Encyclopedia of
TOXICOLOGY

Volume 3   Q–Z

Encyclopedia of
TOXICOLOGY

Volume 3   **Q–Z**

Editor-in-Chief

Philip Wexler

Toxicology and Environmental Health Information Program
National Library of Medicine
National Institutes of Health
Bethesda, Maryland

Academic Press

San Diego   London   Boston   New York   Sydney   Tokyo   Toronto

The following articles are U. S. government works in the public domain:

Carbamate Pesticides
Combustion Toxicology
Medical Surveillance
Nerve Gases
Nitrogen Mustards
Organotins
Risk Assessment, Ecological
Sarin
Soman
Sulfur Mustard
Tabun
VX

This book is printed on acid-free paper. ∞

Academic Press
a division of Harcourt Brace & Company
525 B Street, Suite 1900, San Diego, California 92101-4495, USA
http://www.apnet.com

Academic Press Limited
24-28 Oval Road, London NW1 7DX, UK
http://www.hbuk.co.uk/ap/

Library of Congress Card Catalog Number: 98-84114

International Standard Book Number: 0-12-227220-X (set)
International Standard Book Number: 0-12-227221-8 (volume 1)
International Standard Book Number: 0-12-227222-6 (volume 2)
International Standard Book Number: 0-12-227223-4 (volume 3)

PRINTED IN THE UNITED STATES OF AMERICA
98 99 00 01 02 03 MM 9 8 7 6 5 4 3 2 1

Contents

Contents   vii

Volume 1

Quartz

- CAS: 14808-60-7
- PREFERRED NAME: Silicon dioxide
- SYNONYM: Silicic anhydride
- CHEMICAL CLASS: Inorganic crystalline mineral
- CHEMICAL STRUCTURE:

$$O—Si—O$$

Uses
Quartz is used in the manufacture of glass, water glass, abrasives in grinding and scouring, and components of ceramics and enamels.

Exposure Pathways
Dermal exposure and inhalation are the primary routes of exposure.

Toxicokinetics
Particle size and shape play a crucial role in the toxicogenetic properties of quartz. Its tetrahedrate structure facilitates its fibrogenic potential. Particles $\geqq 3$ μm are usually filtered through the nasal passages. Particles <0.5 μm are readily absorbed and eliminated. Fibrous particles or particles between 0.5 and 3 μm are most likely to settle in the lungs and are most difficult to excrete.

Mechanism of Toxicity
Inhaled quartz particles are absorbed in the alveolar macrophages, the cells responsible for phagocytosis of xenobiotics. This incites the cells' defenses, which over prolonged exposure destroy the macrophages and compromise the immune system. As cell membranes break down and collagen production is activated, fibrosis occurs.

Human Toxicity
Granuloma may result from dermal exposure. Pneumoconiosis is the first sign of deposition of dust in the lung. Chronic inhalation can produce silicosis, a fibrotic lung disease that progresses beyond simple deposition of dust in the lung. Silica workers are also at greater risk for lung and stomach cancer, tuberculosis, and COPD. Quartz is also implicated in autoimmune illnesses such as schleroderma and lupus.

Clinical Management
Bronchial alveolar lavage should be performed to remove excess particulate. Systemic steroids should be administered to reduce inflammatory response.

—Jayne E. Ash and Shayne C. Gad

Related Topics

Asbestos
Respiratory Tract

Quinidine

- CAS: 56-54-2
- Synonyms: Cin-Quin; quinidine sulfate; quinidine gluconate; quinidine polygalacturanate; Quinaglute
- Pharmaceutical Class: Class IA antiarrhythmic; the dextrorotary isomer of quinine
- Chemical Structure:

Uses

Quinidine is used to treat and control atrial fibrillation and atrial flutter. Quinidine is also approved to treat premature ventricular contractions and to treat paroxysmal atrial tachycardia or paroxysmal atrioventricular junctional rhythm. It may also be used to treat malaria, although quinine is preferred.

Exposure Pathways

Ingestion is the most common route of exposure in both accidental and intentional poisonings. Quinidine is also available in intravenous and intramuscular forms.

Toxicokinetics

The bioavailability of quinidine is up to 90%. Peak plasma effects occur in 1–3 hr. Sustained-release preparations produce peak plasma levels in 5 or 6 hr. Quinidine is up to 90% protein bound. The volume of distribution is 2 or 3 liters/kg. Congestive heart failure can lower the volume of distribution to 0.5 liters/kg. Chronic liver disease increases the volume of distribution by 30%. Up to 80% of quinidine undergoes hepatic hydroxylation. The remainder (approximately 20% of a therapeutic dose) is eliminated unchanged in the urine. The average half-life in therapeutic doses is approximately 7 hr.

Mechanism of Toxicity

Quinidine has direct and indirect, or antimuscarinic, effects on cardiac tissue. Quinidine decreases myocardial excitability, conduction velocity, and contractility. As quinidine concentrations increase, conduction velocity progressively decreases. This is evident in an increase in PR interval, an increase in QRS duration, and an increase in QT interval. The effective refractory period is prolonged by quinidine. The anticholinergic effect on the heart is a decrease in vagal tone. In overdose, sinus node automaticity may be depressed. It is the α-adrenergic blocking properties of quinidine that cause vasodilation and hypotension.

Human Toxicity: Acute

Acute overdosage can result in both cardiovascular and neurologic effects. Cardiac effects occur as a result of myocardial depression and depression of atrial, atrioventricular, and ventricular conduction. EKG changes will be evident. These EKG changes include a widening of the QT, PR, and QRS complexes; ST depression; and T inversion. Myocardial depression and vasodilation can cause hypotension to develop. Syncope can result from transient *torsade de pointes* (i.e., bursts of atypical ventricular tachycardia). Ventricular tachycardia and ventricular fibrillation may develop. Possible central nervous system effects include lethargy, seizures, and coma. Signs of toxicity are expected to occur in adults ingesting ≥ 1 g. Therapeutic plasma levels of quinidine range from 1 to 4 $\mu g/ml$. Cardiac toxicity can occur with levels ≥ 14 to 16 $\mu g/ml$.

Human Toxicity: Chronic

With chronic toxicity, gastrointestinal symptoms are common. Nausea, vomiting, and diarrhea are generally seen. The toxidrome known as cinchonism can occur in chronic toxicity. Clinical symptoms of this toxidrome include tinnitus, headache, vertigo, deafness, fever, visual changes, and confusion. Loss of vision can occur. Cinchonism can occur when quinidine plasma levels are ≥ 5 $\mu g/ml$. Loss of vision can occur when levels are ≥ 10 $\mu g/ml$.

Clinical Management

Basic and advanced life-support measures should be used as needed. Induction of emesis is not recommended due to the potential for a decreased level of consciousness, seizures, and arrhythmias. Gastric lavage followed by activated charcoal is recommended. Repeated doses of activated charcoal may enhance elimination. Serum electrolytes should be monitored in all serious exposures. Intravenous administration of sodium bicarbonate may decrease toxicity. Hypotension can be treated with fluids and vasopressors if needed. Ventricular dysrhythmias can be treated with class IB antiarrhythmics such as phenytoin or lidocaine. Persistent bradycardia and third-degree heart block are indications for insertion of a temporary pacemaker. Seizures can be treated with diazepam. If seizures are uncontrolled, phenobarbital or phenytoin can be administered.

Animal Toxicity

Clinical effects similar to those noted in humans are expected to occur in animals following acute exposures to quinidine. Appropriate gastrointestinal decontamination measures and supportive care should be utilized.

—Bridget Flaherty

Related Topics

Cardiovascular System
Quinine

Quinine

- ◆ CAS: 130-95-0
- ◆ Preferred Name: Quinine sulfate
- ◆ Synonyms: Chininum, Quinina
- ◆ Pharmaceutical Class: Antimalarial
- ◆ Chemical Structure:

Uses

Quinine is the drug of choice for treatment of malaria; it is also used in the treatment of nocturnal leg cramps. It is often misused as an abortifacient.

Exposure Pathways

Quinine is available in oral dosage forms. Ingestion is the most common exposure pathway.

Toxicokinetics

Quinine is rapidly absorbed orally. It is metabolized in the liver by oxidation to several polar hydroxy metabolites. The volume of distribution is 1 or 2 liters/kg and protein binding is 70%. Quinine is excreted by the kidneys; approximately 10% is excreted as unchanged drug. The therapeutic half-life of quinine is 11.1 ± 4.1 hr. The half-life can more than double at toxic doses to 26.5 ± 5.8 hr.

Mechanism of Toxicity

The exact mechanism of toxicity is unknown. Quinine acts on all body muscle groups, most notably cardiac, uterine, and skeletal muscles.

Human Toxicity: Acute

Accidental and intentional ingestions have resulted in headache, deafness, blindness, tachycardia, respiratory arrest, and death. Reversible renal failure can occur. The adult toxic dose can be as low as 2 g.

Human Toxicity: Chronic

Cinchonism can arise from cumulative dosing. Clinical symptoms associated with the syndrome of cinchonism include headache, dizziness, tinnitus, rash, cardiovascular effects, intestinal cramping, vomiting, diarrhea, fever, confusion, and seizures. The symptoms resolve with cessation and elimination of the drug.

Clinical Management

Recommended treatment includes gastric decontamination with gastric lavage and repeated doses of activated charcoal. Intensive monitoring of vital signs and the EKG are important. Quinine levels may be useful to confirm exposure. Other treatment may include administering a Stellate block for quinine-induced blindness and the use of vasodilators for residual visual impairment.

—Linda Hart

Radiation Toxicology

R adiation toxicology is a specialized area of toxicology that deals with the health consequences of exposure of biological systems to ionizing radiation. Because everyone is exposed to natural radiation, and most exposure is from man-made sources during their lifetimes, it is important to understand the potential effects from such exposures as well as those that can occur as a result of larger exposures associated with events such as radiological accidents or radiotherapy. Fortunately, this well-researched discipline provides some of the best quantitative estimates of health risks of all the areas of toxicology. This section will provide a brief summary of the current knowledge of radiation health effects together with certain basic elements of radiation toxicology needed to understand the concepts and terminology that are specific to this area.

Basic Physical Concepts

Ionizing radiation is a term applied to radiations that give rise, directly or indirectly, to ionizations, or the ejection of one or more orbital electrons from an atom or molecule, when they interact with matter. Ionizing radiations consist of particulate forms, such as β particles, positrons, electrons, neutrons, protons, α particles, and heavy nuclei, and electromagnetic radiation, which comprises gamma (γ) rays, X rays, and higher energy ultraviolet rays.

Gamma rays and X rays having the same energy (usually given in units of keV) do not differ in their nature or properties. The different terminology reflects the difference in their origin, i.e., γ rays occur as a result of the deexcitation of unstable nuclei seeking a lower nuclear energy state, whereas X rays occur as a result of energy loss of excited atomic electrons. Gamma rays have discrete energies that are characteristic of the nuclide from which they are emitted. When electrons that have been accelerated by a positive, high-voltage field collide with a target such as tungsten (as in an X-ray machine), the electron loses energy in the form of X rays called bremsstrahlung radiation. In this type of interaction, the X rays can assume a continuum of energy values up to a maximum value. When rearrangement of orbital electrons occur as a result of ionization of an atom, the X rays so produced are called characteristic X rays and have discrete energy values unique to the atom from which they arise.

Gamma rays and X rays are among the most penetrating types of radiations. Since they are not charged, they interact with matter primarily by photoelectric effect, Compton scattering, and pair production. In these interactions, an electron is ejected from an atom and carries with it all or part of the energy of the original photon. This "secondary electron" is the agent by which the photon energy is dissipated within the medium, and once created, these electrons interact with the atoms of the surrounding medium identically to β particles (see below).

As photons pass through matter, they lose energy at an exponential rate. The fractional energy expected to remain after traveling a distance d in a medium can be described by

$$\frac{I}{I_0} = e^{-\frac{\mu}{\rho}d}$$

where: I = intensity of the photon beam at distance, d, in the medium

I_0 = incident photon intensity

μ/ρ = attenuation coefficient in the medium for the energy considered (in units of cm^2/g)

d = distance traveled into the medium in g/cm^2 (equal to the distance in cm × density)

The penetrating ability of photons depends on both the photon energy and the composition of the absorbing media, with penetration increasing with increasing energy, and decreasing density and effective atomic number of the medium. The penetrating ability of photons is sometimes described by a half thickness or half-value layer (HVL), which is the thickness of an absorbing medium that decreases the photon intensity by one-half. For example, the HVL for 1-MeV γ rays is about 9 m air, 9.6 cm water, 4 cm Al, and 9 mm Pb.

Of the various particulate emissions, the most important in terms of likelihood for human exposure are the α particle, the β particle, the proton, and the neutron. The first two radiations generally occur as a result of the radioactive decay of unstable atoms, whereas neutrons generally result from nuclear reactions, such as nuclear fission (as in nuclear reactors and fission-based nuclear weapons) and charged-particle activation of target atoms (as with some accelerator-produced radioisotopes). Protons arise from atomic interactions of neutrons.

Alpha particles are identical with helium nuclei and consist of two protons and two neutrons. They result from the radioactive decay of heavy elements such as radium, thorium, uranium, and plutonium. Because of their double-positive charge, α particles have great ionizing power, but their large mass results in very little penetration. For example, α particles from 4 to 10 MeV have ranges in air of 5–11 cm; the corresponding range for α particles in water would be from 20 to 100 μm.

Beta particles are equivalent to electrons but arise from radioactive decay. They are emitted with a continuous spectrum of energies up to a maximum that is characteristic of each radionuclide. Electrons have a greater range and penetrating power but much less ionizing potential compared to α particles. The range of β particles in air is about 4 m per MeV. In water the range (in cm) can be approximated as one-half the maximum β energy (in MeV). For example, the range

of the energetic β particles from ^{90}Y (E_{max} of 2.27 MeV) is about 1.15 cm in water (and similarly soft tissue).

Neutrons are neutrally charged particles with a mass slightly larger than that of a proton. Because they are neutrally charged, they produce ionizations indirectly. In biological material, neutrons can eject protons from the nuclei of hydrogen atoms by nuclear collisions, which in turn are charged and directly ionizing. Neutrons also can activate hydrogen and other elements by neutron capture, which results in the release of γ rays, and sometimes radioactive by-products. Neutrons that are produced by fission or by special sources such as Am-Be sources have a spectrum of energies that range over many orders of magnitude. Since their ranges depend on both neutron energy and the composition of the absorbing material, it often requires complex calculations to describe the ranges of neutrons. However, as simple examples, very low-energy neutrons (called thermal neutrons) have a range of about 30 cm in soft tissue, whereas high-energy or fast neutrons can penetrate tissue to the about the same extent as can highly penetrating γ rays.

Quantities and Units

Radiotoxicology, like other disciplines of toxicology, has specialized quantities that define the relationships between exposure to radiation and the resulting dose received by specific biological units. Some of these quantities are based on measurements and/or calculations; others, particularly those used in radiation protection, consist of quantities that include modifying factors designed to allow comparison of risks to people exposed to a variety of radiation types and with widely varying spatial patterns of dose.

The defining event of a radioactive nuclide is the transformation of its nucleus into the nucleus of another species, i.e., radioactive decay. The number of nuclear transformations occurring per unit of time is called the "activity." The traditional unit of activity has been the Curie (Ci), which is equal to 3.7×10^{10} nuclear transformations per second. The conversion of radiation units to the international system (Systme International d'Unit or SI) is currently under way in the United States. The more fundamental unit of activity, the Becquerel (Bq), equal to 1 nuclear transformation per second, is gradually replacing the Curie. Both units of activity are modified by prefixes such as kilo-, milli-, and micro- to achieve standard multiples of the fundamental unit. A listing of the most commonly used prefixes is given in Table R-1.

In radiation physics, the term "exposure" is used to describe the amount of ionization caused in air by γ and X rays. The unit of exposure is the Roentgen (R), which is equal to 2.58×10^{-4} coulomb per kilogram of air; the corresponding SI unit is coulomb/kg. This quantity is most often used in diagnostic radiology and does not apply to ionizations produced by either particulate radiations or high-energy ($\geqq 2$ MeV) X rays or γ rays. For radiobiological applications, the exposure rate (e.g., in R/min) is most commonly used.

When viewed at the microscopic level of a cell, or smaller biological subunit, the dose D is replaced by what is called specific energy (z). While D is a macroscopic dose, z is a microscopic dose. For a single absorbed dose D to an organ or tissue, there can be many different microscopic doses z to cells in that organ or tissue. Also, a cell may have no dose at all, while another cell in the same tissue may have a very large dose. However, when z is averaged over the tissue or organ, the average value obtained should equal the absorbed dose D as defined previously.

TABLE R-1
Standard Multiples Used with Radiation Units

Prefix	Multiplication factor	Symbol
exa	10^{18}	E
peta	10^{15}	P
tera	10^{12}	T
giga	10^{9}	G
mega	10^{6}	M
kilo	10^{3}	k
hecto	10^{2}	h[a]
deca	10^{1}	da[a]
deci	10^{-1}	d[a]
centi	10^{-2}	c[a]
milli	10^{-3}	m
micro	10^{-6}	μ
nano	10^{-9}	n
pico	10^{-12}	p
femto	10^{-15}	f
atto	10^{-18}	a

[a] It has been suggested that all SI units be expressed in "preferred standard form" in which the multiplier is 10^{3n}, where n is a positive or negative whole number. Consequently, the use of hecto, deca, deci, and centi is to be avoided wherever possible.

The most commonly used quantity describing radiation dose is the absorbed dose (D), which is defined as the mean energy, e, imparted by ionizing radiation to matter of mass m, i.e., $D = e/m$. This quantity is a measurement of the deposition of energy in any substance by all types of ionizing radiation. The traditional unit of absorbed dose is the rad, which is equal to 100 ergs/g or 0.01 J/kg; the corresponding SI unit is the Gray (Gy), which is equal to 1 J/kg (therefore, 1 Gy is equivalent to 100 rad). The absorbed dose should be used in preference to the exposure whenever the former can be measured or calculated.

For radiation protection purposes, several dose quantities have been created that attempt to "normalize" the responses of different tissues and organs of the body from irradiation by different types of ionizing radiation so that uniform radiation protection guidelines can be promulgated that are insensitive to the particulars of any given irradiation scenario. The traditionally used quantity has been the dose equivalent (DE), which is defined as the absorbed dose (D) multiplied by a quality factor (Q) that takes into account the type of radiation and other factors such as a dose distribution factor (f) within a given tissue. This latter factor, however, had been considered so uncertain that a default value of 1 was customarily assigned to it so that the dose equivalent became simply the absorbed dose multiplied by the quality factor. The unit of dose equivalent has been the rem, which is dimensionally the same as the rad; the SI unit is the Sievert (Sv).

The quality factor for the various types of ionizing radiation is based on the linear energy transfer (LET) of the type of radiation. The LET is the average rate at which charged particles transfer their energies to the atoms in a medium and is a function of the energy and velocity of the charged particle. For example, α particles with their large mass, +2 charge, and slow speed impart much more energy per unit of path length than do electrons. The unit generally attributed to LET is keV/μm path length.

Since equal doses of different types of ionizing radiations do not produce equivalent biological effects, a quantity called the relative biological effect (RBE) was developed to allow comparison of effects produced in identical biological systems from different types of radiations. The RBE is customarily defined as the ratio of two doses (a reference dose divided by a test dose) for producing a given level of biological effect under a given condition. The reference is often taken to be X rays. For example, if 90% cell killing is produced by 10 Gy of X rays, but 0.5 Gy of neutrons, then the RBE in this case would be D_x/D_n or 20.

Current radiation protection guidelines are specified in terms of the quantity effective dose (E). This effective dose has associated with it the same probability of occurrence of cancer and genetic effects whether received by the whole body, via uniform irradiation, or by partial-body or individual-organ irradiation. To take into account the observed varying radiosensitivities of the different organ systems of the body and to adjust for nonuniformity of irradiation, a tissue weighting factor, w_t, is used. An additional radiation weighting factor, w_r, is used to adjust for the biological effectiveness of different radiations. The current weighting factors, as stated by the National Council on Radiation Protection and Measurements, are summarized in Table R-2.

TABLE R-2
Radiation and Tissue Weighting Factors Used in Radiation Protection Guidelines[a]

Radiation type and energy range	W_r
X rays and γ rays, electrons and positrons	1
Neutrons, energy	5
<10 keV	10
10–100 keV	10
>100 keV to 2 MeV	20
>2–20 MeV	10
>20 MeV	5
Protons	2
α Particles, fission fragments	20

Organ or tissue	W_t
Gonads	0.20
Red bone marrow	0.12
Colon	0.12
Lung	0.12
Stomach	0.12
Bladder	0.05
Breast	0.05
Liver	0.05
Esophagus	0.05
Thyroid	0.05
Skin	0.01
Bone surface	0.01
Remainder	0.05

[a] W_r is the radiation weighting factor, and W_t is the tissue weighting factor.

Sources of Ionizing Radiation Exposure

Humans are routinely exposed to ionizing radiation. Some of the sources are naturally occurring, and others are due to man-made uses of radiation and radioactive materials. In general the radiation from natural sources includes cosmic radiation, external radiation from radionuclides in the earth's crust, and internal radiation from radionuclides inhaled or ingested and retained in the body. Man-made sources of radiation include X-ray equipment, particle accelerators and nuclear reactors used in the generation of nuclear energy, radionuclides used in nuclear medicine, radionuclides released to the environment as a result of nuclear weapons testing and nuclear accident, and occupational exposure to both external and internal radiation. The magnitude of the exposure to natural sources depends mostly on geographical location, whereas exposure to man-made sources depends on human activities.

Natural Background Radiation

Exposure to natural sources of external ionizing radiation results from the levels of cosmic and terrestrial X and γ radiation present in the environment. Cosmic radiation at the earth's surface is affected by altitude, geomagnetic latitude, and solar modulation. For example, the dose rate at 1800 m is about double that at sea level. Within the United States, the effect of latitude and solar modulation on cosmic ray dose rate is <10%. Because cosmic radiation is highly penetrating, it results in relatively uniform whole body irradiation. The average dose rate from cosmic irradiation in the United States has been estimated to be about 28 mrem/year.

Humans are also exposed to external γ irradiation from concentrations of naturally occurring radioactive materials in soils and rocks. These radioactive elements include uranium and thorium radionuclides plus their radioactive progeny and ^{40}K, and result in widely varying dose rates that depend on the geology of the particular region. Estimates of the annual dose rate for this type of exposure in the United States averages 28 mrem/year.

Internally deposited naturally occurring radionuclides also contribute to the natural radiation dose from inhalation and ingestion of these materials in air, food, and water. Included are radionuclides of lead, polonium, bismuth, radium, potassium, carbon, hydrogen, uranium, and thorium. Potassium-40 is the most prominent radionuclide in normal foods and human tissues. The dose to the total body from these internally deposited radionuclides has been estimated to be about 39 mrem/year.

The major exposure of the population to natural radiation arises from inhalation of the short-lived radioactive progeny of the radioactive noble gas ^{222}Rn, which in turn is a sixth-generation radioactive decay product of natural uranium. The amount of ^{222}Rn present in the air depends on many factors (e.g., gas permeability in soil and rock, relative humidity, and barometric pressure) but is necessarily linked to the geological concentration of the uranium parent radionuclide. There is about an eightfold range of concentrations of uranium in different types of rocks and soils.

Most of the early measurements of Rn levels were made outdoors; however, it has become apparent that the indoor concentrations are generally several times higher than those outdoors. Because people in Western countries spend only about 15% of their time outdoors, most of the exposure to radon therefore occurs indoors. Additionally, the trend toward the construction of more energy-efficient housing (more air-tight) has also enhanced the concentrations of ^{222}Rn indoors.

The average annual radiation dose to the general population due to inhalation of radon and its progeny is estimated to be about 200 mrem. However, this dose can range upward by one or two orders of magnitude in cases in which the indoor Rn concentrations are very high. Because of the short half-lives of the radon progeny, and the fact that the most important radionuclides decay by α-particle emission, their radiation dose is delivered primarily to the tissues of the respiratory tract.

Man-Made Radiation Sources

Several human activities involving the production and use of radionuclides have resulted in releases of radioactive materials to the environment. Such activities include past atmospheric testing of nuclear weapons, production of nuclear weapons materials, production of electricity by nuclear reactors, radioisotope production and use in industry and medicine, and accidental releases of radionuclides at both civilian (Three Mile Island and Chernobyl) and military (Kyshtym and Wind-

scale) nuclear installations. Additionally, there has been a significant increase in the types of quantities of sources of potential radiation exposure from consumer products. These include radioluminescent devices containing ^{3}H, ^{147}Pm, or ^{226}Ra; smoke detectors containing ^{241}Am; static eliminators containing ^{210}Po; and airport X-ray baggage inspection systems. In other cases, radiation emissions are incidental or extraneous to the purpose for which the consumer product was designed, e.g., television receivers, tobacco products containing ^{210}Po and ^{210}Pb, combustible fuels and building materials containing uranium- and thorium-series radionuclides, and gas mantles, camera lenses, and welding rods containing thorium.

A summary of the contributions of the various natural and man-made radiation sources to our radiation background is given in Table R-3. It can be seen that natural sources contribute about 82% of the total, with radon being the largest single source (67% of natural radiation dose). Of the 18% contributed by man-made sources, medical exposure is the most prominent (83%). Attempts to significantly reduce population radiation doses would most likely be focused on the largest contributors, i.e., indoor radon and medical radiation.

TABLE R-3
Radiation Exposure of the U.S. General Population

Radiation source	Per capita annual effective dose equivalent (mrem)
Natural radiation	
Cosmic rays	28
Terrestrial, external	28
Internally deposited radionuclides (except radon)	39
Inhaled radon and progeny	200
Sources due to or enhanced by human activity	
Medical uses	53
Nuclear power[a]	0.05
Consumer products	8
Weapons fallout (averaged to Year 2000)	5
Total	361

[a] Includes contributions from uranium mining and milling, fuel fabrication, power plant operation, reprocessing of spent fuel, and transportation.

Radiobiological Effects

Radiation-induced biological effects include chromosomal aberrations, mutations, cancer, genetic effects, and loss of normal tissue and organ functions. Although the biological effects of concern from exposure to ionizing radiation are described typically at the tissue or organ level, it has long been recognized that an understanding of the mechanisms by which radiation produces effects such as cancer, genetic changes, and tissue destruction are best obtained from studies performed at the cellular and molecular levels. This is appropriate because a single ionizing particle passing through a cell deposits its energy in a random and nonuniform manner on a microscopic scale. This deposited energy creates positively and negatively charged molecules and atoms (called ion pairs) along the path (or track) traveled by the ionizing particle. The density of ion pairs produced along a track varies significantly depending on the ionizing particle and the medium through which it passes and is proportional to the average energy deposited per unit path length, or LET. For example, electrons produced by 200 kV X rays produce about 80 ions per μm path length, protons generated from interactions with 12 MeV neutrons produce about 300 ions/μm, and α particles from the decay of ^{226}Ra produce about 3700 ions/μm; correspondingly, the LETs for typical X rays range from 0.2 to 15 keV/μm, fast neutrons from 8 to 40 keV/μm, and α particles $\geqq$260 keV/μm. As mentioned previously, the efficiency with which a particular type of radiation produces biological effects (RBE) depends strongly on LET.

Direct and Indirect Radiation Effects

At the chemical level, a solute molecule in a biological system (DNA, RNA, and protein) can be affected by radiation in two different ways. When an ionization track passes either directly through a molecule or close enough so that the created ions can drift to and interact chemically with the molecule before they recombine and neutralize in solution, the phenomenon is called a direct radiation effect. On the other hand, since the largest fraction of almost any biological system consists of water (e.g., 70–80% of a typical cell), the most likely radiation interactions will be with water molecules. When this occurs, ion radicals and free radicals are

created. When irradiated, water molecules become ionized in a two-step process:

$$H_2O + radiation \rightarrow H_2O^+ + e^-$$
$$e^- + H_2O \rightarrow H_2O^-$$

However, the charged water molecules are unstable, having lifetimes less than 10–15 sec, and almost immediately dissociate into one smaller ion and a free radical:

$$H_2O^+ \rightarrow H^+ + OH\cdot \text{ (the dot symbolizes the unpaired electron of the OH free radical)}$$
$$H_2O^- \rightarrow H\cdot + OH^-$$

The free radicals thus produced are very reactive and diffuse through the solvent system interacting in a fairly indiscriminate manner with other free radicals, with molecules previously damaged by radiation, or, most important, with intact solute molecules previously unchanged by the radiation. Free radical reactions may also produce other more or less reactive chemical species, such as H_2, H_2O, and H_2O_2, they may react with oxygen to enhance the effect of the radiation, or they may interact with organic molecules creating organic free radicals:

$$H_2O^- + RH \rightarrow R\cdot + H_2O_2$$
$$RH + HO_2\cdot \rightarrow RO\cdot + H_2O$$

This latter phenomenon is called an indirect radiation effect. If RH is an important molecule (e.g., DNA and RNA) then these interactions can affect cell functions.

Effects of Radiation on Cells

Depending on type and quantity, radiation can produce a variety of effects on cells. These effects include cell killing, the production of DNA point mutations or chromosome aberrations, altered cell-cycle regulation, or alterations of metabolic functions. As more studies are done using the tools of molecular biology, more insights into the mechanisms of radiation action are being found. This knowledge plays a key role in the development of biophysically based risk models that are used to extrapolate results from studies performed at relatively high radiation doses and dose rates to the region of low doses and dose rates that are of greatest interest and importance for the protection of people and the environment from radiation damage.

Cell Survival

Cell survival curves are used to describe the relationship between radiation dose and the proportion of cells that survive. The endpoint of survival can have two different meanings depending on whether the cell populations studied have proliferative potential. In the case of non-dividing, terminally differentiated cells, survival is generally related to the ability of the cells to maintain function. In general, relatively large radiation doses are required to inactivate cell function for terminally differentiated cells. For cells that proliferate either in tissue or in culture, cell death or survival is more related to that cell's ability to continue to divide and produce clones of cells. Thus, a cell that is able to undergo no more than one or two cell divisions after irradiation would still be considered "dead." Doses on the order of a few rads are generally required to kill the most sensitive proliferating cells, although there is a range of radiosensitivities to this effect. For example, normal human fibroblasts irradiated with X rays have a D_0 of 120 rads (D_0 is the dose at which survival is reduced to 37% of its original value); in comparison, cells from a patient with the disease ataxia-telangiectasia (AT) have a D_0 of about 50 rads and are therefore more radiosensitive—a hallmark of AT.

Studies using experimental designs in which only certain selected portions of a cell were irradiated have shown that the sensitive sites for radiation-induced cell killing are in the nucleus as opposed to the cytoplasm. Furthermore, there is strong circumstantial evidence that the DNA in the chromosomes is the primary target for radiation-induced cell killing.

Cell and Tissue Radiosensitivity

The sensitivity of various cells and cell lines to radiation-induced lethality can differ significantly, and different organs consist of different cell populations (e.g., connective tissue and vascular, parenchymal). As early as 1906, Bergoné and Tribondeau studied cellular radiosensitivity and postulated that actively proliferating cells were most radiosensitive, that the degree of cellular differentiation was inversely related to radiosensitivity, and that radiosensitivity of cells was proportional to the duration of mitotic and developmental activity. In general, this "law" is valid for different cell types, although there are exceptions. For example, the small lymphocyte, which is highly differentiated and has little if any proliferative potential, is one of the most radiosensitive mammalian cells. One scheme for categorizing the sensitivity of normal cells to cell killing is the following:

Very high
  Lymphocytes, immature hematopoietic cells, intestinal epithelium, spermatogonia, ovarian follicular cells

High
  Urinary bladder epithelium, esophageal epithelium, gastric mucosa, mucus membranes

Intermediate
  Endothelium, growing bone and cartilage, fibroblasts, glandular epithelium of breast, pulmonary, renal, pancreatic, thyroid and hepatic epithelia

Low
  Erythrocytes, muscle cells, mature connective tissue, osteocytes, chondrocytes, ganglion cells

The effects of radiation on more complex organ systems will depend on the effects produced on the different subpopulations of cells that comprise the organ. For example, if the parenchymal cells are most radiosensitive (as in intestinal mucosa), then loss of function of the organ may occur at the lowest doses, followed perhaps by vascular damage at higher doses. If the parenchymal cells are normally nondividing (as in the brain), then the reverse may occur, with damage to the microcirculation predominating at lower doses.

Genetics Effects

Radiation Effects on Inheritance

Radiations such as X rays, γ rays, and β particles can damage genetic material in reproductive cells and cause mutations that can be transmitted from one generation to another. In the 1920s, researchers using fruit flies (*Drosophila*) found that chromosomes could be easily injured by radiation and such injury could lead to mutations that were expressed in subsequent generations. This finding was quickly confirmed in numerous species of plants and animals. Today it is known that relatively small radiation doses (<10 rad) can cause alterations in nucleotides and visible breaks in chromosomes of germ cells that can lead to genomic instability that can be passed on to subsequent generations.

In evaluating the effect of radiation on heredity of germ cells, two specific germ-cell stages are considered important: (1) the stem-cell spermatogonia in males and (b) the oocytes, primarily the immature ones, in the female. The spermatogonia continue to multiply throughout the reproductive life span of an individual. However, oocytes are not replaced during adult life.

Because of the lack of information for humans, most genetic studies have been carried out with experimental organisms, especially mice. Radiation has been found to cause mutations in all nonhuman experimental organisms studied and therefore such effects are also expected to arise in humans.

The genetic effects that could be caused by radiation are too numerous to be considered individually. For nuclear accident risk assessment, genetic disorders have been grouped as (1) dominant and X-linked single-gene disorders, (2) chromosome disorders, and (3) multifactorial disorders.

Dominant and X-Linked Single-Gene Disorders

Most cells from humans contain two sets of chromosomes with matched pairs of genes, one gene from each parent. The matched genes can differ, with one gene being dominant over its recessive counterpart. A recessive gene can only show its effect if both matching chromosomes carry that gene. If an altered gene is present on the X chromosome it will invariably produce an effect in boys, who have only one X chromosome, but will behave as recessive in girls, who have two X chromosomes. Single-gene disorders related to damage to the X chromosome are called X-linked effects.

Dominant gene disorders that could be caused by radiation include traits such as Huntington's chorea, hypercholesterolemia, and achondroplastic dwarfism. The X-linked traits include traits such as muscular dys-

trophy, hemophilia, and agammaglobulinemia. However, there is no direct evidence that these diseases have been induced in humans.

Chromosome Disorders

One of the mechanisms of chromosomal damage that depends on radiation quality (i.e., LET) is the induction of single-strand (SS) and double-strand breaks (DSBs). The two types of damage are considered to be important because SSBs are more easily and accurately repaired by the cell than are DSBs. Thus, DSBs result in damage that is both more lethal and more able to result in chromosome disorders. For low LET radiation, increased production of DSBs is a function of dose rate, as single tracks are so sparsely ionizing that breaking more than one chromosome with a single track is unlikely; therefore, DSBs arise as a consequence of multiple tracks occurring sufficiently close in time and space. On the other hand, high LET radiation produces a high enough ionization density within its tracks that DSBs can occur from single traversals of a cell nucleus. This in part is responsible for the greater RBE for high LET radiation for both cell killing and cell transformation.

Damage that is produced by radiation can be chromosomal or chromatid, depending on whether the cell is in a pre- or postreplication state. In either case, sufficient energy is imparted to break a chromosome or chromatid, usually into a major and a minor fragment. Once this has occurred, (1) the broken ends may rejoin to restore the chromosome's original configuration; (2) a fragment may fail to rejoin, resulting in a deletion, which is sometimes large enough to be scored as a micronucleus; or (3) broken ends may rejoin with other broken ends to yield abnormal forms that are subsequently scored at the following mitosis as rings, dicentrics, anaphase bridges, or symmetric and asymmetric translocations.

Chromosome anomalies and aberrations can influence heredity. Most somatic cells of humans contain 23 pairs of chromosomes, with one member of each pair contributed by the sperm and the other contributed by the egg. When the process of sperm or egg cell production goes awry as a result of radiation damage, abnormal chromosome numbers (aneuploidy) can arise. Aneuploidy is a form of genetic instability.

It has been estimated that in about 90% of cases, aneuploidy will result in spontaneous loss of pregnancy. In the remaining 10% of cases, a severely affected child would be expected because of the inherited genomic instability. Conditions such as Down's syndrome and both Klinefelter and Turner anomalies are the result of genomic instability associated with aneuploidy. These defects are relatively severe—in terms of both life expectancy (about 45 years) and level of disability (about 50%). Persons born with aneuploidy usually are physiologically and morphologically abnormal and do not have children. Thus, their genomic instability tends not to be passed on to other generations.

Chromosomes can be easily broken by radiation which can lead to a structural rearrangement (called a translocation). Translocations are also a form of genomic instability. When translocations occur in germ cells, they can be transmitted to the offspring. Translocations normally yield chromosomes with too little or too much genetic information. If a child is born with a balanced translocation (not too little or too much information) he or she would not normally be affected but could pass on genomic instability to future generations. Those born with such genomic instability could suffer from severe physical and mental disabilities.

Multifactorial Disorders

Multifactorial diseases involve complex patterns of inheritance and represent a very large class of genetic diseases. For such diseases to arise, a specific combination of mutant genes must be present. Environmental factors can also be important. Examples of multifactorial diseases include congenital malformations (e.g., spina bifida and cleft palate), constitutional diseases, and degenerative diseases.

Genetic Effects in Irradiated Populations

Epidemiology has not detected hereditary effects of radiation in humans with a statistically significant degree of confidence. Nevertheless, there can be no doubt of the existence of hereditary effects in man. Following radiation exposure of a large population (e.g. as occurred in the former Soviet Union after the Chernobyl nuclear accident in 1986), an increase in the incidence of genetic disease would be expected to occur. The genetic damage would show up both early (as an increased incidence of birth defects among some children of the exposed population) and late (through latent mutations expressed in their grandchildren, great-grandchildren, and subsequent generations). It has been estimated that about 50% of all genetic damage introduced by radiation exposure following a major nuclear accident will be manifest within the first three to five subsequent generations with the remaining damage dispersed over future generations.

Early and Continuing Deterministic Effects

If a person is exposed to a large amount of radiation (i.e., large radiation dose) delivered to the entire body, cells in tissues can be destroyed in large numbers. Because tissues have important functions, the destruction of significant numbers of cells can lead to impairment in one or more of these functions. The biological effects that arise when large numbers of cells are destroyed by radiation are called "acute somatic effects" if they occur in a relatively short period of time (e.g., within a few weeks). Acute somatic effects are a subset of what is now formally called "early and continuing deterministic effects" (once called nonstochastic effects).

Deterministic effects are those that increase in severity as the radiation dose increases and for which a threshold is presumed to exist. Besides acute somatic effects, deterministic effects also include radiation effects (other than cancer and genetic effects) that continue to occur after an extended period (e.g., years) of chronic exposure. Such chronic exposures can arise from long-lived radionuclides (e.g., isotopes of plutonium and cesium) ingested via contaminated food or inhaled via contaminated air and retained in the body. Chronic exposure can also arise from the continuous intake of contaminated food or air over long periods. Populations in Russia, the Ukraine, and Byelarus continue to ingest and inhale long-lived radionuclides that were released during the 1986 nuclear accident at Chernobyl. Firemen that fought the reactor fire during the Chernobyl accident and plant workers present at the time of the accident were chronically exposed to large radiation doses from inhaled radionuclides.

Examples of deterministic effects are hypothyroidism arising from large radiation doses to the thyroid gland; skin burns arising from exposure of small or large areas of the skin; permanent suppression of ovulation in females; temporary suppression of sperm production in males; growth and mental retardation caused by exposure of a fetus during pregnancy; and death from severe damage to critical organs such as the bone marrow, lung, or small intestine.

Thresholds arise for deterministic effects because large numbers of cells usually must be simultaneously destroyed to produce such effects, which is highly unlikely at low doses. The threshold dose for a specific deterministic effect depends on the type of radiation, on the rate at which the dose is delivered (dose rate), and, for some effects, on other factors.

Factors Affecting the Production of Deterministic Effects

The type of radiation is important because different types of radiation interact with body tissue differently. Gamma rays and X rays can easily penetrate into body tissue and therefore can produce deterministic effects in all body organs if the dose and amount of tissue irradiated are both large enough. Beta radiation can cause skin burns and ulcers when β-emitting hot particles (highly radioactive, very small particles) are deposited on the skin, but little damage is likely to be done to other tissue unless the β-emitting particles are taken into the body in large amounts (e.g., by inhalation of ingestion). Alpha radiation does not cause skin burns or ulcers when α-emitting particles are deposited on the skin because the α radiation does not have enough energy to penetrate the dead layer of tissue that covers the skin surface. However, when taken into the body in large amounts, α-emitting particles can cause deterministic effects.

For total-body exposure to X rays or γ rays, organs and tissue at risk include all organs and tissue in the body. For inhalation or ingestion exposure to β-emitting materials, organs and tissue at risk include the lungs and gastrointestinal tract as well as other sites depending on the metabolic fate of the radionuclide of concern. For example, strontium isotopes preferentially irradiate the skeleton, while iodine isotopes preferentially irradiated the thyroid. When considering possible deterministic effects from inhaled radionuclides, organs other than the lung should also be considered because the radionuclides can translocate from the lung to other organs such as the liver and skeleton.

Radiation dose and dose rate are important because the larger the dose, the larger the amount of potentially destructive radiation energy deposited in tissue, which can lead to extensive cell death and concomitant impairment in important tissue functions. A significant

impairment can lead to morbidity and lethality. Likewise, radiation dose rate is important because when it is sufficiently high, the radiation can overwhelm cell repair mechanisms and organs cannot recover from tissue injury. Most efficient recovery occurs when the radiation dose rate is low and when the amount of tissue that the radiation interacts with is small. In the administration of radiation therapy to cancer patients, physicians try to minimize damage to healthy tissue by delivering the radiation in a number of fractions over a number of days or weeks. This allows damaged normal tissue to recover during the periods between the fractionated exposures. The rate of recovery differs for different organs.

Other factors that can be important in determining the impact of radiation exposure include a person's age and sex, how healthy they are, and the type of medical support received from physicians after being injured by radiation.

Thresholds Doses for Specific Deterministic Effects

For nuclear accident risk assessment, organs of primary interest because of their high sensitivity or their potential for receiving large radiation doses are bone marrow, gastrointestinal tract, thyroid gland, lungs, skin, gonads, and eyes. Table R-4 shows estimates (central, lower bound, and upper bound) of threshold doses for a variety of deterministic effects of exposure to γ rays when the dose is deliver quickly (within 1 hr). Larger doses would apply when the dose is delivered over hours, days, weeks, or longer. For example, the central estimate of the γ-ray threshold for acute lethality from radiation-induced injury to the hematopoietic system is 150 rad (see Table R-4) when the dose is delivered within an hour. However, when the dose is delivered continuously over several years, individuals have survived γ-ray doses as high as 600–1000 rad (which would be fatal if received within a few hours). Nuclear workers in the former Soviet Union (MAYAK workers) that participated, during the late 1940s through mid-1950s, in the production of plutonium for nuclear weapons received large γ-ray doses (up to about 1000 rad in some cases) over several years and survived.

Table R-5 shows estimates (lower bound, central, and upper bound) of thresholds for specific deterministic effects of exposure of the unborn embryo or fetus to X or γ rays delivered quickly (within 1 hr).

TABLE R-4

Threshold γ or X Radiation Doses (Lower, Central, and Upper Estimates) for Specific Deterministic Effects[a]

Effect	Organ/tissue	Lower bound (rad)	Central (rad)	Upper bound (rad)
Vomiting	Upper abdomen	Not estimated	50	Not estimated
Diarrhea	Upper abdomen	Not estimated	100	Not estimated
Erythema	Skin[b]	200	300	400
Moist desquamation	Skin[b]	800	1000	1200
Permanently suppressed ovulation	Ovum in females	20	60	100
Suppressed sperm counts[c]	Testes in males	20	30	40
Cataracts	Lens of eye	0[d]	100	150
Hypothyroidism	Thyroid	Not estimated	200	Not estimated
Radiation pneumonitis	Lung	400	500	600
Hematopoietic death[e]	Bone marrow	120	150	180

[a] Applies to γ rays of X rays delivered to the indicated organ or tissue in less than 1 hr.

[b] For 50–100 cm² area of skin and the dose evaluated at a depth of 0.1 mm.

[c] Two-year suppression of sperm counts in males.

[d] Used to include the possibility that cataracts may be a stochastic effect with no threshold.

[e] Death from lethal injury to the sensitive bone marrow.

Late Somatic Effects

Late somatic effects are those that occur long after exposure to a DNA-damaging agent in progeny of cells other than germ cells. The late somatic effect that is of most concern is cancer.

Induction of Cancer by Radiation

One of the first observations of cancer following irradiation was the appearance of skin cancer on the hands of many of the early workers who used X rays. Since that time, animal and epidemiological studies have shown that radiation can cause an increase in the incidence of specific cancers. They have also shown that cancer does not appear immediately after exposure to radiation but only after a delay (latent period). For humans, the latent period may be quite long (many years) for some cancers.

Mechanism that may be involved in the induction of cancer by radiation have been proposed. These mechanisms include (1) the induction of mutations, (2) the activation of oncogenes, (3) the inactivation of tumor suppressor genes, and (4) the induction of cancer-causing viruses. Although the relative importance of the various mechanisms in the induction of cancer is not clear, more than one mechanism could be involved for a given type of cancer.

For both humans and laboratory animals, one cannot currently distinguish between a radiation-induced

TABLE R-5

Thresholds (Lower, Central, and Upper Estimates) for Deterministic Effects of Exposure of the Unborn Embryo or Fetus[a]

Effect	Time/Period[b]	Lower bound (rad)	Central (rad)	Upper bound (rad)
Small head size	0–17 weeks	5	10	Not estimated
Severe mental retardation	8–15 weeks	0	10	20
	16–25 weeks	0	20	50
Death of embryo or fetus	0–18 days	0	10	50
	18–150 days	20	40	50
	150–term (days)	250	300	350

[a] Applies to X or γ rays delivered within 1 hr.

[b] Refers to time after conception in days or weeks.

cancer and a spontaneously occurring cancer (i.e., from an unknown cause). Therefore, statistical methods are used to determine whether radiation exposure is associated with an increase in cancer in a given study population. There have been several epidemiological studies in which definite dose–response relationships have been established for radiation-induced cancers. The best studied populations include atomic bomb survivors, *Tinea capitis* irradiation patients, ankylosing spondylitis irradiation patients, radium dial painters, radium therapy (^{224}Ra) patients, thorotrast patients, uranium miners, and Chernobyl fallout victims.

Atomic Bomb Survivors

Within a 3-day period in August 1945, atomic bombs were dropped on the Japanese cities of Hiroshima and Nagasaki, killing a total of 64,000 people within 1 km of the explosions as a result of blast, thermal effects, and instantaneous gamma and neutron irradiation. Since that time, a prospective epidemiological study has been conducted by a joint group of U.S. and Japanese scientists (the Radiation Effects Research Foundation; RERF) on about 92,000 survivors who were within 10 km of the center of the respective blasts and about 27,000 others who were not in either city at the time of the explosions. The study includes detailed dose reconstruction for about 76,000 individuals and medical follow-up on as many of the survivors as possible. As the follow-up has continued, the RERF has periodically published updates of the cancer incidence and mortality data for these populations. Data complete as of 1988 have shown a total of 3435 cancers, of which 357 were said to be radiation induced. From these data, excess cancer risks are calculated which form the basis for many of the current radiation risk factors in use today. It should be noted that a large fraction of the atomic bomb survivors are still alive, particularly those who were irradiated as children, so that additional information can be anticipated as this population continues to be studied.

Tinea capitis Irradiation

From 1905 to 1960, X-ray irradiation of the scalp for treating ringworm, *T. capitis*, was regularly performed on as many as 200,000 children worldwide. For a typical series of X-ray treatments, doses of 220–540 rad were received by the scalp, 140 rad to the brain, 380 rad to the cranial marrow and <100 rad to other

organs and tissues of the head and neck. Cancers of the thyroid and skin (basal cell carcinoma) were the major consequences of the irradiation.

Ankylosing Spondylitis Irradiation

About 14,000 patients with the disease ankylosing spondylitis received X-ray therapy between 1935 and 1954 in Great Britain and Northern Ireland. In irradiating the spine, doses of 300–700 rad were received by tissues in the thoracic region. The major radiation-related outcome has been an excess of leukemia due to irradiation of bone marrow progenitor cells within the ribs and vertebrae and, recently, an indication of excess solid tumors in lung, esophagus, and breast. The importance of this study has been in the health effects from partial-body irradiation and in the temporal pattern of appearance of solid tumors.

Radium Dial Painters

Radium, as ^{226}Ra and ^{228}Ra, was used in luminous paints in the period 1920–1950. Large amounts of radium were ingested by painters of watch and instrument dials as they tipped their brushes by mouth to achieve a fine point. The radium, once ingested, behaves chemically like calcium and, therefore, deposits in significant quantities in bone mineral, where it is retained for very long times. Being an α-emitting radionuclide, the radium irradiates bone surface-lining cells and has resulted in an excess incidence of osteogenic sarcomas. Of interest in these patients has been the observation of a "practical threshold" of dose and dose rate from ^{226}Ra, below which bone cancers do not appear to occur. This has also been observed in some experimental animal studies.

Radium Therapy (^{224}Ra)

In Europe, the short-lived radionuclide (3.6 day half-life) ^{224}Ra was used for more than 40 years in the early 1900s in treating tuberculosis and ankylosing spondylitis. Because of its effectiveness as an analgesic in treating debilitating bone pain from the latter, its use has continued. ^{224}Ra, being an α-emitting radionuclide that deposits on bone surfaces, delivers its radiation dose effectively to bone-lining cells, inducing an excess of osteogenic sarcoma, similar to those found in the radium dial painters. Interestingly, no excess in leukemia has been found in this population, even though portions of the hematopoietic precursor cell populations are purportedly within range of the α-particle irradiation.

Thorotrast Patients

From 1928 to the 1950s, a preparation of the radioactive, colloidal thorium dioxide (Thorotrast) was used extensively as an X-ray contrast medium in angiographic studies. Because of the very high density of thorium to X rays and the tendency of the colloidal particles to be taken up by the fixed phagocytes within the liver and spleen, it was effective in diagnostic imaging of these organs. However, because Thorotrast is chemically insoluble *in vivo* and is retained tenaciously for long times, long-term α irradiation of liver, spleen, and bone marrow tissues occurred, with a resultant large increased incidence of various liver carcinomas and sarcomas. In this case and unlike the results from Ra exposure, an excess incidence of leukemia has been observed.

Uranium Miners

As part of the radioactive decay series of uranium, ^{222}Rn, a radioactive noble gas, emanates from geological deposits. During underground mining, this gas was released to the work space, and miners inhaled both this gas and its radioactive progeny in significant amounts. Epidemiological studies have been done on mining populations from the United States, Canada, Australia, Czechoslovakia, France, China, and Sweden. Their results have shown conclusively that inhalation exposure to radon and progeny is a strong risk factor for lung cancer, both with and without concurrent exposure to cigarette smoke. This database is used to project lung cancer risk for exposure of the general population to radon in indoor environments.

Chernobyl Fallout Victims

The nuclear reactor accident that occurred in Chernobyl in April 1986 released large quantities of radionuclides to the environment. The contamination was highest near the reactor, with significant fallout also occurring in the western part of the former Soviet Union and spreading to many parts of western Europe. At this point, 10 years after the accident, the medical follow-up of the populations that lived near the reactor has found only one significant disease attributable to the radiation from the accident, i.e., thyroid cancer in persons who were children at the time of the accident. The radiation dose to the thyroid was due to inhalation and ingestion of radioactive iodine isotopes released when the reactor core was breached; estimates indicate that the doses to children's thyroids ranged upward to as high as 1000 rad. The relatively high incidence of thyroid cancer is significant in that it was not expected based on extrapolation from the results of the atom bomb survivors study.

Excess radiation-induced cancers have also been demonstrated in well-controlled studies using laboratory animals (e.g., mice, rats, and dogs). The data from animal studies are being used to supplement the dose–response information obtained from epidemiological studies in humans and recently are providing model systems for the investigations of the mechanisms of radiation-induced diseases such as cancer.

Models Used to Demonstrate Excess Cancers in Populations

Specific risk-assessment models are used to demonstrate an excess in radiation-induced cancer by relating the risk of cancer induction to radiation dose and to other variables and factors such as sex, genetic makeup, the presence of cigarette smoking, and the type of radiation considered. For example, smokers exposed in uranium mines to α radiation from inhaling radon and progeny have a higher risk of lung cancer than do nonsmokers. Also, α radiation is about 20 times more effective than γ rays in producing lung cancer.

Important variables used in risk-assessment models include radiation dose and dose rate, age, and follow-up time. For example, very high dose rates of γ or X rays are thought to be about two times more effective in causing cancer in humans than are very low dose rates. There is also some evidence that a very low doses rate of α radiation can be more effective in producing lung cancer than somewhat higher dose rates. However, this phenomena may be related to changes in the susceptibility of the lung cancer induction with age. It is now known that the ability to repair DNA damage declines with increasing age.

Application of Absolute and Relative Risk Models

Two types of models are usually used for conducting statistical analysis of cancer risks: (1) absolute-risk models and (2) relative-risk models. With absolute-risk models, the excess risk due to exposure to radiation does not depend on the normal risk that would arise when there is no radiation exposure. With relative-risk models, the relative risk is a multiple of the normal risk. Unlike absolute risk, which is measured on a scale that starts at 0 and goes to 1, relative risk values begin at 1 and go to infinity (i.e., very large numbers). A value of 1 for the relative risk means that there is no excess risk.

As an example of application of absolute risk, if the normal risk over the lifetime is 0.001 for a specific type of cancer, and radiation adds an additional risk of 0.01, then the absolute risk of cancer over the lifetime is 0.001 + 0.01, or 0.011.

The relative risk takes into consideration how the normal risk changes with age. For example, if the normal risk of developing a given type of cancer between the ages of 50 and 51 years is 0.001, and radiation exposure leads to a relative risk of 2; then, the relative risk is used to multiply the normal risk so one has to calculate the product 2 × 0.001, or 0.002. Thus, instead of having a normal risk of 0.001 for cancer in the age interval 50 to 51 years, the risk is increased to 0.002 because of the radiation exposure. Similar calculations are carried out for other age intervals depending on the age of the person at exposure and the latent period for the cancer type of interest. The risk for the different age intervals would then be added to obtain a lifetime risk. However, no radiation-related risk would be counted during the latent period. Currently used lifetime risk estimates for cancer induction are largely based on either relative-risk or absolute-risk models.

Current Lifetime Risk Estimates

On the basis of available evidence, the Committee on the Biological Effects of Ionizing Radiations (called BEIR V Committee) has recommended use of a lifetime excess risk (i.e., normal risk has been subtracted) of 0.08 per 100 rad for death from γ-ray- or X-ray-induced cancer. This risk applies to the average person in the U.S. population (all ages considered) exposed to doses up to 10 rad, when delivered in a short time (e.g., minutes to a few hours). When the same dose is delivered over weeks or months, the risk is expected to be reduced, possibly by a factor of 2 or more. The

risk for exposure during childhood is estimated to be about twice as large as that for adults. Males and females are judged to have similar risks. However, all of the cited risk estimates should be regarded as uncertain. These same risks would apply to other radiation sources (e.g., neutrons, β particles, and α particles) if the absorbed dose in rads or Gy were replaced by effective dose in rem or Sv (see Quantities and Units for an explanation of effective dose). The cited risk estimates do not apply to the known subpopulations that are highly sensitive to radiation.

Further Reading

Abrahamson, S., Bender, M., Book, S., Bunchner, C., Denniston, C., Gilbert, E., Hahn, F., Hertzberg, V., Maxon, H., Scott, B., Schull, W., and Thomas, S. (1989). *Health Effects Models for Nuclear Power Plant Accident Consequence Analysis,* Document No. NUREG/CR-4214. U.S. Nuclear Regulatory Commission, Washington, D.C.

Hall, E. J. (1988). *Radiobiology for the Radiologist,* 3rd ed. Lippincott, Philadelphia.

Mettler, F. A., Jr., and Upton, A. C. (1995). *Medical Effects of Ionizing Radiation,* 2nd ed. Saunders, Philadelphia.

National Academy of Sciences, National Research Council (1990). *Health Effects of Exposure to Low Levels of Ionizing Radiation—BEIR V.* National Academy Press, Washington, DC.

United Nations Scientific Committee on the Effects of Atomic Radiation (1993). *Sources and Effects of Ionizing Radiation.* United Nations, New York.

—Raymond A. Guilmette and Bobby R. Scott

Related Topics

Carcinogenesis
Chromosome Aberrations
Developmental Toxicology
Gastrointestinal System
Indoor Air Pollution
Metals
Molecular Toxicology
Occupational Toxicology
Respiratory Tract
Risk Assessment, Human Health
Skeletal System
Skin

*Radon**

- ◆ CAS: 10043-92-2
- ◆ SYNONYMS: Radon-222; nitron; alphatron
- ◆ ATOMICAL CLASS: Unstable radioisotope
- ◆ CHEMICAL STRUCTURE: ^{222}Ra

Uses

Radon is a naturally occurring radioactive gas and environmental contaminant resulting from the radioactive decay of radium (see Radium). It is a colorless, odorless, and extremely dense gas which phosphoresces when condensed into liquid. The half-life of ^{222}radon is about 3.8 days, decaying to ^{218}polonium and ^{214}polonium among other materials. Radon is used in cancer treatment, as a tracer in leak detection, in flow rate measurement, in radiography, and in chemical research.

Exposure Pathways

Radon and its decay products enter the body by inhalation, dermal absorption, and ingestion. The extent to which the population is exposed to ^{222}radon and its daughters (^{218}polonium and ^{214}polonium) in the air, especially indoors, has recently received increased attention. Indoor ^{222}radon and daughter concentrations arise from outside air, building materials, water supplies, and the soil and rock underlying the building. Ventilation rates may be altered to obviate unacceptable levels of radon.

Persons working with radium and its compounds are also exposed to radon.

Toxicokinetics

Radon is transported in the air by absorption on dust particles that are easily deposited in the bronchiolar areas of the pulmonary system. Deposition on the sticky surface of the bronchial epithelial tissue allows for the irradiation of that tissue with α particles and consequent transformation to cancer tissue. Radon daughters are also easily absorbed on solid surfaces, especially colloids and dust particles present in the atmosphere.

* Information Source: Hazardous Substance Data Bank, National Library of Medicine, Bethesda, MD, July 1996.

Short-lived and long-lived radon daughters, produced within the atmosphere and the body, may become selectively distributed to various organs via the bloodstream.

The major systemic threat of these materials is to the kidneys from biotransformed radon daughters. Radon transported by the blood reaches various tissues and organs. Its distribution depends chiefly on the fat content of organs and tissues since it is lipid soluble. From 50 to 90% of the radon body burden is located in the fatty tissues. Radon daughters taken in become localized largely in active deposits in the lungs, to which they represent a grave threat.

Radon is eliminated mainly in exhaled air (about 90% in the first hour and the remainder within 6 or 7 hr), whereas radon daughters are eliminated mainly by excretion in feces and urine.

Mechanism of Toxicity

Radon gas has demonstrated carcinogenicity attributed chiefly to its radioactive properties. Radon gas has been implicated in the occurrence of lung cancer in individuals engaged in mining ores. Miners who smoke cigarettes are at higher risk, indicating a possible synergistic effect between ore dust, radiation, and cigarette smoking. This situation leads to a high risk of cancer in the respiratory tract. Occupancy of radon-containing homes, particularly in the lower floor levels, might also be a cause of lung cancer. Deliberate or inadvertent intake of radioactive elements or their compounds that concentrate in certain organs or tissues may be a cancer risk.

Human Toxicity

Inhalation of dust particles contaminated with radon and its daughters represents the major hazard to human health from these materials. The absorbed material is deposited in the bronchial area of the lung. Before the dust can be cleared from the lung, some of it is absorbed and all of it has irradiated the epithelial surface of the bronchial region of the lung with α particles, creating a significant risk of cell transformation to cancer foci. An increased risk of lung cancer has been associated with radon exposure in uranium miners. This increased risk of the development of respiratory cancer has been well documented. An additive, rather than multiplicative, model has been gaining support to illustrate the connection between smoking and radon daughter-induced lung cancer. Mostly bronchogenic cancers are produced, including squamous cell carcinomas, mixed adenocarcinomas, and, in miners, mostly oat cell carcinomas.

Other human consequences of radon exposure include cataracts, nephritis, and dermatitis. Congenital malformations and spontaneous abortions have also been reported in miners exposed to significant concentrations of radon.

Chromosomal aberrations in peripheral lymphocytes from underground miners have also been reported at significantly increased incidence levels. Peripheral lymphocyte chromosomes from 80 underground uranium miners were studied. Significantly more chromosomal aberrations were observed among workers with markedly atypical bronchial cell cytology, suspected carcinoma, or carcinoma *in situ* than among miners with regular or mildly atypical cells.

Radon and its daughters have been classified as carcinogenic to humans based on extensive data in humans and animals.

Animal Toxicity

Sprague–Dawley rats were exposed to radon progeny up to 82 days. The lung cancer incidence in rats was directly proportional to the lifetime cumulative exposure to radon progeny. Mixed adenosquamous carcinomas, bronchiolar/alveolar carcinomas, and squamous cell carcinomas were observed in treated animals and were significantly elevated above control animals. Exposed hamsters also showed increased incidences of squamous cell carcinomas. Chromosomal aberrations have also been demonstrated in animals exposed to radon. Radon and its daughters are considered to be carcinogenic in animals.

—R. A. Parent, T. R. Kline,
and D. E. Sharp

Related Topics

Indoor Air Pollution
Pollution, Soil
Radiation Toxicology
Respiratory Tract

Ranitidine

- CAS: 6637-35-5
- SYNONYM: Zantac
- PHARMACEUTICAL CLASS: An H2 receptor antagonist. The drug has a furan ring and is structurally similar to histamine.
- CHEMICAL STRUCTURE:

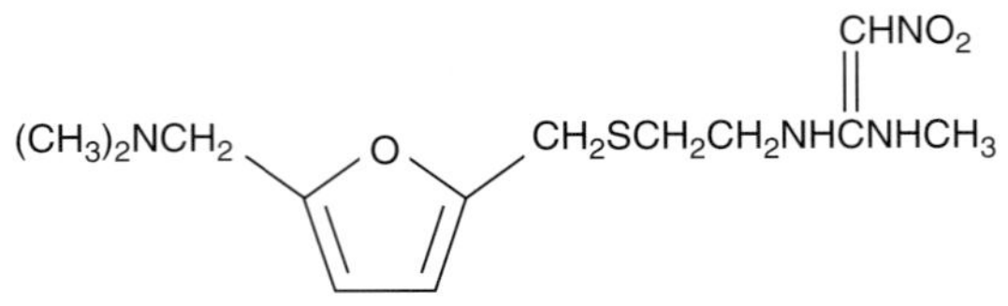

Uses

Ranitidine is indicated for the short-term and maintenance therapy of duodenal ulcers. The drug is also approved for the active treatment of benign gastric ulcers, gastroesophageal reflux disease, pathological hypersecretory conditions (i.e., Zollinger-Ellison syndrome), and erosive esophagitis.

Exposure Pathways

Ingestion or injection are the routes of both accidental and intentional exposure to ranitidine.

Toxicokinetics

Ranitidine is absorbed rapidly from the gastrointestinal tract and undergoes extensive first-pass metabolism. The absolute bioavailability of orally administered ranitidine is approximately 50%. Mean peak serum concentrations occur within 2 or 3 hr following oral doses of 150 mg. Ranitidine is metabolized in the liver to N-oxide, desmethyl ranitidine, and ranitidine s-oxide. Ranitidine is widely distributed throughout the body and is 10–19% protein bound. The apparent volume of distribution is 1.2–1.9 liters/kg. Ranitidine is excreted principally in urine via glomerular filtration and tubular secretion. Approximately 30% of an oral dose and 70% of the parenteral dose is excreted unchanged in the urine. The elimination half-life of ranitidine is

2.5 hr in healthy adults and is prolonged in patients with renal failure (5.9–8.9 hr).

Mechanism of Toxicity
Isolated cases of cardiac toxicity have been reported following rapid parenteral and oral administration of ranitidine.

Human Toxicity
Ranitidine reportedly has less central nervous system (CNS) penetration, endocrine effects, and cardiovascular effects than cimetidine. CNS effects, including hallucinations, delirium, and headaches, have been reported with ranitidine. Bradycardia with dyspnea, tachycardia, atrioventricular block, asystole, and ventricular premature complexes have been reported rarely. Hematologic effects (i.e., leukopenia, granulocytopenia, and thrombocytopenia) and increases in liver function tests have been reported as being associated with chronic and acute ranitidine ingestions. Sexual impotence has occurred in one patient and gynecomastia in another following chronic ranitidine therapy. Acute oral ingestion of up to 18 g has been associated with only transient adverse effects.

Clinical Management
Basic and advanced life-support measures should be utilized as necessary. Appropriate gastrointestinal decontamination procedures should be administered based on the history of the ingestion and the patient's level of consciousness. Electrocardiogram monitoring should be considered along with liver function tests and a complete blood count.

—*Carla M. Goetz*

Red Dye No. 2

- ◆ CAS: 915-67-3
- ◆ Preferred Name: Amaranth

- ◆ Synonyms: Amaranth; FD&C No. 2; 3-hydroxy-4-[(4-sulfo-1-naphthalenyl)azo]-2,7-naphthalenedisulfonic acid trisodium salt
- ◆ Chemical Class: Azo dye
- ◆ Chemical Structure:

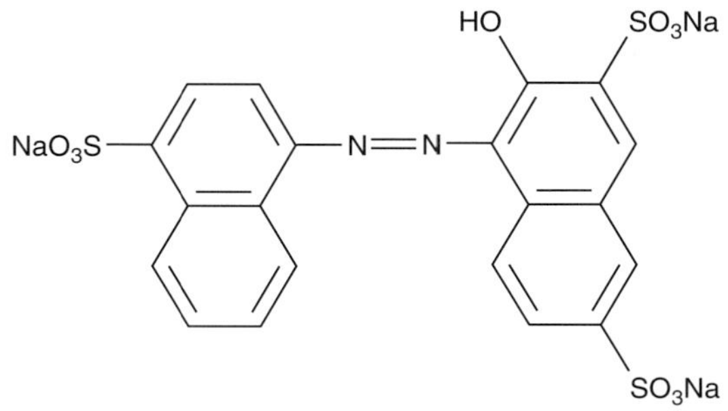

Uses
Red Dye No. 2 was formerly used in food, drugs, and cosmetics but was banned by the U.S. FDA in 1976. It is currently used for dyeing wool, silk, and other textiles as well as paper, wood, and leather products. It is also used as an indicator in hydrazine titrations and is used in color photography and in the manufacture of phenol-formaldehyde resins.

Exposure Pathways
Exposure may occur through oral or dermal routes. Inhalation routes of exposure are unlikely.

Toxicokinetics
When given orally, 2–8% of the amaranth is absorbed from the intestinal tract. Intestinal flora have a reducing effect on amaranth, and azo reduction is also mediated by the hepatic monooxygenase system. Metabolites, including 1-amino-4-naphthalene sulfonic acid and 1-amino-2-hydroxy-3,6-naphthalene disulfonic acid, are excreted in the urine and bile.

Mechanism of Toxicity
Dietary amaranth in animals results in an exfoliating or solubilizing effect on the brush border membrane of the small intestine. Amaranth stimulates *in vitro* RNA synthesis by causing the dissociation of chromatin. Amaranth has also been shown to increase kidney malate dehydrogenase activity after intramuscular dosing.

Human Toxicity

There are no data on the carcinogenicity of amaranth in humans. Some persons are sensitive to azo dyes, with reactions including recurrent urticaria. Children with a sensitivity to amaranth have exhibited behavioral changes.

Clinical Management

Patients exhibiting toxicity or sensitivity should be treated symptomatically. Phenobarbital has been shown to increase plasma disappearance and biliary excretion of amaranth but is not recommended for clinical treatment on a routine basis.

Animal Toxicity

Amaranth did not cause sensitization in guinea pigs, nor was there any significant dermal or systemic toxicity related to dermal treatment in rabbits. An allergic response was observed after intradermal stimulation in the guinea pig. The acute toxicity of amaranth is low: in rats, the intraperitoneal and intravenous LD_{50} is $\geqq 1000$ mg/kg, and the oral LD_{50} in mice is $\geqq 10,000$ mg/kg.

Numerous chronic and transgenerational studies have shown no statistically significant increase in reproductive, developmental, or teratogenic effects due to dietary amaranth. Chronic feeding studies in rats have shown increased mortality, growth inhibition, vacuolar dystrophy, granular deposits in the intestinal tract, increased kidney weight, decreased vitamin A content of the liver, and fatty degeneration of liver cells. However, no histopathological or other effects were noted in beagle dogs fed amaranth for 7 years. Carcinogenicity studies on amaranth have mixed results, with some studies showing skin carcinoma, intestinal carcinoma, lymphosarcoma, mammary tumors, hepatoma, and adenofibroma. Many other studies showed no statistically significant increase in tumors. Amaranth has not tested positive in a variety of *in vitro* mutagenicity studies; however, its metabolites have tested positive in some assays.

—*Janice M. McKee*

Related Topics

Food Additives
Food and Drug Administration

Red Tide

"Red tides" are situations in which minute marine organisms (primarily dinoflagellates) that produce toxins cause red coloring of large areas of the ocean, fish kills, and, potentially, human health effects.

Among the small marine organisms (protistans) are the various protozoans, algae diatoms, bacteria, yeasts, and fungi. The marine protistans are widely distributed throughout neuritic waters and in the world's seas from the polar oceans to the tropics. At least 80 species are known to be toxic to humans and other animals. Most of the toxic organisms are of the order *Dinoflagellata*, of which there are more than 1200 species. Protistans have been shown to contain or release a toxin that (1) gives rise to paralytic shellfish poisoning through the food chain, (2) produces respiratory or gastrointestinal distress or dermatitis in humans, (3) causes mass mortality of marine animals, or (4) has been identified as toxic in laboratory experiments.

Blooms of protistans sometimes occur and result in the phenomenon commonly referred to as red tide or "red water." However, such blooms may appear yellowish, brownish, greenish, bluish, or even milky in color, depending on the organism involved and other factors. Such blooms usually become visible when 20,000 or more of the organisms are present in 1 ml of water. However, some blooms may contain 50,000 or more organisms. The color of red tides probably is due to peritenon, a canthophyl.

Paralytic shellfish poison, variously known as saxitoxin, *Gonyaulax* toxin, dinoflagellate poison, mussel or clam poison, or mytilotixin, is a toxin or group of toxins found in certain mollusks, arthropods, echinoderms, and some other marine animals that have ingested toxic protistans and have become "poisonous." Paralytic shellfish poisoning through the food chain is well-known in both domestic animals and humans (see Shellfish Poisoning, Paralytic).

The amount of poison in the shellfish or other organism is dependent on the number of toxic protistans filtered by the host animal. Off California, mussels become dangerous for human consumption when 200/ml protistans or more are found in the coastal waters. As the count rises, the mussels become more toxic. Within 1 or 2 weeks, in the absence of the toxic protistans, the mussels become relatively free of the poison. The toxin has been studied by extractions from shellfish, from dinoflagellates secured from natural blooms, and recently from laboratory cultures. Paralytic shellfish poison can be obtained from all three sources in similar form.

There is some question about how many toxins exist in the complex of paralytic shellfish poison. In earlier works it was considered a single poison, but it must now be thought of as a complex of toxins. In the dinoflagellate *Alexandrium* (*Gonyaulax*) *tamarensis* there are several other toxins in addition to saxitoxin, and they differ from saxitoxin only in their weak binding ability on carboxylate resins. Further studies on organisms obtained from red tides along the New England coast have resulted in the isolation of two other toxins: gonyautoxin II (GTX2) and gonyautoxin III (GTX3). Another toxin, neosaxitoxin, has also been isolated from *A. tamarensis*.

While a number of pharmacological and toxicological studies on shellfish poisons were carried out in the late 1800s, it was not until 1928–1932 that reports showed that the poison from the mussel *Mytilus californians* was slowly absorbed from the gastrointestinal tract and rapidly excreted by the kidneys. It was said to depress respiration, activity in the cardioinhibitory and vasomotor centers, and conduction in the myocardium. Subsequent studies showed that saxitoxin had a marked effect on peripheral nerve and skeletal muscle in the frog. The "curare-like" action was attributed to a mechanism that prevented the muscle from responding to acetylcholine. The toxin produced progressive diminution in the amplitude of the endplate potential in the front nerve–muscle preparation. It also depressed mammalian phrenic nerve potentials, suppressed indirectly elicited contractions of the diaphragm, and often reduced the directly stimulated contractions. With respect to the cardiovascular system, the toxin was shown to have a direct effect on the heart and its conduction system. It produced changes that ranged from a slight decrease in heart rate and contractile force with simple PR interval prolongation or ST segment changes to severe bradycardia and bundle branch block or complete cardiac failure. The poison provoked a prompt but reversible depression in the contractility of isolated cat papillary muscle.

In 1967, it was demonstrated that this toxin blocks action potentials in nerves and muscles by preventing, in a very specific manner, an increase in the ionic permeability that is normally associated with the inward flow of sodium. It appeared to do this without altering potassium or chloride conductances. It was also shown that in cats, mussel poison blocked transmission between the peripheral nerves and the spinal roots. The large myelinated sensory fibers were blocked by intravenous doses of 4.5–10.3 μg/kg, whereas the large motor fibers were not blocked until this dose was increased by 30–40%. It was suggested that one of the layers in the connective tissue sheath of peripheral nerve is impermeable to saxitoxin, whereas the leptomeninges covering the spinal roots are either deficient in or lack this layer.

During the 1950s, the Canadian–United States Conference on Shellfish Toxicology adopted a bioassay based on the use of the purified toxin. The intraperitoneal minimal lethal dose of the toxin for the mouse was approximately 9.0 μg/kg body weight. The intravenous minimal lethal dose for the rabbit was 3.0 or 4.0 μg/kg body weight, whereas the minimal lethal oral dose for humans was thought to be between 1.0 and 4.0 ng.

—*Shayne C. Gad*

Related Topics

Cardiovascular System
Neurotoxicology: Central and Peripheral

Reproductive System, Female

Introduction

The need for a better understanding of female reproductive toxicology is driven by issues associated with women's health in the context of today's changing society and our current knowledge of women's general health. Modern technology now creates and applies new chemicals faster than they can be tested for their potential adverse health effects. At the same time, more women work outside the home in expanding job classifications, many of which have a potential for chemical exposure that is increased above the typical exposures in the home. In addition, the "reproductive years" for women today have been widened with the development of new medical technologies that permit older women to bear children and younger women to delay their pregnancies. This, in turn, has broadened the age that must be considered as risks for reproductive impairments. In addition, health risks associated with reproductive failure in women are now recognized to extend beyond the issues of family planning and fertility. Reduced bone accumulation in the second and third decade, bone loss, and increased risk of heart disease in later life are just three of the non-reproductive problems which may result from abnormal ovarian function.

A woman's reproductive function is often considered to be more sensitive to environmental perturbations than that of a man's. This concept is supported by studies that demonstrate that reproductive failure in women can be induced by strenuous exercise, marginal nutrition, and acute physical or emotional stress. Despite concerns that the women may be more adversely affected than men when confronted by the same exposures, very little information is available in terms of female reproductive toxicology. Recent reference books on reproductive toxicology generally provide much more information relating to male- than female-related issues because more information is available for male compared to female toxicology. The reason for this disparity is most likely attributable to the practical considerations in experimental design and the availability of critical research materials that make the studies of males more attractive to research programs. For example, in many experimental designs, viable gametes are the ultimate object of evaluation and spermatozoa are plentiful and easily collected compared to ova. Although semen collection and analyses are not simple procedures, sperm production provides a constant objective and quantifiable endpoint for male fertility. Ova, in contrast, are irregularly produced and seldom available for study except in animal studies or under complex clinical management protocols. Even when ova are made available for scientific study, there is little technology available to evaluate their quality compared to spermatozoa.

While it is becoming increasingly evident that gender differences probably do exist in reproductive toxicity, progress in documenting male–female differences has been slow. Perceived and real difficulties in using female animal models in controlled studies have been the major impediment to progress in this area of research. The result of this difficulty is that much more information exists for the male than the female in terms of reproductive toxicology and the inequities in the database will change only when methods are developed which permit the female system to be studied as efficiently and effectively as the male is studied at the present time. Fortunately, toxicologic investigations of female reproduction are becoming more common, new techniques are being developed to monitor women's reproductive health, and the information relating to female reproductive toxicology is growing. As a consequence, the gender gap in information relating to the effects of toxic exposures is closing.

Reproductive Epidemiology

Infertility for humans is usually defined as the failure to conceive following natural attempts to achieve pregnancy for a year or more. Individuals and couples not wishing to have children are seldom evaluated or even surveyed in terms of their reproductive health. In poor economic times the desire for children decreases in developed countries and reproductive health in terms of fertility becomes a less important issue. Thus infertility rates may be higher than current estimates due to insufficient and/or inaccurate information. There is a general lack of epidemiologic information regarding reproduction function in human populations and most estimates relating to fertility are biased toward the clinic population. Most of the existing information concerning human reproduction is obtained from clinical records and does not adequately represent the general population. Only recently have tools been developed to design and evaluate population-based prospective studies assessing basic reproductive physiology.

Recognized exposures of humans to reproductive toxicants have been infrequent and, when detected, were eliminated as quickly as possible rather than studied. Controlled and/or prospective experiments with humans have moral and ethical ramifications; therefore, most of the information relating to the mechanism of action of recognized or putative toxicants on human reproduction is limited. The lack of information relating to real-world exposures combined with the limitation of adequate animal models for human reproductive processes make the study of spontaneous reproductive failures an attractive approach to predict sites and mechanisms of action of putative reproductive toxicants. This approach assumes that induced reproductive failures resulting from toxic exposures will mimic spontaneous events since there is a finite number of control mechanisms that can fail. The spontaneous failures are thought to reveal "weak links" and are the most susceptible targets for the adverse effects of toxic exposure.

The primary problem in studying human reproduction is that many aspects of the reproductive process occur without the knowledge of either the woman or her physician. Ovulation, fertilization, and implantation all occur as "concealed" events. Reproductive biologists and epidemiologists, however, have recently develop new and incisive tools to associate exposures to reproductive health. While it will be years before the "baseline" information is available to be able to apply these broadly, the promise for the future is that reproductive health can be monitored as well as any other aspect of life. Meanwhile, toxicologists use animal models in invasive or terminal experiments to demonstrate the effects of documented and putative toxicants to gain insight regarding their basic mechanisms of action. Because of species differences in the expression of reproductive function, the direct application of these kinds information and the validity of certain endpoints to human reproductive health are not altogether clear.

What are the risks to women in regard to reproductive toxicants at home or at work? Overall increases in reported cases of infertility are difficult to evaluate. The increasing median age of most societies and delays in family planning lead to an increase in age-related fertility which is a natural phenomenon. However, fertility rates of young couples have deceased during the past 10 years and experts speculate that as much as 37% of the reproductive failure seen in modern society could be related to environmental factors. There is no direct evidence that women have been affected more or less than men but this is often assumed. Sperm counts have been assessed and show a decline over this same time period; however, no such similar assessment can be made on the potential of female fertility for comparison. It is doubtful, therefore, that the entire decline in overall fertility can be attributed to men. Traditional epidemiologic studies that are used to monitor menstrual function do not provide information to permit an assessment of fecundity or explain the trends in female fertility. This lack of information regarding female fecundity does underscore the basic theme of this review: Female reproductive toxicology has been, perhaps, one of the most neglected area of toxicology.

Problems Associated with Female Reproductive Toxicology

Female reproductive toxicology is a challenging study area for several reasons. One of the purposes of this review is to delineate some of the problems that reproductive toxicologists face when attempting to investigate putative female reproductive toxicants. This will be approached by focusing on three broad aspects or qualities of female reproduction that most limit progress in this field. These areas include the sensitivity of the female reproductive system to environmental factors, the complexity of the female reproductive system compared to that of the male, and the species specificity in regard to ovarian function. Since it is essential that human and animal female reproductive physiology be understood for the effects of toxic exposures to be studied, each overview of female reproductive toxicology must provide a brief yet detailed description of normal female reproductive physiology and types of reproductive failures. Finally, examples of known reproductive toxicants and their mechanism of action are presented.

Sensitivity of Female Reproduction to Environmental Stressors

The most obvious quality of female reproductive system that influences the way it must be studied is its sensitivity to environmental influences. As mentioned previously, the female reproductive system appears to be more sensitive to environmental perturbations and more complex in its organization than most of the physiologic systems or when compared to male reproductive physiology. Reproductive failure in female animals is often considered to be the first and only detriment resulting from non-lethal adverse environmental conditions and the capacity to reproduce (fecundity) is considered the essential quality of that animal's fitness in its current location. Chemical or physical stressors which are not adequate to perturb other physiologic functions can interrupt or delay reproductive function. In times of acute stress or in response to chronic challenges to the survival of the organism, reproductive function may be selectively suppressed as a short-term adaptive process. Thus, as a non-essential mechanism for short-term survival, reproductive function is sacrificed as an immediate fail-safe strategy for long-term survival. This preferential selection to curtail reproduction in response to non-lethal stressors and the sensitivity of reproduction to nonspecific physical, chemical, and emotional influences make it difficult to identify nonreproductive toxins as being distinctly separate from specific reproductive toxicants. Agents which cause reproductive failure directly and may not be specific reproductive toxins can lead to reduced fertility. In both males and females (but probably to a higher degree in females) the general health of the individual is likely to be reflected in reproductive capacity. Toxicants which have nonreproductive organs as targets may affect sexual development or influence reproductive processes sooner and more noticeably than they affect other physiologic processes.

Complexities of Reproductive Processes

The second issue to be discussed regarding female reproductive toxicology is that of the complexities within the female reproductive system. The degree that female reproductive system is considered to be more complex, compared to the male, is not necessarily the issue since this supposition is debatable. Female reproduction may be recognized to be complex because it has been studied in greater detail and many more aspects of the female reproductive system are defined than for the male. Certainly the female reproductive system is overtly more dynamic and, perhaps because of this dynamicism, more susceptible to physical, chemical, and emotional stressors. The discrete series of events of the ovarian cycle which requires precise coordination between the central nervous system, hypothalamus, and pituitary in order for gametogenesis and ovulation to take place provides the opportunity for environmental changes to adversely influence normal processes. If these events are delayed or altered appreciably, some form of short-term infertility will most likely result. When this is compared to the male, the relatively monotonous production of hormone and gametes is not as likely to be overtly influenced by short-term events.

Female reproductive function, in general, is intermittently expressed, cannot be completely assessed in a single individual, and is often influenced by normal environmental factors. In contrast, most other organ functions are expressed continuously, can be appraised equally well at any point on the time line, and respond predictably to changes in the environment. Most other physiologic processes can be studied in an individual and therefore can be characterize in terms of the biological variation within one individual. In contrast, reproduction can only be evaluated completely when pairs of individuals are studied for prolonged time periods and in some cases when more than one generation is studied serially. Reproductive failure can vary from reduced sexual drive to complete sterility. Infertility can be the result of either functional or organic defects and sub-fertility may be the result of defects at one of several levels of reproductive function (e.g., menstrual dysfunction, anovulation, early fetal loss, and pregnancy loss).

Both the nervous system and the endocrine system are involved in reproductive processes and any number of metabolic processes are essential for normal reproductive function. Both the synthesis and metabolism of neural transmitters and endocrine messengers (glycoprotein and steroid hormones) are critical for normal fertility. Reproduction is a process that includes growth and development of organ systems, gametogenesis, courtship behavior, coitus, gamete transport/interaction, internal fertilization, implantation, gestation, and nurture. Perturbations and/or derangements at any stage in this process can reduce or eliminate fertility. The complete reliance on the endocrine mechanism makes the reproductive system susceptible to the downstream effects of vascular, hepatic, and renal dysfunction.

While reproductive biology is a progressive research field, many of the physiologic mechanisms involved with gamete transport fertilization, implantation, and gestation are still poorly defined. As much as 20% of the clinically described sub-fertility is classified as unexplained and may indicate that a large portion of infertility may be the result of yet undefined environmental hazards. Sub-fertility in human populations is estimated at frequencies as high as 20% in married couples of child-bearing age and over 10% of married non-contracepting women between 20 and 35 years of age.

Experimental designs which involve the female reproductive tract must consider the influence of changing hormonal events. Portions of the female reproductive tract are sequentially modified under the influence of pituitary and ovarian hormones. Many female tissues are induced to proliferate and then differentiate in response to steroid and protein hormone patterns. There are clearly time periods of sensitivity and time intervals that are specific for different toxins or the same toxicant in different species. The concept of precise "sensitive" periods for toxic effects is well established for developmental toxins (teratogen) and this same concept is likely to be true for toxicants which impact reproductive functions such as follicle recruitment, folliculogenesis, gamete transport, and endometrial maturation in the adult female. Very few studies have addressed these issues directly due to the complexities of the experimental design as well as concerns relating to the adequacy of the animal models that are available to study. Long-term testing with sub-lethal doses is currently the approach used to ensure that exposures are delivered at all possible sensitive time periods. This kind of design may have very little relevance to real-world exposures which generally occur acutely.

Historically, toxicologist have viewed the developmental aspects of reproductive toxicology (e.g., teratogenicity and growth retardation) as the fundamental or basic component of reproductive toxicology. The effects of toxicants on adult reproductive processes and organs (those aspects which limit or perturb fertility) are often considered as less important issues. Because of this oversight, much of female reproductive toxicology has been focused on effects of agent conception and pregnancy. In terms of the potential for life-threatening exposures and our responsibility to safeguard the fetus, the emphasis on teratology is understandable. However, it should be recognized that the opportunity for exposure and the potential for adverse effects on reproductive processes is as great if not greater for the nonpregnant woman as it is for the fetus. Currently, the amount of scientific information available and the degree that we understand developmental reproductive toxins are greater than those for nondevelopmental reproductive toxins. For this reason, the following discussion will be limited to perturbations to reproduction success in the adult human female.

Species Specificity of Female Reproductive Physiology

The third issue is that of species specificity of reproductive physiology. The great variation in reproductive function between species creates the greatest challenge for reproductive toxicologists who study the female. Whereas the basic events of female reproductive cycles can be compared between species, the organization of the components of these events is more varied than any other of the physiologic systems. When compared to the kinds of differences observed between species for the other systems (e.g., the cardiovascular, digestive, integument, muscle–skeletal, immune, and respiratory), the physiologic mechanism and expression of reproductive function is more diverse than any other. Even if an abundance of good and practical models existed for human reproductive toxicology, there would still be concerns regarding species specificity of toxicants because of differences in metabolic mechanisms through other organ systems, such as the liver or kidney, and in the expression of reproduction function, i.e., reproductive performance.

The ability to extrapolate data relating to ovarian function from females of one species to females of another is limited. Each species of mammal (there are more than 4000 species) has developed and retained a unique organization of the physiologic processes that make up the complex set of processes which are essential for reproduction to take place. Of the more than 4000 patterns of ovarian cycle organization that we might expect, less than 100 have actually been characterized and this represents the limited numbers of species which have been domesticated or adapted to captivity. It is important to remember that most of our domestic and laboratory species have been artificially selected for reproductive performance and may be more tolerant of environmental influences on their reproductive processes. Very little is understood regarding the effects of multiple stressors on any physiologic system and for most system this can be justified because of the substantial independence of most systems from others. The reproductive system is clearly one for which this simple logic does not apply, particularly when animal modeling is involved.

When the most basic components of female reproductive physiology are compared, such as neural and pituitary control of ovarian function and ovarian morphology and endocrinology of the ovarian cycle, the diversity becomes clear. Differences between species are most easily discussed in terms of the higher nervous center control over gonadal function and organization of the ovarian cycle. Changes in photoperiod, temperature, conditions of the substrate, nutrition, and even nonspecific stressors can modulate gonadal activity through highly species-specific control mechanisms at the level of the hypothalamus. Each species has adapted to reproduce optimally in response to unique environmental conditions which artificial enclosures cannot duplicate. These factors make interspecific comparisons of reproductive performance within a controlled setting difficult.

The laboratory macaque represents the best potential model for the human female; however, expense of maintaining the monkey model, difficulties in handling and manipulating mature monkeys, insufficient baseline data, a limitation of animal resources, the time required to perform multi-generation studies, and inadequate experimental tools make the use of monkeys as a model for female reproductive toxicology severely limited. Since less than the ideal model is usually used, model selection for human reproductive toxicology must be built on a solid understanding of female reproductive physiology and modern trends in reproductive medicine and pharmacology.

Female Reproductive Development and Physiology

Development

Unlike the male phenotype, the female mammal requires little additional directing force beyond the correct genotype. Thus, fetal development in the female is similar in the presence or absence of the normal fetal gonads. However, since the same somatic substrates are present in male and female fetuses, the introduction of androgenic substances to a female fetus will produce the inappropriate development of male-type secondary sex characteristics. This can occur as a result of endogenous adrenal production of weak androgens (congenital adrenal hyperplasia) or exogenous androgenic agents (anabolic steroids) which can cause the development of ambiguous or male-type genitalia and, in the most severe cases, infertility. This ability to respond inappropriately to androgens persists throughout life and females can be virilized at any time, although the sequelae of virilization is generally decreases with age (see Androgens.)

All aspects of ovarian function and adult reproductive normalcy in the female are ultimately dependent on the process of germ cell maturation in the adolescent and adult. Ovarian steroids are responsible for the de-

velopment of the secondary sex characteristics as well as the function and maintenance of the reproductive tract. The ovary can be compared to an undifferentiated organ which retains its embryonic capacity to differentiate throughout the reproductive years. Ovarian stroma cells derive their function only under the direction of a competent hypothalamic–pituitary drive and the presence of developing primary oocytes. If the oocytes are depleted, by accident or age, the ovary ceases to function and all aspects of reproductive function that depend on sex steroid support will regress. The preponderance of reproductive functions are either driven or modulated by sex steroid hormones. For this reason a clear understanding of steroid hormone production and action is essential to understand either reproductive physiology or reproductive toxicology.

Germ Cells

Germ cell numbers are finite in females and are present in the early embryo; having multiplied by mitosis and migrated to the genital ridge, they initiate but do not complete meiosis. Unlike spermatogonia, which continue to be replenished throughout adult life, all of the germ cells are present in a resting stage from the early fetal period to the end of the reproductive life of the female. Usually the oocytes remain in a suspended stage of meiosis which is complete just prior to fertilization. If these original germ cells are lost they cannot be replaced. If they are all lost then ovarian function and all reproductive function will irreversibly cease. Unlike the testis, in which the endocrine and gametogenic activities of the gonad are physically separated, the individual ovarian follicle comprises the combined endocrine and gametic functional unit of the ovary. This is an important difference between the sexes and is reflected in the approaches that are used to assess reproductive health in each.

The "resting" stage of germ cells (primary oocytes) and their vestments of follicle cells are tightly clustered in the cortical portion of the ovary in what is termed the germinal epithelium. The counterpart to this in the male would be the lining of the seminiferous tubules behind the "testis–blood barrier." While the presence of the ovary and its hormonal products is not essential for embryonic development of the female phenotype up to the neonatal stage, the presence of viable germ cells together with their surrounding differentiated gonadal tissue is essential for complete sexual maturation.

Just prior to sexual maturity and under the control of higher nervous centers which control gonadotropin secretion, increased pituitary secretion of gonadotropins stimulates some of the resting oocytes and their surrounding primitive follicle cells to mature. Both the resting oocytes and the undifferentiated follicular cells must be present in the germinal epithelium for this earliest phase of normal ovarian function to occur. In the absence of viable oocytes follicle cells will not develop and, as a consequence, there will be no response to gonadotropin stimulation.

Ovarian Function

At the onset of puberty, the undifferentiated ovarian stromal cells in the immature ovary, in response to gonadotropin secretion and their close proximity to a viable, resting primary oocyte, differentiate and develop the capacity to produce sex steroid hormones (primarily androgens and estrogens). These sex steroids are directly responsible for the development and maturation of the secondary sex organs and complete the process of sexual maturation (puberty) at the appropriate time. The follicular events associated with this follicle activation require the differentiation of the previously undifferentiated primitive follicle cells into two separate cell types; the theca and the granulosa cells. The theca cells produce androgens (androstenedione and testosterone) from acetate and circulating cholesterol and depend on the granulosa cells to convert the androgens to estrogens. Increased production of androgens from the theca cells or decreased ability to aromatize these androgens by the granulosa calls can lead to virilization (masculinization) and infertility.

In the earliest stages of puberty, ovarian follicles develop to the stage of producing estrogen but do not ovulate. This period of ''adolescent sterility'' in primate species is associated with adequate estrogen stimulation for development of the secondary sex characteristics, general sexual development, and sexual maturity. Inappropriate release of gonadotropins (luteinizing hormone and follicle stimulating hormone from the anterior pituitary gland) prior to the age of normal puberty will induce follicular development and precocious sexual development since all of the components of adult ovarian function are present at birth and only lack gonadotropin stimulation for complete, normal function.

The Ovarian Cycle

The hormonal events associated with ovulation are incompletely understood. Simplistically, a feedback loop is established between the hypothalamus–pituitary and the ovary at the onset of sexual maturity. Pituitary gonadotropins function to stimulate primitive follicles to secrete estrogen, progesterone, and peptide hormones which modulate both the pituitary and the hypothalamic control of reproduction through positive and negative feedback loops. Collectively this is referred to the hypothalamo–pituitary–ovarian axis (HPO axis). The maturing follicle secretes increasing amounts of estrogen prior to ovulation, which modulates the secretion of gonadotropins, orchestrates sexual behavior (in some species), and prepares the reproductive tract for mating, gamete transport, and potential conception. Ovulation occurs during a limited time period within the complete ovarian cycle and the extruded ova are fertilizable for only a short time period. Thus, there is a great deal of importance regarding the synchrony of ovulation and mating. Precise synchrony of these events is achieved by the action of estrogen from the maturing follicle acting upon the central nervous system, the pituitary, as well as the reproductive tract.

Gametogenesis in the female (folliculogenesis and ovulation) is intimately associated with hormone production patterns and can be monitored by measuring changes in hormone production rates. Although the peri-ovulatory LH surge requires increasing estrogen in order to be elicited, the final release mechanisms vary between species. For some rodents the hypothalamic release of neural peptides (catecholamines, indolamines, and specific gonadotropin releasing factors) is closely associated with the diurnal light–dark exposure; in other species copulation is an absolute requirement; and in others, such as primates and most domestic and laboratory animals, it occurs as a result of follicular maturation only. For all species, the coordination between the higher nervous centers, the pituitary, and the ovary is mediated through hormone signals from the maturing follicle cells. Thus, in the female, hormone patterns which represent HPO activity are precise and appropriate indicators of reproductive status and potential fertility. This should be contrasted to the male in which the gamete itself is usually evaluated and endocrine parameters have limited clinical significance.

Ovulation is associated with a surge of gonadotropin release which is the result of estrogen positive feedback.

This massive release of gonadotropins probably functions primarily as a fail-safe mechanism to complete the ovulatory process which was initiated through follicle maturation and the synergism of gonadotropins and estrogen acting upon the mural granulosum. The LH surge secondarily functions to convert the original vestments of the oocyte (follicle cells) into a different cell type (the corpus luteum) which will secrete another sex steroid (progesterone) during the pre-gestational period. If conception does not occur, progesterone secretion by the ovary is limited to the time interval following ovulation and may be produced for as short as 2 days or as long as 3 weeks depending on species. Progesterone action serves to prepare the reproductive tract, primarily the lining of the uterus (endometrium), for the embryo implantation and also acts centrally to prevent addition follicles from developing during this time interval.

Following ovulation the ova that are extruded onto the surface of the ovary (either into the body cavity or within a bursa) are picked up by the oviducts. The oviducts serve as conduits which transfer both fertilizable ova toward the uterus and allow selected spermatozoa from the lower reproductive tract to meet and fertilize the ova. Following fertilization the resulting zygote is transported to the uterus where successful implantation can take place. The function of the oviducts, which are responsible for gamete and zygote transport, is controlled by the ovary through estrogen and progesterone production throughout the ovarian cycle and is dependent on pituitary gonadotropin support. When implantation does occurs, it is usually 2–6 days following ovulation and fertilization; however, delays of implantation can be as long as several months in species such as bears, seals, most mustelids, and some edentates.

The embryo survives unattached to the uterine lining for 5 days to 7 weeks in laboratory and domestic species. During this pre-implantation period, nutrient requirements are absorbed from the materials within the uterine lumen. After this time some form of stable attachment is formed between the trophoblast (primitive placenta) and the endometrium. This can range from a very superficial apposition with many cell layers separating maternal and fetal circulations to true implantations with only one cell layer separating the two vascular beds.

Hormone Action

Neuropeptides and polypeptides derived from neural tissue are responsible for pituitary function. The neurotransmitters (catechols, indoles, endorphins, and dopamine) act directly or indirectly to cause the release of the gonadotropins or prolactin and exert their action through synaptic junctions to alter neural activity in the hypothalamus and other areas of the brain. The polypeptide hormones (gonadotropin hormone releasing hormone, adrenocorticoid releasing hormone, and thyroid releasing hormone) act through membrane receptors and transduce their signal by intracellular second messengers such a cAMP, calcium, and/or phosphoinositol. The fetal pituitary is capable of responding to higher nervous centers by mid-gestation but does not do so until these centers "awake" at the time of puberty long after birth. Premature "awakenings" of these centers lead to premature sexual development and arrest of the normal somatic growth pattern through the action of sex steroids on bone growth. The absence of the awakening of the central nervous centers that control pituitary function leads to a failure to undergo sexual development.

The primary pituitary hormones that influence female reproductive function are two glycoproteins (luteinizing hormone and follicle stimulating hormone) and one protein hormone (prolactin). The glycoprotein hormones (LH and FSH) are stimulated to be synthesized and released by a single polypeptide neurohormone, gonadotropin releasing hormone (GnRH), and inhibited by both ovarian steroid (estrogen and progesterone) and one ovarian peptide (inhibin) hormone through negative feedback loops which impinge both at the level of the hypothalamus and the pituitary itself. Derangements which lead to spontaneous or early release of the gonadotropins are rare as the GnRH "drive" is an essential stimulation. The opposite is true for prolactin because it is controlled primarily for the negative actions of dopamine. Any action which decreases the domaminergic drive to the pituitary will lead to hyperprolactinemia and reproductive dysfunction relating to prolactin excess. Failure of the pituitary to release gonadotropins is common particularly in women and can be the result of inadequate hypothalamic support or inability of the pituitary to manufacture and release gonadotropins. When this occurs prior to puberty delayed maturity and infantilism are the result.

If this occurs following puberty, then ovarian function and menstrual periods cease.

All aspects of female reproduction are regulated by ovarian sex steroids. These small lipids, which are ubiquitous to all species, act to both develop and differentiate all secondary sex characters. The principal female sex steroids are estradiol and progesterone. Estradiol is mitogenic in estrogen-sensitive tissues and is responsible for end organ growth and proliferation and acts through estrogen receptors that are constitutive in all estrogen-sensitive tissue. Progesterone can be mitogenic and/or differentiating, depending on the tissue. In the uterine endometrium progesterone differentiates the "proliferated" endometrium and decreases the action of estrogen by decreasing estrogen receptors. In the breast progesterone compliments the proliferative action of estrogen. All progestational cells require the antecedent action of estrogen in order to express progesterone receptors. Exogenous compounds, either natural or synthetic, can mimic endogenous hormones and cause infertility, inappropriate somatic changes, or induce hyperplastic disease. Examples include the synthetic steroid hormones and oral contraceptives, which have been produced to regulate fertility. These compounds were created to be able to be easily ingested or absorbed and possess unusually long biological half-lives. If an exogenous ligand has androgenic activity then the results of female exposure are similar to those of excessive adrenal androgen production as described previously. Estrogenic or progestational agents would have different, nonmasculinizing effects but still can result in infertility by interfering with the normal signals which are sent by the ovary to the hypothalamus, pituitary, and reproductive organs.

The effects of steroids are limited to cells that contain steroid hormone receptors. Each steroid has its own specific receptor but each steroid receptor has a strong structural and functional relationship to other compounds with a similar structure. While each steroid has specific effects on specific target organs, these effects can general be divided into three categories: mitogenic (proliferative), differentiating, or regulatory. Estrogen has all three effects, is the principle female sex steroid hormone, and will be discussed in detail.

Because of the pivotal importance of estrogens it is somewhat surprising that many naturally occurring compounds, other than hormonal estrogens, have estrogenic potential due to structural similarities to the steroidal estrogens (estrone, estradiol, and estriol) and their ability to bind the estrogen receptor. It is important that toxicologists understand that many xenobiotics have estrogenic capacities for the same reason. Since estradiol is the primary estrogenic hormone for all species, the estrogen receptor that transduces the "estrogenic message" has been conserved to a great degree. Conservation of both the ligand and the receptor should permit a great deal of uniformity of estrogenic activity of different compounds between species. This, however, is not the case. Estrogens of all types have different effects in different species. In terms of toxins, for example, the "clover disease" in sheep caused by phytoestrogens found in certain legumes and tubers does not occur in other species even though the same phytoestrogen is found at the same circulating concentrations.

In humans, estrogens are primarily responsible for the development of female sex characteristics. The development of vagina, uterus, and fallopian tubes as well as the breasts, fat deposition for body contours, the pubertal growth spurt, pubic/axillary hair, and pigmentation of the genital region and areolae are all also a part of estrogen action, although androgens are probably also involved. Some overlap exists between the action of estrogen and androgen in terms of anabolic effects, but in general estrogens oppose or have opposite effects of androgens. The loss of estrogen at menopause leads to decreased bone deposition, decrease turgor of the skin, and sclerosis of the blood vessels. The relative immunity to coronary disease and gout is also dependent on estrogen and is lost at menopause. In animals, estrogens are psychogenic and the desire to mate or become receptive to males is a direct action of estrogens. The effects of estrogens on emotions are not well defined in humans, although changes in emotion can be striking following the menopause.

Toxic effects of estrogens are generally considered to be the adverse effects observed when estrogen is given in supra-physiologic doses or for inappropriate time intervals. Since the normal pattern of estrogen production is quite variable and the effects of even normal patterns have a wide range of effects in different individuals, it is difficult to separate some adverse estrogenic effects from "normal effects," i.e., swelling and soreness of the breast, morning sickness in early pregnancy, and dysfunctional uterine bleeding. In very high doses estrogens can cause water retention, thus edema related to heart failure or renal disease could be accentuated by extremely high doses of estrogen. The effect of estrogen on liver function varies tremendously between species. In birds and fish, for instance, estrogens mobilize large amounts of lipids for egg production to the point that the serum become milky in appearance. In humans, estrogens change the pattern of circulating lipids (they can be considered "protective" for circulatory diseases) and this effect is of great interest to those who study heart disease and atherosclerosis.

In women, menstrual function disturbances can be an early and accurate indicator of estrogen imbalance. In other species which do not slough the endometrium, other endpoints need to be assessed. Assessments of estrogen deficiency can be particularly difficult to detect particularly for the effects of estrogen antagonists or antiestrogens. Estrogen-induced changes are not only different between species but also different within the same species at different levels of the reproductive system. In some species a carcinogenic action of estrogens has been described as for diethylstilbestrol (a nonsteroidal estrogen). In most studies the ability of estrogens to cause tumors is largely attributed to a genetic predisposition for tumor formation and has caused unnecessary fears for its use therapeutically; the exception, however, is diethylstilbestrol, which, when taken by pregnant women, leads to hyperplastic disease in the female children. Estrogens are both mitogenic and tend to increase there own receptor numbers and these effects are counteracted by progesterone. Thus, prolonged exposure to estrogens in the absence of progesterone can cause abnormal growth of proliferative tissues like the endometrium.

Progestins are the generic term for compounds that exert an effect similar to that of progesterone. Progesterone is the second most important steroid hormone in the female reproductive function. Progesterone receptors are induced by the action of estrogen, thus progesterone has little effect in the absence of estrogen priming. The primary role of progesterone in the uterus is to create an implantation site for the embryo. Insufficient progesterone leads to an inadequate implantation site and implantation failure. Progesterone plays a key role in pregnancy by reducing uterine contractile tone and preserving the pregnancy. Within the uterus progesterone causes a decrease in estrogen receptors, thus attenuating the mitogenic effect of estrogen on this tissue. Similarly, progestins reverse the actions of estrogen on cervical mucus, labial color and swelling, and

sexual behavior. In the breast, however, progesterone acts to augment the mitogenic effect of estrogen and acts to proliferate the epithelium of the ducts in preparation for milk production.

Female Reproductive Failure

Reproductive failure can be induced by environmental hazards at any level, although all levels have not been linked to environmental factors. Since it is not ethically possible to observe or impose reproductive toxic exposures on human subjects, much of what we believe about the effects of putative toxins are based on the assumption that well-defined spontaneous reproductive failures are accurate surrogates for environmentally induced defects. Spontaneous reproductive failures, together with results from animal experiments, are used as models to predict or understand the impact of reproductive toxins on the human system.

Ovarian senescence due to increasing age (menopause) is another form of infertility that is a normal consequence of the aging process and can be compared to the effects of an ovarian toxin. As oocytes are depleted through the normal process of atresia, the reproductive system responds in much the same way as it would to an ovarian toxin that acts to destroy oocytes. In both cases, the absence of gonadal hormones results in increased pituitary drive in compensation for the decreased negative feedback from the ovaries which, without germ cells, cannot produce steroid hormones. The increased gonadotropins (hypergonadotropism) cannot, however, compensate for the irretrievable gonadal deficiency (hypogonadism).

Defects which act to suppress hypothalamic or pituitary function also lead to infertility but are expressed differently. Kallman's syndrome (anosmia with isolated gonadotropin deficiency), for example, is a form of hypothalamic deficiency in which the pituitary is not stimulated to release gonadotropins due to a defect at the level of the hypothalamus and higher nerve centers. The gonads of affected individuals remain in a preadolescent condition and sexual maturity is never achieved. In such case the disease simulates a central nervous system toxin which leads to reduced gonadotropin release (hypogonadotropic) and a subsequent reduced ovarian activity (hypogonadism). Since the pituitary is not compromised in this condition the appropriate administration of the hypothalamic factor which causes the release of gonadotropins (GnRH) will restore pituitary and subsequently gonadal function as well. Physical, nutrition, or even emotional stress can lead to different degrees of "hypothalamic amenorrhea," which is a general term for hypogonadotropic hypogonadism in women. Professional dancers, athletes, and over-zealous dieters can exhibit this reversible form of infertility at any stage in life and, as a consequence, this type of reproductive failure occurs relatively frequently. Such cases can be used to model theoretical toxicants which block normal hypothalamic or pituitary function.

Many kinds of organic or functional defects can lead to post-ovulatory reductions in fertility. For instance, anatomical impediments to gamete transport can prevent fertilization. Poorly developed or insufficient endometria due to end organ insensitivity to steroids, insufficient steroid hormone production, or impediments to steroid action at the level of the endometrium will not adequately support the implantation site of an otherwise healthy embryo. Previous reproductive tract infections are responsible for the majority of these kinds of reproductive failures; however, developmental defects and alterations in organ function caused by inappropriate stimulation or response also contribute.

Reproductive Toxicants

In recent years public concern regarding toxic exposure has focused on the potential of reproductive hazards resulting from agricultural and industrial chemicals. This has led to the suggestion that a significant amount of the recognized reproductive failure among humans and animals can be attributed to increased toxic exposures. The increasing number of female workers in industry as well as the recent recognition of hazards to female reproduction in the workplace has heightened concerns relating to female reproductive toxicology. Such concerns, however documented, have resulted in an increase in risk assessments of both putative and real reproductive toxicants as well as in the creation of regulations concerning disposal of and exposure to xenobiotics. Progress in this area, however, has been slow for a number of reasons.

Reproductive toxicants obey the rules of other toxins and their effect can usually be linked to some interruption of normal physiologic mechanism such as galactose and its interaction with ovarian transferases and ki-

nases. They can act directly by inducing a change through their inherent chemical activity such as the purine analogs, which interfere with the normal process of oogenesis and predictively have greatly different effects at different stages of reproductive development. Some reproductive toxicants mimic or block hormone action by virtue of their structural similarity and mimic or antagonize endogenous messengers. Other toxicants act through their own receptors and interact with hormone transduction signals within the cell. Some toxicants act directly while others must be metabolized to an active form before they can exert their adverse effects. Other reproductive toxicants act indirectly after being metabolized from an inert compound to a form that is chemically or biologically active. The polycyclic aromatic hydrocarbons exert their effects indirectly by inducing hepatic and ovarian enzymes, which govern steroid production and metabolism, and act by transducing adverse signals or signals that impede normal physiologic functions.

The identification of chemical compounds of high concern as human reproductive and developmental toxicants was provided by a report from the Government Accounting Office (GAO) in 1991. That report reviews the evidence that identifies compounds as male, female, or developmental toxicants and what safeguards are in place to protect the public. The compiled list reveals 30 compounds, 21 of which have adverse reproductive effects in women or female animals. These compounds include industrial solvents (toluene, ethylene glycol monoethyl, and monomethyl ethers); metals (cadmium and lead), pesticides, fungicides, and fumigants (clordecone and its metabolite mirex, DDT, ethylene dibromide, ethylene oxide, hexachlorobenzene, and the pesticide contaminant dioxin); halogenated hydrocarbons (vinyl chloride, PBBs, and PCBs); products of combustion (carbon disulfide, carbon monoxide, and tobacco smoke); as well as arsenic, diethylstilbestrol, and warfarin.

The GOA list of reproductive toxicants does not include some putative reproductive toxicants which are currently highly regulated or banned for industrial use such as benzene, benzamine, chloroprene, formaldehyde, styrene, and xylene. Of the toxicants listed as female reproductive, approximately half have strong evidence of direct adverse effects on human (or nonhuman primates) female reproduction separate from their action as developmental toxicants and teratogens. A number of compounds such as the glycol ethers are very likely reproductive toxicants but reports demonstrating this effect are only now appearing in the literature. Thus, the list of compounds for which there is strong evidence of adverse effects on fertility, menstrual function, or other gynecological disorders in nonpregnant women can be condensed to the 14 individual or groups of compounds. These are listed in Table R-6 along with the adverse effect that was associated with each. The actual list of compounds which have adverse effects on female reproduction is undoubtedly much longer than this list indicates, and many of the compounds which are documented for adverse male effects are likely female toxicants as well. However, until effects of exposures of women are documented they cannot be included.

Table R-6 not only demonstrates the relatively small number of documented human female reproductive toxicants but also underscores the difficulty of investigating exposures to reproductive toxicants in human populations. There is a lack of specific knowledge in terms of the targets and mechanism of action of most reproductive toxicants because the end-point for recognizing the adverse effect is "downstream" of the actual target. The literature lists more than half of the adverse effects as only "menstrual dysfunction," which provides little help in identifying a specific site or action. This general outcome is reported because it is the only relevant endpoint that is usually available during the

study periods which usually follows the actual exposure. Assessment fertility, for instance, would need to include a relatively large number of women who were simultaneously exposed to the possibility of pregnancy over an interval of time which would permit adequate pregnancies to occur and be completed.

Regardless of the number of women exposed and the time of exposure, most reproductive toxicity data are collected in retrospect and using the subjects' recall as the source of information relating to reproduction. Practical endpoints for assessing the target of toxicity other than the woman's menstrual calendar have not been available historically and only general symptoms such as menstrual function can be recalled and reported. As reviewed earlier, menstruation is the normal result of an ovulatory ovarian cycle and ovulatory cycles can be quite variable in length and regularity. Irregular menstrual cycles may be typical for some women and not for others and a woman's recall of her previous "regularity" may not be accurate. In addition, vaginal bleeding for other reasons which may have characteristics of true menstruation may occur in the absence of ovulation, e.g., breakthrough bleeding as a result of unopposed estrogen stimulation. While the listing of adverse effects as menstrual dysfunction may be adequate to indicate that female reproduction has been perturbed, it provides very little information as to the target or mechanism of action, nor does it provide information to the health risk except in the most severe cases.

A deeper understanding of the site of toxicity and the mechanism of action can come only from controlled animal studies in which basic hypotheses are tested using laboratory rodents or primates. As indicated previously, there are concerns of species specificity in terms of sensitivity or response to reproductive toxicants, routes of exposure, and relevant dosage that make this less than a perfect science. However, knowledge of the similarities and differences in the reproductive physiology of the model species to human function as presented earlier in this section, together with a knowledge of human reproductive health and disease, permits a great deal of information to be obtained from laboratory animal studies. The complete understanding of basic reproductive physiology allows the toxicologist to focus on specific targets of toxic action. It is from understanding of basic reproductive physiology, the experiments of nature provided by spontaneous reproductive diseases, and laboratory experiments with animal mod-

TABLE R-6
Condensed List of Human Female Reproductive Toxicants
and Their Adverse Effects

Toxicant	Effect
Benzene	Menstrual dysfunction
Benzamine	Menstrual dysfunction
Chloroprene	Menstrual dysfunction
Formaldehyde	Menstrual dysfunction
Mercury	Menstrual dysfunction
Halogenated hydrocarbons	Menstrual dysfunction
Anesthetic gases	Infertility
Toluene	Menstrual dysfunction
Styrene	Menstrual dysfunction
Diethylstilbestrol	Inferility
Ethyl oxide	Abortion
Lead	Abortion
Vinyl chloride	Ovarian dysfunction
Polyaromatic hydrocarbons	Infertility

els that targets of toxicity on functional and anatomical bases are appreciated. These targets are defined in the following sections.

Central Targets

The organs, nuclei, and organelles that are required for normal pituitary secretion of gonadotropins are considered to be the central targets. They are generally divided into the neural tissues, nerve tracts, specific nuclei in the brain, and their organelles (including the hypothalamus and higher nervous centers) and the anterior pituitary gland. In some cases the pineal gland would also be considered because of its direct effect on pituitary function.

Hypothalamus and Higher Brain Centers

Toxicants which disrupt the synthesis of GnRH or its normal pulsatile release will cause reproductive failure by way of pituitary dysfunction. There are two general mechanisms for this to occur. The direct effect is one in which neural transmission is altered by other neurotransmitters or their analogs. Anesthetics, anticonvulsant and recreational drugs of abuse, are examples of agents that can cause hypothalamic dysfunction which in most cases decreases in GnRH pulse and amplitude. These kinds of toxicants can reduce neuronal firing rate and reduce either the baseline gonadotropin secretion or block the midcycle periovulatory surge in laboratory animals. In human subjects decreased nutrition, increased exercise, as well as physical or emotional stress can lead to similar derangements of the hypothalamic–pituitary axis by increasing catecholamine, indolamine, and endorphin levels with oligomenorrhea or amenorrhea as a result. Some compounds such as the ergot derivatives can mimic dopamine action and reduce prolactin secretion. The indirect effect is one in which the normal "long-loop" feedback mechanisms are altered. Bioactive steroid hormones or their analogs can inappropriately increase or decrease hypothalamic drive leading to alterations in pituitary gonadotropin secretion. Increased adrenal glucocorticoid, for example, is thought to decrease gonadotropin secretion although the precise mechanism in not known. Diethylstilbestrol is a model for a toxicant which might decrease hypothalamic drive because it is a potent estrogen agonist. In contrast, tamoxifen or clomiphene citrate, which are estrogen antagonists, would have the opposite effect.

Some of the halogenated hydrocarbons are thought to act as estrogen antagonists and may influence hypothalamic function by acting through estrogen receptors.

Anterior Pituitary

The anterior pituitary can also be adversely influenced by two separate but general mechanisms. It can be directly affected by changes in the stimulatory effect of GnRH from the hypothalamus and it can be modulated by ovarian steroid and peptide hormones from the ovary (as discussed previously). Perturbations of the hypothalamus are transduced directly to the pituitary through the primary GnRH signal; thus, normal pituitary function is unlikely when the hypothalamic drive is perturbed. Because of the location of the hypothalamus and pituitary at the base of the brain and the intimate vascular and neuronal connections between them, it is difficult to separate toxic actions which occur at the hypothalamic or pituitary level and are often considered together as simple "central" as opposed to actions at the level of the ovary, reproductive tract, or related organ.

Inappropriate circulating levels of bioactive steroid hormones or their analogs can lead to perturbations of pituitary function. Increased blood concentrations of bioactive estrogen, progesterone, androgen, or their analogs will lead to decreased secretion of gonadotropins. Steroid antagonists will open this feedback loop and cause increased amounts of gonadotropins to be secreted. The therapeutic basis of oral contraception and one aspect of fertility enhancement are based on these principles. Many halogenated hydrocarbons are thought to act as estrogen analogs and act to either transduce false signals through the estrogen receptor or block endogenous estrogen from exerting its normal action. The latter case is well defined for DDT, which causes thin egg shells in exposed birds. PCBs, for instance, have been suggested to have agonistic and antagonistic action in different animal species and different organs within the same species. It is not clear how many chlorinated biphenyls have estrogenic effects or if any of these compounds are serious hazards to women. It is difficult to separate actions which occur at the hypothalamus and pituitary; therefore, these are often considered collectively as "central" effects as opposed to effects that occur downstream such as at the level of the ovary.

New evidence is now emerging that some halogenated hydrocarbons exert their effects through a receptor other than the estrogen receptor. This receptor, which has been termed the Ah receptor (Ah = Aryl hydrocarbon), has no known physiologic role but acts much like the receptors of the steroid hormone super family of receptors. The binding of the Ah receptor to its ligand, which can be dioxin or related coplanar, chlorinated biphenyls, elicits transcription of new proteins and/or blockage of other proteins such as estrogen receptors.

Ovarian Targets

Ovarian tissue can be compared to embryonic tissue in that most of its functional elements are still in various stages of development. All of the endocrine aspects of the ovary are differentiated at the time that a subpopulation of germs cells mature. Both the endocrine and the germ cell populations are transient populations which must be renewed with each reproductive cycle. The ovarian targets of toxicity are therefore ever-changing populations of different cell types. For this reason it has been difficult to identify cytotoxic agents which have specific ovarian cell types as their unique target. In general, ovarian targets can be divided into two categories. The most important category is the germ cells, which are primarily primary oocytes in a resting stage of meiosis. The second category is represented by the cells which produce steroid and peptide hormones.

Germ Cells

Unlike the testes, in which steroid production can proceed in the absence of spermatogenesis, the ovary can function as an endocrine organ only if viable germ cells are in residence. Toxicants which eliminate the resting germs cells automatically eliminate all endocrine function. Since all aspects of female reproduction are dependent on ovarian steroids, the growth, development, and integrity of the entire reproductive system will be disrupted by loss of the germ cells. A complete loss of ovarian function would ensue and in humans, menstruation would cease as it would with complete hypothalamic–pituitary dysfunction. In contrast, toxicants which adversely affect only the oocytes which have ended their resting phase and begun to mature will interrupt only the current ovarian cycles as additional oocytes can be recruited from the resting germs cells. Such compounds may be "silent" hazards having the effect of delaying conception only slightly. Exposure to such toxicants would most likely be recognized through

menstrual dysfunction, long menstrual cycles, and possibly as a delay to conception.

Ovarian Steroid Secreting Cells

Steroidogenic cells within the ovary are also transient cell populations. While the development of gonadotropin receptors and steroidogenic machinery is dependent on the proximity and continued viability of a healthy oocyte, the mature cell will survive only the length of the reproductive cycle and possibly through one pregnancy. The steroidogenic cells of the ovary are recruited each cycle from undifferentiated ovarian stroma. Agents which arrest differentiation, block the expression of gonadotropin receptors, or block the production of steroid hormones will have adverse effects on ovarian function. Steroidogenesis can be blocked by either blocking the transport and availability of cholesterol, compromising the reducing capacity of the cell, or by direct block of steroidogenic enzymes. Such disruptions would be recognized as menstrual dysfunction ranging from irregular menstrual cycles to complete amenorrhea if steroidogenesis is completely stopped.

Reproductive Tract Targets

The female reproductive tract is completely dependent on the functioning ovary to provide estrogen and progesterone for its growth, development, and function. Reduced steroid production, increased clearance of circulating steroid hormones, or antagonism of steroid action at the level of the steroid hormone receptor will lead to decreased size and function of all aspects of the female reproductive tract. Increased circulating concentrations of sex steroids or their agonist generally leads to hypertrophy, hyperplasia, and dysfunction. Sex steroids or their analogs at relatively high circulating concentrations will disrupt pituitary function through the long-loop feedback. However, low levels of steroid analogs for prolonged time periods may have adverse effects on the reproductive tract without disrupting the HPO axis. Such theoretical toxicants could cause infertility with no other overt signs. An example of this kind of toxicant is a low-dose progestin therapy which, in some women, is an effective contraception although relatively normal menstrual cycles are observed, suggesting adverse effects at the level of the endometrium while exerting no demonstrable effect at the level of the HPO axis. Similarly, weak sex steroids could act locally to alter cervical secretion, reducing sperm survival and transport through the reproductive tract as observed in sheep exposed to plant estrogens.

Over 300 different plants contain either compounds with estrogenic activity or precursors for the formation of nonsteroidal estrogens. Coumestrol, equol, and zearalenone are examples of phytoestrogens that are found in legumes, tubers, and fungi that infest grains. These substances clearly act as reproductive toxins in sheep (equol in clover disease) and pigs (zearalenone in moldy corn syndrome). Evidence is not as convincing for carnivores fed commercial diets with plant "fillers." In humans there is some evidence that Asian diets act as a protectant for some forms of hyperplastic disease and some claims for precocious puberty being the result of contamination with environmental estrogens have been made. Although there are structural similarities between the parent phytoestrogen molecules and DES, it is speculated that the only phenolic metabolites of these compounds are active compounds since pretreatment with carbon tetrachloride (to inhibit the mixed function oxidase in the liver) reduces the estrogenic *in vivo* potency of *o,p′*-DDT in rats. However, *in vitro* studies indicate that *in vitro* competition of compounds, such as *o,p′*-DDT and methoxychlor, with estradiol for binding the rat uterine estrogen receptor, is positively correlated to *in vivo* estrogenicity. The fact that the estrogenicity of either the parent compound or its metabolite competitively competes with estradiol for receptor binding may limit the action of circulating steroidal estrogens. By limiting the action of the more potent steroidal estrogens the weaker nonsteroidal estrogens may act as antiestrogens. In addition, some DDT analogs such as *p,p′*-DDT are thought to act by inducing liver enzymes which metabolize endogenous steroidal estrogen, thus reducing normal estrogen delivery to the target tissue.

Substances which increase or decrease smooth muscle activity can cause adverse reproductive effects. Nicotine, for example, acting through epinephrine and oxytocin can influence tubal and uterine contractions. Theoretically such agents could cause mis-timing of gamete and/or embryo transport and failure of fertilization or implantation, respectively. Hemotoxic agents can alter menstrual flow and result in menstrual irregularities (as indicated previously) at the level of the endometrium without having any effect on the reproductive system directly.

Non-reproductive Organ Targets

Key non-reproductive organs are essential for normal reproduction. An example of this kind of interaction is the production of binding proteins by the liver which are essential for steroid hormone transport. Hepatotoxins such as ethanol can limit binding protein production and adversely alter the ability of sex steroids to be transported to their binding sites. The liver also plays the primary role in deactivating and eliminating steroid hormones. Hepatotoxins such as the halogenated hydrocarbons, barbiturates, and anticonvulsants which alter enzymes that either conjugate or metabolize steroid hormones can also adversely affect reproductive function. Normal thyroid function is important for normal reproduction. Thyroid hormone is essential for normal cell function in general and thyroid disease is often associated with reproductive failure.

Summary and Conclusion

In summary, in the broadest view reproductive toxicants can impinge on the female system through changing normal development, cytotoxic effects on gametes, dysfunction of reproductive organs, interference with the differentiation of cell types, or through interrupting the hormone messages through which the processes of hormone synthesis, transduction, or metabolism occur. Reproductive toxins can influence reproductive performance by affecting sexual or social behavior, embryo survival and development, as well as affecting reproduction indirectly by influencing general health. The primary difficulties faced in identifying reproductive toxicants are the sensitivity of female reproductive processes to normal environmental change, the lack of baseline data, the complexities of the ovarian cycle, and the species-specific nature of female reproductive physiology.

A great deal of progress is currently being made in this discipline in response to pressures exerted by the public. Real concerns are now being express that environmental factors are causing an increase in female infertility and more women are exposed to chemicals through the workplace. Perhaps the greatest impediment to progress is adequate animal models or an *in vitro* screening test for human sensitivity to the large number of chemicals being produced.

Further Reading

Barlow, S. M., and Sullivan, F. M. (1982). *Reproductive Hazards of Industrial Chemicals.* Academic Press, New York

Dixon, R. L. (1980). Toxic response of the reproductive system. In *Casarret & Doull's Toxicology: The Basic Science of Poisons* (J. Doull, C. D. Klaassen, and M. O. Amdur, Eds.). McMillan, New York.

Matteson, D. R., and Ross, G. T. (1982). Oogenesis and ovulation. In *Methods for Assessing the Effects of Chemicals on Reproductive Dysfunction* (J. Voulk and K. Sheenan, Eds.), pp. 217–247. Wiley-Interscience, New York.

Scialli, A. R., and Zinaman, M. J. (1993). *Reproductive Toxicology and Infertility.* McGraw-Hill, New York.

U.S. General Accounting Office (GAO) (1991, October). *Reproductive and Developmental Toxicants,* GAO/PEMD-92-3. U.S. GAO, Washington, DC.

Working, P. K. (1989). *Toxicology of the Male and Female Reproductive Systems.* Hemisphere, New York.

Zielhuis, R. L., Stijkel, A., Verberk, M. M., and van de Poel-Bot, M. (1984). *Health Risks to Female Workers in Occupational Exposure to Chemical Agents.* Springer-Verlag, Berlin.

—*Bill L. Lasley*

Related Topics

Carcinogen–DNA Adduct Formation and DNA Repair
Chromosome Aberrations
Developmental Toxicology
Dose–Response Relationship
Endocrine System
Environmental Hormone Disrupters
Epidemiology
Mutagenesis
Reproductive System, Male
Risk Assessment, Human Health
Sister Chromatid Exchange
Toxicity Testing, Developmental
Toxicity Testing, Reproductive

Reproductive System, Male

Both the public and the scientific community have become increasingly aware of the potential for chemical exposure to cause adverse effects on the male reproductive system. It has been estimated that, in the United States, the rate of infertility in couples could be as high as 15%. For about 40% of those couples, infertility is associated with the male. Recent reports have suggested that, during the period from 1938 to 1990, the average sperm density of human semen has declined and that testicular abnormalities have increased. It has been proposed that exposure to environmental estrogens may be having an adverse effect on the male reproductive system. However, the influence of environmental exposure to chemicals on human male reproductive capabilities is still unclear. In general, a direct causal relationship between chemical exposure and damage to the human male reproductive system has been apparent only when reproductive dysfunction has been severe. For example, it was clearly established that pesticide formulators who had been exposed to high levels of the nematocide dibromochloropropane exhibited testicular atrophy and infertility.

As awareness of the potential for chemicals to cause male reproductive harm has grown, so has an appreciation of the gaps in our knowledge about the events underlying toxicity and the adequacy of our testing procedures. Reproduction is biologically complex and chemicals could have adverse effects at many different levels. Since perturbations of any of the biological processes necessary for integrated function could lead to infertility or failure to produce viable offspring, adequate evaluation of potential reproductive effects must consider how chemical interactions at various sites could result in reproductive dysfunction. In the following description of male reproductive toxicology, first the basic processes of the male reproductive system are described. Current information about sites and mechanisms of male reproductive toxicity are then presented. Procedures for assessing adverse effects on male reproductive capability are also described.

The Male Reproductive System

For a male to produce viable sperm which are capable of fertilization and production of normal offspring, a series of tightly orchestrated events must take place (Fig. R-1). The process of spermatogenesis in the testis is subject to neuroendocrine controls (I) via the hypothalamic–pituitary axis as well as to indirect modulatory influences (II) arising from nutritional status, liver metabolism, and vascularization. Within the testis (III), endocrine, autocrine and paracrine controls are re-

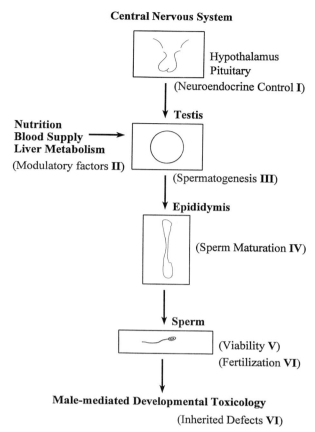

Central Nervous System

Hypothalamus
Pituitary

(Neuroendocrine Control **I**)

Testis

**Nutrition
Blood Supply
Liver Metabolism**

(Modulatory factors **II**)

(Spermatogenesis **III**)

Epididymis

(Sperm Maturation **IV**)

Sperm

(Viability **V**)

(Fertilization **VI**)

Male-mediated Developmental Toxicology

(Inherited Defects **VI**)

FIGURE R-1   Overview of the male reproductive system.

quired for the proliferation and differentiation of the stem cell spermatogonia into the mature spermatid which is released into the lumen of the seminiferous tubule. The released spermatozoa travel through the rete testis and efferent ducts to the head (caput) of the epididymis. As the spermatozoa pass through the middle (corpus) and tail (cauda) of the epididymis, they undergo maturation (IV) and gain motility as well as the ability to fertilize. Toxicants could affect any of the previous processes as well as have direct effects on sperm cell viability (V) or on ability to penetrate and fertilize the oocyte (VI). As yet, little is known about male-mediated developmental toxicity (VII) in which the male gamete transmits inherited defects, but this possibility has lately received attention.

There is a considerable body of both basic biological knowledge and toxicological information about the hypothalamic–pituitary control of testicular function and testicular cell biology. Less is known about epididymal maturation of spermatozoa. Although sperm cell biol-

ogy is a well-researched area, few toxicants are known to directly interact with this cell type. Therefore, the areas of greatest emphasis will be those about which most toxicology is known.

Hypothalamic–Pituitary– Gonadal Axis

Neuroendocrine control of gonadal function is regulated through the hypothalamus in the brain and the closely associated anterior pituitary gland (Fig. R-2). Gonadotropin releasing hormone (GnRH) is released from the hypothalamus in a pulsatile manner and carried in the blood supply directly to the anterior pituitary gland. Under the influence of GnRH, the pituitary is stimulated by a receptor-mediated process to secrete the gonadotrophins, luteinizing hormone (LH), and follicle stimulating hormone (FSH). LH and FSH are carried to the testis, where they play a central role in regulation of testicular function. Like GnRH, LH and probably FSH release are pulsatile in nature.

In the testis, the cellular targets for these hormones are the Leydig cell for LH and the Sertoli cell for FSH. LH binds to receptors on the Leydig cell and stimulates steroidogenesis. Testosterone production is episodic, coincident with the pulsatile release of LH. The Sertoli cell is the testicular target for FSH. Sertoli cells are the "nurse" cells of the testis surrounding and supporting the developing germ cells. The complete role of FSH in spermatogenesis is as yet unknown. However, FSH

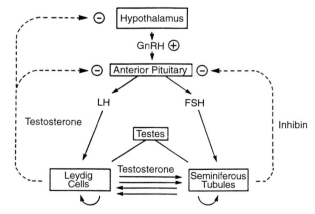

FIGURE R-2   Neuroendocrine control of male reproductive system (from J. J. Heindel and K. A. Treinen, Toxicol. Pathol. 17(2), 411–445 1989. Copyright the Society of Toxicologic Pathologists).

is necessary for spermatogenesis presumably due to its involvement in Sertoli cell functions.

Hormonal regulation is through a series of feedback mechanisms taking place at both central and peripheral sites. To complete the endocrine feedback loops, testosterone regulates LH production, and inhibin and other Sertoli cell products regulate FSH secretion. These feedback loops modify/modulate the release of GnRH from the hypothalamus as well as LH and FSH from the pituitary. Within these loops, factors which perturb one component will alter regulatory influences on another. For example, where Leydig cell damage decreases steroidogenic capability, testosterone production would be diminished and, in response to low circulating levels of testosterone, LH levels would increase in an attempt to restore testosterone production.

Spermatogenesis

The testis is made up of the tunica albuginea, a tough fibrous capsule within which are tightly packed seminiferous tubules surrounded by vascularized interstitium (Fig. R-3). The two major functions of the testis are spermatogenesis, which takes place within seminiferous tubules, and steroidogenesis, which is carried out by Leydig cells in the interstitium. Steroidogenesis is essential both for spermatogenesis and for secondary sexual characteristics. Spermatogenesis is the process

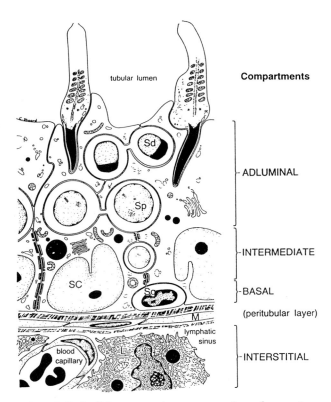

FIGURE R-4 *Diagrammatic representation of a portion of a seminiferous tubule. L, Leydig cell; M, myoepithelial peritubular cell; SC, Sertoli cell; Sg, spermatogonium; Sp, spermatocyte; Sd, spermatid (from J. C. Lamb, IV, and P. M. D. Foster, Physiology and Toxicology of Male Reproduction, Academic Press, San Diego, 1988).*

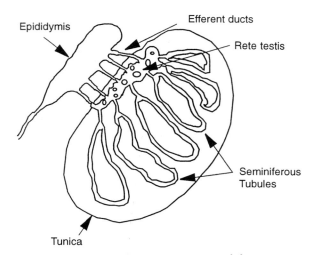

FIGURE R-3 *Schematic representation of the structure of the testis (from P. K. Working, Toxicology of the Male and Female Reproductive Systems, Hemisphere, New York, 1989).*

by which undifferentiated spermatogonia develop through spermatocytes and spermatids into specialized spermatazoa. The immature germ cells develop from the basal area at the circumference of the tubule and progress toward the central tubular lumen into which they are released (Fig. R-4). The Sertoli cell is the "nurse cell" of the testis, supplying support, nutrients, and protection to the developing germ cells which it surrounds. A "blood–tubule" barrier is formed by Sertoli cell tight junctions which maintains the tubular milieu for germ cell development and acts as a barrier to the entry of blood-borne materials.

Broadly, there are three germ cell types represented in the process of spermatogenesis: the spermatogonia, which are outside the blood–tubule barrier; spermatocytes at various points of meiosis; and the postmeiotic spermatids, which are undergoing elongation prior to release as spermatozoa. Spermatogonia undergo mitotic proliferation and differentiation passing through types A_1–A_4, intermediate, and type B spermatogonia before dividing to produce primary spermatocytes which enter meiosis. During the preleptotene phase of meiosis I, the spermatocyte passes through the blood–tubule barrier and proceeds through meiotic prophase—leptotene, zygotene, pachytene, and diplotene—producing two secondary spermatocytes at the end of meiosis I. The secondary spermatocytes enter meiosis II and rapidly divide to produce a total of four haploid round spermatids.

The metamorphosis of round spermatids into spermatozoa is described as spermiogenesis. Initially, the round spermatid develops an acrosome derived from the intracellular Golgi complex. The acrosome starts as a small vesicle and develops into a pronounced nuclear cap. It is necessary for oocyte fertilization because it contains the lysosomal enzymes required to penetrate the vestments surrounding the egg. Early in spermiogenesis, a microtubule-containing flagellum begins to develop. Nuclear DNA undergoes condensation and is no longer synthesized before the nuclei elongate around a microtubule structure called the manchette. As the spermatid elongates, mitochondria are arrayed in a sheath behind the sperm head and around the flagellum in what will be the middle piece of the spermatozoa. These are the organelles which will supply the energy for sperm movement. Release of the mature spermatid into the tubular lumen (spermiation) is accompanied by loss of spermatid cytoplasm. The residual cytoplasm from the spermatid is engulfed by the Sertoli cell to form residual bodies within the Sertoli cell cytoplasm. The spermatozoon is a cell ideally designed to transport DNA from the male to the oocyte with little cytoplasmic baggage, a good mitochondrial engine, and a large tail for propulsion.

In the rat, it takes about 54 days for the spermatogonia to traverse spermatogenenesis and be released as a testicular spermatozoa. Since spermatogonial differentiation is initiated at approx 12.9-day intervals in each segment of the seminiferous epithelium, germ cells from successive generations and at different phases of development are present in a seminiferous tubule cross section. These associations of developing germ cells are very specific and, in the rat, have been defined as 14 individual stages of spermatogenesis based on nuclear morphology and the appearance of the spermatid acrosome. The seminiferous epithelium not only cycles through the stages of spermatogenesis in a time-dependent manner but also the different stages follow one another along the length of the seminiferous tubule in what is known as the wave of spermatogenesis. The "wave" is necessary to maintain continuous sperm production. If there were no wave, spermatogonia throughout the testis would all enter spermatogenesis at the same time, and fertility would be episodic with a 12.9-day interval coincident with a spermiation event occurring throughout the testis. Obviously, this is not the case and different parts of the seminiferous epithelium undergo spermiation at different times.

The characteristic stages of spermatogenesis differ between species. This would be predicted when the duration of spermatogenesis and the length of the repeating cycle differs between species. For example, there are 6 specific cellular associations defined as stages of spermatogenesis in the human in comparison with 14 in the rat. Interestingly, in the human male there is no clearly defined wave of spermatogenesis arranged consecutively along the length of the seminiferous tubule. Instead, the arrangement of stages seems to result from development along a helical as well as a longitudinal axis. From a toxicological point of view, cells in different stages may have both a different biochemical milieu and different susceptibility to toxicity.

Epididymal Maturation of Sperm

Spermatozoa leave the testis via the excurrent duct system first passing through the rete testis then the efferent ducts to reach the epididymis. The efferent ducts (4–20 depending on species) join to form the epididymis, which is a long, highly coiled, single tubule lying above and connected to the testes (Fig. R-5). The epididymis is divided into three regions: head (caput), body (corpus), and tail (cauda). In the epididymis, spermatozoa undergo maturation. Only the mature spermatozoa found in the caudal region show forward motility and are capable of fertilization. Mature spermatozoa are stored in the cauda epididymis and, on ejaculation, leave the epididymis via the vas deferens. Spermatozoa are discharged through the ejaculatory duct. The major portion of ejaculate volume is made up of products secreted by the accessory sex glands: seminal vesicles, the prostate, and bulbourethral glands. Rodents also have coagulating glands and preputial glands. It has been demonstrated in rodents, using mature spermatozoa collected from the cauda epididymis, that the secretions of the accessory glands are not important for successful *in*

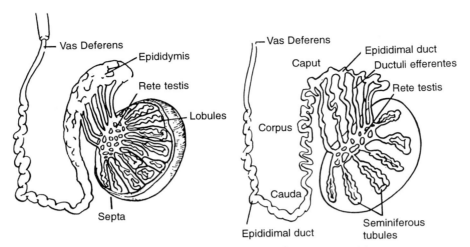

FIGURE R-5 *Structural relationships between the testis and the epididymis (from L. J. D. Zaneveld and R. T. Chatterton (Eds.), Biochemistry of Mammalian Reproduction, Wiley, New York, 1982).*

vitro fertilization. However, there is a reduction in fertility *in vivo* when accessory gland products are not present indicating they do play a role in successful reproduction.

Testicular spermatozoa are immotile and not capable of fertilization until they undergo epididymal maturation. It has recently been demonstrated in humans that the immature testicular spermatid, if directly microinjected into an oocyte, will result in successful pregnancy and birth. However, in unassisted reproduction, the epididymis confers essential spermatozoa characteristics of motility and the ability to bind to and fertilize an oocyte. Many changes in epididymal fluid composition occur with epididymal transit. The epididymis has important absorptive and secretory functions. For example, spermatozoa entering the caput epididymis are in a dilute suspension and are sparsely seen in the lumen of the tubule. However, in the cauda epididymis they are highly concentrated and packed tightly. Carnitine levels in epididymal fluid are 2000 times those found in blood. Spermatozoa characteristics which change with passage through the epididymis include some minor structural remodeling of the head and midpiece region of the spermatozoon, changes in surface membrane composition, as well as many biochemical changes which may be important for the acquisition and maintenance of motility. Unlike caudal spermatozoa, caput spermatozoa have minimal ability to swim in a forward direction. Spermatozoa usually reach the tail of the epididymis 1 or 2 weeks after leaving the testis and this is called epididymal transit time.

Male Reproductive Toxicants

Many compounds have been implicated as male reproductive toxicants but their sites and mechanisms of action are not well understood. The difficulty of understanding mechanisms of male reproductive toxicity are exemplified by the nematocide, dibromochloropropane (DBCP), which as yet has no clearly defined mechanism of male reproductive toxicity. To confound matters, it seems that DBCP can affect numerous cell types and DBCP metabolites also have the potential to cause reproductive dysfunction. Interestingly, the mouse is insensitive to the effects of DBCP, unlike the rat and the human. Thus, animal studies in the mouse would not have raised concerns about potential reproductive hazards associated with human exposure to DBCP. Compounds which are known to cause adverse effects on male reproduction in animal studies include DBCP, ethylene dibromide, cadmium, lead, polychlorinated biphenyls, phthalate esters, 2-methoxyethanol, 2,5-hexanedione, ethanol, nitrobenzenes, and nitrotoluenes. For some of these agents and others, what is known about mechanisms and sites of toxicity will be described later.

Metabolic Mechanisms in Male Reproductive Toxicity

Numerous studies have documented that enzymatic metabolism to toxic reactive metabolites plays an important role in toxicity. The liver is the organ in which there is the greatest concentration of xenobiotic metabolizing enzymes. There are numerous examples of liver toxicants where toxicity is mediated by reactive metabolites. Often, the cytochrome P450 mixed function oxidase enzyme system has been implicated in metabolic activation. Other organs also contain appreciable amounts of metabolizing enzymes. In particular, those cell types which show high levels of cytochrome P450 activity have been found to be susceptible to damage via formation of toxic metabolites (e.g., the Clara cells of the lung and the cells of the proximal tubule in the kidney).

Metabolism to toxic metabolites has also been implicated in testicular toxicity caused by the phthalate esters, 2-methoxyethanol, nitrobenzenes, and DBCP. The site of metabolic activation can be the testis or the liver. If liver metabolism was important for formation of a toxic metabolite, the metabolite would have to be sufficiently stable to reach the testis. Conversely, the testis has been shown to have measurable metabolizing capabilities albeit at significantly lower levels than the liver. Nevertheless, the close proximity of these testicular enzyme activities to the developing germ cell could be of particular importance for toxicity.

Classification of Male Reproductive Toxicants

Various approaches have been taken to the classification of male reproductive toxicants. Mattison proposed a schema in which toxicants were divided into those which were direct acting and those which were indirect acting. The definition suggested that a direct-action toxicant interacted directly with vital cellular components based either on inherent chemical reactivity or on action at specific receptor sites. Conversely, an indirect-action toxicant caused toxicity either through disruption of endocrine homeostasis or via metabolic activation to a direct toxicant. Similarly, Working utilized the definitions of direct and indirect male reproductive toxicants. However, he simplified the situation by removing any distinction based on a requirement for metabolism. He classified an indirect reproductive toxicant as one which acts at a non-germ cell site (e.g., the hypothalamic–pituitary axis or the Leydig cell) to alter hormonal control of the testis. Conversely, a direct toxicant would affect the testis without endocrine mediation.

While the classification of male reproductive toxicants as direct or indirect toxicants is very useful, the terms utilized could be further clarified to more clearly define the primary site of toxicity (Fig. R-6). An indirect toxicant would have a primary action on hypothalamic–pituitary neuroendocrine controls or on extragonadal systems. A direct toxicant would produce its primary effect on testicular cells, the duct system of the male reproductive tract (e.g., epididymis), or on the mature spermatozoa. Direct toxicity to testicular cell types which are not germ cells (i.e., the Leydig cell and the Sertoli cell) would be included in the definition of action for a direct toxicant. Since the testis is subject to hormonal control and feedback loops, ultimately the action of toxicants which damage testicular cell types will result in a secondary disruption of endocrine homeostasis. When toxicants are classified by the processes altered, distinctions based on whether a parent compound or a metabolite is eliciting toxicity become unimportant.

Reproductive Toxicity Testing

A substantial literature exists which emphasizes male reproductive toxicity testing procedures using laboratory animals, usually rodents, in mating trials and generational studies to determine the ability of chemicals to affect reproductive performance. Multigeneration reproductive toxicity studies are the established paradigm in reproductive toxicity testing. Both males and females are exposed before and after mating. The parents and litters are evaluated for adverse effects. End-

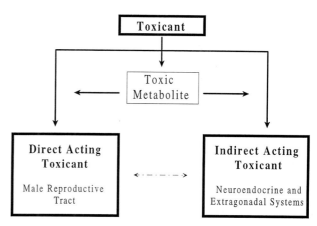

FIGURE R-6 *Classification of male reproductive toxicants as direct or indirect acting.*

points used to assess toxicity include fertility and histopathology in the parent generation, litter size and offspring weight, external abnormalities, and subsequent growth. In order to determine if *in utero* exposure alters reproductive capability, the offspring go on into a second-generation breeding trial. Such work is necessary for regulatory purposes, but it cannot supply detailed mechanistic information.

In the risk assessment process, which determines the potential for chemical exposure to cause toxicity in humans, much of the data used for regulatory purposes are derived from laboratory animal toxicity testing protocols such as those described previously. An approach is to obtain dose levels at which no-observable-adverse-effect occurs or lowest-observable-adverse-effect level in the animal model. To determine an exposure level acceptable for the human population a safety factor (often ×1000) is incorporated. The EPA has published specific guidelines for reproductive toxicity risk assessment.

While the process of reproduction in humans and in the rodent animal models routinely used in toxicology studies could be expected to have broad similarities, many species differences do exist. For example, it is well documented that rodents have sperm reserves which are much greater than those of man. In the rat, epididymal sperm counts can be reduced by as much as 90% without a significant affect on fertility. Since rodents have such large sperm reserves, rodent breeding studies may not detect marginal changes in reproductive capacity. To complement breeding studies, information about the effect of toxicant exposure on sperm numbers, motility, and morphology would be desirable. Mechanistic insights about causative events underlying changes in sperm parameters and their relationship to fertility would improve our ability to devise sensitive and specific toxicity tests which predict those chemicals most likely to cause reproductive harm.

Male Reproductive Toxicants: Sites of Action

The majority of information about male reproductive toxicants has been obtained from animal studies carried out in the rat either *in vitro* or *in vivo*. Animal studies allow controlled experimentation where events underlying toxicity can be better understood, allowing more rationale extrapolation to the human situation. The following sections describe some of the compounds which cause male reproductive damage and summarize possible sites and mechanisms of toxicity. For the majority of compounds the data are incomplete.

Agents Affecting the Hypothalamic–Pituitary–Gonadal Axis

Lead

Animals exposed *in utero* to lead acetate, administered in maternal drinking water, had decreased sperm counts and diminished male sexual behavior. When these *in utero* exposed animals grew to adulthood, secretion of LH and FSH from the pituitary was found to be altered suggesting development of the hypothalamic–pituitary–gonadal axis had been affected.

A relatively recent human study found decreased levels of testosterone which correlated with lead exposure and were associated with the expected feedback-mediated increase in LH secretion from the pituitary. Since the gonadal–pituitary feedback loops were apparently complete, this study suggested that the effects of lead were directly on the testis. However, a series of animal studies in rats would seem to contradict this conclusion. When rats were exposed to levels of lead sufficient to give blood levels found after human occupational exposures, levels of testosterone were decreased similarly to those found in the human study. In contrast to the human data, LH levels in the rat study did not increase concomitant with low testosterone level and FSH levels were also decreased. This study suggested that lead was acting at the level of the hypothalamus and/or pituitary rather than directly on the testis.

Antiandrogens

Agents that interact with the androgen receptor can compete with endogenous testosterone and thus disrupt feedback regulation. By blocking the action of testosterone, there is an inappropriate release of LH and disruption of gonadal function. Examples of compounds capable of antiandrogen activity are the drugs digitoxin, spironolactone, and cimetidine. Recent reports have also indicated that DDE, the major metabolite and breakdown product of DDT, can act as an antiandrogen (see Androgens).

Estrogens and Antiestrogens

Administration of estrogens to both humans and animals causes a decrease in serum levels of both LH and testosterone, suggesting that estrogens can disrupt normal feedback mechanisms controlling LH production. However, it has also been shown that estrogenic compounds can directly inhibit the steroidogenic capacity of the Leydig cell. Tamoxifen, a drug used in the treatment of breast cancer, has been shown to have tissue-specific effects as an estrogen, partial estrogen, or estrogen antagonist. Other estrogenic drugs are the synthetic estrogen, diethylstilbestrol, and the antiestrogen clomiphene. Examples of environmental agents which compete for the estrogen receptor are DDT, chlordecone (Kepone), and polychlorinated biphenyls (PCBs). Exposure of wildlife to PCBs has been associated with feminization of populations and decreased levels of testosterone in males.

Central Nervous System-Active Agents

Agents which alter central nervous system (CNS) control of the hypothalamic release of GnRH also have the potential to disrupt the hypothalamic–pituitary–gonadal axis. Release of endogenous neurotransmitters and neuropeptides, important for normal neuronal function, may be altered by chemical exposure. Hypothalamic release of GnRH is stimulated by α-adrenergic receptors; therefore, agents which alter α-adrenergic function may alter GnRH release. Chlordimeform, an insecticide, may decrease GnRH release through an adrenergic mechanism. Conversely, the endogenous opioids inhibit GnRH release and morphine and morphine-like drugs can suppress GnRH-mediated secretion of LH. Many CNS-active drugs have the potential to affect GnRH secretion.

Extragonadal Modulation of Reproductive Function

Nutritional Factors

Dietary insufficiency and nutritional status can disrupt spermatogenesis. In animal studies, levels of circulating hormones are decreased by malnourishment. For effects to be manifest on spermatogenesis, deprivation has to be severe. Specific vitamins, for example, vitamins A and E, and trace minerals such as zinc are essential for spermatogenesis. The mechanistic basis for disruption of spermatogenesis induced by vitamin or mineral deficiencies is not clear.

Liver Metabolism

The liver is the primary organ responsible for the metabolism of fat-soluble compounds into more water-soluble, readily excreted metabolites. Therefore, compounds which act as inducing agents to increase the metabolic capabilities of the liver could potentially alter enzymatic steps important in the formation and further metabolism of steroidogenic compounds. A major liver enzyme system is the cytochrome P450 mixed function oxidase, which can be induced by many compounds including the barbiturate, phenobarbital, and chlorinated hydrocarbons such as chlordane and DDT. Cytochrome P450 induction can lead to enhanced metabolism of testosterone and a decrease in circulating levels.

Vascularization

The testis is particularly vulnerable to changes in blood supply. This is thought to be due to the fact that it is a relatively anoxic organ with low blood flow and low oxygen tension. The effects of cadmium on testis vasculature have been most extensively studied but there are indications that lead and cobalt could also affect blood supply. The testis is not adversely affected by chronic low-dose exposures to cadmium although toxicity is apparent in other organs with this dosage regimen. However, acute exposure to cadmium causes testicular damage at dose levels which do not affect other tissues. Damage is characterized by increased permeability of the testicular vasculature, with first increased then decreased blood flow.

Agents Affecting Particular Testicular Cell Types

Leydig Cell Target

The Leydig cell has a central role in the synthesis and secretion of testosterone. LH, released from the pituitary, stimulates Leydig cell testosterone production. Testosterone is synthesized from cholesterol via reactions catalyzed by four different enzymes: (1) cholesterol ≫ pregnenolone, (2) pregnenolone ≫ progesterone, (3) progesterone ≫ hydroxyprogesterone ≫ androstenedione catalyzed by the same enzyme, and (4) androstenedione ≫ testosterone. Cytochrome P450-containing enzymes catalyze cholesterol side chain cleavage (reaction 1) and the synthesis of androstenedione from progesterone (reaction 3).

Many chemicals alter Leydig cell function and some can directly cause Leydig cell death. Often, it is unclear whether agents which decrease testosterone levels are acting directly at the level of the Leydig cell or are related to disruption of the hypothalamic–pituitary–gonadal axis. There is evidence for both central and peripheral effects on testosterone production induced by ethanol, cannabis, and cyclosporine, an immunosuppressant used in organ transplantation. As indicated earlier, for lead-induced decreases in testosterone levels, there are data to support both a direct effect on the Leydig cell and an indirect effect via the hypothalamic–pituitary–gonadal axis.

Many and diverse chemicals can have an affect on testosterone production. Some of these compounds are listed by broad mechanistic class in the following sections.

Inhibitors of Enzymes Involved in Testosterone Biosynthesis

The antifungal agent, ketoconazole, inhibits testosterone biosynthesis apparently by inhibiting cytochrome P450 enzymes involved in steroidogenesis. The environmental pollutant, 2,3,7,8-tetrachlorodibenzodioxin, has been shown to inhibit Leydig cell steroidogenesis. However, this compound has many other actions in the whole animal. Numerous compounds which possess estrogenic activity have been shown to directly inhibit Leydig cell testosterone production by an effect on the cytochrome P450 enzyme required for the conversion of progesterone to androstenedione.

Leydig Cell Cytotoxicants

Ethylene dimethanesulfonate (EDS) belongs to a class of agents used in the treatment of cancer. EDS was unusual in that it was found specifically to cause Leydig cell death. In laboratory studies, EDS has proved to be a valuable chemical tool to explore the role of the Leydig cell in spermatogenesis. The mechanism of EDS-induced cell death is as yet unclear.

Agents Causing Leydig Cell Hyperplasia

The hyperplastic response could be viewed as a precursor of neoplasia or tumor formation. However, this relationship is not clearly delineated. A plethora of compounds, including some hypolipidemic agents, estrogens, and antiandrogens, induce hyperplasia but the toxicological importance of this effect is open to debate.

Sertoli Cell Target

The Sertoli cell performs a pivotal role in spermatogenesis, orchestrating and nurturing the development of the germ cells. The Sertoli call has many functions. It serves a protective role through Sertoli–Sertoli cell tight junctions which compartmentalize the developing germ cell away from the extratesticular milieu. However, this barrier means that the nutritive and hormonal requirements of the germ cell must pass through or be generated by the Sertoli cell. The Sertoli cell cytoskeleton performs both specialized transport and support functions. Microtubule networks track through the cell carrying a multitude of hormonal and nutritive factors essential for germ cell development. The elements which comprise seminiferous tubule fluid are carried on this network. The cytoskeleton also provides the scaffolding which serves as a physical support for the developing germ cell.

The Sertoli cell has become a focus for mechanistic studies investigating the cellular events involved in testicular toxicity. Two processes will be used to illustrate how toxicants could affect Sertoli cell function.

Agents Altering Microtubule Assembly

As indicated previously, microtubules play an important role in both support and transport processes as well as seminiferous fluid secretion. 2,5-Hexanedione has been studied extensively as an agent capable of altering testicular microtubules. This compound causes neurotoxicity as well as testicular toxicity. The mechanistic basis for both toxicities is thought to arise through the same chemical characteristics which result in alterations of microtubule assembly. Hexanedione is reactive per se and capable of binding to amino groups on protein to form pyrroles. Pyrrole formation has been correlated with both neurotoxicity and testicular toxicity. 2,5-Hexanedione apparently acts to cross-link and stabilize microtubules causing toxicity in the brain and the testis. The function of both of these organs is highly dependent on an undamaged microtubule network. Other compounds which act on testicular microtubules are the benzimidazole fungicides, benomyl and carbendazim, and the anti-inflammatory agent colchicine. These prevent the assembly of testicular tubulin into microtubules.

Agents That Can Alter Cellular Metabolism

A good indicator of toxicity in isolated Sertoli cell preparations is increased secretion of lactate. Why this would signal Sertoli cell dysfunction is unclear since lactate is a required substrate for germ cell metabolism. However, since Sertoli cells play a key metabolic role, compounds which disrupts Sertoli cell metabolism would be expected to cause testicular toxicity. A possible candidate for this mechanism is 1,3-dinitrobenzene. Various nitroaromatic compounds, including nitrobenzene and di- and tri-nitrotoluenes, cause testicular toxicity apparently by disruption of Sertoli cell function. 1,3-Dinitrobenzene has been most extensively studied as a model nitroaromatic. Nitroaromatics can undergo reductive metabolism to aniline derivatives. Nitroreduction can give rise to various reactive and potentially toxic intermediates. It has been proposed that the nitroso intermediate may be involved in testicular toxicity although it is possible that other metabolic intermediates may be important.

Other Sertoli Cell Toxicants

Phthalic acid esters have been widely used as plasticizers, imparting flexibility to plastic products. Diethylhexyl-phthalate undergoes metabolism to monoethylhexyl-phthalate which causes testicular toxicity. Numerous hypotheses exist as to how these compounds act. Tri-*o*-cresyl phosphate, which was used as a plasticizer and lubricant, is both a neurotoxicant and testicular toxicant. The saligenin metabolite of tri-*o*-cresyl phosphate apparently causes toxicity by an as yet unknown mechanism.

Germ Cell Target

The germ cells of the testis are continually and consistently changing in order to accommodate the biological process with which they are involved. No other cell in the body does so much over it's life span. On entering spermatogenesis, spermatogonia undergo mitosis and increase their cell number by rapid division; then they enter meiosis, becoming part of the spermatocyte population, undergoing extensive DNA synthesis, and doubling their chromosome number. After two meiotic divisions, spermatocytes form the haploid spermatids which elongate and tightly package DNA prior to release. Because of the diversity of the events occurring in spermatogenesis, the germ cells of the testis could have very different susceptibilities to the action of toxicants. However, the complexity of the interactions between cell types makes it often very difficult to discern which cell type was first affected. Compounds described here will be those which are known to affect particular germ cell types. Where possible, relationships will be drawn between germ cell characteristics and toxicity. For example, agents which affect microtubules involved in cell division would be expected to disrupt both mitosis in spermatogonia and meiosis in spermatocytes.

Spermatogonia

The rapidly dividing spermatogonia would be expected to be susceptible to toxicity induced by agents which affect cell division. For example, radiation and some anticancer drugs, such as the DNA alkylating agent busulfan and procarbazine, have been shown to cause toxicity to spermatogonia. Since these agents are used because of their potential to damage dividing tumor cells, it is not surprising that they adversely affect reproductive capabilities.

After DBCP was found to cause testicular atrophy in pesticide formulators it's use was halted. It has been difficult to clearly implicate one cell type as being specifically affected by DBCP. However, the spermatogonia seem to be especially sensitive to DBCP.

Spermatocytes

Ethylene glycol monomethyl ether (EGME) is used in the semiconductor industry and has been shown to elicit a relatively specific toxicity to late pachytene spermatocytes. However, the actual chemical species responsible for spermatocyte damage is not EGME but 2-methoxyacetic acid, a metabolite of EGME generated by the liver. It has been proposed that the methoxyacetic acid specifically affects one carbon oxidation pathways necessary for synthesis of purine bases. The stage of spermatogenesis susceptible to EGME toxicity is one in which there is a high level of meiotic RNA synthesis.

Spermatids

Few agents have specifically been implicated as causing toxicity to testicular spermatids. After exposure to methyl chloride, which was once used as a fumigant, the initial lesion was identified as delayed release of mature spermatids. Spermatids were present at much later stages than would be expected. Another discontinued fumigant, ethylene dibromide, has been shown to have a direct effect on spermatids but other germ cell types were also affected.

In terms of fertility, the response to and recovery from injury associated with damage to a particular germ cell type would be expected to occur at different times after toxicant exposure depending on which stage of germ cell development had been affected. For example, spermatocyte toxicity induced by EGME would be seen as decreased fertility 4 or 5 weeks after dosing in the rat. This delay reflects the time that would have been required for the damaged cell population to complete spermatogenesis and pass through the epididymis. Similarly, fertility changes would be manifest 2 or 3 weeks after administration of a toxicant which acts on the spermatid. These decreases in fertility would be of short duration and would be reversible as long as the stem cell population was capable of dividing and replacing the affected cell types. For spermatogonial toxicants, fertility changes would have the greatest delay in onset (6–8 weeks) and may not be reversible if the stem cell population was severely damaged.

Agents Affecting the Epididymis

The epididymis is an androgen-dependent organ; therefore, agents which alter testosterone production also cause a decrease in epididymal weight and disturb epididymal events necessary for sperm maturation. Agents known to directly affect the epididymis include α-chlorohydrin, which is thought to affect epididymal vasculature causing sloughing of epithelial cells. Inter-

estingly, α-chlorohydrin is a metabolite of DBCP and may be responsible for DBCP-induced epididymal damage. DBCP is metabolized first to epichlorohydrin (also an epididymal toxicant) before forming the presumably ultimate toxicant, α-chlorohydrin. EDS, previously presented as a Leydig cell toxicant, also has direct effects on the epididymis besides those that would be anticipated by testosterone depletion.

Agents Directly Affecting Sperm

α-Chlorohydrin not only causes epididymal damage but also directly affects spermatozoa with a resultant diminution in motility. Since this direct effect was reversible, at one time α-chlorohydrin was considered as a male contraceptive. However, this possibility was no longer contemplated after the irreversibility of epididymal toxicity was noted. The sulfonamide drug, sulfasalazine, causes male infertility possibly through a direct action on epididymal sperm. However, the antifertility effect of sulfasalazine has been noted primarily in patients being treated for inflammatory bowel disease and the possibility has been raised that reproductive dysfunction may be a consequence of the disease state rather than the drug.

Agents Affecting Fertilization

Changes in fertilization capabilities have been difficult to study in vivo where events occur in the female reproductive tract. Recent in vitro technologies should markedly increase knowledge of these events and the toxicants which could perturb them. In patients receiving sulfasalazine, sperm are unable to undergo the acrosome reaction which is necessary for egg penetration.

Male-Mediated Developmental Toxicants

There are few studies which have investigated the possibility that paternal exposure to toxicants can result in developmental abnormalities. However, it is well-known that toxicants can induce genetic changes in the male germ cell; therefore, it is plausible that such changes could be paternally transmitted and have a deleterious effect on offspring development. Human epidemiology studies have found positive associations between paternal occupational exposure to, for example, vinylchloride, anaesthetic gases, and DBCP, and developmental endpoints. However, the causality of the associations remains to be clarified.

Further Reading

Lamb, J. C., IV, and Foster, P. M. D. (Eds.) (1988). *Physiology and Toxicology of Male Reproduction.* Academic Press, San Diego.

Scialli, A. R., and Clegg, E. D. (Eds.) (1992). *Reversibility in Testicular Toxicity Assessment.* CRC Press, Boca Raton, FL.

Witorsch, R. J. (Ed.) (1995). *Reproductive Toxicology,* 2nd ed., Target Organ Toxicology Series. Raven Press, New York.

—*Marion G. Miller*

Related Topics

Carcinogen–DNA Adduct Formation and DNA Repair
Chromosome Aberrations
Developmental Toxicology
Dose–Response Relationship
Endocrine System
Environmental Hormone Disrupters
Epidemiology
Mutagenesis
Reproductive System, Female
Risk Assessment, Human Health
Sister Chromatid Exchange
Toxicity Testing, Developmental
Toxicity Testing, Reproductive

Reserpine

- CAS: 50-55-5
- SYNONYMS: Reserpinum; Serpasil

- PHARMACEUTICAL CLASS: Rauwolfia alkaloid hypotensive agent; a peripherally acting adrenergic antagonist
- CHEMICAL STRUCTURE:

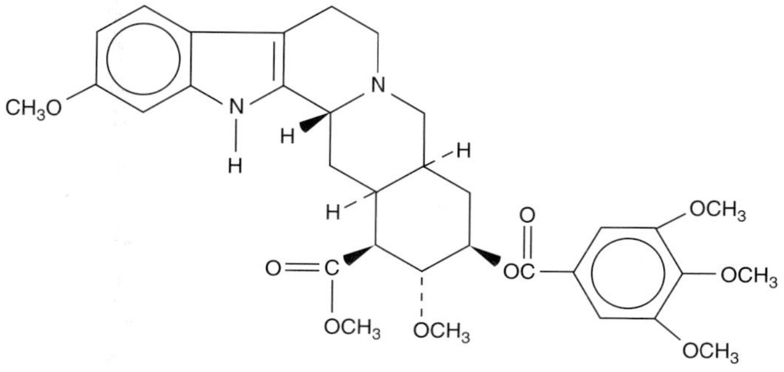

Uses

Reserpine is used in the management of mild to moderate hypertension, the symptomatic treatment of agitated psychotic states where other antipsychotic agents are not tolerated, the management of thyrotoxicosis resistant to propranolol, and the management of Raynaud's phenomenon.

Exposure Pathways

Ingestion is the most common route of both accidental and intentional exposure to reserpine. It is available in an oral dosage form either alone or in combination with a thiazide diuretic.

Toxicokinetics

Reserpine is readily absorbed with peak blood levels being reached within 1 or 2 hr; however, the onset and duration of reserpine's pharmacologic effects are not related to drug concentrations in the blood or brain. Reserpine is extensively metabolized in the liver to inactive metabolites. It is widely distributed into tissues, especially adipose tissue. A volume of distribution has not been determined. Reserpine crosses the placenta and appears in breast milk. Protein binding is 40%. The elimination half-life is biphasic. The half-life of the first phase is 4.5 hr, and the half-life of the second phase is 271 hr (11.5–16 days). The half-life may be prolonged in obese patients.

Mechanism of Toxicity

Reserpine depletes postganglionic adrenergic neurons of norepinephrine, both peripherally and centrally, by inhibiting its re-uptake into storage vesicles and by blocking the re-uptake of dopamine. Unimpeded parasympathetic activity from adrenergic inhibition results in nausea, vomiting, diarrhea, hyperchlorhydria, and miosis. Depletion of serotonin, dopamine, and norepinephrine in the brain results in sedation and depression.

Human Toxicity: Acute

Manifestations include central nervous system depression, coma, and ataxia. Tachycardia may precede bradycardia, and hypertension may precede hypotension. Nausea, vomiting, and diarrhea may occur. Therapeutic doses for adults are 0.1–1 mg daily and for children 0.01–0.02 mg/kg/day in two divided doses. Nonfatal poisonings have been reported in children ingesting up to 1000 mg. No fatalities have been reported in either children or adults; all patients have fully recovered.

Human Toxicity: Chronic

Depression, sometimes severe, can develop within 2–8 months after onset of therapy; peptic ulcer activation and gastrointestinal hemorrhage may occur.

Clinical Management

Induction of emesis may be contraindicated due to the potential for rapid loss of consciousness. Reserpine is adsorbed by activated charcoal. Treatment is largely symptomatic and supportive with basic and advanced life-support measures being utilized as necessary.

Animal Toxicity

The LD_{50} (intravenous) for dogs is 500 μg/kg. Oral doses of 5 mg/450 kg have produced colic-like behavior in horses, whereas parenteral doses of 12.5 mg have resulted in marked toxicity.

—*Elizabeth J. Scharman*

Resistance to Toxicants

In toxicology, the term resistance may be defined as an inherent genetic capability of an organism to oppose any adverse effects, manifest in either potency or dose, of a toxicant. It is important to distinguish the phenomenon of resistance from tolerance, which is the ability of an organism to adapt to the adverse effects of a toxicant with each successive dose of that toxicant. Resistance can also be a relative term with regard to the population or species that may oppose a toxic effect. For example, in a typical toxicology study the number of animals responding to a range of doses of a chemical usually reveals a small percentage of the population showing adverse effects in the lower dose range and a small percentage of the population showing no adverse effects in the higher dose range. The animals responding at the lower doses are typically categorized as susceptible individuals, whereas the animals showing little or no response at the higher doses are categorized as resistant individuals. Similarly, some species and/or strains of bacteria are resistant to penicillin, whereas others are susceptible.

Microorganisms are probably the best example for illustrating the phenomenon of resistance. Although the science of toxicology generally addresses higher levels of organisms such as fish or mammals, bacteria may serve as a good illustration of resistance to toxic effects because antibiotics, generally derived from microorganisms, evolved in nature as a form of "toxic warfare" allowing one microorganism to gain a competitive advantage over another. Humans have taken advantage of this by developing drugs that are effective in curing infectious diseases based on the structure of these natural antibiotics. Bacteria may be resistant to certain antimicrobial agents because (1) the drug fails to reach its target, (2) the drug is detoxified, or (3) the intended target is changed in a way that the drug cannot affect it. Some bacteria have cell walls that will not allow a drug to cross it, thus providing resistance. Other species or strains have enzymes on or within the cell wall that are capable of inactivating the drug. The physical and/or chemical composition of the cell wall may also resist the diffusion of a drug that may be dependent on certain environmental conditions such as pH or the presence of oxygen. Because bacteria can produce hundreds to thousands of generations within a very small time frame, they can acquire resistance through natural selection, i.e., a small mutation may change a cellular process to allow resistance to a drug, and the subpopulation, cloned from the cell that acquired the mutation, now has an advantage to oppose any drug treatment and can cause infection in the presence of the drug. Other means of acquiring resistance by bacteria include a transfer of a resistant gene to another stain or even a different species. This can occur through conjugation (direct transfer of genes through a sex pilus or bridge), transduction (transfer via a bacteriophage), or transformation (envelopments and incorporation of resistant-encoded DNA that is free in the environment into the bacteria).

Resistance is a phenomenon that can also be observed in the higher animals, although here differences arise between scientists as to relativity (i.e., whether an animal is simply less sensitive vs. more resistant). Factors that may impart resistance include age, sex, species, and/or strain; of these, species is probably the most common denominator imparting resistance to a toxicant. For example, a human cannot eat acorns because of the presence of toxic alkaloids present in the meat of the nut. Squirrels, however, are resistant to the toxic effects of these alkaloids because they possess liver enzymes capable of detoxifying these natural toxins. Another example is the classic resistance of certain strains of mice to oppose the effects of cadmium on the male reproductive system. It has been known for decades that most strains of mice will show severe testicular hemorrhage, followed by necrosis and sterility, after the parenteral injection of small amounts of cadmium chloride. Some strains, however, are remarkably resistant to this toxic phenomenon, being able to endure lethal doses with little or no effect on the testis. This resistance is also seen in some species of animals that have testis that are located within the abdominal cavity.

An age-related effect of resistance to metal toxicity can be seen following exposure to lead in humans. Although adults usually have higher blood lead concentrations than children, they are apparently more resistant to the neurotoxic effects of lead poisoning than a child. This is probably due to age-related differences in neurological development, as well as the permeability of the gastrointestinal tract and the blood–brain barrier to lead. Some may argue that children are simply more susceptible to lead poisoning than adults, but the change in resistance with age deserves some attention.

—*Stephen Clough*

Related Topics

Immune System
Modifying Factors of Toxicity

Resource Conservation and Recovery Act

- TITLE: RCRA
- AGENCY: U.S. EPA
- YEAR PASSED: 1976
- GROUPS REGULATED: Chemical industry

Synopsis of Law

Several statutes administered by U.S. EPA regulate land disposal of hazardous materials. The principal law is the Resource Conservation and Recovery Act (RCRA), enacted in 1976. RCRA established a comprehensive federal scheme for regulating hazardous waste. Directed to promulgate criteria for identifying hazardous wastes, U.S. EPA has specified these criteria as ignitability, corrosivity, reactivity, and toxicity. The agency has identified accepted protocols for determining these characteristics and established a list of substances whose presence will make waste hazardous.

RCRA directs U.S. EPA to regulate the activities of generators, transporters, and those who treat, store, or dispose of hazardous wastes. Standards applicable to generators, transporters, and handlers of hazardous wastes must "protect human health and the environment." U.S. EPA's regulations applicable to generators and transporters establish a manifest system that is designed to create a paper trail for every shipment of waste, from generator to final destination, to ensure proper authority over persons who own or operate hazardous waste treatment, storage, or disposal facilities. Pursuant to RCRA, U.S. EPA issued regulations prescribing methods of treating, storing, and disposing of waste; governing the location, design, and construction of facilities; mandating contingency plans to minimize negative impacts from such facilities; setting qualifications for ownership, training, and financial responsibility; and requiring permits for all such facilities.

—*Shayne C. Gad*

Related Topics

    Clean Air Act
    Clean Water Act
    Environmental Toxicology
    Hazardous Waste
    Toxic Substances Control Act

Respiratory Tract

Introduction

The route by which a chemical enters the body is a major factor in determining whether a substance is toxic. More than 100 years ago it was noted that the air we exhaled was less dusty than the air we inhaled, demonstrating that airborne substances were removed from the inhaled air and deposited in the respiratory tract. When toxic chemicals are inhaled and deposited on sensitive tissues, normal respiratory functions required to maintain the morphological and physiological viability of the respiratory system may be significantly impaired, increasing an individual's risk of disease. With each breath, our body is potentially exposed to numerous gases, vapors, and airborne viable and nonviable particles that could adversely effect the vital function of this system. The lung is a most vulnerable target organ since it has nearly four times the total

surface area interfacing with the environment as does the total combined surface area of the gastrointestinal tract and the skin. Because of this large surface area (70 m^2), inhalation becomes a major route for entry into the body of toxic substances from occupational and environmental exposure. It has been calculated that at rest, the average adult breathes about 15 kg of air each day. This is significantly more than the daily intake of food and water, which is about 1.5 and 2.0 kg per day, respectively. Breathing is a function that must be continuous on a minute-by-minute basis, whereas extended intervals without exposure occur between periods of water and food intake. Also, the dose of polluted air reaching the respiratory tract is dependent on the state of exercise with minute ventilation varying by up to a factor of 30 between sleeping and exercise.

To maintain its primary function as an organ of gas exchange, the mammalian respiratory system must be able to defend itself from constant assault of hazardous agents that enter the body by this route of exposure. When these normal pulmonary defenses are compromised, inhaled toxic substances have the potential for initiating or aggravating existing lung disease. The health effects associated with airborne contaminants are not limited to the respiratory tract. This route of exposure may also be the portal of entry for substances that can then be translocated from the respiratory tract to systemic sites. Because the blood leaving the lung is rapidly distributed to all parts of the body, deposited contaminants may be transported to the entire body. To produce an effect that is beyond the pulmonary system, it is necessary that the chemical, its metabolite(s), or a reactive product(s) be transported to some specific susceptible target site. There is also evidence that lung tissue can be damaged when toxic chemicals enter the body by other routes and are then transported by the bloodstream to the lung. For example, interperitoneal injection of butylated hydroxytoluene or ingestion of the pesticide, paraquat, produces acute lung damage.

This entry presents a discussion of the principles of respiratory toxicology including (1) an historical perspective, (2) approaches used to evaluate respiratory responses to inhaled chemicals, (3) classification of airborne chemicals, (4) concepts of dose–time relationships, (5) factors influencing toxicity of airborne substances, (6) the basic biology of the respiratory system with emphasis on those structures and functions that are involved in toxicological responses, (7) biomarkers of pulmonary effects, (8) toxicological response associated with inhaled chemicals, and (9) assessing the human risk of airborne chemicals.

Historical Perspective: Respiratory Morbidity and Mortality

The consequences of breathing contaminated air have long been known. The public awareness of and concern for the nature and degree of health and welfare risk associated with exposure to airborne chemicals have varied considerably over history. It has been suggested that concern for air pollution may have begun with human's first use of fire for heating and cooking and was accelerated following the wide use of coal as an energy source. As early as the thirteenth century the public began to complain of impaired visibility, soiling, odor, and health effects associated with coal smoke. As a consequence of the need for more energy to support the industrial revolution, a number of serious air pollution episodes began to occur. In 1930, pollution reached levels in the Meuse Valley of Belgium sufficient to cause over 60 deaths and hundreds of illnesses. In 1952, the high levels of air pollution in London combined with fog, resulting in an atmosphere that caused over 4000 deaths. Other serious air pollution problems occurring in Donora, Pennsylvania, in 1948, in Tokyo in 1970, and in New York in 1953 and 1963, brought public attention to the hazards associated with uncontrolled emissions. These most serious life-threatening episodes usually were of an acute nature and produced the most serious effects among the old, infirm, and those with respiratory disease. Meterologic conditions (inversions) over the polluted area typically led to an increase in mortality and morbidity. The 1950s and 1960s were a period during which the public became increasing aware of environmental pollution with industrial chemicals. Rachel Carson's book *Silent Spring* was a milestone in arousing public concern over environmental contaminants that produced human health effects. Increased health risk from accidental release of massive amounts of extremely hazardous substances into the environment from chemical spills, industrial explosions, fires, or accidents involving railroad cars and trucks transporting these chemicals have recently become a major concern. People living in communities surrounding these spills are at considerable risk of being exposed. The possibility of such sudden exposure at sites where hazardous substances are produced, stored, or used became very evident following the release of methyl isocyanate from a pesticide manufacturing plant in Bhopal, India, in 1984. The incident caused over 2,000 deaths and approximately 20,000 more suffered irreversible damage to their eyes and lungs. In another case in which the exposure was chronic rather than acute, levels of beryllium released from a manufacturing plant were sufficient to cause beryllium disease in people residing near the plant. These most serious incidents brought an increased public and scientific awareness of air pollution and its sources, the health and welfare effects associated with such exposure, and the need to develop sound strategies for elimination of such risk. Guidelines have now been developed by the National Research Council to be used to develop community emergency exposure levels for such extremely hazardous airborne substances (see Emergency Response). While today's health effects due to such air pollution episodes may be less dramatic, the scientific community is greatly concerned that other natural and man-made substances may be released into the environment at very low levels that may still result in serious long-term effects. For some chemicals, a threshold level for effects might not exist. Four of the leading causes of death (cancer, pneumonia and flu, chronic obstructive pulmonary disease, and emphysema) involve the respiratory system and may be related to exposure to airborne chemicals. It has been estimated that at least 11%, and possibly as much as 21%, of the lung cancers may be attributed to air pollution. Air monitoring studies have revealed a large number of contaminants, including carcinogens such as benzene, vinyl chloride, and chloroform. Such airborne pollutants may be associated with the occurrence of a higher incidence of lung cancer in urban populations. In the United States, like most other developed countries, regulations have been established to control pollutant concentrations in outdoor air and in the workplace. New pollutants are regularly introduced into the environment and identifying and understanding the association between such contaminants and the resulting disease states remains a challenge for the toxicologist.

Exposure to airborne contaminants is not limited to the outdoor air or to the working environment, but may also take place voluntarily through personal activities such as cigarette smoking and certain hobbies. Exposure to hazardous air pollutants can also occur in the home. The health effects associated with the indoor environment, where an individual may spend as much as 90% of his/her time, have become a major concern. Families of workers have developed documented illness associated with contact with clothing that is contaminated with industrial dusts. The U.S. Environmental Protection Agency (U.S. EPA) has ranked the American home as fourth on their list of serious health hazards. New building materials, emissions from wood and gas stoves, heaters, furnishings, air-conditioning systems, insulation, tobacco smoke, and household products such as pesticides that are used indoors have been linked to serious health risk. Nonspecific symptoms in occupants of modern office buildings, often referred as sick building syndrome (SBS), have been widely reported but have not been clearly linked to a specific air contaminant. SBS symptoms include (1) eye, nose, and throat irritation; (2) sensation of dry mucous membranes; (3) erythema (skin irritation and redness); (4) mental fatigue and headaches; (5) high frequency of airway infections and cough; (6) hoarseness and wheezing; (7) itching and unspecific hypersensitivity; and (8) nausea and dizziness. Young children may be unusually sensitive to the toxic effects of these chemicals.

Approaches Used to Evaluate Respiratory System Response to Airborne Chemicals

Supporting data for evaluation of adverse biological responses to chemical exposure and subsequent prediction of human health risk of a particular level and pattern of exposure are generated using epidemiology, studies of controlled clinical exposures, laboratory animal toxicology, and *in vitro* studies. Each category of study has certain intrinsic advantages and limitations and, in general, a database including results from multiple study categories is required to overcome the individual shortcomings (Table R-7).

Epidemiological studies may show an association between exposure and mortality, morbidity, or a specific disease and may allow direct inference of human risks since actual human exposure conditions, such as the presence of appropriate chemical mixtures, are involved. Examples of strong associations include lung cancer with cigarette smoking or with inhalation of asbestos or metallic compounds of arsenic, chromium,

or nickel; liver tumors with occupational exposure to vinyl chloride; and leukemia with occupational exposure to benzene. Studies involving environmental exposures of the general population have the advantage of including sensitive subpopulations. For example, the elderly and individuals with preexisting cardiopulmonary disease were a sensitive cohort during the previously described London air pollution episode of 1952 and, recently, children were found to be more sensitive than adults to NO_2 emissions from gas cooking stoves.

A prominent shortcoming of most such studies is limited or incomplete exposure information, including both the exact chemicals involved and the airborne concentrations. As a result, evaluation of dose–response relationships and acceptable exposure limits is difficult. Even when strong associations can be demonstrated for high levels of exposure, as in the case of benzene exposure, the low statistical sensitivity of epidemiological methods makes it difficult to assess the risk to individuals with a history of long-term exposure at lower levels. Confounding variable bias is usually a significant problem since exposure histories typically include multiple chemicals. This is particularly important for both cancer and nonneoplastic disease of the respiratory tract. For example, inaccurately reported cigarette smoking or work history can greatly distort findings. It has been suggested that such bias may be involved in the highly controversial finding of an association between environmental tobacco smoke exposures and lung cancer. Studies of worker populations in the synthetic rubber, plastic resin, and coatings industries provide additional examples of confounding exposures. Synthetic rubber and thermoplastics workers may have complex exposure histories with prominent exposures to styrene, butadiene, and, in some cases, acrylonitrile. Coatings industry workers typically have histories of exposure to a broad spectrum of chemicals including styrene, epoxy compounds, acrylic monomers, isocyanates, and anhydrides. Finally, exposure to wood dust may have been a confounding factor in studies reporting an association between formaldehyde exposure and sinonasal tumors.

Epidemiology also suffers from the fact that effects are generally counted when significant disease, morbidity, or mortality has occurred and, thus, protection from disease is not addressed. The use of validated biomarkers for early effects may improve the utility of these studies. As an example, studies have found a clear link between occupational beryllium exposure and

TABLE R-7
Advantages and Limitations of Study Approaches Used to Assess Pulmonary Response to Inhaled Toxicants

	Epidemiological studies	*Controlled clinical studies*	*Animal studies*
Exposure conditions	+ Realistic concentrations + Real chemical interactions − Definition of exposure difficult − Confounding agents interfere	+ Well-defined exposures − Limited to low concentrations	+ Well-defined exposures + Wide concentration range possible + Easy exposure manipulation
Exposure time frame	+ Realistic, acute to chronic	− Short-term only	+ Acute to chronic − Relevance of pattern/length of exposure is questionable
Toxicologic effects	− Limited to severe or crude effects (mortality, morbidity) − Insensitive, two-fold change required to detect effect − Disease present, prevention not addressed	+ Subtle, less severe effects measurable − Only mild, reversible effects, questionable toxicological significance	+ Wide range of responses may be evaluated − Relevance of subtle effects to human is uncertain
Population characteristics	+ Measured in humans + Large population size possible + Full range of sensitive subpopulations possible	+ Measured in humans − Limited number of subjects + Possible to study sensitive subpopulations	− Extrapolation to humans + Large group size possible − Homogeneity of animal model population and environmental factors-relevance to human
Utility	− Assessment of dose response is difficult − No information on mechanism of action − Costly and time-consuming	+ Dose response may be tested (limited range) − Limited information on mechanism of action − High cost	+ Dose response may be tested over a wide range + Possible to investigate mechanism of action + Relatively lower cost

Note. +, Advantage; − limitation.

the presence of a sensitization-dependent, progressive, and incurable granulomatous disease of the lung (chronic beryllium disease). Development of beryllium-dependent transformation tests for bronchoalveolar lavage and peripheral blood lymphocytes may lead to early diagnosis of sensitization and subsequent removal from further exposure and/or corticosteroid treatment to limit progression of the disease.

Controlled clinical studies using volunteers have most frequently been used to evaluate human effects of exposure to low levels of air pollutants, including sulfur dioxide, nitrogen dioxide, ozone, carbon monoxide, and acid aerosols of sulfates and nitrates. Major advantages of this approach are that humans make up the exposure population and that it is possible to closely define and control the exposure concentration. To a limited extent, sensitive subpopulations may be tested. For example, airway hyperreactivity to sulfur dioxide and sulfuric acid aerosols have been demonstrated in asthmatics (asymptomatic at time of testing). Individuals with heart disease are especially at risk to carbon monoxide exposure. When patients with histories of angina pectoris were exposed to low levels of carbon monoxide, they experienced reduced time to onset of chest pain as a result of insufficient oxygen supply to the heart muscle. Since the safety of the experimental subjects must be a primary concern, only short-term exposures to low concentrations that produce only mild and transient responses may be used. The effects of chronic exposures cannot be tested and the range of endpoints that can be assessed is often limited to pulmonary function measurements and blood clinical chemistry assays. In some university hospital settings, additional evaluation procedures are used, including examination of bronchoalveolar or nasal lavage fluids and evaluation of effects on mucociliary clearance and epithelial permeability using inhalation of radio-labeled aerosols. In general, the reversible changes that are observed following single human exposures are of uncertain clinical significance in predicting long-term effects.

Assessments of chemicals or chemical mixtures for safety risk to workers and/or the general population clearly require a database obtained from intact, living organisms. *In vitro* methods cannot be used to model the complex interactions and feedback processes between cells, tissues, and organ systems of a functioning mammalian organism or the complex deposition, uptake, and clearance processes of the respiratory system. Alternatively, *in vitro* studies can be very useful for screening a large number of chemicals for a specific effect, for example, genotoxicity or cytotoxicity, and for development of information on mechanism of action. In terms of risk assessment, *in vitro* study data may provide information that aids in the interpretation of the database derived from animal and human exposure.

For data to be used in regulatory decisions, the primary standard *in vitro* methods are the genotoxicity assays, including the Ames test and the mouse lymphoma cell mutagenesis assay. Animal and human respiratory tract cells or tissues in culture are frequently used for screening and mechanistic studies. For example, alveolar type II cells in culture have been used to evaluate xenobiotic chemical metabolism, alveolar macrophages in culture have been used to test for cytotoxicity, macrophage activation, and the effects of exposure on macrophage function (e.g., phagocytosis and bacterial or virus inactivation), and tracheal explant cultures have been used to model the preneoplastic action of airway carcinogens.

The driving philosophy behind an aggressive strategy of toxicity testing in laboratory animals is the conviction that human beings should not have to suffer from avoidable, debilitating, or lethal chemical-induced toxicity or cancer when the chemical's effect can be demonstrated in a test animal species. Historical examples, such as benzene, asbestos, and vinyl chloride, for which animal models were developed after the association of disease with exposure was demonstrated in humans, show the need for well-designed safety testing in animals. Animal studies allow maximal flexibility in choice of chemical agents, exposure concentrations and regimens, biological endpoints, and test species. Exposure conditions can be tightly controlled and readily manipulated and exposures can be acute, subchronic, or chronic. Studies can be designed to help elucidate mechanism of action and the existence and basis for species differences in response. A broad range of biological responses can be evaluated, including target organ histopathology, changes in hematological and blood chemistry parameters, changes in organ system function, changes in immunological responses, effects on neurobehavioral parameters, and reproductive/developmental effects. Of particular importance to evaluate

respiratory tract toxicity are histopathology, lung function, and bronchoalveolar lavage fluid cytology and chemistry.

Many examples of the use of animal exposures to study the respiratory tract toxicity of inhaled chemicals are discussed in portions of this entry describing indicators of respiratory tract response. Examples cited here demonstrate ways in which animal studies are used to help protect human populations and guide assessment of human risk. For most chemicals for which there is a potential risk of worker exposure by inhalation, there is insufficient human data to set safe occupational exposure limits. Using inorganic nickel compounds as an example, epidemiological data indicate an increased risk for nasal and lung cancer in workers involved in nickel sulfide ore smelting and refining processes and lung tumors have been found in rodents chronically exposed to nickel compounds. The current occupational Threshold Limit Value (TLV) for soluble nickel salts has been set based on studies in which nonneoplastic lesions, including epithelial hyperplasia, inflammation, and fibrosis have been evaluated in laboratory animals.

Animal models have been developed for chronic pulmonary effects of asbestos fibers and silica. The insolubility and cytotoxicity of the chemicals and the inability of alveolar macrophages to normally phagocytize and clear the chemicals are important features of the models of asbestos-induced fibrosis and cancer and silica-induced fibrosis. These validated animal models have been used to evaluate the potential for man-made fibers to cause cancer or other crystalline materials to cause fibrosis. Tested using this approach, exposure to glass fibers has produced both positive and negative results. Using interpleural and intraperitoneal installation of very thin glass fibers, researchers in Germany produced tumors in rats and hamsters. In a recent article supporting the need for inhalation testing, data cited indicated that chronic whole body inhalation of glass fibers failed to induce lung cancer in rats, even at very high fiber loads in the lung. In parallel studies using chrysotile asbestos, 18.9% of the animals had tumors and about half of those were malignant or carcinomas. From these studies, the author concluded that even with levels of exposure 1000 times higher than seen in a typical exposure situation, there was no evidence of tumors. Such information indicates the importance of using a natural route of exposure when assessing the risk of inhaled substances.

Effects of long-term exposure to air pollutants are difficult to evaluate using human data since, in epidemiology, exposure history and confounding factors cannot be controlled or, in clinical studies, only short-term exposures are possible. Long-term exposure studies using laboratory animals provide information that can be used to predict human effects with several models suggesting changes that correspond to well-documented human disease states. Long-term exposure of rats to sulfur dioxide produces thickening of the tracheal mucous layer and hypertrophy of goblet cells, both features of human chronic bronchitis. Repeated sulfur dioxide exposure of rats also interferes with clearance of inert particles. Rabbits that have been exposed to sulfuric acid aerosols have a slowing of mucociliary clearance, goblet cell hyperplasia, decreased pH of intracellular mucous, decreased airway diameter, and increased airway reactivity to acetylcholine. This pattern of response is similar to the pathology observed in patients with chronic bronchitis and asthma. U.S. EPA scientists have exposed rats to ozone using a diurnal concentration pattern (range, 0.06–0.25 ppm), producing alveolar epithelial hyperplasia within 12 weeks, which resulted in a slowing of the clearance rate of asbestos after 6 weeks and functional changes indicative of a stiffer lung after 12 months. Use of such an exposure regimen considered to be realistic (for an urban area of high pollution) suggests possible relevance to human toxicity.

The use of data derived from laboratory animal exposures to assess human risk is complicated by issues concerning extrapolation from animals to humans. Differences between animal and human biochemical and pharmacokinetic processes may diminish or negate the relevance of a particular animal model. Xenobiotic metabolizing capacities and patterns of distribution of these activities within the respiratory tract may differ between species. A biochemical that is specific for male rats (and is not found in humans), α_{2u}-globular protein, appears to be required for susceptibility to renal tubular nephropathy and tumors induced by inhalation of unleaded gasoline vapors. In the respiratory tract, species differences in three-dimensional airway structure may result in differences in toxic effects. For example, the complexity and relative surface area of the nasal turbinates are very different in rodents and humans. Respiratory tract detoxification processes may also differ between species. In addition, the genetic homogeneity of laboratory animal strains and the closely controlled environmental conditions (e.g., diet) used in laboratory animal studies may affect the relevance of such studies to humans. In laboratory rodent carcinogenicity studies, high background incidence rates for certain tumors appear to be related to unrestricted food availability. There is concern that this rodent model might also have heightened susceptibility to chemically induced tumors or that resultant life-shortening for the model might interfere with detection of tumors. All of these potential differences highlight the importance of animal and *in vitro* studies to provide pharmacokinetic data and information on mechanism of action.

Extrapolation from high dose exposures in animals to realistic human exposure levels is also a serious concern for risk assessment. For example, metabolic and detoxification processes may be dependent on exposure level. A commonly cited example involves the increased incidence of lung tumors in rats following particulate exposure regimens that produce high lung particle loads (e.g., chronic, high level diesel exposure). It has been suggested that macrophage-based clearance mechanisms become overwhelmed with chronic exposure at high concentrations and that this may be associated with tumor development that would not be seen at ambient levels. The relevance of tumor incidence under these conditions to prediction of human risk has been questioned.

Classification of Airborne Chemicals

Airborne substances that are of interest to inhalation toxicology include gases, vapors, aerosols, and complex mixtures in various combinations. Aerosols may exist as mists, fogs, smokes, fumes, or dusts. Physical properties of airborne chemicals are most frequently used by the inhalation toxicologist for primary classification, with the first division based on whether the material is a gas, vapor, or aerosol (particulate material). For materials that are not highly reactive, movement and behavior in the respiratory airstream, the sites of deposition an/or uptake, the fraction retained, and the rate of interaction with airway tissues and cells are highly dependent on the physical state (see Factors Affecting Toxicity). In addition, this approach provides the inhalation toxicologist with a convenient breakdown for consideration of the nature of the material to which a population at risk is exposed and for design of appropriate methodology for generation of test atmospheres of a material for toxicological studies. Although classification by chemical type is used for applications such as industrial hygiene, this approach has important limitations. Materials with very different chemical structures may have similar toxic effects and materials with similar chemical properties or even a single chemical may have different toxic effects depending on whether the form inhaled is a gas or an aerosol/gas mixture.

Gases and vapors are usually grouped together since a vapor is the gaseous fraction of a chemical that is a liquid at ambient temperature and atmospheric pressure. Two properties, solubility and chemical reactivity, are particularly important determinants of the toxic actions of inhaled gases and are also used for classification. In general, the solubility of a gas/vapor is important in determining the primary sites of deposition and injury (see Factors Influencing Toxicity). Reactive gases interact chemically with components of cells at the site of deposition, producing direct injury that is typically followed quickly by inflammation and edema but can progress to cause a variety of toxic effects. Following deposition, nonreactive materials may undergo activation to a reactive intermediate or may interact with cellular oxidation–reduction machinery to be activated or to deplete critical cellular reducing substances or antioxidants (e.g., NADPH and glutathione).

An aerosol may be defined as a suspension in air of solid particles, as in dusts, fumes, and smokes, or liquid droplets, as in fogs, mists, and liquid aerosols of organic materials. Dusts are formed by milling or grinding of larger masses of a parent material, while fumes and smokes are formed by combustion, sublimation, or condensation usually with a chemical change in the material. In fibrous aerosols, the solid particles have a length along one axis that is at least three times greater than that along either of the other two axes (i.e., an aspect ratio of greater than $3:1$). Examples include asbestos, glass and plastic fibers, and mineral wool. Mists and fogs are typically formed by condensation of water on microscopic particles. Liquid aerosols are also produced by nebulization or spraying in the use of man-made products (e.g., pesticides and paints). Particles in aerosols may also consist of viable agents, including bacteria and viruses, as well as fungal spores and pollen. Thus, inhaled biological aerosols may produce infectious diseases, such as influenza, viral and bacterial pneumonia and tuberculosis, and allergic reactions. Other properties of inhaled aerosols that are used for classification include particle size, which is

the primary determinant of regional airway deposition, electrical charge, solubility, and rate of dissolution in aqueous media, and hygroscopicity.

Most human inhalation exposures in the workplace, home, or outdoor environment involve airborne mixtures of chemicals, which frequently include both gaseous and particulate material. In addition, gases and vapors may be adsorbed onto the surface of aerosol particles and be carried to potential sites of injury in the lungs by the respirable particles. Three environmental mixtures of continuing concern for air pollution are photochemical smog, diesel exhaust, and environmental tobacco smoke. Smog is typically a complex mixture of gaseous combustion products, including oxides of carbon, sulfur and nitrogen, ozone, hydrocarbons, reaction products of ozone, and other pollutants and particulate aerosols of carbon and various metal oxides. Diesel exhaust and environmental tobacco smoke are also mixtures of particulate and gaseous combustion products. Of particular concern are potential carcinogens, including polycyclic aromatic hydrocarbons and tobacco-specific nitrosamines, that may reach sensitive regions of the airways adsorbed to particles.

Exposure, Concentration, and Dose–Time Concepts

Confusion often occurs with the use of the terms "exposure," "concentration," and "dose." Dose is the amount of contaminant that is deposited or absorbed in the body of an exposed individual over a specific duration. Dose occurs as a result of exposure. Concentration is that level of contaminant present in the air potentially available to be inhaled. The atmospheric concentration of a chemical by itself does not define the total dose of a chemical delivered nor the specific sites of potential injury. For a substance present in inhaled air to be toxic, a significant dose must first be removed from the inhaled air and be deposited on sensitive tissue. Knowledge of the dose to initial target sites provides a critical link between exposure and the subsequent biological response. Understanding the disposition of inhaled xenobiotics is complex and, due to space limitations, cannot be described in detail here. However, certain basic concepts need to be presented to provide information on the various factors related to exposure, dose, and response that are fundamental to understanding the potential human risk from inhaled chemical agents.

The prediction of biologic effects from inhaled pollutants is often based on the study of concentration–time. However, in inhalation toxicology, the concept of dose is most important to the understanding of the relationship between exposure concentration and the body's response. Actually dose can be portioned into two components: internal dose and biological effective dose. Internal dose is the amount of a contaminant that is absorbed into the body over a given time. Biological effective dose is the amount of contaminant or its metabolites that has interacted with a target site over a given period of time so as to alter a physiological function. The consequence of the chemical reaching the target tissue is governed by its pharmacokinetic behavior, which includes the processes of absorption, distribution, metabolism, and elimination. The effective dose to the respiratory tract, for example, for inhaled particles, is proportional to particle retention and integrated particle retention is derived from the balance of two processes: deposition and clearance.

Because of the difficulties in determining actual dose in inhalation studies, the toxicologist must assess the extent to which the concentration (C) of a given chemical and the duration of exposure (T) interrelate to determine the magnitude of the biological response (K). Often the formula, $C \times T = K$ will be used to relate the toxic effect of certain inhaled substances to its concentration and time of exposure. This formula, referred to as Haber's law, has been shown to be valid only for certain combinations of concentrations and exposure time and for only a limited number of substances. While it may be necessary to use Haber's formula in certain conditions, caution must be exercised in using the general expression, $C \times T = K$, when comparing exposure conditions that are to be used in extrapolating from effects seen in laboratory animals to humans. A more appropriate general expression for estimating $C \times T = K$ would be given by $C^a \times T^b = K$, where the exponents a and b are estimated from the data. Such a formula allows for the fact that C and T do not always contribute equally to the observed toxicity. Haber's law may be inappropriate for certain materials such as ammonia and nitrogen dioxide which have been shown to be more toxic with high concentration over shorter exposure periods.

Factors Influencing Toxicity

Scientists who seek an understanding of the toxicological hazards associated with inhalation of airborne substances need a basic knowledge of the structure and normal functioning processes of the respiratory system. This information is essential to understanding how this system responds to inhaled substances and the possible health consequences resulting from exposure to toxic substances. Various regions of the respiratory system can be sensitive target sites for inhaled xenobiotics. However, the potential hazard associated with such exposure will depend on many interacting factors that will be discussed in the following section. In each region of the respiratory tract there are certain defense systems capable of coping with an insult. However, when these defenses are compromised or overwhelmed, the potential for disease is significantly increased. A number of factors can significantly influence the deposition, retention, and redistribution of these inhaled substances, which in turn can directly affect the toxicity of the inhaled substance.

The factors and processes affecting the deposition of airborne substances in all regions of the respiratory tract can be broadly categorized as those related to the (1) structure of the respiratory system, (2) chemical and physical properties of the airborne substance, and (3) ventilatory functions including route of breathing (nasal, oral, and oronasal).

The morphology of the specific respiratory tract region at both the gross anatomical and the microscopic levels are important factors. In extrapolating animal effects to the human, one must be aware that the respiratory tract structure will vary both within individuals and between species at each level of anatomy.

In all regions of the respiratory tract, the specific anatomy, dimensions, composition, flow, and thickness of the mucous or fluid lining layers and regional differences in tissue types and metabolic capabilities all have a major effect on that region's dosimetry. Dosimetry refers to estimating or measuring the amount of a compound or its metabolite or reactive product that reaches a specific target site after exposure to a given concentration.

Whenever the airborne substance is deposited on the linings of the respiratory tract, its new biological environment will react to it. For inhaled particles, a major factor that influences deposition is size. A particle's characteristics may alter its size; for example, if the particle is hygroscopic, it can be expected to grow substantially while still airborne within the respiratory tract and will be deposited based on its hydrated size. The deposition probability for particles with geometric diameter ≥ 0.5 μm is governed largely by their equiva-

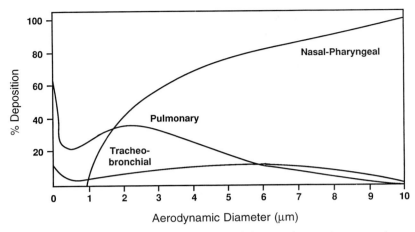

FIGURE R-7   *Regional deposition of inhaled aerosols as a function of particle size (reproduced with permission from A. W. Hayes (Ed.), Principles and Methods of Toxicology, 2nd ed., p. 364, Raven Press, New York, 1989).*

lent aerodynamic diameter. Smaller particles are deposited based on their actual diameter. Since particles are generally inhaled as aerosols rather than as a single particle, the mass median aerodynamic diameter is the most appropriate parameter to use with aerosols in which the particles have actual diameter ≥ 0.5 μm. Aerosols containing particles with diameters less than this should be expressed in terms of diffusion diameter or geometric size. Figure R-7 shows the range of deposition variations in the various respiratory regions. Particle deposition at various sites within the respiratory tract is dependent on several mechanisms. These include impaction, sedimentation, Brownian diffusion, interception, and electrostatic precipitation (Fig. R-8). The most important are impaction, sedimentation, and diffusion. Impaction is the inertial deposition of a particle onto an airway surface. It is the main mechanism by which particles having a diameter ≥ 0.5 μm are deposited in the upper respiratory tract. The probability of impaction increases with increasing air velocity, rate of breathing, and particle size and density. Sedimentation is deposition due to gravity and is an important mechanism for particles with a diameter $\geq$ to 0.5 μm that penetrate to those airways when air velocity is relatively low. Submicrometer-size particles are deposited due to a random motion owning to their bombardment by surrounding air molecules (Browning diffusion) that results in particle contact with the nearest airway wall. This is a major mechanism in airways where the airflow is very low (e.g., bronchioles and alveoli).

It should be noted that respirable fibers may be quite long and extend beyond 50 μm. However, the most important factor in the deposition of fibers, such as asbestos, is the diameter of the fiber, not the length. Fibers of small diameter (-0.5 μm) will remain suspended in the airway and drift with the airflow to be deposited in the airspace.

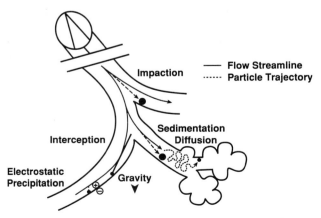

FIGURE R-8   *Mechanisms by which particles may be deposited in the respiratory tract [adapted from R. B. Schlesinger, Deposition and clearance of inhaled particles, in Concepts in Inhalation Toxicology (R. O. McClellan and R. F. Henderson, Eds.), p. 164, Hemisphere, New York, 1989].*

Particle size, lengths, and configurations not only influence the site of deposition, which in turn affects the mode by which the particle is cleared, but also influence the metabolic fate of the chemical. All segments of the respiratory tract, from the nasal cavity to the periphery of the pulmonary compartment, contain enzymes that are capable of metabolizing xenobiotic compounds. These enzymes are capable of metabolizing some compounds to products that are less toxic, while other metabolites may be more toxic than the original inhaled chemical. There are significant differences in rates of metabolism at the different sites in the respiratory tract. In general, the nasal and the pulmonary regions have a higher metabolic activity than other regions. Metabolic capability is an important factor that plays a crucial role in defining species susceptibility to toxicants.

The solubility of an inhaled contaminant influences the disposition of gases, vapors, and particulates. In general, those substances that are highly water soluble, such as ammonia, formaldehyde, and hydrogen chloride, will be removed by the upper respiratory tract. Formaldehyde is concentrated in the nasal mucosa and has been shown to be a nasal carcinogen in the rat. Chemicals with intermediate solubility, such as halogens and ozone, deposit in both the upper respiratory tract and the lung, while chemicals with low solubility, such as phosgene and nitrogen dioxide, deposit in and affect mainly the lung. To understand the kinetics related to solubility and to predict the toxic response one must be able to establish the solubility of the chemical not only in water but also in different media including mucus, blood, or tissue. Some particles, such as fogs, mists, and therapeutic aerosols, are aqueous droplets that rapidly merge with the mucus or liquid lining layer, greatly increasing their bioavailability for absorption.

The absorption of gases is dependent on the solubility of the gas in the blood. For example, chloroform has high solubility and is nearly completely absorbed. Respiration rate is the limiting factor. However, ethylene has low solubility and only a small percentage is absorbed—blood flow—limited absorption. It is of interest to note that as a generalization, there is a pattern of relative absorption rates that extends between the different routes of exposure. This order of absorption (by rate from fastest to slowest and in degree of absorption from most to least) is intravenous ε inhalation ε intramuscular ε intraperitoneal ε subcutaneous ε oral intradermal ε other dermal. It should be remembered that because of the arrangement of the body's circulatory system, compounds inhaled and absorbed initially enter the systemic circulation without any "first-pass" metabolism by the liver.

The depth and rate of breathing influences the dose and site of deposition of airborne substances. The process of ventilation is controlled by a variety of internal and external physical and chemical stimuli which can be affected by airborne chemicals. For many inhaled agents, deviation from normal breathing pattern serves as the earliest indicator of response. Assessing the breathing patterns, lung volumes, and lung mechanical properties are frequently used techniques in evaluating the toxicology of inhaled materials. These tests can provide useful information on whether or not pulmonary function has been impaired, the type of impairment, and the extent or magnitude of the function loss. These are not only excellent methods for indicating toxicity but are also useful in characterizing the pathogenesis of lung disease and for extrapolation of such data from animal to humans. There have been numerous studies documenting the importance of measuring respiratory parameters such as respiratory rate and tidal volume in animals exposed to inhaled toxicants. Significant alterations (depression) in these functions have been associated with exposure to methyl chloride, methylene chloride, methyl bromide, and formaldehyde. In such cases, the predicted delivered dose would have been overestimated had respiratory measurements not been recorded.

Structural Factors Influencing Toxicity

Because of the complexity of the respiratory system, it is frequently described by dividing the system into three general regions or compartments based on the anatomical structure and the corresponding physiological functions attributed to that region. Figure R-9 is a schematic showing these various compartmental areas.

Nasopharyngeal Region

The nasopharyngeal (NP) region is the most proximal region of the respiratory system and is the first potential target for airborne substances. The specific structures making up this region include the anterior nares, the turbinates, the epiglottis, the glottis, the pharynx, and the larynx. The nose is the normal portal of entry for

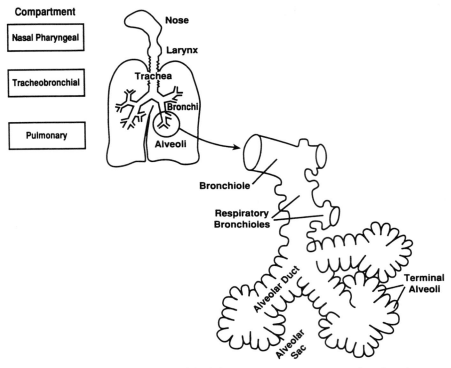

FIGURE R-9   Compartment model of the respiratory tract (reproduced with permission from P. Witorsch and S. Spagnolo (Eds.), Air Pollution and Lung Disease in Adults, p. 22, Boca Raton, FL).

all inhaled material. In addition to being an organ for smell, the nose has other functions including the conditioning and transporting of inhaled air and providing an effective filtering system that serves to protect the upper respiratory system against toxic chemicals and biological agents. This mechanical barrier, while being fundamentally nonspecific, can be quite effective. For example, under normal conditions nearly 100% of the sulfur dioxide, 20–80% of the ozone, and 73% of the nitrogen dioxide drawn through the nose are trapped in this region, preventing the pollutant from reaching the lower areas of the lung. However, under certain conditions, such as at times of high physical activity, individuals resort to mouth breathing and, as a result, the inhaled air bypasses these defenses. This significantly changes the deposition pattern of the inhaled gases or particulate and possibly their toxicity. The concentration of inhaled pollutants at the NP region can be expected to be higher than the level delivered to the lower respiratory airways and is most similar to the ambient concentration.

While the removal of airborne contaminants by the nose is effective, this action also renders this organ susceptible to toxic damage. The behavior of the inhaled substances in the NP airways and the ultimate determination of whether they are deposited or exhaled depends on numerous factors, for example, breathing patterns that influence nasal airflow rates and the chemical and physical properties of the airborne material, such as size, shape, water solubility, and reactivity. Soluble particles may, once deposited, rapidly enter the blood circulation and be transported systemically. Thus, the effective dose of toxicant delivered to the target tissue depends on factors other than the environmental concentration.

In the anterior one-third of the nose, where particles larger than 5 μm are deposited, the principal means of clearance is by blowing, sneezing, or wiping. However, following some exposures, certain particles may actually remain in this area for several days after deposition. Such retention patterns may be responsible for serious health effects. For example, nasal cancer identified in machinists was shown to be related to the high nasal collection efficiency for those particles above 10 μm in diameter. Such large-size particles are made airborne by various grinding and sandblasting activities. Particles approximating 3 μm in diameter are deposited in the NP region primarily through inertial impaction that occurs at airway branching. Breathing patterns involving higher air flow rates would tend to increase deposition of particles as small as 1 or 2 μm in diameter. In all cases, deposition can be expected to increase with duration of breathing and depth of breathing. Deposition occurs during both inhalation and expiration. Chemical and biological agents deposited in this region may lead to inflammation (rhinitis), congestion, impairment of the sense of smell, ulceration, and cancer. To prevent a buildup or uptake of these deposited substances and possible long-term health effects, it is important that they are removed quickly. The mucus layer lining the NP epithelium plays an important role in clearance of deposited material by providing a moist, sticky surface that entraps inhaled particles and gases. Particles are then transported mouthward, due to the beating of the underlying cilia, where they are swallowed or expectorated. Various disease states, such as chronic sinusitis, bronchiolectasis, rhinitis, and cystic fibrosis, adversely alter mucociliary clearance from this region. It has been estimated that normal physical clearance from the NP region has a half-life of approximately 4 min.

Nasal metabolism also plays a role in response of this organ to xenobiotics. The nose contains a large number of enzymes including cytochromes P450, dehydrogenases, esterases, transferases, and hydrolases. Nasal metabolism can be responsible for both the protection of and the damage to the nose. For example, the breakdown of formaldehyde by dehydrogenases may be protective, whereas the toxicity of certain nitrosamines has been attributed to metabolic activation to toxic metabolites by nasal cytochromes P450.

Because the nose is directly exposed to a wide variety of infectious and antigenic agents, the nasal immune system also plays an important role in defending this region from such agents. Nasal secretions contain locally produced antibodies that can be monitored in both humans and test animals by nasal lavage. This is a technique that permits the detection of both immune mediators and the influx of inflammatory cells in this region, providing useful biomarkers of effects.

Tracheobronchial Region

Inhaled air together with all of the airborne substances not removed in the NP region enter the tracheobronchial (TB) region. The TB region of the respiratory system consists of the trachea and the bronchial tree down to and including the terminal bronchioles. In the mammalian tracheobronchial trees, there are two forms of branching, monopodial and regular dichotomous. In most nonprimate species the branching is monopodial; that is, there are long, tapering airways with small lateral branches that come off of the main airway at an angle of about 60°. In a human lung the branching is symmetric (dichotomous) involving the division of a tube into two daughters, each with nearly equal diameters and nearly equal angles of branching with respect to their parent tube. The major function of this TB region is to transport the inhaled air into the pulmonary region for gas exchange and to remove it during exhalation. At the entry into this region, the proximal end of the trachea is continuous with the larynx and extends distally into the thoracic cavity to the carina where it bifurcates to form two primary branches called bronchi. The tracheal airway is maintained during breathing by cartilaginous rings that prevent it from collapsing. As the bronchi continue to divide into smaller and smaller diameters there is a point where the cartilage is no longer present and the airways are composed of smooth muscle and loose connective tissue. At this region they are referred to as nonrespiratory bronchioles. There are several generations of such bronchioles. The most distal conducting and nonrespiratory airways in the respiratory tract are the terminal bronchioles. It is this distal end of the conducting airways that connects to the respiratory bronchioles.

Since the TB region contains both very large and very small conductive airways, substances of all sizes and chemical compositions can be expected to be deposited in this region. The many bifurcations in the TB region are vulnerable sites of high regional deposition. For example, the centriacinus, which is the site of the junction between the most distal conducting airways and the gas exchanges area, is a common site of injury from a variety of airborne chemicals including diesel exhaust, oxidant air pollutants, and asbestos fibers. With physical exertion and mouth breathing, the beneficial defenses of the NP region are significantly reduced and a greater number of larger particles can be expected to be deposited in this region. Toxic substances deposited in this region in sufficient quantities may lead to bronchospasm, allergic reactions, congestion, bronchitis, and cancer. The more rapidly these materials are cleared, the less time for injury.

The primary clearance mechanism for this region is similar to that of the NP region; that is, by mucociliary transport to the glottis with subsequent expectoration or swallowing. The TB region is equipped with both ciliated and secreting cells for removal of deposited material. The airway epithelium has secretory capabilities capable of synthesis and release of mucus, ions, and water. In certain diseases (e.g., bronchitis, cystic fibrosis, and asthma) an excess of secretions may result, causing airflow obstruction, increasing the residence time of inhaled substances, and thus increasing the dose to the airways. The ciliated cell is one of the major cell types and is probably a nonproliferative, terminally differentiated cell. The ciliated cells have on their mucosal surface about 200 cilia per cell. Their length is about 6 μm in the large airways and 5 μm in the smaller airways. Groups of cilia beat spontaneously with metachromal waves in a coordinated fashion independent of nervous control. The nonciliated cells consist of various secretory cells (mucous, clara, and serous) depending on species and a nonsecretory basal cell. The mucous and clara cells may differentiate into ciliated cells. It has been suggested that the ciliated cells can actually retract their cilia and become a secretory cell. Because of the difference in the airway size, the clearance rate differs significantly within the region. Since smaller sized particles are deposited deeper in the airways, the clearance of such particles from the TB tree is slower when compared to those larger sized particles deposited higher up in the airways. It has been estimated that within the large airways, the half-time for clearance is about 0.5 hr. For the intermediate airways the half-time is 2.5 hr and in the finer airways the removal half-time is 5 hr. Exposure to inhaled toxicants and viral and bacterial microorganisms can significantly decrease the efficiency and speed of TB clearance and hence increase the potential health risk. Certain human disease states are also associated with alteration of clearance from this region. Bronchial mucus transport may be impaired in people with bronchial carcinoma, chronic bronchitis, asthma, and acute infections.

Pulmonary Region

Ultimately, the air with its remaining contaminants reaches the pulmonary region, which is the most distal and includes the respiratory bronchioles, alveolar ducts, alveolar sacs, and alveoli. The main purpose of the alveolar architecture of the mammalian lung is to expose blood to gas over a large surface area within a comparatively small volume.

The pulmonary region includes the functional gas exchange sites of the lung. The terminal bronchioles of the TB tree branch to form the respiratory bronchioles. The major structural elements of the parenchyma of this region of the lung include the alveolar ducts, alveolar sacs, alveolar capillaries, and pulmonary lymphatics. The walls of the tubular alveolar ducts are covered with alveoli. As these branch they exhibit increasing number of alveoli opening into their lumina. The human alveolus (about 300 million) is a polyhedral structure about 250–300 μm in diameter. These alveoli are thin walled and surrounded by blood capillaries for ease of gas exchange. The total thickness of this air to blood interface has been demonstrated by electron microscopy to range from 0.36 to 2.5 μ in the human. Gas exchange in the lung must be very efficient since the average human consumes about 2.5 liters of O_2 per minute. To be transported through this air–blood barrier and reach a red blood cell, a molecule such as oxygen must penetrate this tissue at the alveolar surface, transverse the tissue barrier, enter the capillary blood at the capillary surface, move through the blood to a red blood cell, and finally penetrate the red blood cell to bind with hemoglobin. The greater the exchange surface, the more area available for oxygen to diffuse. Also, the thinner the air–blood barrier, the smaller the resistance to the oxygen diffusion, and the more blood flow, the greater the amount of oxygen that can be bound to hemoglobin. The air–blood tissue barrier is sealed toward both blood and the air space by two continuous cell layers, the blood capillary endothelial cell layer and the alveolar epithelium. The endothelial cells function to control the passage of fluid, proteins, and other blood components from the vessel lumen into the interstitium and the air spaces of the lung. These endothelial cells also function in a wide variety of specific metabolic activities important in the pulmonary processing of vasoactive substances—for example, enzymes inhibitors, receptors, and transport systems of these endothelial cells can determine the level of biogenic amines, kinin, angiotensins, and prostaglandins entering the circulation. Airborne toxic substances are capable of causing many types of injury to this gas exchange region, possibly increasing the thickness of the epithelial lining or changing the permeability and resulting in an influx of cellular and acellular fluids into the alveolar spaces. Such changes can have an adverse effect on normal gas exchange.

The most prominent cell making up the epithelial lining layer in the pulmonary region is the type I cell. It makes up more than 93% of the alveolar surface area. These squamous epithelial cells line the alveolar surface and are only about 0.1–0.3 μm or less in thickness, minimizing the barrier for gas exchange. The thicker type II cell is cuboidal and covers only approximately 7% of the alveolar surface area. The airway surface of these cells is covered with microvilli, greatly increasing the surface area. Together, these two cell types function as a permeability barrier to limit the movement of molecules between the alveolar space and the interstitium. The type II cells are the progenitors of the type I cell and proliferate to reestablish the epithelial surface when the type I cells are injured. The type II cells also function as a source of an essential alveolar lining fluid, surfactant. A deficiency of surfactant may lead to alveolar collapse, resulting in hypoxemia and decrease in lung compliance. It is speculated that the alveolar macrophages release certain factors that may promote growth of the type II cells.

Due to its anatomic location, the alveolar epithelial surface is often directly exposed to inhaled gases and particulates. Injury to these cell types has been associated with a number of pulmonary toxicants, for example, silica, ozone, NO_2 herbicides, trace metals, and a number of organic vapors. It is of interest that the chemicals paraquat and diquat both cause type I cell damage but only paraquat damages type II cells, indicating that these two cells may differ in their sensitivity to chemical insults. This alveolar region is the site of several pathological lesions including centrilobular emphysema, fibrosis, and a variety of cellular injuries due to oxidant gases. Loss of or damage to alveolar tissue adversely affects the efficiency gas exchange in this region.

The alveolar macrophages are large, nucleated cells that are found on the surface of the alveoli. These cells are not a fixed part of the alveolar epithelial wall but are mobile and possess the ability to engulf (phagocytize) and remove foreign material from the region. Macrophages, in addition to being responsible for clearing the lung of debris, play a major role in initiating and modulating the primary immune response in this area of the lung and are also effective in maintaining the sterility of the lung by killing or inactivating viable microorganisms. These cells locate the material by ei-

ther random motion or are directed to the site by certain chemotactic substances. Under normal conditions the number of macrophages is usually estimated to be about 3% of the total alveolar cells, but this number can be significantly increased with an increase in deposition of particles in the lung. These cells can ingest more than 10 times their weight in particles without any measurable loss in mobility or phagocytic ability. A number of host and environmental factors can modify the rates of pulmonary clearance by this mechanism. Individuals with chronic obstructive lung disease, viral infections, asthma, interstitial fibrosis, and inflammation, as well as individuals that have been exposed to numerous inhaled gases and particulates, have been shown to have reduced numbers and impaired function of these cells, resulting in a concomitant increase risk of pulmonary disease. Certain kinds of particles may be difficult to clear due to their particular shape. Long fibers such as asbestos may be cleared more slowly and may induce certain biochemical changes that can ultimately be toxic to the macrophage.

Once loaded with particles the macrophage may clear them from this region by a number of pathways. The primary route out of the alveoli is via the mucociliary escalator. The macrophages reach the distal terminus of the mucus blanket and then are swept distally by ciliary beating within the airways. Macrophages may also migrate within the interstitium to the lymphatic system. There is also evidence that macrophages may enter the blood directly where they, together with their engulfed particles, can travel to extrapulmonary sites. If the ingested substance is toxic to the macrophage, it may be lysed while still in the alveolar region and the particle released to be taken up by another macrophage. Such cytotoxic substances may thus remain in the lung for considerable time. Clearance kinetics would indicate that the successful removal of insoluble particles by macrophages consists of two phases. The first has a half-life measured in days and the second in hundreds of days. Clearance routes and kinetics are a function of lung burden, the physicochemical and toxicological properties of the material to be transported, and the health of the individual.

An actively functioning pulmonary immune system is critical for defense of the lung. Immune activity has been shown in both the conducting airways and in the lung parenchyma. All major immunoglobulins—IgA, IgG, IgM, and IgE—are present in the bronchial secretions. These are derived from local synthesis and by transudation from serum. In the parenchyma of the lung, the pulmonary macrophage participates in the generation, expression, and regulation of the immune response. These cells serve as antigen-presenting cells, as effector cells for T cell immunity, and as regulatory cells that modulate either as a promoter or suppressor of pulmonary immune response. The mechanism of the immune response in the lung is complex and beyond the scope of this entry.

Briefly, antigenic material deposited on pulmonary tissue initiate and stimulate the immune process. The antigen is taken up by and processed by the macrophage. Antigens in alveolar spaces that escape this phagocytic action and other clearance mechanisms may still gain access to the pulmonary interstitium, where they may be subsequently transported to nearby lymphoid tissue where immune stimulation can occur. The pulmonary macrophage presents the antigen to local lymphatic tissue that ultimately will produce cell-mediated or humoral immune response. Lymphatic tissue and lymphocytes are present at or near the air–tissue interface at all levels of the respiratory tract from the nasopharynx to the alveolar spaces of the pulmonary racemus. These tissues are important in ensuring pulmonary immune response since they contain antigen-presenting cells and the full repertoire of antigen-reactive T and B lymphocytes needed to react with the antigen.

Cell-mediated response begins with the macrophages but is then mediated through the thymus-derived lymphocytes (T cells). These T cells regulate the immune system. T cells interact with B cells for antibody production. T cells not only can kill cells presenting antigen but also can release cytokines that modulate the immune response. Humoral immune responses are the end result of antigen interacting with marrow-derived or bursal cell-equivalent lymphocytes (B cell). B cells secrete antibodies that inactivate antigens in the body. The B cell function is regulated by two subpopulations of T cells: helper T cells that are required for optimal production of antibody and suppressor T cells that are active in modulating the humoral response once initiated. A wide array of substances that can affect the immune response are discussed in the section on effects on pulmonary defenses. Table R-8 list examples of immunomodulation by various inhaled chemicals.

Macrophages also release substantial amounts of diverse substances that exhibit a broad range of bio-

logical activities. Examples of such mediators include (1) interleukins, which play an important role as mediators of inflammation, are chemotactic for neutrophils, promote the differentiation of natural killer (NK) cells, and function in the maturation of helper T cells; (2) monokines, which regulate the growth and activation of other cells such as fibroblasts and endothelial cells; and (3) interferon, which represents a group of antiviral proteins that function to inhibit the intracellular replication of many viruses and the proliferation of malignant cells and promote NK cell functions. These NK cells do not play a role in the antigen-specific antibody response but are critical components of the general, nonspecific immune defenses.

From this brief description it is evident that the immune system is very complex and to function properly depends on the interaction of several components. Each of these steps is a potential target for a toxic chemical.

Biomarkers of Pulmonary Effects

There is a growing need for development of sensitive assays that can be used in inhalation toxicology as biological markers of adverse health effects associated with pulmonary injury. A pulmonary biomarker should be able to reflect a change in a biological system that can be related to a specific effect or an exposure for a specific toxic substance. Such a marker should be an indicator of early biological response of the respiratory system, indicating alterations in cellular, biochemical, or immunological processes or functional or structural changes. An ideal biomarker of an effect should be unique to a specific disease, capable of quantitatively relating to a particular stage of the disease, reproducible, sensitive to small changes due to an exposure, and specific to a particular test substance. A number of such markers have been developed and are discussed in the 1989 National Research Council monograph, *Biological Markers of Pulmonary Toxicity*. The major focus of this section is to provide examples of biomarkers of pulmonary response and to discuss how these indicators can help to improve our understanding of the respiratory system in normal and disease states.

Markers of physiological effects can be useful in identifying early changes in respiratory functions of the lung due to inhaled material. Biomarkers are available to measure lung mechanical properties, ventilation, intrapulmonary gas distribution, alveolar–capillary gas

TABLE R-8
Examples of Immunomodulation by Various Inhaled Chemicals

Classification	Symptoms	Chemical agents
Immediate (typle I) hypersensitivity	Bronchial asthma, asthmatic bronchitis, urticara, rhinitis, atopy	Beryllium, chloramine, ethylenediamine, ethylene oxide formaldehyde, isocyanates, platinum, nickel
Cytolytic (type II) hypersensitivity	Chemically induced hemolytic anemia, bone marrow depression, thrombocytopenia	Trimellitic anhydride, mercury
Arthus–immune complex (type III) hypersensitivity	Hypersensitivity pneumonitis, rheumatoid disease, sarcoidosis, vasculitis	Trimellitic anhydride, mercury
Cell-mediated (type IV) hypersensitivity	Contact dermatitis, sarcoidosis, anergy, delayed hypersensitivity	Beryllium, chromium, isocyanates, mercury, phthalic anhydride, trimellitic anhydride
Immunosuppression	Altered immune responses and host resistance following inhalation exposure	Asbestos, silica, metals, toluene, oxidant gases, tobacco smoke, benzene, toluene
Irritancy or nonimmunological	Pseudoallergic symptoms of bronchial asthma and asthmatic bronchitis	Formaldehyde, isocyanates, ethylenediamine

exchange, and perfusion. Such measurements have been used to test the effects of exposure to an array of inhaled toxicants. These assays can reveal functional manifestation of structural changes in the respiratory system, whether these changes are transient, resulting from bronchoconstriction, inflammation, or edema, or irreversible, such as from fibrosis, emphysema, or chronic obstructive lung disease. While the current functional tests are useful in evaluating clinical lung disease, they, by themselves, are not sensitive markers of the lung injury. The lung responds to air contaminants in much the same way regardless of the specific toxic nature of the toxicant. Increased efforts are being devoted to the development of functional tests that will be better indicators of specific alteration, focusing on certain regions of the respiratory system, such as the terminal bronchioles and respiratory bronchioles. Alterations at these sites, which are likely targets of several types of airborne toxicants, would be indicative of small airway disease.

Measuring airway hyperreactivity is a useful marker that involves measuring increased bronchoconstriction (i.e., contraction of airway smooth muscle). Hyperreactivity can be measured in the pollutant-exposed subject by following with a challenge of (1) a variety of pharmacological chemicals such as methacholine, carbachol, histamine; (2) a physical stimuli such as cold or dry air or exercise; or (3) air pollutants such as sulfur dioxide. An exposed individual may develop bronchoconstriction after inhaling a lower concentration of a provoking agent than is needed to cause a similar degree of change in the airway in a normal subject. Airway hyperreactivity has proven to be useful in assessing airway responsiveness following exposure to a low concentration of pollutants, such as ozone, nitrogen dioxide, sulfuric acid aerosols, allergens, and certain irritant gases. Evidence indicates that these tests constitute markers that are useful for detecting risk of accelerated loss of lung function, which may be indicative of the development of chronic lung disease.

Since the mechanisms of clearance of particles from the respiratory tract have been shown to be similar in most mammals, markers measuring alterations in the effectiveness of these defenses have been used to predict respiratory tract disease and for extrapolating animal data to humans. Both human and animal studies have shown that exposure to certain gases and particulate matter may significantly alter bronchial mucociliary clearance rate. Relating these effects to specific health effects remains speculative. However, there is a predisposition to respiratory infections (e.g., chronic bronchitis), with retarded clearance from the airways. By increasing the residence time of carcinogens, altered mucociliary clearance may also be a factor in development of bronchial cancer.

More nonevasive markers are needed for assessing early alterations in lung structure. The cells of the nasal, tracheal, and bronchial regions can be relatively accessible with bronchoscopy, brushing, and biopsy. In the tracheobronchial region, markers of differentiated phenotypes are useful in providing a direct indication of cellular damage. Mucous glycoproteins are markers for alterations of mucous cells and specific histochemical staining techniques are used to characterize secretory cell products. A low-molecular-weight protein appears to be a specific marker of clara cell secretory products. The presence of dynein appears to be a good marker for structural changes in the ciliated cells. Other biochemical and immunological markers (keratin expression, transglutaminase, and sulfotransferase) may be capable of indicating differentiation of tracheobronchial epithelial cells. Measuring such changes could provide early indicators of pathologic changes in easily accessible airway lining cells.

Many types of lung injury caused by inhaling toxic chemicals can cause selective injury to the more proximal portions of the gas exchange region of the lung. Markers focusing on specific focal patterns of injury that may be caused by different pollutants would be useful. For the alveolar region, specific markers of injury or disease are even less developed. Because an early response to cell injury from airborne pollutants is likely to result in proliferation of airway cells (e.g., epithelial, fibroblast, and macrophages) markers have been used to measure these responses following exposure to cigarette smoke, asbestos fibers, and oxidant gases. Using morphometrics, the total number of cells in the lung and the distribution of cells among the various types of alveolar cells have been determined in both humans and animals. Such techniques, although difficult to perform, offer promise that such assays will become sensitive markers of early structural changes.

Cellular and biochemical markers have been widely used to detect changes in the acellular and cellular content of nasal, bronchial, or bronchoalveolar lavages. The response of these regions to several inhaled substances, such as ozone, nitrogen dioxide, fibrogenic material, and several trace metals, can be measured by examining the lavage fluid to determine how it varies from normal. Indicators being used include the presence of blood neutrophils and mast cells (markers of permeability changes and influx of inflammatory cells), influx of eosinophils and basophils (indicators of allergic reaction), serum protein (marker of increased permeability of alveolar–capillary barrier), lactate dehydrogenase (marker of cytotoxicity), and lysosomal enzymes (marker of activation or lysis of macrophages). Other markers of effect have also been measured in lavage fluid, including growth factor, interleukins, arachidonate metabolites, and increase in prostaglandins. There is still a need to develop reliable markers to detect specific cell responses at the molecular level. Molecular type markers to characterize changes in DNA and RNA, changes in DNA sequences, and changes in the extent or pattern of gene expression would be of most value since they might aid the scientist in identifying individual susceptibility to pulmonary disease.

Toxicological Response to Inhaled Chemicals

The toxicology literature is extensive in the documentation of many human and animal studies that have been conducted to detect the health effects associated with airborne pollutants. Causal relationships between exposure to an agent and various forms of toxicity can be readily established using controlled animal studies. Animal studies suffer the obvious drawbacks when trying to extrapolate these responses to humans. Such studies are nevertheless commonly used to identify toxic properties of chemical agents because of the shortcomings of human epidemiological and clinical studies. Unfortunately, many of the available animal studies were designed and conducted to study the responses at relatively high concentrations, making it difficult to directly relate such responses to the relatively low levels found in the ambient environment. It is not the intent of this entry to provide a complete overview of all treatment-induced effects associated with inhaling airborne chemicals; instead, this entry provides a toxicity profile that is focused on an array of health effects caused by exposure and relates these observed responses to the potential health risk of the population.

The objective of any toxicological study is to determine a relationship between a measured biological response in a susceptible species by the most valid and sensitive technique following an appropriate exposure. Current research continues to focus on identifying that portion of the respiratory system that experiences the greatest effect of an inhaled toxicant. However, the point of maximal injury can be expected to vary with the nature of the toxicant, its concentration and duration of the exposure, the effectiveness of local defense mechanisms, and the inherent susceptibility to damage of the cells at risk. The size and complexity of the respiratory system in humans and animals provides numerous sites of potential damage caused by inhaled gases and particulates. To fully understand the toxicological consequences resulting from exposures, testing procedures must apply multiple endpoints and varying durations of exposure and must evaluate a variety of target tissues for injury. In evaluating the significance of the available database, the toxicologist needs to understand the various relationships that may exist between the measured response and the exposure. With multiple endpoints of toxicity and given concentration of the agent, an infinite number of linear and curvilinear relationships could be generated. The dose–effect relationship may be steep, indicating that a small increase in concentration (dosage) elicits a dramatic increase in the effect, or the slope may be shallow, indicative of only a small change in the altered state accompanying a large increase in the dosage of the toxicant. Frequently, a toxicant may elicit effects on more than one target organ, giving rise to dose–effect curves of different configurations. Such information is vital in predicting dose–response relationships.

Irritation and Inflammatory Response

Although irritation often suggests a relatively mild, transient effect, respiratory tract irritation is one of the most significant airway responses for the inhalation toxicologist. Irritation is frequently the first observable adverse response of the airways following exposure to airborne materials. In addition, irritation often occurs at relatively low concentrations that may be realistic for typical human exposures. The number of chemicals and common mixtures that are known to be respiratory irritants is far greater than that for any other respiratory system response. Many common components of air pollution, including sulfur dioxide, H_2SO_4, nitrogen dioxide, ozone, and various metal oxides, are respiratory irritants. This, along with the fact that many people have personal experience with such irritation, for example, by household ammonia, cigarette smoke, or photochemical smog, produces a high public awareness and concern for the irritancy of airborne chemicals. Agents that produce an irritant response on contact with airway tissues are termed direct irritants. The responses may be mild to severe, with typical concentration dependence, and they are usually reversible. Many organic vapors that are potential workplace hazards are sufficiently reactive to produce irritant injury to the airways. Examples include aldehydes (e.g., acrolein), epoxy compounds (e.g., ethylene oxide and propylene oxide), halogenated alkanes (e.g., bromotrichloromethane), aliphatic isocyanates (e.g., methyl isocya-

nate), and aliphatic nitro compounds (e.g., tetranitromethane). Many of these chemicals are also capable of producing respiratory tract neoplasms in laboratory animals. Respiratory irritancy is the most frequently used basis for setting occupational exposure limits, such as ACGIH TLVs.

The mouse respiratory depression model of Alarie, which is described in more detail in the section on physiological assessment, provides a lung function-based system for classifying and describing the relative potency of respiratory tract irritants. Upper respiratory tract irritants, the "sensory" irritants of the Alarie model, are usually water-soluble chemicals, such as formaldehyde, ammonia, sulfur dioxide, and acrolein. The early effects produced by such chemicals, including burning sensations of the eyes and upper airways and the cough and bronchoconstriction caused by irritation of conducting airways including the larynx, as well as the decreased respiratory frequency in mice are neurally mediated reflex responses. Irritant receptors in the conducting airways also respond to mediators, such as histamine, serotonin, and prostaglandins, and produce bronchoconstriction via a reflex increase in vagal efferent activity. Some human populations, such as asthmatics and the young, may be especially sensitive to the effects of upper airway irritants, responding at lower concentrations than the general population. Irritants that penetrate to the deeper regions of the lung, the pulmonary irritants of the Alarie model, are generally less water soluble or, in the case of aerosols, have small particle diameters. Examples include ozone, nitrogen dioxide, phosgene, and oxides of metals such as cadmium and beryllium. Again, the early responses—cough, chest tightness, and substernal soreness in humans, rapid, shallow breathing in rats, and respiratory depression in mice—appear to be neurally mediated reflexes.

Although the initial responses to irritants are reflexes mediated by irritant nerve endings, prolonged and/or repeated exposures result in cellular and tissue injury, edema, and inflammation. Such irritant-induced structural effects have been demonstrated for most sensory and pulmonary irritants, including chlorine, sulfuric acid, methyl isocyanate, formaldehyde, ozone, and nitrogen dioxide. It is generally believed that materials that produce primary respiratory irritation have the potential to produce long-term effects following repeated exposure. An important question concerns the potential role of the irritant response in the pathogenesis of chronic disease and cancer.

Under normal conditions, the alveolar epithelial and endothelial cell layers that make up the air–blood barrier under normal conditions control the passage of fluids and cells between the air spaces of the lung and the interstitium. Damage to this delicate barrier can causes an inflammatory response and the impairment of lung function. Changes in the permeability of the alveolar–capillary barrier lead to an infusion of proteinaceous serous fluid (edema) and blood cells (neutrophils, macrophages, and eosinophils). This influx of cells usually peaks within the first 3–7 days of the inflammatory response. If the inflammation is sustained it is usually accompanied by a specific immune response mediated by pulmonary lymphocytes.

This is the normal reaction and may be the lung's first response against the insult. However, after entering the lung, inflammatory cells can actually enhance the effect of the original insult and may be casually related to certain chronic lung diseases. These cells respond to injury by producing a number of potent chemicals, such as cytokines, chemotactic factors, prostaglandins, lysosomal enzymes, active oxygen radical species, and leukotaxines. Involvement of oxygen radicals has been hypothesized for a number of pulmonary diseases related to exposures to numerous agents, including asbestos, paraquat, cigarette smoke, ozone, nitrogen dioxide, and ionizing radiation. In normal circumstances, the generation of oxidants by defense cells is essential for effective host defense against invading microoganisms. If the inhaled substance causes subsequent lysis of these cells, these highly active cellular products would be released into the lung where they could act directly on the pulmonary tissue. Macrophages, for example, have been shown to release proteolytic enzymes that can degrade intercellular components of lung connective tissue and also interact with certain constituents of serum such as complement. These agents may, alone or in combination, cause functional impairment of epithelial cells, mesothelial cells, and fibroblasts, resulting in disease.

The analysis of isolated bronchoalveolar lavage fluid is an effective means for detection of inflammatory responses in the lung. In both animals and humans, cell counts and cell distributions can be determined, along with measurements of protein and bioactive mediators.

Asphyxiation

By definition, asphyxiants are chemicals that deprive the tissues of oxygen when inhaled. Any physiologically inert gas, including hydrogen, nitrogen, helium, and methane, that is inhaled at a high enough concentration to exclude an adequate concentration of oxygen acts as a simple asphyxiant. Chemical asphyxiants such as carbon monoxide, cyanide, hydrogen sulfide, and nitrites block the use of oxygen, causing asphyxiation when inhaled along with an adequate concentration of oxygen. Carbon monoxide is an odorless and tasteless by-product of incomplete combustion of carbonaceous materials. Carbon monoxide poisoning continues to be a significant public health concern both because of its use in suicides and because of accidental poisonings caused by faulty ventilation of home heating devices. Since the binding affinity of red blood cell hemoglobin is 200 times greater for carbon monoxide than for oxygen, carboxyhemoglobin formed at a relatively low concentration of this gas can block oxygen transport by a large proportion of the hemoglobin. Full dissociation of carbon monoxide from hemoglobin occurs following removal from the carbon monoxide-containing environment. Therefore, the poisoning is not cumulative. Carbon monoxide is an air pollutant and component of cigarette smoke, and smokers, parking garage workers, and traffic policemen are repeatedly exposed at low levels. Although asphyxiation is not a concern with such exposures, transient neurobehavioral deficits may develop and there may be an increased risk to individuals with heart disease. Other chemicals, such as sodium nitrite, interfere with transport of oxygen by oxidizing the iron moiety of hemoglobin, producing methemoglobin. Cyanide does not block oxygen transport but is a classic tissue-level poison, inhibiting cytochrome oxidase and blocking energy production.

Morphological and Structural Effects

Morphological studies are often the cornerstones of toxicity experiments. Pathological evaluation of exposed tissue permits the identification and characterization of structural damage to the respiratory system. Animal studies have been effective in improving our understanding of the pathologic sequelae of chemical deposition at specific sites in the respiratory system. The difference in the structure of the respiratory system of humans and experimental animals may complicate but does not necessarily prevent qualitative extrapolation of risk to humans. Since the lesions resulting from a particular exposure can be shown to be similar in several mammalian species of test animals, it would appear likely that the biological processes responsible for the lesions in animals could also occur in humans. However, it should be understood that different exposure levels may be required to produce a similar response in humans. The concentration at which effects become evident in humans can be influenced by a number of factors such as preexisting disease, dietary factors, combination with other pollutants, and the presence of other stresses.

A wide variety of morphological changes have been associated with inhalation of airborne contaminants. Both acute and chronic exposures have been shown to directly affect the structural integrity of the respiratory system. Acute studies are conducted primarily to define the intrinsic toxicity of the chemical, to identify the target organs, and provide information for design and the selection of doses for longer term studies.

The epithelium of the conducting airways represents a tissue that is uniquely sensitive to a number of inhaled toxicants and that shows early histopathological damage when injured. Such injury in turn often elicits a variety of acute inflammatory responses. The ciliated cells, which are distributed throughout much of the length of the conducting airways, often exhibit morphological damage causing ciliary dysfunction, slowing of transport rate, and excessive mucus production. It appears that these ciliated conducting airway cells are the most sensitive to direct-acting toxicants and that cells with the most secretory capacity are less sensitive (i.e., mucous and clara cells). Cilia may be reduced in length or diameter and exhibit reduced density, and the cells may exhibit a variety of cytoplasmic changes including dilated endoplasmic reticulum, swollen mitochondria, and condensed nuclei. Tests for clearance of marker substances have been used to demonstrate that morphological effects on the cilia can result in a significant reduction in mucociliary clearance. Cigarette smoke, sulfur dioxide, alcohol, H_2SO_4, ozone, nitrogen dioxide, trace metals, and certain bacterial infections are toxic to the cilia and lead to impairment of mucociliary clearance. Individuals with bronchial carcinomas, cystic fibrosis, chronic bronchitis, and certain infectious diseases, such as influenza, atypical pneumonia, and tuberculosis, have impairment of lung clearance. Disruption or impairment of this defense system may result

in greater accumulation of and potential injury by various airborne substances and increase the susceptibility to bacterial and viral infections. Continued chemical exposure can cause necrosis and the subsequent sloughing off of ciliated epithelial cells. The epithelial tissue may be repaired by the proliferation of the secretory cells. In areas of repair, the nonciliated cells often appear to be more numerous. In this type of injury, the repair process is initiated soon after the test animals are removed from the exposure atmosphere.

The respiratory alveolar epithelial response to toxic injury can be rapid, resulting in necrosis and subsequently sloughing of the sensitive type I cells. This type of response is seen with exposure to such toxicants as ozone, nitrogen dioxide, and butylated hydroxytoluene. This injury stimulates the proliferation of the more resistant type II cells. This proliferative response typically peaks at about 48 hr after onset of the initial injury to the type I cells. The increase in number of type II cells can be expected to alter diffusion capacity of the pulmonary region through population of this membrane with these thicker cells.

Following lung injury, recovery depends on prompt and orderly repair. The type and extent of the injury determine whether cell replication results in the restoration of the normal structure or in abnormal remodeling that may lead to profound anatomic distortion due to an exuberant fibroproliferation response. Reepithelization of any damaged respiratory area is critical to maintenance of normal lung function. For example, shortly after an injury, the alveolar surface may be denuded with only type II cells remaining. These type II cells begin to replicate, resulting in the repopulation of alveolar basement membrane. Eventually, the replacement cells flatten and begin to acquire the morphological features of the type I cells as the air–lung interface is reconstituted. However, it is also possible that in this repair process, a rapid migration of fibroblasts into the damaged area may occur. When this happens these cells begin to replicate and deposit connective tissue. This obliterates the air space architecture, resulting in alveolar fibrosis. Pulmonary fibrosis results in decrease in diffusion capacity and a decrease in lung volume and compliance. Inhaled agents causing such fibrosis in humans include silica, asbestos, organic dust, cadmium fumes, paraquat, and some infectious microorganisms. The proliferation of epithelial cells, fibroblasts, and other lung cells following exposure can be measured *in vivo* and is useful in studying the pathogenesis of pulmonary disease. Cell proliferation assays are designed to quantify the relative rates of cell division within such target tissues using specialized immunohistochemical staining techniques to detect proliferating cells.

Alterations in capillary permeability are often associated with structural injury to endothelial cells. The endothelial defects are less evident at low concentrations but include cell swelling and disruption of the basement membrane. It has been suggested that the difference in the extent of epithelial and endothelial damage can be explained by the different repair potential of these two lining layers rather than the dissimilar reaction to the injury. The pulmonary endothelium has been shown to be susceptible to injury by oxygen-based free radicals. They are especially sensitive to the effects of high oxygen tension. Numerous studies have shown that such lung oxygen damage is the result of a direct toxic effect through intracellularly generated O_2 intermediates and not solely by the recruited polymorphonuclear cells. Paraquat, nitrofurantoin, cyclophosphamide, and bleomycin are among substances known to injury endothelial cells.

Three-dimensional reconstruction of cells and tissue is now being used to study subtle changes in intracellular organelles and cell-to-cell relationships that are affected by exposure. Developing such techniques has required advances in computer processing power to supply the memory and appropriate algorithms necessary to make this process technically feasible. Together with time-lapse photography and high-voltage electron microscopy, computer-time reconstructions can be used to study the effects of chemicals on cell function and cell regulation.

The main types of noncarcinogenic response of lung cells to chronic exposure are hyperplasia, hypertrophy, and metaplasia. In human studies, it is difficult to identifying the chemical(s) causing a chronic pulmonary disease that is associated with morphological alterations due to the long latency periods involved. In many cases, the disease symptoms may fail to be evident until after 20 or more years of exposure. Chronic lung disease can be conveniently classified into three broad groups: restrictive lung disease, chronic obstructive lung disease, and cancer.

Both restrictive and obstructive lung disease are associated with serious impairment of the flow of gases into the gas exchange regions of the lung. Pulmonary function tests are used to distinguish between these two diseases. Chronic obstructive pulmonary disease (COPD) includes three major types—asthma, chronic bronchitis, and emphysema. In humans, a clear distinction between emphysema and bronchitis is not possible. Most patients that have chronic bronchitis also have emphysema. The resulting gas trapping and persistent slowing of airflow make expiration difficult. The bronchial wall thickness may be 50–100% greater than normal. Individuals with COPD can be recognized by their difficulty in performing more than light to moderate exercise and nonuniform distribution of ventilation. They frequently have associated cardiovascular disease and chronic cough and recurrent expectoration.

Chronic bronchitis is a major health problem that is associated with long-term cigarette smoking, dusty environments such as grain elevators and coal mines, trace metal exposure (vanadium, arsenic, and iron oxide), phosgene exposure, and the exposure to ambient air heavily polluted with sulfur oxides and combustion products. Chronic bronchitis is clinically evident as excessive bronchial mucus production. Histological examination of human bronchial airways shows hypertrophy of mucus glands in the large bronchi; chronic inflammatory changes, including cellular infiltration and an accumulation of fibroblasts and connective tissue; edema; and possibly increases in smooth muscle in the airways. In the early stages, these effects are potentially reversible, but advanced stages are irreversible. Additional features of chronic bronchitis include inflammation of the mucous membranes of the bronchial airways and a reduction in number of ciliated cells together with increased secretions having abnormal physicochemical properties. These effects ultimately result in grossly impaired mucociliary transport. Cough aids as the clearance mechanism for the excess mucus.

Asthma is defined clinically by recurrent episodes of airway obstruction that reverse either spontaneously or with bronchodilator therapy. The airway obstruction is accompanied by increase in airway resistance due to bronchospasm, inflammation, and excessive mucus production. Bronchoconstriction, airway closure, and gas trapping may eventually lead to respiratory failure. Hyperresponsiveness is considered a hallmark of asthma, making these individuals uniquely sensitive to exposure to airborne chemicals such as isocyanates.

Emphysema differs from the other two conditions in that there is evidence of anatomic alterations of the lung characterized by abnormal, uneven, permanent enlargement of the air spaces distal to the terminal bronchioles, resulting from the destruction/distension of the alveolar walls. Airway restriction or collapse results from loss of supporting tissue that normally maintains airway patency. Such structural changes are associated with various pulmonary functional abnormalities related to loss in lung elasticity and decreases in normal diffusion capacity and forced expiratory volume. Emphysema has been associated with long-term exposure to coal dusts, cigarette smoke, osmium tetroxide, cadmium oxide, and some common atmospheric pollutants (e.g., ozone). When such chemicals are inhaled they cause cell injury and an inflammatory response. During this process, proteases, lysosomal enzymes, and oxidants are released during phagocytosis, cell injury, and cell death. To maintain structural integrity under such conditions, the lung can respond with biochemical modifiers such as antiproteases. A balance between these two responses must be maintained since these reactive substances can degrade pulmonary elastin and collagen, resulting in a destruction of the supporting structure of the alveoli. With this destruction of lung tissue, there is a subsequent loss in total lung surface area and reduction of the lung's ability to meet gas exchange demands.

Restrictive lung disease occurs when the elastic properties of the lung are so impaired that the lung becomes stiff as in fibrotic diseases related to silicosis, pneumonia, and asbestosis. This disease condition is characterized by increased lung recoil and a decrease in lung volumes, such as vital and total lung capacity. Such restriction decreases the lung's normal ability to expand, making inflation of the lung more difficult.

When the lung is chronically exposed to a contaminant that is not easily removed or degraded, the lung may undergo a process referred to as granuloma formation. This lesion is characterized by accumulation of mononuclear cells (macrophages, lymphocytes, and giant cells) into a relatively discrete structure. These granulomas may distort the interstitial architecture, interfering with the normal process of gas exchange, and can cause tissue damage and fibrosis that may result in permanent dysfunction and morbidity. Granulomas are dynamic structures in that freshly recruited monocytes are continually entering the lesion and replacing mature cells. Ultimately, these granulomas may resolve or become fibrotic due to the influx and proliferation of fibroblasts. These lesions may be initiated by infectious agents (mycobacteria, fungi, and viruses) and inorganic

substances like beryllium. A common property of all such agents is their low biodegradability and persistence, often within the macrophage. Individuals exposed to beryllium fumes may develop acute pulmonary edema and pneumonia. While most of these individuals recover, some develop chronic granulomatous lesions appearing years after the initial exposure. Generally, such chronic disease results from prolonged exposure to low concentrations of beryllium.

Fibrotic lung disease is directly associated with chronic inflammation in which the inflammatory process in the lower respiratory tract injures the lung and modulates the proliferation of mesenchymal cells to form a fibrotic scar. Practically any chronic injury that is capable of sustaining a continued inflammation will produce some degree of interstitial fibrosis. Such effects involve a chronic, ongoing process and may ultimately involve the entire organ. The fibrotic process involves damage to the normal alveolar architecture, which in turn leads to activation of the macrophage and release of potent growth factors. These factors cause the mesenchymal cells to proliferate and produce large amounts of collagen that then accumulates in the interstitial space. This excess collagen deposition leads to pulmonary fibrosis. It is interesting to note that in the postexposure period, the fibrotic process tends to continue. Once fibrosis occurs within a group of alveoli, it is unlikely that those alveoli will ever recover. Fibrosis-producing agents include inorganic particulates (silica, beryllium, coal dust, iron oxide, chromium, and asbestos), toxic gases (ozone, nitrogen dioxide, and high concentrations of oxygen), cigarette smoke, paraquat, and a variety of immunologic insults.

Pulmonary Function: Physiological Assessment

Although pulmonary injury by a toxic agent and/or disease process is normally defined by morphological change, the functional manifestations of these structural effects have proven to be sensitive indicators of toxic response and lung disease. Pulmonary function testing provides a safe, noninvasive approach for clinical evaluation of the presence, type, and severity of pulmonary impairment. When workplace conditions include a risk of inhalation exposure to toxicants, preemployment and periodic, repeated lung function testing can be a key element in health effects screening and disease prevention. For many lung function tests, repeated testing is also possible in laboratory animals and progression of or recovery from disease may be evaluated in individuals. Evaluation of pulmonary function in both humans and animals complements evaluation of structural changes caused by inhaled chemicals. In addition, specific lung function tests may detect significant respiratory tract effects or disease states that do not produce lasting or detectable structural changes. Finally, a large body of experimental evidence from animal models of specific pulmonary diseases and animal toxicology studies suggests that similar lung insults and/or structural changes produce similar functional effects in humans and animals. Therefore, effects observed in animals may be used to predict human pulmonary effects. Table R-9 provides

examples of pulmonary function measurements that have been used for evaluation of impairment by airborne toxicants.

In interpreting pulmonary function data, several key points should be understood. (1) A specific functional effect is not diagnostic for a single structural change. For example, reduced vital capacity or compliance may be caused by several structural changes, including fibrosis, edema, hemorrhage, cellular hyperplasia, and heavy particle loading. (2) The respiratory system has a large functional reserve. Therefore, a relatively diffuse and extensive lung lesion may be required to produce a detectable effect on lung function. (3) A useful approach to pulmonary function testing is the use of a battery of measurements to develop patterns of func-

TABLE R-9
Common Measurements for Assessment of Changes in Pulmonary Function

Test category	Individual test/parameter	Definition/functional significance
Ventilatory pattern	Respiration rate	Breathing frequency (breaths/min)
	Tidal volume	Volume of breath
	Minute volume	Total volume inspired/expired per minute
Static lung volumes	Vital capacity	Maximum volume that can be expelled from the lungs by forced effort following maximum inspiration
	Total lung capacity	Volume of gas in lungs at end of maximum inspiration
	Residual volume (RV)	Volume of gas in lungs at end of maximum expiration
	Functional residual capacity (FRC)	Volume of gas remaining in lungs at end of tidal expiration
	Inspiratory capacity	Maximum volume of gas that can be inhaled from FRC level
	Expiratory reserve volume	Maximum volume of gas that can be expired below FRC level
Respiratory mechanics	Air flow resistance	
	Total lung flow resistance	Flow resistance of airways
	Static lung compliance	
	Dynamic lung compliance	Stiffness (elasticity) of the lung
	Maximum forced expiratory maneuver	"Stress" test for obstruction of airflow in peripheral airways
	Forced vital capacity (FVC)	
	Forced expiratory volume	
	Peak expiratory flow rate	
	Expiratory flow at 50, 25, and 10% of FVC	
Distribution of ventilation	Single and multiple breath nitrogen washout	Homogeneity of ventilation in lungs—Airflow obstruction and gas trapping causes greater variability of ventilation
	Closing volume	Volume difference from RV representing onset of closure of small airways; increases with air flow obstruction
Diffusion	Carbon monoxide diffusing capacity	Measurement of efficiency of alveolar gas exchange; decreases with thickening of alveolar blood–air barrier
Blood gases	Measurements of arterial pO_2, pCO_2, and pH	Evaluates adequacy of ventilation; changes typically require severe functional deficits
Pulmonary circulation	Edema: marker radioisotope movement to airways; wet/dry lung weight ratios	Evaluation for transudation of fluid into airways
	Cardiovascular pressures	
	Cardiovascular volumes, flow resistance	Hyper- or hypotension in vascular system Cardiovascular function

tional change that are consistent with a particular disease such as fibrosis or emphysema. (4) Restrictive lesions (e.g., fibrosis) are characterized by a lung that is less elastic, while obstructive lesions (e.g., emphysema) are characterized by changes that obstruct the movement of air in the airways. (5) Methods used for animals often require the use of anesthesia or restraint and potential interference with measurements must be considered.

The lung function tests that are most frequently used in animals evaluate breathing pattern, lung volumes, lung mechanical properties (including compliance, airway resistance, and flow rates), and diffusing capacity. Parameters for breathing pattern include respiratory frequency, tidal volume (the volume of a single, normal breath), and minute ventilation. A useful screening approach for evaluation of acute respiratory irritancy of inhaled chemicals has been developed by Alarie and co-workers. Using a head-only exposure system for mice, the effect of chemical exposure on respiratory frequency is monitored and, to allow comparison of irritancy between chemicals, the concentration that depresses the frequency by 50% is calculated. Two patterns of irritancy have been described. "Sensory" or upper airway irritants are usually highly water-soluble chemicals, such as formaldehyde and ammonia, and cause a reflex depression in respiratory frequency with a slow expiratory phase. Pulmonary or peripheral airway irritants are typically less soluble chemicals, such as phosgene, ozone, and nitrogen dioxide, and cause a respiratory depression marked by pauses between breaths. It should be noted that in rats, pulmonary irritants produce tachypnea (rapid shallow) breathing.

Measures of breathing pattern tend to be relatively insensitive to early restrictive or obstructive lesions, but with advanced chronic disease, restrictive lesions produce rapid, shallow breathing and obstructive lesions cause slow, deeper breathing. Fibrosis produced by subchronic inhalation exposures to metal oxides of cadmium or vanadium produce tachypnea. "Stress test" methods have been developed to enhance the sensitivity of these ventilatory endpoints in unanesthetized animals by exercise-induced or carbon dioxide-induced hyperventilation. The latter approach, which has been used in guinea pigs and restrained rats, has been used to detect lung injury by several agents, including methyl isocyanate, sulfuric acid, quartz dust, cotton dust, and wood smoke.

Figure R-10 depicts the physiologically defined lung volumes used for humans and animals. Inhalation exposures that cause restrictive lung lesions, including exposure to silica, cadmium compounds, ozone, and diesel exhaust, produce decreases in total lung capacity and vital capacity. Obstructive lesions lead to breathing at higher lung inflation (due to gas trapping), with increased total lung capacity, residual volume, and functional residual capacity. Inhalation exposure to ozone and acrolein may produce this type of response.

Mechanical properties of the lung may be tested using static or dynamic testing modes. The static test can be conducted in living animals or using excised lungs and involves information derived from the pressure–volume curve produced during lung deflation. Static compliance, a measure of lung elasticity derived from this curve, is decreased following exposure to fibrogenic agents, such as mineral dusts, or agents that cause edema, inflammation, and cellular hyperplasia, such as oxidants and irritants, and is increased following exposures that cause emphysema-like lesions, such as subchronic ozone and nitrogen dioxide exposures. Tests for dynamic lung mechanics require monitoring of flow, volume, and pressure and provide measures of total lung resistance, which is most dependent on large airway obstruction, and dynamic compliance, which is a measure of elasticity that is also sensitive to peripheral airway obstruction. Because of their sensitivity to bronchoconstrictors, guinea pigs are frequently used for these measurements with applications including evaluation of irritant effects, nonspecific airway hyperreactivity to bronchoconstrictors (e.g., histamine), and immunologically determined airway hypersensitivity. Many irritants that cause decreased dynamic compliance and/or lung resistance, such as ozone, sulfur dioxide, sulfuric acid, acrolein, and toluene diisocyanate, also cause nonspecific increased airway reactivity. Specific immunoglobulin-dependent sensitization to inhaled proteinaceous materials (e.g., ragweed pollen, animal dander, and grain dust) has been demonstrated in animals and humans using lung function measurements. Asthmatic responses with pulmonary sensitization to low-molecular-weight chemicals (haptens), such as isocyanates and anhydrides, have been observed in workers and have been modeled in guinea pigs using lung function tests. The role of the immune system in responses to many haptens remains in question since it has not been possible to consistently demonstrate the presence of an antigen-specific antibody.

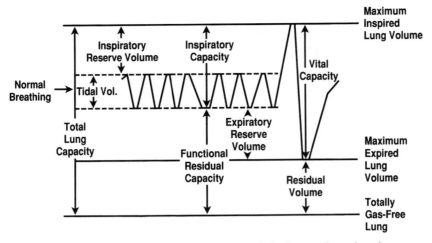

FIGURE R-10   *Capacities and lung volumes of the lung (adapted with permission from R. O. McClellan and R. F. Henderson (Eds.), Concepts in Inhalation Toxicology, p. 364, Hemisphere, New York, 1989).*

The approach most commonly used to evaluate effects on distal airways in clinical and occupational medicine is the maximum forced expiratory maneuver, which allows measurement of airflows as a function of lung volume from total lung capacity to residual volume. Typically, the forced vital capacity (FVC) and the forced expiratory volume at 1 sec (as a % of FVC) (FEV$_1$) are measured. Decreased airflow rates are seen with emphysema, chronic bronchitis, and following acute exposures to bronchoconstricting irritants or agents that produce asthmatic responses following previous sensitization. When exercise is superimposed on certain pollutants such as with an exposure of humans to certain pollutants, such as ozone, decrements for FVC and FEV$_1$ have been observed at realistic levels for air pollution (0.08–0.12 ppm). Animal models for forced expiratory maneuvers require the use of anesthetics and/or paralytic agents with the forced expiration produced in an apneic animal by applying a steady negative airway pressure following inflation to total lung capacity. FVC, peak expiratory flow rate, mean mid-expiratory flow, and expiratory flow at 50, 25, and 10% of FVC may be derived from the maximum expiratory flow–volume curve. Using such methods, it has been demonstrated that the qualitative patterns of flow–volume curves are similar in humans and small laboratory animals when values are normalized for lung volume. As expected, decrements in airflow rates are seen in rats with elastase-induced emphysema. However, for demonstration of effects of more realistic

exposures to toxicants and/or more subtle injury, the utility of measurements from forced expiratory tests in small animals may be limited.

Since the efficiency of the diffusion of O$_2$ and carbon dioxide at the alveoli is directly related to the primary function of the lung, tests of diffusing capacity are an important component of a lung function testing battery. Diffusing capacity for carbon monoxide (DL$_{CO}$) is usually measured and normalized for lung volume. Diffusing capacity is reduced whether the cause is related to structural changes in the alveolar region, as in thickening of the air–blood barrier (restrictive disease/ fibrosis) or destruction of alveolar epithelium, or to an effective reduction in alveolar surface, as in obstruction disease. Thus, decrements in DL$_{CO}$ may be difficult to interpret without correlation with other functional endpoints or histopathological evidence. In addition, rodents appear to compensate with hypertrophic or hyperplastic lung changes resulting in unexpectedly normal DL$_{CO}$ values.

Pulmonary Carcinogenicity

Testing for the potential carcinogenic effects of airborne chemicals has received high priority in an effort to protect human health. Lung cancers are the most rapidly increasing cancers in Western Europe and North America. In the United States, cancer of the lung or bronchus were diagnosed in more than 170,000 people in 1993. Because of the long latency period associated with chemical carcinogenesis, it is often difficult to identify, in the human population, the specific causative agent. Although there are a number of short-term tests for determining genotoxicity, these are generally used in screening and have not replaced the need for long-term animal testing. The basic premise of carcinogenesis research is that a substance that affects animal cells in such a way as to cause cancer is highly likely to affect human cells in the same way. Positive results from such testing are useful in that they demonstrate that a specific chemical is carcinogenic to animals under the conditions of the test. This information can be used as an indicator for predicting the potential carcinogenic hazard to humans. Pulmonary cancers, like other forms, can be caused by both external factors (chemical, radiation, viruses, and diet) or internal factors (hormones, inherited genes, and immune conditions) or the interaction of these factors. Evidence of pulmonary carcinogenicity can be based on (1) an increase in the incidence of a specific tumor type, (2) the development of a specific tumor type earlier than seen in controls, (3) the presence of types of tumors normally not seen in control groups, and (4) an increase in multiplicity of tumors. In the absence of adequate human data, it appears reasonable and appropriate to predict that, with sufficient evidence of carcinogenicity in animals, the chemical presents a similar risk to humans. Nearly all known human carcinogens have been shown to be carcinogenic in some animal model. Recent review on chemical carcinogenesis, mechanisms of carcinogenesis, and pathophysiology of induced tumors provides current information on all aspects of carcinogenesis in each organ system.

Carcinogens can be divided into two general types: those that act directly and those that act indirectly. Direct-acting carcinogens are those that interact with cellular constituents such as protein, lipids, and nucleic acids. There are relatively few direct-acting carcinogens [e.g., bis(chloromethyl)ether, ethylene oxide, and nitrogen mustard].

The indirect-acting carcinogens are all substances that require metabolic activation before they interact with cellular macromolecules. These agents are often referred to as pro- or precarcinogens and include certain cyclic and polycyclic aromatic hydrocarbons [benzene and benzo(*a*)pyrene], aliphatic hydrocarbons (methylene chloride and pesticides), nitrosamines, and other chemicals such as urethane, formaldehyde, and acrylonitrile. Both the indirect- and direct-acting car-

cinogens can ultimately react with the genetic material of the cell. Such substances are referred to as genotoxic carcinogens, which differ from nongenotoxic or epigenetic carcinogens that do not appear to bind with the DNA of the cell but have other mechanisms of action. Agents included in this later group include fibers, trichloracetic acid, and certain plasticizers. These chemicals do not damage DNA nor are they mutagenic in the standard short-term screening assays.

Examples of inorganic carcinogens include arsenic, asbestos, chromium, and nickel. The chemical form is important in determining the dose response. For example, beryllium sulfate is more carcinogenic than beryllium oxide. The difference may relate to the solubility of these compounds in the lung and the actual dose of the chemical to the target tissue. Although beryllium, lead, cadmium, and silica have also been shown to be carcinogenic in animals, the evidence in humans is less substantial. Epidemiological studies have implicated nickel as a carcinogen for cancer of the nasal cavity, lung, and possibly larynx. Carcinogenicity of chromium is associated with slightly soluble chromates (CR^{+6}), while the insoluble and very soluble salts of chromic acid or trivalent forms show little or no carcinogenicity. Mesotheliomas, tumors of the pleural lining of the lung and thoracic cavity, result from inhaled asbestos. Quartz and silica are generally not considered to be carcinogenic in humans, but some recent studies have demonstrated that these substances may cause lung cancer in rats. Multiple mechanisms may be involved in the carcinogenicity of asbestos. Asbestos is considered to be a cocarcinogen or tumor promoter. The hypothesis is that asbestos may cause generation of active oxygen species that lead to lipid peroxidation and DNA strand breakage. High levels of arsenic increase risk of lung cancer. Trivalent arsenic appears to be the most active form and affects DNA synthesis and repair.

The induction of nasal carcinomas following inhalation exposure to several chemicals has been reported, including benzene, acetaldehyde, diallylnitrosamine, formaldehyde, hydrazine, and vinyl chloride. It is of interest that some chemicals, such as epichlorohydrin and bis(chloromethyl)ether, that produce nasal carcinomas in rats also produce lung cancer in humans. However, there is no epidemiological evidence indicating an increased incidence of nasal cancer in workers exposed to these industrial chemicals. Also, agents such as nitrosamines that are delivered by other routes of exposure can also induce tumors in the nose. For example, the consumption of salted fish having high concentrations of volatile nitrosamines may cause nasal tumors in experimental animals. Consumption of alcohol has also been associated with laryngeal cancers in humans.

Tumors of the nasal passages can vary from small papillomas and adenomas to large carcinomas that have the potential to metastasize to other parts of the body. While most of these neoplasms arise within the epithelium of the nose, chemical carcinogens have also induced mesenchymal and neuroectodermal neoplasms. Nasal tumors are encountered most frequently. Laryngeal tumors are relatively rare but have been reported following exposure to smoke and acetaldehyde vapors.

Numerous chemicals have been identified as capable of causing pulmonary cancer in both animals and humans. The International Agency for Research on Cancer (IARC) states that there are adequate experimental inhalation studies in animals for several chemicals. Table R-10 lists chemicals that cause lung neoplasia in laboratory animals and humans following inhalation. Studies to investigate the carcinogenic potential of airborne chemicals usually focus on the morphological examination of tissue to determine the number of various types of tumors, the number of tumor-bearing animals, the number of tumors per animal, and the onset of the tumor. In such studies, few biochemical or physiological endpoints are investigated except for periodic hematological assays.

Inhalation of certain durable natural mineral fibers of amphibole asbestos, such as amosite and crocidolite, can lead to the development of inflammation, fibroproliferation, and pulmonary neoplasms and cancer of the serosal lining of the body cavities or mesothelioma. Administration of fibrous particles to laboratory animals has included, in addition to inhalation, intratracheal instillation and intracavitary implantation and instillation. Inhalation studies are difficult to conduct due to the problems associated with the generation and characterization of the fibers during all phases of the assay.

While many studies have been conducted using rats, there has been concern that the rat may not be an appropriate model for studying particulate-induced pulmonary tumorigenesis. One problem involves the finding that tumors can be induced under conditions of so-called "pulmonary overload" even with "inert"

particles. These overload tumors may arise via mechanisms distinct from those normally associated with pulmonary carcinogenesis.

Inhalation studies conducted with rats have also been criticized due to a lack of sensitivity for the induction of fiber-inducted neoplasms, particularly mesotheliomas. Rats develop a low incidence of such tumors following exposure to amosite or crocidolite, which are known to cause tumors in humans.

In an effort to increase sensitivity, investigators have used other exposure methods including intratracheal, intrapleural, and intraperitoneal methods. There is significant concern over the induction of cancers by these nonphysiological exposure routes. Cancers induced by intracavity instillation may be due more to chronic inflammation and fibrosis from the "bolus effect" rather than to the mechanisms of fiber-induced proliferative disease that normally occurs following the inhalation route of exposure. With intratracheal instillation, the distribution in the lung is not uniform and the resulting lesions differ from those reported in inhalation studies. It is generally agreed that long-term rodent inhalation studies provide the most definitive animal data for extrapolation to human assessment.

While many carcinogenicity studies on individual chemicals have been conducted, the study of complex mixtures presents a formidable scientific challenge for the toxicologist. One of the most difficult tasks is related to finding the primary causative agents of the effects. Examples of complex mixtures that have been studied for carcinogenicity include tobacco smoke and diesel engine emissions.

Cigarette smoke has been extensively studied due to its association with human lung cancer. However, it is not an impressive inducer of lung tumors in experimental animals. Lung tumors have been observed following long-term exposure using special strains of mice. Strains of mice such as A strain are known to have a high incidence ($\geq 70\%$) of spontaneous tumors. While there have been many studies of tobacco smoke using the laboratory rat, only one study showed an increase in lung tumors. Syrian hamsters exposed to whole smoke have developed laryngeal cancer but not lung tumors. This may be the result of an unusual increase in deposition of the inhaled smoke at this site in the hamster.

Several studies have shown that diesel exhaust is carcinogenic to the rat following long-term exposure. In these studies two basic types of tumors were found—bronchoalveolar tumors and squamous cell tumors

TABLE R-10
Carcinogenic Agents Associated with Lung or Pleural Cancer in Laboratory Animals
and Humans

Agents causing lung tumors	*Agents associated with human lung cancer*
Organic chemicals	
Gases	Industrial processes
Benzene[a]	Aluminum production
Bis(chloromethyl)ether[a]	Coal gasification
Bromomethane (ethyl bromide)[a]	Coke production
1,3-Butadiene[a]	Hermatite mining, underground with exposure to radon
1,2-Dibromo-3-chloropropane	Iron and steel founding
1,2-Dibromoethane	Painter, occupational exposure
Dimethyl sulfate[a]	Rubber industry
1,2-Epoxybutane	Chemicals for which exposure has been occupational
Ethylene oxide[a]	Asbestos[a]
Formaldehyde[a]	Bis(chloromethyl)ether[a]
Methylene chloride	Chromium compounds, hexavalent[a]
3-Nitro-3-hexene	Coal tars[a]
1,2-Propylene oxide[a]	Coal tar pitches
Tetrachloroethylene	Mustard gas[a]
Tetranitromethane	Nickel and nickel compounds[a]
Urethan	Soots[a]
Vinyl chloride[a]	Talc containing asbestiform fibers[a]
Particles	Vinyl chloride[a]
Benzo(*a*)pyrene[a]	Environmental agents and cultural risk factors
Polyurethane dust	Erionite
Inorganic compounds	Radon and its decay products
Metallic	Tobacco smoke[a]
Antimony compounds	
Berryllium compounds	
Cadmium chloride[a]	
Chromium dioxide[a]	
Nickel compounds[a]	
Titanium compounds	
Nonmetallic	
Asbestos fibers[a]	
Zeolite fibers	
Ceramic aluminosilicate fibers	
Kelvar aramid fibers	
Silica[a]	
Oil shale dust[a]	
Quartz	
Volcanic ash	
Radionuclides	
α-emitting radionuclide particles	
β-emitting radionuclide particles	
Radon and its decay products[a]	
Complex mixture	
Cigarette smoke[a]	
Diesel engine exhaust	
Gasoline engine exhaust	
Coal tar aerosols[a]	
Artificial smog	

[a] Identified by IARC as chemical causing lung cancer in humans and respiratory cancers in animals (from D. E. Gardner, J. D. Crapo and R. O. McClellan, (Eds.), *Toxicology of the Lung,* Raven Press, New York, 1993).

both arising from the alveolar parenchyma. Diesel exhaust represents complex mixtures of numerous organic and inorganic chemicals as well as various gases that may be toxic or carcinogenic. The complex interaction of organic hydrocarbons and carbon particles may be responsible for the tumors seen in these studies. It has been hypothesized that the organic hydrocarbons present initiate the process and the particles, with absorbed hydrocarbons, promote the initiated cells. A number of epidemiologic studies in London, the United States, and Canada did not detect significant health risk or indicated only a small increase in lung cancer incidence among workers exposed chronically to diesel exhaust. However, the animal studies together with supporting *in vitro* (e.g., mutagenic to bacteria and mammalian cells) data taken in aggregate led to the conclusion that diesel engine exhaust is a potential human carcinogen but probably represents a low level of risk.

While various types of lung cancer have been noted in humans, all known human pulmonary carcinogens are taken into the body by inhalation. It is equally important that inhaled chemicals can cause neoplasms at sites elsewhere in the body. For example, exposure to vinyl chloride, butadiene, acrylonitrile, and ethylene oxide by inhalation may produce a significant increase in cancer incidence in other organs (liver, brain, and blood) as well as in the lung. Often, the neoplasms in other organs have a higher incidence and are of a more serious health risk than the lung tumors.

Effects on Normal Pulmonary Defenses

The host defense system is one of the prime targets for which function has been shown to be adversely affected by exposure to a wide range of environmental chemicals. During the air pollution episodes of this century (Meuse Valley, Belgium, London, and Donora, Pennsylvania) excess deaths were recorded from lower respiratory tract infections. The American Thoracic Society has published guidelines on what constitutes an adverse respiratory health effect. Among the five most important adverse respiratory effects are a greater incidence of lower respiratory infections. Because of the importance and the complexity of this system many *in vivo* and *in vitro* assay systems have been used to assess the integrity and biological activity of both the cellular and acellular components of the lung defenses. Any breach in these defenses should be considered as a possible

indicator of an increased risk of pulmonary disease. This section is intended to familiarize the reader with the various defense system responses that have been studied, the measurements made, and the gaps in the information database. The host defense parameters which have been used most widely to examine the association between airborne toxicants and lung disease include mucociliary clearance dysfunction, functional and biochemical activity of the alveolar macrophages, immunological competency, and susceptibility to infectious disease. Increases in respiratory morbidity and impairment of lung clearance have been shown to occur at ambient levels of air pollution and associated with susceptibility to pathogenic microorganisms.

As discussed earlier, a major component of the respiratory defense system is the mucociliary clearance mechanism of the conducting airways. Mechanisms for clearance of deposited substances appear to be quite similar in most mammals, including humans. The effectiveness of this defense has been determined by measuring the rate of transport of deposited particles, the frequency of ciliary beating, the integrity of the ciliated cells, the physical–chemical properties of the mucus blanket, and the rate of mucus production and transport. Exposure to a variety of inhaled agents, such as formaldehyde, cigarette smoke, ozone, nitrogen dioxide, airborne particulates including trace metals (cadmium and nickel), and sulfuric acid, causes ciliary damage and dysfunction, such as slowing of the frequency of ciliary beating, resulting in a significant reduction in transport rates.

Impairment of alveolar macrophage function alters the ability of the cell and/or the lung to (1) maintain sterility within the gas exchange regions of the lung, (2) provide an effective clearance mechanism from the lung for inhaled particles and cellular debris phagocytized by these cells, (3) interact with lymphocytes, and (4) release immunologically active soluble mediators. To fully meet these functional responsibilities, these cells must maintain mobility, a high degree of phagocytic activity, an integrated membrane structure, and a well-developed functional enzyme system.

As the first line of defense, the resident macrophage must (1) isolate ingested particles by phagocytosis, (2) act as a vehicle for physical movement from the lung, and (3) inactivate or detoxify inhaled and ingested microbes or chemicals. The sequence of events that must take place for this defense system to function is complex and involves a number of intricate and interrelated biological functions. Chemicals can interfere with this function at many sites. Alterations in the ability of any of these functions could be expected to significantly increase the host's risk of pulmonary disease. A number of assays have been developed to identify functional changes in alveolar macrophages and have been used to demonstrate effects following exposure to agents such as carbon, diesel exhaust, PbO, nickel chloride, CO, Pb_2O_3, cigarette smoke, cotton dust, and quartz. These chemicals also promote the influx of new macrophages into the lung. These new cells may be derived from (1) an influx of interstitial macrophages, (2) proliferation of interstitial macrophages with subsequent migration of the progeny into the airspace, (3) migration of blood monocytes, or (4) division of free lung macrophages. While such an accumulation of macrophages may appear to be a necessary response to the immediate insult, a possible consequence of this mass recruitment may be the development of chronic pulmonary disease, as was discussed earlier.

Not just macrophages migrate into the lung during pulmonary insults. Polymorphonuclear leukocytes (PMN) also accumulate following exposure to such agents as diesel exhaust, ozone, nitrogen dioxide, iron oxide, cotton dust, cigarette smoke, HCl, and cadmium chloride. A large pool of PMN normally remains within the microvessels and few are found in the air spaces. However, following injury these migrate out of the vascular space. These cells migrate through the endothelium to the inflammatory site, where they attempt to phagocytose and destroy foreign material and may sometimes damage the host tissue in the process. While in the lung, powerful oxidants (oxygen radicals) and enzymes can be released and produce tissue injury. For example, there is evidence that cigarette smoke can activate the PMNs and produce a shift in favor of proteolysis by the release of elastase from its lysosomal granules and generating oxidants with nadph oxidase system and myeloperoxidase. It has been postulated that such a proteolytic imbalance may cause lung tissue destruction leading to emphysema.

Not all exposure to chemicals results in an increase in number of available macrophages in the lung. Exposure to lead sequioxide, silica, asbestos, Pb_2O_3, cadmium fumes, MnO_2, Mn_3O_4, ozone, crysotile, amosite, cadmium oxide, acrolein, and nickel chloride actually causes a reduction in the number of these defense cells. These chemicals are cytotoxic and result in a lysis of the macrophage upon exposure. Some chemicals, such

as fly ash, carbon monoxide, and certain trace metals, may affect cellular viability but not cause lysis. In some cases, the same chemical may, at different concentrations, elicit a variety of measurable and significant effects. For this reason, the most appropriate approach is to use a battery of functional assays. Parameters such as total number, stability, viability, morphology, phagocytic and bacterial function, and biochemical metabolism are useful measurements of total functional capacity of the macrophage.

The efficiency of the phagocytic and lytic system of the macrophage determines the sterility and health of the lung. Marked changes in phagocytic efficiency of these cells is found following exposure to nickel chloride, nitrogen dioxide, ozone, sulfur dioxide, CH_2O, cadmium chloride, and cigarette smoke. Depressed bactericidal function of macrophages has been reported following exposure to many of the previous chemicals and to H_2SO_4, ethanol, and lead chloride.

The activity of a number of macrophage enzymes (e.g., acid phosphatase, β-glucuronidase, β-N-acetylglucosaminidase, peroxidase, and lysozyme), which function to combat infectious disease, has been significantly depressed following exposure to a number of toxicants. Depression in the ability of the macrophage to produce interferon, a substance that is involved in host defense against viral infection, has been identified following exposure to ozone, irradiated auto exhaust, and nitrogen dioxide.

Alterations in the previous respiratory defenses would be expected to make the lung more vulnerable to infectious disease. Animal models have served to demonstrate the effects of airborne chemicals and to associate these effects with actual increases in susceptibility to respiratory disease. These *in vivo* models combine the adverse effects of the toxicant plus the added stress induced by an infectious microorganism to measure the effectiveness of the host defenses after exposure to the toxicant. The test animals, and in some cases humans, are challenged with a laboratory-induced respiratory infection (bacterial, viral, or mycoplasma) following the exposure to the test chemical. If the host defense mechanisms are functioning normally, there is a rapid inactivation of inhaled organisms that have been deposited in the lung. However, if the pollutant exposure has caused a dysfunction(s) in these defenses, the microbes will proliferate rapidly and a measurable increase in pulmonary infection can be identified. While exposure to a test substance alone may not be life-threatening, association with other environmental stresses, such as infection, could prove critical in the promotion or exacerbation of a particular disease.

The lung is an active immunologic organ which, when exposed to toxicants, can have specific local immunologic effects and play a role in systemic alterations, such as changes in circulating immunoglobulins. Wheezing, chest tightness, rhinitis, and asthma are symptoms of a sensitization response to a foreign material by the pulmonary immune system. Immunity can be defined as all of the physiological mechanisms that enable an individual's body to recognize materials as foreign and to neutralize, eliminate, or metabolize them without injury to its own tissue. Over the past decade, data have been accumulated to clearly substantiate cases in which lung immunoregulatory functions of humoral and/or cell-mediated immunity have been compromised by inhaled chemicals.

While the immune system is highly regulated by complex interactions, both between components of the system and between immune and nonimmune organ systems, xenobiotics can modulate the immune system effecting either "up" or "down" regulation of the process. Inhaled chemicals may provoke a variety of different responses, including (1) reduction of normal immune response—immunosuppression resulting in an increase incidence of infection or tumors (e.g., benzene, malathion, lead, cadmium, nickel, and nitrogen dioxide); (2) overactivation of the immune system or exaggeration of the response causing hypersensitivity reactions (beryllium, mercaptans, chromates, diiocyanates); and (3) promoting of an autoimmune reaction, a pathological condition, in which there is a failure of the body to distinguish between "self" and "nonself" and production of structural and/or functional damage to tissues and organs (e.g., mercury, cadmium, vinyl chloride, and methyl cholanthrene). Over 100 xenobiotics have been associated with such autoimmunological effects.

Inhaled substances have been shown to exacerbate various immune-mediated disorders including asthma, hypersensitivity, pneumonitis, allergic rhinitis, and workers' pneumoconiosis. Table R-11 gives examples of the chemical agents that, when inhaled, are capable of eliciting an immunotoxic effect.

It is of interest that the same person may have an immediate-onset response on one occasion, a delayed-onset reaction on another, and, under other exposure conditions, exhibit a dual response with immediate-

TABLE R-11
Examples of Immunotoxins

Halogenated aromatic hydrocarbons
    Polychlorinated biphenyls
    Polybrominated biphenyls
    Dioxins
Pesticides
    Organophosphates
    Organochlorides
    Carbamates
Polycyclic aromatic hydrocarbons
    Benzo(a)pyrenes
    Methylcholanthrene
    Dimethylbenz(a)anthracene
Solvents
    Benzene
Heavy metals
    Beryllium
    Manganese
    Nickel
    Cadmium
    Platinum
Air pollutants
    Ozone
    Nitrogen dioxide
    Cigarette smoke

onset symptoms that resolve within an hour followed several hours later by a second set of symptoms. The underlying mechanisms for such effects are not known. However, clinical and experimental evidence has indicated that this process is like many other toxicologic effects in that the response is related to concentration, duration, and frequency of exposure.

Assessing the Risk of Airborne Chemicals

Risk assessment has been defined as the process whereby the most relevant biological, dose–response, and exposure data are used to identify and characterize risk. This information is used to produce a qualitative and/or quantitative estimate of the probable hazard to human health resulting from exposure to the airborne chemical(s). This process was significantly improved when the National Research Council evaluated the role of risk assessment as it relates to toxicology and developed uniform guidelines for federal agencies to use in assessing risk. They categorized the process into four steps: hazard identification, dose–response assessment, exposure assessment, and risk characterization. The process, summarized in Fig. R-11, is now widely used in evaluating the risk of inhaled chemicals. The important question in risk analysis is not simply what is the specific toxic response to some chemical, but rather what is the probable risk that the chemical may actually produce, under conditions of human exposure, significant health effects. There are multiple reasons for conducting risk assessment for airborne material. In addition to using the information for establishing federal, state, and local governmental regulations necessary to protect the worker or the general public, the risk assessment is also of value in identifying data gaps and planning future research and is a useful integration of our existing knowledge database. Since such risk analysis is being used to establish air standards by policymakers, the toxicologist must play a key role in this process. Because of the public health and economic implications of risk assessment, the U.S. Congress has requested a survey and analysis of the risk literature and a federally funded program on health risk assessment.

While each assessment is unique, there are certain basic principles that are common to all. This review concentrates on those that are appropriate for airborne chemicals. In analyzing the data for risk, certain assumptions have to be made. For noncarcinogenic effects, it is assumed that adverse effects will not occur below a certain level of exposure, even if the exposure continues over a lifetime. This threshold effect is supported by the fact that the toxicity of many chemicals, including airborne materials, is manifested only after the depletion of a physiological reserve and that the various host biological repair and defense capabilities can accommodate a certain degree of damage. In such cases, the objective of the toxicological risk assessment is to be able to establish, with best scientific certainty, a threshold dose below which adverse health effects are not expected to occur. For carcinogenic effects, especially those considered to be due to genotoxic events (e.g., mutations), a threshold may not exist. Regulatory agencies consider that exposure to carcinogens pose a finite risk at all doses and that the probability of developing cancer increases with increased dose. The U.S. EPA, in assessing the risk of carcinogens, assumes that the same total daily body burden will give the same tumor incidence, regardless of the route of exposure. This approach does not consider that some tumors at the site of contact (e.g., following inhalation) may be site specific or that the dose to a target organ may be modulated by the route of exposure. An important implication of this is that all levels of exposure, however small, add to the background risk and thus the experimental data are usually never adequate to exclude the possibility of added risk for the exposed population.

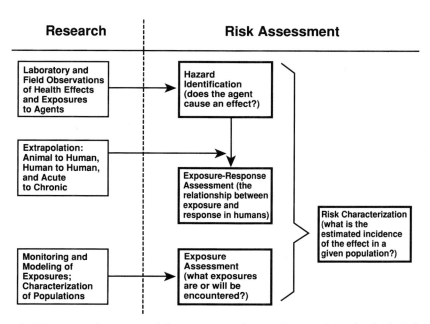

FIGURE R-11   *Summary of the process used in evaluating the risk of inhaled chemicals.*

Effectively predicting the human health risk from exposure to airborne contaminants is complex and requires reliable data for hazard identification, an understanding of the dose–response relationships, and an analysis of the human exposure. Significant advances have been made in inhalation toxicology in accumulating useful data appropriate to supporting hazard identification and in dose–response assessment. The development of sensitive analytical instrumentation permits the exposure and monitoring of chemicals at increasingly lower concentrations. This is important not only in conducting laboratory studies but also is of value in developing techniques useful in assessing total human exposure of individuals or populations. This includes the identification of the contaminant, contaminant sources, environmental media of exposure, chemical and physical properties of the airborne substance, and the intensity and frequency of the exposure. Important improvements have been made in developing and validating reliable biomarkers of health effects and of exposure. Traditional analysis of body samples, such as urine, blood, exhaled air, and tissue, is useful for determination of dose. Detailed toxicokinetic studies in animals have provided information on dose–response relationship. Physiology-based mathematical modeling of toxicokinetic parameters measured in animal studies has allowed easier interpretation and extrapolation of these animal data to human exposures.

The first step in risk assessment is to identify the potential deleterious effects of the substance. Frequently, risk assessment is conducted based on limited data. In cases in which the most relevant data are not available, attempts are often made to extrapolate other toxicological information regardless of the route of administration or the concentration/dose used to produce a certain effect. Using such information may not ensure the protection of human health from inhaling the substance and may result in either overregulating or underregulating the exposure risk of the chemical in question. The most reliable and scientifically defensible risk assessment analysis should be based on toxicological data collected under exposure conditions that are realistic and relevant to the human exposure; that is, by the same route of exposure, for similar duration, and in quantities that mimic the expected human exposure. All these factors are known to modulate the dose of the inhaled pollutant and/or its metabolites and hence the toxicological effects resulting from the exposure. Currently, scientists are attempting to develop mathe-

matical models that would permit appropriate extrapolation of high-dose studies to low-dose studies, short-term effects to lifetime health risk, and the ability to make appropriate route-to-route and species-to-species extrapolation from one compound to another and from *in vitro* to *in vivo*.

Developing reliable dose–response data from well-conducted inhalation studies is an essential step in the overall risk assessment process. With inhalation exposure, more so than with other routes of exposure, special attention needs to be paid to the difference between exposure and dose. In assessing health effects, exposure is often used as a surrogate for dose. However, when this is done important factors may significantly modify the predicted effect (these factors include physical–chemical characteristics of the material, protective mechanisms, metabolism, and biological characteristics of the subject). Defining the appropriate dose resulting from certain exposure becomes a more difficult task when the exposure involves studies with complex mixtures of airborne material such as automobile emissions, cigarette smoke, and atmospheric pollutants.

Being able to determine whether people exposed to airborne chemicals are at significant risk and the magnitude of the risk requires meaningful exposure assessment analysis. This analysis should include all exposures a person has to a specific contaminant, regardless of environmental medium (air, water, food, or soil) or the route of entry (inhalation, ingestion, or dermal contact). Exposure assessment can be useful in providing information for describing the distribution of the contaminant exposure within a population, for estimating the dose received, and for determining routes of entry into the body. Exposure can be assessed by using personal monitors near the breathing zone of the individuals. Passive samplers for air contaminants frequently use diffusion or permeation to concentrate the airborne material on a collecting medium, which is then returned to the laboratory for analysis. These have been used for volatile organics, nicotine, formaldehyde, nitrogen dioxide, and carbon monoxide. Active samplers use small pumps to draw the contaminated air through some collecting medium for analysis or through some form of direct-reading detector. When appropriate biomarkers are used in combination with the personal exposure data, an indication of internal dose can be estimated. For example, blood and urinary cotinine levels can be linked to air nicotine concentration and blood carboxyhemoglobin levels can be related to air carbon monoxide concentrations. The National Academy of Science has published an extensive discussion of the various approaches being used for assessing human exposure to airborne pollutants.

The final step, risk characterization, involves the integration and analysis of the existing database to provide a numerical estimate of the incidence of the adverse effect in a given population, assuming specific conditions of exposure.

The existing methods available for scientifically defensible risk characterization are not yet ideal since each step has an associated uncertainty due to data limitation and incomplete knowledge on exact mechanism of action of the toxic chemical on the human body. For noncancer endpoints, safety factors or uncertainty factors are applied since these effects are assumed to have a threshold below which no adverse effect is expected to be observed. U.S. EPA has used the concept of a reference concentration (RfC) to estimate acceptable daily human exposure from hazardous air pollutants. The RfC was adapted for inhalation studies based on a reference dose (RfD) method previously used for oral exposure assessment. The RfC differs from the RfD paradigm by using certain dosimetric adjustment to score the exposure concentration for animals to a human equivalent concentration. Both are estimates, with uncertainty spanning perhaps an order of magnitude, of a daily exposure to the human population, including sensitive subgroups, that would be without appreciable risk of a deleterious effects over a lifetime.

The RfC is estimated based on available knowledge on the toxic response of both humans and animals.

Appropriate uncertainty factors (UF) and modifying factors (MF) are incorporated into the equation:

$$RfC = NOAEL/(UF) \times (MF)$$

The NOAEL is the no-observed-adverse-effect level. Table R-12 indicates how these factors are used in deriving appropriate risk characterization.

For carcinogens, risk is estimated based on human and experimental animal data and other supporting evidence of carcinogenicity (e.g., structure–activity correlations, kinetics, and *in vitro* data). Decisions on the carcinogenicity of chemicals in humans need to be based on considerations of all relevant data, whether they are indicative of a positive or negative response, and should embody sound biological and statistical principles.

However, because animal carcinogens are not the same with respect to potency, target organs, mechanism, and so forth, and thus are not equally relevant to humans, hazard evaluation is on a weight-of-evidence basis. The weight-of-evidence evaluation of carcinogenic hazard to humans provides a basis for carcinogen classification. These assessments involve fitting mathematical models to experimental data and extrapolating from these models to predicting risk at doses well below the experimental range. A range of risks can be produced using different models and assumptions about dose–response curves and the susceptibility of humans and animals to the test agent. Both IARC and U.S. EPA have established a weight-of-evidence classification for carcinogens. They are similar but the IARC method does not address the potency of carcinogens, whereas

TABLE R-12
Application of Uncertainty Factors in Deriving RfC

Type	Magnitude	Purpose
Interindividual	10	Intended to account for the variation in sensitivity among the human population
Interspecies	10	Used to account for uncertainty in extrapolating results from animals to average human population
Subchronic to chronic	5–10	Used to account for uncertainty in extrapolating less than chronic exposure results on animals or humans when no long-term human data are available
LOAEL to NOAEL	5–10	Accounts for the uncertainty inherent in extrapolation downward from LOAEL to a NOAEL
Incomplete to complete data	10	Used when experimental data are incomplete; this factor is intended to account for the inability of any single study to adequately address all possible adverse effects in human; depends on scientific judgment of the uncertainties of the study and database

the U.S. EPA approach offers a means for developing quantitative estimate of carcinogen potency.

Further Reading

Bates, D. V., Dungworth, D. L., Lee, P. N., McClellan, R. O., and Roe, F. J. C. (Eds.) (1989). *Assessment of Inhalation Hazards.* Springer-Verlag, New York.

Brain, J. D., Beck, B. D., Warner, A. J., and Shaikh, R. A. (Eds.) (1990). *Variations in Susceptibility to Inhaled Pollutants: Identification, Mechanisms and Policy Implications.* Johns Hopkins Univ. Press, Baltimore, MD.

Calabrese, E. J., and Kenyon, E. M. (1991). *Air Toxic and Risk Assessment.* Lewis, Chelsea, MI.

Gardner, D. E., Crapo, J. D., and McClellan, R. O. (Eds.) (1993). *Toxicology of the Lung.* Raven Press, New York.

National Research Council (1991). *Human Exposure Assessment for Airborne Pollutants: Advances and Opportunities.* National Academy of Science, Washington, DC.

—*Donald E. Gardner and
Daniel T. Kirkpatrick*

Related Topics

Absorption
Ames Test
Animal Models
Biomarkers, Human Health
Clean Air Act
Combustion Toxicology
Dose–Response Relationship
Emergency Response
Indoor Air Pollution
International Agency for Research on Cancer
Mouse Lymphoma Assay
Occupational Toxicology
Pharmacokinetics/Toxicokinetics
Photochemical Oxidants
Pollution, Air
Polycyclic Aromatic Hydrocarbons
Radiation Toxicology
Risk Assessment, Human Health
Risk Characterization
Sick Building Syndrome
Threshold Limit Values
Tissue Repair
Toxicity Testing, Inhalation

Rhododendron Genus

- ◆ REPRESENTATIVE SPECIES: Azalea, rhododendron
- ◆ SYNONYMS: *Rhododendron catabiense*, Ericaceae (heath) family; catawba rhododendron; mountain rosebay; purple laurel

Exposure Pathway

Ingestion of plant parts (and sometimes honey contaminated by bees) is the route of exposure.

Mechanism of Toxicity

Grayanotoxins exert their effect in the cell membrane by binding to sodium channels. Nerve and muscle cells are kept in a state of depolarization.

Human Toxicity

All parts of the plant are toxic although the severity of symptoms differs among species. A form of contamination involves the ingestion of honey contaminated by bees frequenting rhododendron leaves. Consumption of the plant may cause abdominal pain, vomiting, watering of the eyes and mouth, ataxia, and weakness. Hypotension and bradycardia along with progressive paralysis of the limbs and convulsions may be noted.

Clinical Management

Minimal ingestions usually do not require treatment. Larger ingestions may require gastric decontamination with syrup of ipecac or lavage followed by administration of activated charcoal and a cathartic. Symptomatic care should follow.

Animal Toxicity

All animal species appear to be affected. Symptoms ranging from lethargy and ataxia in dogs to tachycardia, dyspnea, and paralysis in donkeys have been reported.

—*Rita Mrvos*

Rhubarb

- SYNONYMS: *Rheum rhaponticum;* pie plant; garden rhubarb; wine plant
- DESCRIPTION: Rhubarb is a perennial plant with stalks that grow 1–3 ft in length and become reddened when ripe. The leaves are large and wrinkled with wavy margins.

Exposure Pathways
The routes of exposure are ingestion and dermal contact.

Mechanism of Toxicity
Soluble oxalates in the leaves are absorbed via the gastrointestinal tract. Once absorbed, oxalates bind with calcium producing secondary hypocalcemia. The insoluble oxalate salts precipitate in the renal system resulting in kidney malfunction and electrolyte imbalance.

Human Toxicity
Rhubarb stalks are edible. Soluble oxalates are found in the leaf blades and in a much lower concentration in the stalk. Anthraquinone glycosides are not present in rhubarb grown in the United States, therefore exposures do not result in the expected cathartic effects. Fatal poisonings due to ingestion of leaves are rare. Symptoms include mouth and throat irritation. Because it contains a less irritating oxalate, large amounts of the leaf must be ingested before symptoms develop. With ingestion of large amounts, symptoms include abdominal pain, nausea, vomiting, weakness and drowsiness, seizures, possible liver damage, and kidney damage. Because digestion of the plant is slow, effects may be delayed several days. The ingestion of stalks and small leaf exposures are unlikely to cause serious problems.

Clinical Management
If presentation is early, treatment with syrup of ipecac, activated charcoal, and a cathartic may be useful. Low calcium levels and tetany should be treated with intravenous calcium gluconate. Serum blood urea nitrogen and creatinine should be measured, and urine should be checked for the presence of oxalates. The patient should be rehydrated with fluid and electrolyte therapy as needed. Hemodialysis may be indicated if anuria develops.

Animal Toxicity
Oxalate-containing plants can be a source of poisoning for grazing animals. In ruminants, a large acute exposure results in hypocalcemia and death. In chronic exposures, renal damage and urolithiasis from calcium oxalate deposition result. Other animals display gastrointestinal symptoms and renal damage. Treatment includes activated charcoal and intravenous electrolyte and calcium treatment. With rest, animals become ambulatory after less than 8 hr of treatment. Notable gastrointestinal tract mucosal edema and hemorrhage, abdominal ascites, and hyperemia have occurred.

—*Regina Wiechelt*

Riboflavin

- CAS: 83-88-5
- SYNONYMS: Vitamin B_2; beflavin; flavaxin; lactoflavin; isoalloxazine; 7, 8-dimethyl-10-(d-ribo-2,3,4,5-tetrahydroxyphentyl)
- PHARMACEUTICAL CLASS: Water-soluble vitamin
- MOLECULAR FORMULA: $C_{17}H_2ON_4O_6$
- CHEMICAL STRUCTURE:

Uses

Riboflavin is a nutritional supplement used during periods of deficiency (ariboflavinosis). Riboflavin needs are increased during chronic debilitative stress to the body such as malabsorption diseases of the small intestine, liver disease, hyperthyroidism, alcoholism, and during pregnancy and lactation. Neonates undergoing phototherapy for hyperbilirubinema also have increased need.

Exposure Pathway

The route of exposure is oral. Dietary sources of riboflavin include broccoli, spinach, asparagus, enriched flour, yeast, eggs, milk, cheese, fish, poultry, liver, and kidneys.

Toxicokinetics

Riboflavin is readily absorbed from the gastrointestinal tract mainly in the duodenum. It is hepatically metabolized, moderately protein bound, and widely distributed to tissue; however, little is stored in the liver, spleen, heart, and kidneys. Riboflavin is excreted renally almost entirely as metabolites. All riboflavin in excess of daily body needs is excreted unchanged in the urine.

Human Toxicity

Acute toxicity is unlikely following even 100 times the recommended daily allowance. Chronic exposure to large doses of riboflavin may cause yellow discoloration of the urine.

Clinical Management

Acute ingestions seldom require treatment. Chronic excessive use should be discontinued and any toxic effects will resolve.

Animal Toxicity

Acute toxicity is not expected, and it would be unlikely for animals to be given chronic riboflavin overdoses.

—*Denise L. Kurta*

Rifampin

- CAS: 13292-46-1
- SYNONYM: Rifampicin

- PHARMACEUTICAL CLASS: Antibiotic
- CHEMICAL STRUCTURE:

Use

Rifampin is used as an antibiotic. It is a semisynthetic derivative of rifamycin B, a macrocyclic antibiotic produced by the mold *Streptomyces mediterranei*.

Exposure Pathway

Ingestion is the most common route of exposure. Rifampin is available in oral and parenteral forms.

Toxicokinetics

Rifampin is rapidly and nearly completely absorbed from the gastrointestinal tract. Peak serum levels are seen within 2–4 hr. Food and aminosalicylic acid interfere with absorption and delay peak levels. Massive ingestions in the overdose setting may also delay absorption. Protein binding is 75–90%. The volume of distribution is approximately 1 liter/kg. Rifampin undergoes hepatic deacetylation. Both rifampin and its deacetylated metabolite are excreted into the bile. Rifampin, and to a lesser extent its deacetylated metabolite, undergo enterohepatic recirculation. The half-life of therapeutic doses of rifampin is 1.5–5 hr. The half-life is shortened after regular use due to induction of hepatic enzymes. Chronic liver disease increases the half-life. The kinetics are not well described in the overdose setting. In one case, the half-life was 4.4 hr.

Mechanism of Toxicity

In the acute overdose setting, the mechanism of toxicity is not defined. A number of the toxic reactions occurring with intermittent dosing schedules or on reexposure are postulated to be due to the presence of antirifampin antibodies.

Human Toxicity: Acute

Intentional overdoses of rifampin rarely lead to significant morbidity and fatalities are exceedingly uncommon. The few deaths that have been associated with rifampin have all been in individuals with a history of alcoholism or concomitant ethanol ingestion. Acute overdose with rifampin may cause a red to orange discoloration of the skin, "the red man syndrome." Body fluids are also discolored and urine, feces, sweat, tears, and saliva may exhibit a red to orange discoloration. Symptoms associated with rifampin overdose include headache, abdominal pain, nausea, vomiting, and flushing. Pruritus, which may be limited to the scalp, may be seen, and a cutaneous burning sensation may be noted. Lethargy and obtundation have been reported. Facial or periorbital edema may be seen. Minor and transient elevations of hepatic transaminases, bilirubin, and amylase have been reported. The bilirubin level may be elevated due to inhibition of bilirubin excretion. With large overdoses, rifampin may interfere with the bilirubin assay. An acute ingestion of 60 g was fatal in an alcoholic. Overdoses of 12 g in otherwise healthy individuals have been tolerated, as has 2 g in an 18-month-old. Because of the small number of cases, correlation of dose with severity is not possible, and serum levels are not useful.

Human Toxicity: Chronic

Rifampin used daily at therapeutic doses is associated with facial flushing and itching in less than 5% of patients. More rarely, hepatotoxicity is seen. The risk of hepatotoxicity is increased with chronic liver disease, alcoholism, and old age. Acute renal failure, nephrogenic diabetes insipidus, and thrombocytopenic purpura are very rare complications of continuous daily use. The use of rifampin on an intermittent dosing schedule, two or three times weekly or less, is associated with a higher incidence of toxic side effects. These include a flu-like syndrome that lasts up to 8 hr following each dose of rifampin. More serious toxic effects associated with an intermittent dosing schedule include hemolytic anemia, thrombocytopenia, hepatitis, nephritis, acute renal failure, and shock. These reactions are believed to be hypersensitivity reactions and related to antirifampin antibodies.

Clinical Management

Acute overdoses of rifampin are rarely serious and good supportive care, gastric decontamination, and activated charcoal are all that is usually necessary. Given the extensive enterohepatic circulation of rifampin, repeated doses of activated charcoal may enhance elimination. Serious systemic toxicity associated with the chronic administration of rifampin is an indication to discontinue the drug.

—*Michael J. Hodgman*

Risk Assessment, Ecological

Introduction

The ecological problems facing environmental scientists and decision makers are numerous and varied. Growing concern over potential global climate change, loss of biodiversity, acid precipitation, habitat destruction, and the effects of multiple chemicals on ecological systems has highlighted the need for flexible problem-solving approaches that can link ecological measurements and data with the decision-making needs of environmental managers. Increasingly, ecological risk assessment is being suggested as a way to address this wide array of ecological problems.

Ecological risk assessment has been defined as "the process that evaluates the likelihood that adverse ecological effects may occur or are occurring as a result of exposure to one or more stressors" (U.S. EPA, 1992a). Exposure may be defined as co-occurrence or contact between stressors and the ecological components of concern (e.g., organisms, populations, communities, or ecosystems). Characterization of ecological effects and exposure are two major components of the ecological risk assessment process.

The broad potential scope for ecological risk assessment is shown in Fig. R-12. Stressors may be chemical (e.g., toxic chemicals or nutrients), physical (e.g., habitat alteration or destruction and changes in hydrologic or temperature regimes), or biological (e.g., introduced

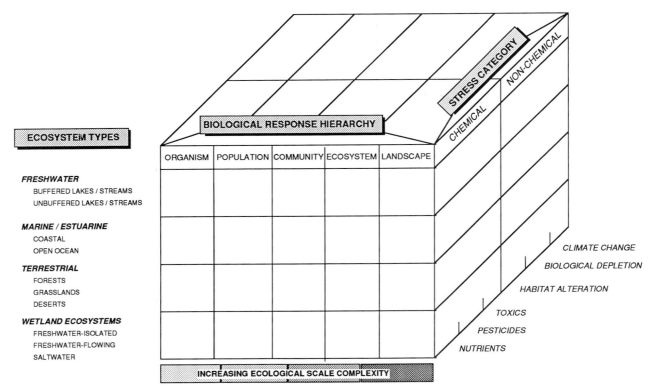

FIGURE R-12 *Ecological risk assessment "cube" (adapted from U.S. Environmental Protection Agency, Ecological Risk Assessment Guidelines Strategic Planning Workshop, EPA/ 630/R-92/002, Risk Assessment Forum, Washington, DC, 1992).*

species and genetically engineered organisms). Adverse ecological effects may be evaluated at levels of biological organization ranging from individuals of a single species to ecosystem structure and function, and ecological systems affected may be aquatic, terrestrial, or wetlands. Although not reflected in Fig. R-12, there is also a continuum of spatial and temporal scales, which may vary from the short-term, localized effects of a small spill of a readily degraded chemical to the long-term, global effects that may result from stratospheric ozone depletion.

Ideally, risk is expressed in quantitative, probabilistic terms. However, because of the scale and complexity of ecological system responses, it is not always possible to express risks quantitatively. Available scientific models may be inadequate given the inherent variability of natural systems, the potential for the direct effects of a stressor to lead to indirect or "cascading" ecological effects, or the potential for nonlinear stressor–response relationships. In such situations, it may be appropriate to estimate risk using a semiquantitative or qualitative

comparison of effects and exposure information. Whatever approach is used, incomplete information and uncertainties will be present in nearly every case, so sound professional judgment and scientific expertise are essential.

The ecological risk assessment process, as defined by the U.S. EPA, is shown within the bold lines in Fig. R-13. There are three phases: problem formulation, analysis, and risk characterization. The problem formulation phase involves evaluating the potential stressors, ecological effects, and ecosystems at risk, selecting appropriate endpoints, and developing hypotheses that will be the focus of the assessment. Data on ecological effects, stressor–response relationships, and exposure to stressors are evaluated in the analysis phase. In risk characterization, exposure and effects information are integrated, and all of the evidence is brought together to reach final conclusions about the likelihood and consequences of effects. Effective risk characterizations acknowledge uncertainties and assumptions and separate scientific conclusions from policy judgments.

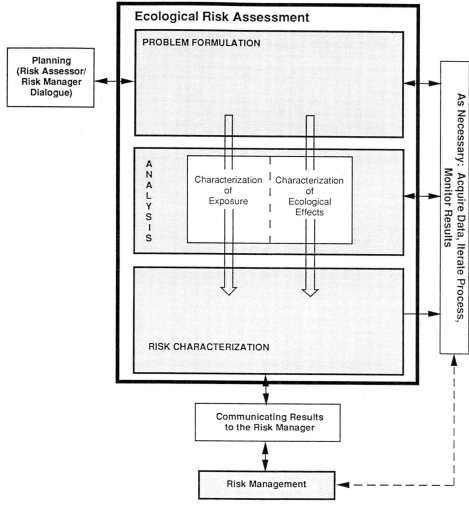

FIGURE R-13 *Overall framework for ecological risk assessment (from U.S. Environmental Protection Agency, proposed guidelines for ecological risk assessment, Fed. Regist. 61, 47552–47631, 1996).*

Other "frameworks" for ecological risk assessment have been suggested that are basically similar to EPA's framework except for slight differences in terminology. The National Research Council (1993) used a slightly modified version of a previous human health risk assessment paradigm (NRC, 1983) and suggested that their new version was suitable for both ecological and human health risk assessments. Suter (1993) used essentially the same diagram as the NRC (1993) for predictive risk assessments but rearranged it slightly for retrospective assessments that evaluate the risks of ongoing events (e.g., ecological effects associated with previous dumping of chemicals at a Superfund site).

Discussions here will focus principally on the terminology and principles found in the EPA's framework report and proposed guidelines (U.S. EPA, 1992a, 1996).

The science-based analysis of information conducted during risk assessment takes place within the context of risk management activities (Fig. R-13). Typically, a risk manager must decide between various management options by using the scientific results from the risk assessment in conjunction with economic, social, legal, and political factors. For example, if a risk assessment demonstrates that use of a pesticide on a certain crop may cause some level of adverse effect on aquatic organisms in nearby streams, the risk manager may weigh potential ecological damage against the economic and social problems that may be caused by banning use of the pesticide or the ecological damage that may result from using other pesticides in place of the banned substance.

Initial discussions between the risk assessor and the risk manager, concerned members of the public, and other "stakeholders" in the assessment can help ensure that the output of the risk assessment will be relevant to making decisions on the issues under consideration and that all relevant ecological concerns are addressed. Also, the regulatory context of the assessment may be important in determining the resources and time available for the evaluation. Interactions between the risk assessor and risk manager at the end of the risk assessment are important to provide the risk manager with a full and complete understanding of the assessment's conclusions, assumptions, and limitations.

The box on the right-hand side of Fig. R-13 represents data collection and monitoring. Data form the basis of risk analyses; these data may have been collected specifically for a risk assessment or developed for another purpose. In the latter case, assessors need to critically evaluate data to ensure that they can support the assessment. Risk assessment results can provide important input to monitoring plans used to evaluate the efficacy of a risk management decision. For example, if the decision was to mitigate risks through exposure reduction, monitoring could help determine whether the desired reductions in exposure (and effects) were achieved. Monitoring is also critical for determining the extent and nature of any ecological recovery that may be occurring and whether adjustments to the management approach are needed (Holling, 1978). Finally, experience obtained by using focused monitoring results to evaluate risk assessment predictions can help improve the risk assessment process.

The ecological risk assessment process is frequently iterative. For example, it may take more than one pass through problem formulation to complete planning for the risk assessment, or information gathered in the analysis phase may suggest further problem formulation activities such as modification of the endpoints selected. To maximize efficient use of limited resources, iterations of ecological risk assessments are frequently performed in a structured order (i.e., in tiers). Tiers often proceed from simple, relatively inexpensive evaluations to more costly and complex assessments. Early tiers may be designed to identify risks that are very low

(which may be dropped from further analysis) or very high (which would be expedited for risk management). Subsequent tiers may be designed to answer more difficult questions, such as the extent and nature of risks, or to evaluate alternative management options. Because a tiered approach can incorporate standardized decision points and supporting analyses, it can be particularly useful for multiple assessments of similar stressors or situations.

Each of the three phases of the ecological risk assessment process are described in more detail.

Problem Formulation Phase

Entry into the problem formulation stage of an ecological risk assessment may be due to the observation of an existing stressor (e.g., rise in global carbon dioxide levels) or effect (e.g., reduction in a sport fishery), may be in anticipation of the introduction of a stressor (e.g., manufacture of a new chemical or filling of a marsh), or may be because of concern over a valued resource (e.g., a pristine lake). As noted previously, preliminary risk assessor/risk manager discussions are important for defining the regulatory context and management goals for the assessment (the planning box shown in Fig. R-13). The management goals may influence the selection of assessment endpoints (explicit expressions of environmental values to be protected) along with ecological considerations. Problem formulation (Fig. R-14) includes evaluations of stressor characteristics, the ecosystem potentially at risk, and the ecological effects expected or observed. The spatial and temporal scales of the assessment are determined, endpoints are selected, and a conceptual model is constructed that describes the causal linkages and pathways through which identified stressors might affect the identified assessment endpoints. When problem formulation is complete, the risk assessor should have a clear focus for the risk assessment and a plan for the analysis phase.

Evaluation of the stressors that are important to a particular assessment may or may not be a difficult issue. If the goal of the assessment is to predict the effects of a new chemical, the stressor is known, but information is required concerning the sources of the chemical, amount produced, and environmental fate and transport. If the concern is for loss of a commercial fishery in an estuary, many possible stressors could be considered, such as toxic chemicals, oxygen depletion

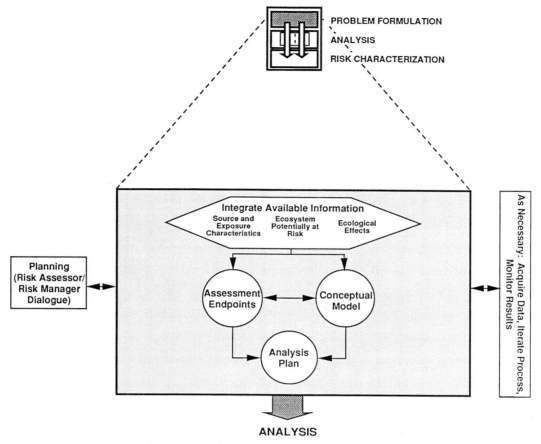

FIGURE R-14 *Problem formulation phase (from U.S. Environmental Protection Agency, proposed guidelines for ecological risk assessment, Fed. Regist. 61, 47552–47631, 1996).*

and habitat loss as indirect effects of nutrient loading, disease, overfishing, and the occurrence of natural events such as strong storms or droughts that may have contributed to adverse water quality conditions. Note that in addition to anthropogenic stressors, the types and historical patterns of natural disturbances in the ecological system should be considered. For example, in a watershed naturally adapted to strong spring floods, the absence of flooding may itself constitute a stressor to the system.

At the problem formulation stage, it is best to consider all possibilities at first, then narrow the focus to a manageable number. Some of the characteristics of a stressor that may be important to consider at this stage include intensity (concentration or magnitude), duration (short or long term), frequency (single event, episodic, or continuous), timing relative to important biological cycles, and spatial scale and heterogeneity

(EPA, 1992a). The need to evaluate spatial and temporal distribution and variation is inherent in many of these example characteristics. Such information is especially useful for determining potential exposure; that is, where there is co-occurrence of or contact between the stressor and ecological components.

An important goal for evaluating stressors at this stage is to establish qualitative linkages or pathways between sources of a stressor, its spatial and temporal distribution in the environment, and the ecological components of concern. This may require consideration of secondarily created stressors. For example, nutrients may exert effects on a commercial fishery indirectly by leading to algal blooms that reduce light levels to the point that higher aquatic plants important as habitat for juvenile fish cannot survive. When the algal blooms die, they may lower dissolved oxygen levels to the point that fish kills result. In this case, habitat loss or low dissolved oxygen are the stressors causing immediate damage to the fish, but knowledge of the whole chain of events would be necessary to plan a useful risk assessment.

If a risk assessment is initiated due to concern over a stressor, the spatial and temporal distribution of the stressor can help determine the ecological systems at risk. In the case of a new chemical, information on the planned manufacture, use, and disposal of the chemical can be useful. In the case of a commercial fishery in an estuary, the area of concern could be expanded upstream into the watershed (if these areas were spawning grounds for the fish) or into the ocean (if the fish spent part of the year outside of the estuary). On the other hand, if the risk assessment is triggered by observed effects, these effects can directly indicate ecosystems or ecological components that should be considered in the assessment. Value-initiated risk assessments are driven by goals for the ecological values of concern. These goals provide the basis for identifying more specific endpoints for the assessment and then for identifying an array of influential stressors and relevant effects.

Evaluation of stressors and ecological effects must be done based on knowledge of the structural and functional properties of the ecological systems potentially at risk. Relevant factors include aspects of the abiotic environment (such as climatic conditions and soil or sediment properties), ecosystem structure (including the types and abundances of different species and their trophic-level relationships), and ecosystem function (such as the ecosystem energy source, pathways of energy utilization, and nutrient processing).

An important component of problem formulation is developing the link between endpoints that we can measure (e.g., survival of fish in a laboratory assay) and a management goal (e.g., preventing fish kills in streams). This link is aided by defining assessment endpoints which are intermediate between broad management goals and specific measurements. The more useful assessment endpoints have a clear unambiguous definition. In the estuary example, "viable commercial fisheries" may be an appropriate management goal but is difficult to work with as an assessment endpoint. To minimize ambiguity, an assessment endpoint should include both an entity (such as striped bass) and a property of the entity (such as abundance or yield) (Suter, 1993). If feasible, additional definition of the assessment endpoint is desirable (e.g., percentage change in abundance).

While ecological risk assessments have a large potential scope (Fig. R-12) that precludes a comprehensive listing of assessment endpoints, several criteria for an appropriate assessment endpoint may be suggested, including ecological relevance, susceptibility to known or potential stressors, and relevance to management goals (EPA, 1996; Suter, 1993):

• Ecological relevance: Ecologically relevant endpoints should reflect important characteristics of the ecological system, so their selection requires an understanding of system structure and function. For example, an assessment endpoint might consider changes in the population of a species that has a major effect on the abundance and distribution of many other species. In the estuary example, one assessment endpoint might be maintaining (or restoring) a certain areal extent of aquatic plant populations in the estuary. The ecological relevance to larval fish and other organisms may compel the inclusion of this endpoint even though the value of the plants to society is not immediately evident.

• Susceptibility to the Stressor: An assessment endpoint is susceptible to a stressor if the ecological component is both exposed to a stressor and sensitive to that exposure. For example, if the stressor is reduced transparency of the water in an estuary due to algal blooms, then the survival and growth of aquatic plants in the estuary may be good components for an assessment endpoint given the likelihood of their exposure and their dependency on certain light levels for growth.

• Relevance to management goals: Most risk assessments are conducted as a response to some legal or regulatory requirement and to be useful must produce a result that can support an environmental decision. Risk manager concerns and public values can range from protection of species that are endangered or have commercial or recreational importance to preservation of ecosystem attributes for functional reasons (e.g., flood water retention by wetlands) or aesthetic reasons (e.g., visibility in the Grand Canyon). While the ecological significance of these values may vary, their importance to society justifies their consideration in the selection of assessment endpoints.

During problem formulation, available information on stressors, ecological systems potentially at risk, ecological effects, and endpoints is assembled into a conceptual model of the problem (U.S. EPA, 1992a). The conceptual model should describe the linkages between stressor sources and pathways to the potentially affected ecological components, including those specifically enumerated in the assessment endpoints. The various exposure scenarios defined by the conceptual model may represent either direct effects or indirect effects (e.g., loss of a key prey organism for a species of concern). Although it is helpful to include many possible exposure/effect hypotheses in the conceptual model, only those that are considered most likely to contribute to risk should be selected for further evaluation in the analysis phase of the risk assessment.

A diagram of the conceptual model showing exposure pathways and scenarios can be useful in planning the risk assessment. A description of the conceptual model should include both the hypotheses selected for further analysis and a rationale for those that were not chosen. For those hypotheses that are to be the focus of the assessment, the description should include the data available or required for subsequent analysis and the types of models and methods to be employed. In addition, major uncertainties and data gaps should be identified.

The conceptual model and assessment endpoints together provide input to the analysis plan, which delineates the assessment design and the data needs, measures, and methods for conducting the analysis phase of the risk assessment. The analysis plan shows how the assessment relates to management decisions that must be made and indicates how data and analyses will be used to estimate risks.

Analysis Phase

In the analysis phase (Fig. R-15), both the ecological effects of and the exposure to stressors are evaluated based on the risk hypotheses and conceptual model developed during problem formulation. The products of the analysis phase are an exposure profile that quantifies the magnitude and spatial/temporal variations of exposure and a stressor–response profile that relates effects on ecological components to varying stressor levels. These profiles and information on associated assumptions and uncertainties are carried forward into risk characterization.

Consideration of both biotic and abiotic ecosystem characteristics is important when evaluating exposure and effects information. Examples of questions that may be asked include the following:

- What are the spatial and temporal distributions of the ecological components of concern? For example, to evaluate the effect of a localized area of chemical contamination on an eagle population, it may be important to consider the total area in which the eagles feed relative to the area of contamination to determine the likelihood of exposure.

- How might characteristics of the ecosystem modify the nature and distribution of the stressor? Chemical stressors can be modified by a wide range of environmental fate processes, including biotransformation by microorganisms, photolysis, hydrolysis, and sorption. Environmental properties such as sediment organic content can influence the bioavailability of chemical stressors and thus their exposure to organisms. Physical stressors can also be affected by ecosystem characteristics. For example, the effects of altered stream flow regimes caused by water diversion will depend on the amounts and patterns of precipitation as well as the frequency and intensity of the diversion.

The interplay between the ecosystem, exposure, and effects indicated by the dotted line and two way arrows in Fig. R-15 can be illustrated by consideration of the potential effects associated with nutrient enrichment in an estuary. High nutrient levels may lead to increased growth of phytoplankton and epiphytes. This initial effect can expose submerged aquatic vegetation to decreased light penetration. Subsequent loss of submerged aquatic vegetation may reduce habitat or certain fish species, causing a reduction in their populations. In addition, death and decomposition of the phytoplankton may create yet another stressor for organisms in the estuary—low dissolved oxygen. This example demonstrates the potential for cascading chains of exposure and effects in complex systems.

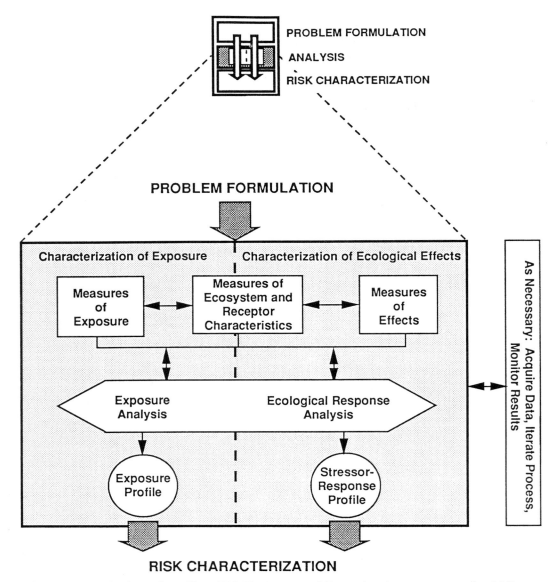

FIGURE R-15   *Analysis phase (from U.S. Environmental Protection Agency, proposed guidelines for ecological risk assessment, Fed. Regist.* **61**, *47552–47631, 1996).*

Characterization of Exposure

Characterization of exposure (the left half of Fig. R-15) evaluates the interaction of the stressor with the ecological components of concern. Exposure must consider the spatial and temporal distributions of both the ecological components of concern and the stressors. Stressor characterization involves determining the stressor's distribution or pattern of change. For chemical stressors, a combination of monitoring and modeling may be used to evaluate environmental fate, transport, and bioavailability. For physical stressors such as

habitat destruction, other approaches such as measuring areal extent may be appropriate. In this case, co-occurrence of physical alteration (e.g., dredging) with the ecological components (e.g., riparian wetlands) would result in exposure (e.g., direct or secondary loss of habitat for species of concern). Exposure assessment of biological stressors such as introduced pathogens must also consider the ability of organisms to reproduce, become dormant, and disperse.

Sampling and analysis approaches for stressors should consider the potentially heterogeneous spatial distribution of the stressor (e.g., contaminated sediments or soils at a hazardous waste site). The timing of a stressor presence in an ecological system can also be of critical importance since diurnal, seasonal, or episodic occurrences may affect the species and life stages exposed as well as the effect of the system of the stressor (e.g., seasonal changes in microbial biotransformation of chemicals). Even short-term events may be significant if they coincide with critical life stages. Periods of reproductive activity may be especially important because early life stages often are more sensitive to stressors, and adults also may be more vulnerable at this time.

For exposure to occur, both the stressor and the ecological component of concern must be present. Exposure is commonly evaluated by measuring or modeling concentrations or amounts of a stressor and combining these stressor estimates with assumptions about the pathways for exposure. For example, measurements of chemical concentrations in water may be adequate to assess direct exposure of fish, but other pathways may also be important. If contaminated food may be consumed, concentrations of chemicals in food items can be measured or estimated and combined with ingestion rates to estimate dietary exposure. If chemical exposure is ongoing at a field site, tissue residues may be used to measure exposure directly, but the contribution of various potential sources of the chemical may require further analysis.

Information obtained from the exposure analysis is summarized in an exposure profile that quantifies the magnitude and spatial and temporal patterns of exposure for the scenarios developed during problem formulation. The exposure profile is integrated with the stressor–response profile in risk characterization, so it must be expressed in a form consistent with the output of ecological effects characterization. For example, if organisms are exposed to chemicals for a significant portion of their life cycle, the corresponding stressor–response profile should provide information on chronic toxicity.

Assumptions and uncertainties associated with the exposure analysis should be included in the exposure profile. Exposure assessments may have significant data gaps or may rely on data of questionable or unknown quality. While some uncertainties can be quantified, a risk assessor may have to rely on assumptions and qualitative estimates based on professional judgment. It is important that the assessor characterize the major sources of uncertainty and carry them forward to the risk characterization so that they may be combined with a similar analysis conducted as part of the characterization of ecological effects.

Characterization of Ecological Effects

The relationship between the stressor and the endpoints identified during problem formulation is analyzed in ecological effects characterization (the right half of Fig. R-15). Stressor–response data and cause–effect relationships are evaluated. Extrapolations are conducted as necessary during this phase. The product is a stressor–response profile that quantifies and summarizes the relationship of the stressor to the assessment endpoint and is used along with the exposure profile in risk characterization.

The type and amount of ecological effects data available will vary depending on the nature of the stressor, the assessment endpoints, the availability of published data, and the resources available for conducting the risk assessment. The evaluation process relies on professional judgment, especially when few data are available or when choices among several sources of data are required. If available data are inadequate, new data may be needed before the assessment can be completed. Some of the sources of data used in ecological risk assessments include the following:

• Quantitative structure–activity relationships (QSARs). QSARs relate physical and chemical properties of a chemical to its biological effects. For example, the octanol–water partition coefficient of a nonionic organic compound can be related to its toxicity and bioconcentration potential for aquatic organisms.

• Laboratory toxicity tests: Acute and chronic toxicity data are available for many species and chemicals. There are substantially more data for aquatic than terrestrial species. As noted below, extrapolating the results of single-species laboratory tests to effects in complex field situations can be problematic.

• Field tests: Controlled field tests (e.g., field enclosures or mesocosms) can provide strong causal evidence linking a stressor with a response and allow evaluation of effects in a more "realistic" setting than laboratory tests. Observational field studies (e.g., comparison of contaminated sites with reference sites) also provide environmental realism lacking in laboratory studies, but the presence of confounding factors (e.g., multiple stressors and variations in habitat quality) can make it difficult to attribute observed effects to specific stressors.

Data derived from these tests are used to quantify the stressor–response relationship and to evaluate the evidence for causality. A variety of techniques may be used, including statistical methods and mathematical modeling. When the assessment endpoint cannot be measured directly, additional analyses may be necessary. Examples of extrapolations that may be required in ecological risk assessments include those

- Between taxa (e.g., bluegill sunfish mortality to rainbow trout mortality);

- Between responses (e.g., acute LC_{50} to chronic no-observed-effect level);

- From laboratory to field (e.g., mortality of bluegills in laboratory tests to mortality of bluegills exposed to the same stressor under field conditions); and

- From field to field (e.g., from the results of a pond mesocosm test to effect in a lake in a different area).

In addition to these extrapolations, evaluation of particular assessment endpoints may require consideration of indirect effects resulting from changes in competition among species, predation, disease prevalence, or utilization of resources, as in the nutrient/estuary example discussed earlier. Compensatory mechanisms present in ecological systems may make it difficult to extrapolate changes at one level of biological organization to another (e.g., from effects on individuals of a species to population level effects). The risk assessor's task is also complicated by the existence of nonlinear stressor–response relationships and the lack of equilibrium conditions that may characterize ecological systems (Reice, 1994). These complexities make it difficult to predict how and when a system may recover following the cessation of a stressor.

Evaluation of the causal association between a stressor and effects is particularly important for retrospective risk assessments. For example, at a Superfund site, the problem might be to associate observed ecological effects with gradients of contaminants present at the site. In this case, both field observations and laboratory

toxicity data could be used to help establish a causal relationship. Concepts applied in human epidemiology may be useful for evaluating causality in observational field studies (Hill, 1965); an example of ecological causality analysis was provided by Fox (1991) and Woodman and Cowling (1987).

The results of ecological effects characterization are summarized in a stressor–response profile that describes the stressor–response relationship, any extrapolations and additional analyses conducted, and evidence of causality (e.g., field effects data). Ideally, the stressor–response relationship will relate the magnitude, duration, frequency, and timing of exposure in the study setting to the magnitude of effects. When possible, use of stressor–response curves are preferable to point estimates of an endpoint such as a 96-hr LC_{50} since such values provide no information about the slope or shape of the stressor–response curve. Stressor–response curves allow expression of the change of the magnitude of effect with changing stressor levels and may be especially useful in risk characterization when the exposure profile is similarly expressed as a distribution of values rather than a single point estimate.

As with the exposure profile, the stressor–response profile should clearly describe and quantify (where possible) the assumptions and uncertainties involved in the evaluation. The description and analysis of uncertainty in characterization of ecological effects are combined with uncertainty analyses from the exposure profile and problem formulation during risk characterization.

Risk Characterization Phase

In the risk characterization phase (Fig. R-16), the risk assessor integrates exposure and effects information into an estimate of risk; summarizes supporting information as well as major assumptions, limitations, and uncertainties; and presents the risk assessment results in terms of the assessment endpoints identified in problem formulation. Clear presentation of the strengths, limitations, and conclusions of the risk assessment will greatly enhance the assessment's usefulness in decision making.

Several different approaches may be used to integrate exposure and effects information. The following are examples:

- Field observational studies: Field observational studies (surveys) can serve as risk estimation tech-

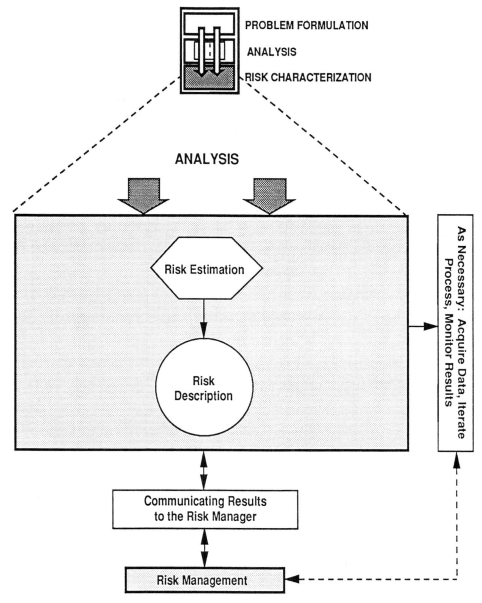

FIGURE R-16   *Risk characterization phase (from U.S. Environmental Protection Agency, proposed guidelines for ecological risk assessment, Fed. Regist. **61**, 47552–47631, 1996).*

niques by providing empirical evidence linking exposure to effects. Field surveys measure biological changes in natural settings through collection of exposure and effects data related to the assessment endpoints identified in problem formulation. A major advantage of field surveys is that they can be used to evaluate multiple stressors and complex ecosystem relationships that cannot be replicated in the laboratory. Field surveys are

designed to delineate both exposures and effects (including secondary effects) found in natural systems, whereas estimates generated from laboratory studies generally delineate either exposure or effects under controlled or prescribed conditions.

• Categories and rankings: When exposure and effects data are limited or are not easily expressed in quantitative terms, professional judgment or other qualitative evaluation techniques may be used to rank risks using categories such as low, medium, and high or yes and no. For example, the U.S. Forest Service assessed the risk of pest introduction from the importation of logs from Chile using qualitative categories owing to limitations in both the exposure and the effects data for the introduced species of concern as well as the resources available for the assessment (USDA, 1993).

• Single-point exposure and effects comparisons: When sufficient data are available to quantify exposure and effects estimates, the simplest approach for comparing the estimates is a ratio. Typically, the ratio (or quotient) is expressed as an exposure concentration divided by an effects concentration. Risk is presumed if the resulting quotient is >1; low risk is indicated by a quotient below 1. Quotients are commonly used for chemical stressors, where reference or benchmark toxicity values are widely available. Simple and quick to use, quotients provide an efficient way to identify high- and low-risk situations but are less useful for evaluating incremental risk (e.g., a quotient value of 4 does not necessarily indicate twice as much risk as a quotient value of 2).

• Comparisons incorporating the entire stressor–response relationship: If a curve relating the stressor level (e.g., chemical concentration) to the magnitude of response (e.g., percentage mortality) is available, then risks associated with many different levels of exposure can be evaluated. The slope of the effects curve shows the magnitude of change in effects associated with incremental changes in exposure, and the capability to predict changes in the magnitude and likelihood of effects for different exposure scenarios can be used to compare different risk management options. Uncertainty can be incorporated by calculating confidence bounds on the stressor–response or exposure estimates. Comparing exposure and stressor–response curves provides a predictive ability lacking in the quotient method.

• Comparisons incorporating variability in exposure and/or effects: When data are available on variability in exposure or effects, many different risk estimates can be calculated. Variability in exposure can be used to estimate risks to moderately or highly exposed members of a population being investigated, while variability in effects can be used to estimate risks to average or sensitive population members. A major advantage of this approach is its ability to predict changes in the magnitude and likelihood of effects for different exposure scenarios and thus provide a means for comparing different risk management options.

• Application of process models: Process models are mathematical expressions that represent our understanding of the mechanistic operation of a system under evaluation. One example is a model that assesses risks to bottomland forest wetlands associated with long-term changes in hydrologic conditions (Brody *et al.*, 1989). This model linked the attributes and placement of levees and corresponding water level measurements (exposure) with changes in forest community structure and wildlife habitat suitability (effects). A major advantage of using process models estimating risk is the ability to consider "what if" scenarios and to forecast beyond the limits of observed data that constrain techniques based solely on empirical data. The process model can also consider secondary or indirect ecological effects, unlike other risk estimation techniques such as the quotient method or comparisons of exposure and effect distributions. However, since process models are only as good as the assumptions on which they are based, they should be treated as hypothetical representations of reality until appropriately tested with empirical data.

In addition to these risk estimation techniques, the risk assessor should consider other available lines of evidence in compiling an overall estimate of risk. Using several lines of evidence and comparing risk estimates may increase confidence in conclusions of a risk assessment. When evaluating lines of evidence, the risk assessor may consider the adequacy and quality of data, the degree and type of uncertainty associated with the evidence, and the relationship of the evidence to the focus of the risk assessment. In conveying this information to the risk manager, the risk assessor should also discuss the underlying uncertainties and the significance of any adverse effects on the assessment endpoints.

Uncertainties originating in both the problem formulation and analysis phases should be described. During problem formulation, conceptual model development may account for one of the most important sources of uncertainty in a risk assessment. If important relationships are missed or specified incorrectly, the risk characterization may misrepresent actual risks. Uncertainty arises from lack of knowledge about how the ecosystem functions, failure to identify and interrelate temporal and spatial parameters, omission of stressors, or overlooking secondary effects. When little is known, a useful approach is to estimate exposure and effects based on alternative sets of assumptions (scenarios). Each scenario is carried through to risk characterization, where the underlying assumptions and the scenario's plausibility are discussed.

Some common sources of uncertainty encountered during the analysis phase of ecological risk assessment are described (U.S. EPA, 1996):

- Data gaps: Nearly every assessment must treat situations in which data are unavailable or available only for parameters other than those of interest. Examples include using laboratory data to estimate a wild animal's response to a stressor or using a bioaccumulation measurement from a different ecosystem. These data gaps are usually bridged with a combination of scientific analyses, scientific judgment, and perhaps policy decisions. Large data or knowledge gaps should be noted and discussed in risk characterization.

- Uncertainty about a quantity's true value: Uncertainty about a quantity's true value may include uncertainty about its magnitude, location, or time of occurrence. This uncertainty can usually be reduced by taking additional measurements.

- Variability: Variability describes a characteristic's true heterogeneity, such as variability in soil organic carbon, seasonal differences in animal diets, or differences in chemical sensitivity in different species. While variability can be described, heterogeneity may not reflect a lack of knowledge and cannot usually be reduced by further measurement.

- Model structure uncertainty (process models) and uncertainty about a model's form (empirical models): Uncertainty in process or empirical models can be quantitatively evaluated by comparing model results to measurements taken in the system of interest or by comparing the results of different models. Process model descriptions should include assumptions, simplifications, and aggregations of variables. Results of comparisons between model predictions and data collected in the system of interest may be useful in evaluating process model uncertainty. Empirical model descriptions should include the rationale for selection and model performance statistics (e.g., goodness of fit).

After potential effects on the assessment endpoints have been determined and uncertainties evaluated, the next step is to interpret whether these effects are considered adverse. Adverse ecological effects, in this context, represent changes that are undesirable because they alter valued structural or functional attributes of the ecological entities under consideration. The following are criteria for evaluating adverse changes in assessment endpoints that may be used in an ecological risk assessment:

- Nature of effects and intensity of effects: Consideration of these criteria is important to distinguish adverse ecological changes from those within the normal pattern of ecosystem variability or those resulting in little or no significant alteration of biota. Natural ecosystem variation can make it very difficult to observe (detect) stressor-related perturbations. For example, natural fluctuations in marine fish populations are often large, with intra- and interannual variability in population levels covering several orders of magnitude. Thus, a lack of statistically significant effects in a field study does not automatically mean that adverse ecological effects are absent. Rather, risk assessors should then consider other lines of evidence in reaching their conclusions.

- Spatial and temporal scale: While effects occurring over large spatial scales or extending over long periods of time can be highly adverse, adverse effects do not always have to occur on a large scale. For example, loss of a small wetland area may be highly adverse if it represents the only habitat available in an area for migratory waterfowl.

• Potential for recovery: Recovery is the rate and extent of return of a population or community to some aspect of its condition prior to a stressor's introduction. Risk assessors should consider the potential irreversibility of significant structural or functional changes in ecosystems or ecosystem components when evaluating ecological adversity.

Evaluating the degree of adversity is often a difficult task and is frequently based on the risk assessor's professional judgment. After the risk assessor discusses the risk assessment results with the risk manager, other factors, such as the economic, legal, or social consequences of ecological damage, are considered. The risk manager will use all of this information to determine whether a particular adverse effect is acceptable and may also find it useful when communicating the risk to interested parties.

Presentation of the risk assessment results to the risk manager is a final, critical stage of the risk assessment process. The risk manager needs to clearly understand the risk predictions in terms of the originally identified assessment endpoints; the major uncertainties, limitations, and assumptions of the assessment; and the degree of adversity of the anticipated risks as discussed previously. A presentation that provides risk estimates associated with a range of stressor levels can be especially useful for a manager who may need to choose among risk management options that have varying capabilities for stressor reduction.

Overall, ecological risk assessment can be a useful, flexible tool for organizing and analyzing exposure and effects data over a wide range of spatial, temporal, and ecological scales. As such, it can be invaluable for risk managers who must make decisions using available information in the face of considerable uncertainties. The usefulness of the risk assessment is enhanced when the risk assessor and risk manager have a common understanding of the scope, goals, and assessment endpoints at the beginning of the process and a clear view of the findings when the assessment is completed.

Further Reading

Brody, M., Conner, W., Pearlstine, L., and Kitchens, W. (1989). Modeling bottomland forest, and wildlife habitat changes in Louisiana's Atchafalaya Basin. In *Freshwater Wetlands and Wildlife* (R. R. Sharitz and J. W. Gibbons, Eds.), U.S. Department of Energy Symposium Series, No. 61, CONF-8603100. Office of Science and Technical Information, U.S. Department of Energy, Oak Ridge, TN.

Fox, G. A. (1991). Practical causal inference for ecoepidemiologists. *J. Toxicol. Environ. Health* **33**, 359–373.

Hill, A. B. (1965). The environment and disease: Association or causation? *Proc. R. Soc. Med.* **58**, 295–300.

Holling, C. S. (Ed.) (1978). *Adaptive Environmental Assessment and Management.* Wiley, Chichester, UK.

National Research Council (1983). *Risk Assessment in the Federal Government: Managing the Process.* National Academy Press, Washington, DC.

National Research Council (1993). *Issues in Risk Assessment.* National Academy Press, Washington, DC.

Oreskes, N., Shrader-Frechette, K., and Belitz, K. (1994). Verification, validation, and confirmation of numerical models in the earth sciences. *Science* **263**, 641–646.

Orr, R. L., Cohen, S. D., and Griffin, R. L. (1993). *Generic Non-Indigenous Pest Risk Assessment Process,* Draft Report. Animal and Plant Health Inspection Service, U.S. Department of Agriculture, Hyattsville, MD.

Reice, S. R. (1994). Nonequilibrium determinants of biological community structure. *Am. Sci.* **82**, 424–435.

Suter, G. W., II (1989, March). Ecological endpoints. In *U.S. EPA. Ecological Assessments of Hazardous Waste Sites: A Field and Laboratory Reference Document* (W. Warren-Hicks, B. R. Parkhurst, and S. S. Baker, Jr., Eds.), EPA 600/3–89/013. U.S. EPA, Washington, DC.

Suter, G. W., II (1990). Endpoints for regional ecological risk assessments. *Environ. Manage.* **14**(1), 19–23.

Suter, G. W., II (1993). *Ecological Risk Assessment.* Lewis, Chelsea, MI.

U.S. Department of Agriculture (1993). *Pest Risk Assessment of the Importation of Pinus radiata, Nothofagus dombeyi, and Laurelia philippiana Logs from Chile,* Forest Service Publication No. 1517.

U.S. Environmental Protection Agency (1992a). *Framework for Ecological Risk Assessment,* EPA/630/R-92/001, Risk Assessment Forum, Washington, DC.

U.S. Environmental Protection Agency (1992b). *Ecological Risk Assessment Guidelines Strategic Planning Workshop,* EPA/630/R-92/002, Risk Assessment Forum, Washington, DC.

U.S. Environmental Protection Agency (1996). Proposed guidelines for ecological risk assessment. *Fed. Regist.* **61**, 47552–47631.

Woodman, J. N, and Cowling, E. B. (1987). Airborne chemicals and forest health. *Environ. Sci. Technol.* **21**(2), 120–126.

—*William H. van der Schalie and Susan B. Norton*

Related Topics

Bioaccumulation
Bioconcentration
Biomagnification
Biomarkers, Environmental
Ecological Toxicology
Effluent Biomonitoring
Environmental Processes
Environmental Toxicology
Hazard Identification
Pollution, Air
Pollution, Soil
Pollution, Water
Population Density
Risk Assessment, Human Health
Risk Characterization
Risk Communication
Risk Management
Sensitivity Analysis
Toxicity Testing, Aquatic
Uncertainty Analysis

Risk Assessment, Human Health

Risk assessment (risk analysis) is a process that can be used to qualitatively and/or quantitatively evaluate the potential for an event or events to occur. The process is useful for characterizing risks associated with a broad range of events that can affect humans and the environment. Events may include risk of adverse health effects occurring after chemical or radiation exposure; risk of injury and/or death during air, highway, or rail travel; and risk of certain catastrophic events occurring, such as nuclear accident, industrial accident, or earthquake. When used to evaluate the kind and degree of hazard posed by agents, the assessment includes consideration of the extent that a particular group may be exposed and the potential for the agent to cause harm. Used as a predictive tool, the risk assessment process

generates information that can support risk management and decision making. Typically, risk management is regarded as distinct from the risk assessment process, and decision making is based on an integration of the results of the risk assessment and other considerations. Other considerations may include engineering data; potential social, economic, and political impact; general feasibility; and cost–benefit analysis.

Evaluating the risks associated with chemical exposure can be complex. Almost everything in our environment (e.g., soil, water, air, food, and drink) is composed of chemicals, and all chemicals have the potential to cause harm under certain conditions. The likelihood of harmful effects occurring due to the presence of a particular chemical is dependent on a chemical's inherent toxicity. Poisons are generally considered to be chemicals that cause harm at very low doses and usually have effects that are seen very quickly.

Humans evolved and thrived in harmony with the chemicals commonly present in the environment. However, there are clearly hazards associated with the naturally occurring agents (e.g., asbestos, lead, arsenic, cyanide, and radioactive species), and there are instances in which the natural environment can pose a significant hazard. Concerns commonly addressed by the risk assessment process, however, are associated with exposures due to human activities. People contact environmental media (air, soil, and water) or ingest food (e.g., fruits, vegetables, and meats) and drinking water that contain chemical residues, may use prescription and over-the-counter drugs, use consumer products, and may be exposed to a variety of chemicals in the work environment. In each instance, the risk assessment process can be used to get a clearer understanding of potential risks and identify conditions in which risk may be unacceptable. For example, a persistent pesticide may remain as a residue in the fruits and vegetables it was applied to during farming. An estimated dose of the pesticide can be made based on the amount of each type of fruit and vegetable typically eaten and the amount of residue likely to remain on each. Typically, criteria is expressed as a daily dose that occurs over a period of time (e.g., lifetime) or a media concentration that developed assuming a time frame over which exposure may occur. Toxicity data can be used to develop a dose that is expected to be safe for people (criteria) and then compared to the dose estimate. If the estimated dose is below the criteria, it does not trigger concern. If it exceeds the criteria, then further evaluation and/or action is appropriate. If the chemical of concern is present in a media in which there is a potential for inhalation, ingestion, and absorption through the skin, the potential dose needs to be estimated based on the sum of the chemical likely to enter the body through all potential routes of exposure.

There is no simple way to compile a list of chemicals that would include everything that is hazardous or likely to carry unacceptable risks under all conditions. People may come in contact with environmental media (air, soil, and water) or ingest food and drinking water that contains chemical residues (e.g., pesticides) and contaminants. In addition, people may use prescription or over-the-counter drugs, use consumer products, and be exposed to a variety of hazardous chemicals in the workplace environment. Typically, a regulatory mandate will dictate in what circumstances and which chemicals need to be evaluated and controlled. For example, at the federal level U.S. Environmental Protection Agency (U.S. EPA) administers the Safe Drinking Water Act (SDWA), which provides for limits on certain chemical contaminants in drinking water, the Clean Air Act, which regulates chemical emissions into the ambient air, and the Resource Conservation and Recovery Act, which regulates hazardous waste handling, storage, and disposal. The Occupational Safety and Health Administration (OSHA) administers the Occupational Safety and Health Act (OSHAct) and regulates human exposure to chemicals in the work environment.

The risk assessment process is commonly used by both the regulating and regulated community to predict the consequences of human chemical exposure in those circumstances in which it has been deemed likely that risks have the potential to be unacceptable. Typically, the process integrates science, science policy, and specific methodologies as defined under policy or regulatory mandate to identify and characterize risks. Approaches to evaluating and managing risk vary with regulatory mandate as well as with circumstances that are being evaluated.

National Academy of Science's Paradigm

The risk assessment process, as used to evaluate the impact of chemicals on human health, was formally defined in the National Academy of Science (NAS)

1983 report, *Risk Assessment in the Federal Government: Managing the Process* (the "Red Book"). [The study on which the report was written was carried out by a committee of the National Research Council (NRC) Commission on Life Sciences with support from the Food and Drug Administration. The NRC is a principal operating agency of the NAS.] The report was developed to provide the federal government with a systematic approach for evaluating risks to human health associated with chemical exposure. The NAS framework was designed to strengthen the reliability and objectivity of the scientific basis of risk assessment as well as ensure that the best scientific data were integrated into the process and to ensure that there was consistency in the approach used by federal agencies. It was intended to minimize controversy and allow for more consistent and rational decision making with regard to human health. The process was defined in broad and general terms and has also been applied to evaluation of risks associated with other organisms (e.g., wildlife) and the environment. The NAS divided the process into four major steps: hazard identification, dose–response assessment, exposure assessment, and risk characterization.

- Hazard identification is the step in the risk assessment that qualitatively characterizes the inherent toxicity of a chemical.

- Scientific data are evaluated to establish a possible causal relationship between the occurrence of adverse health effects and exposure to a chemical. This step includes characterization of acute, subchronic, and chronic effects; the potential for local versus systemic effects; the influence of the route of exposure.

- The relevance, to humans, of effects seen in animals; an evaluation of the biological importance of the observed effects.

- The likelihood of the effects occurring under certain conditions; and the potential implications for public health.

- This step should be based on a thorough review of all the data that may provide information that is relevant to evaluating the potential chemical hazard. This may include data describing the effects on a variety of test animals, *in vitro* studies that characterize mechanisms of toxicity, metabolism, physiologically based pharmacokenetic studies, structure–activity relationships, short-term human studies, and epidemiological studies. Animal studies may focus on particular types of effects and may include reproductive studies, immunotoxicity studies, neurotoxicity studies, genotoxicity studies, and cancer bioassays. Each study must be evaluated with respect to quality, design, interpretation of the data, and statistical considerations to ensure that conclusions are valid before integration into the assessment. This step integrates information that is used in the dose response.

• Dose response assessment characterizes the quantitative relationship between exposure (usually determined in toxicity studies) and the occurrence of adverse health effects. Typically applied or administered dose, rather than effective tissue dose, is used to develop the dose–response relationship. As a rule, the higher the dose, the greater the frequency or intensity of the adverse reaction to a chemical. Often, different effects are observed at high dose and low dose. The approaches used to extrapolate the dose–response relationship from high experimental doses administered to relatively few animals used in laboratory animal studies to relatively low-dose human exposure anticipated to occur in the environment (e.g., via ambient media) vary and are critical to assessing potential risks. Commonly, the lowest dose or doses at which no adverse effects are identified (or, if not available, the lowest dose at which adverse effects are observed) and are used as the basis for setting what are anticipated to be reasonably safe exposure levels (doses associated with acceptable risk) under certain conditions. Exposure criteria (doses at which the risk of adverse effects occurring are determined to be acceptable) are developed and provided as standards or guidelines. A number of agencies develop criteria that are useful in different circumstances. These include OSHA (permissible exposure levels), American Congress of Governmental Industrial Hygienists (Threshold Limit Values), Agency for Toxic Substances and Disease Registry (minimal risk levels), U.S. EPA (maximum contaminant levels),

reference doses (RfDs), reference concentrations (RfCs), Health Canada, and the World Health Organization's International Programme on Chemical Safety. Criteria are often expressed in terms of dose (e.g., milligrams of chemical per kilogram of body weight of the animal per day) or as a media concentration (e.g., milligrams of chemical per kilogram of soil, liters of water, or cubic meters of air).

• Exposure assessment qualitatively and quantitatively characterizes the potential for exposure to occur in particular circumstances and includes an estimate of dose when possible. The assessment includes an estimation or measurement of chemical concentration in the contact media (e.g., soil, water, and air), an estimation of the length of time over which contact will occur, characterization of potential routes of exposure (inhalation, ingestion, and skin contact), and the likelihood for a chemical to be absorbed through those routes. In certain circumstances, direct measurements or fate and transport modeling may be used to estimate chemical concentrations in ambient media. For certain assessments, a quantitative estimate of the total dose of a chemical over a particular time frame and in the given circumstances is made.

• Risk characterization provides for both a qualitative and a quantitative description of risk. The step involves integrating the results of the hazard identification, dose–response assessment, and exposure assessment to characterize risk. Often a direct comparison between exposure criteria developed in the first two steps and the results of the exposure assessment (concentration in the environmental media or the estimated dose, as appropriate) provide a basis for determining whether risks are acceptable. Typically, if criteria are exceeded, the risk is not acceptable. What is defined as acceptable, as well as the way risk is expressed, is often a function of the agency, law, and/or regulation that drives the analysis. Risk may be expressed as excess cancer risk, hazard index, or in terms of a margin of safety. The risk characterization must incorporate considerations of the uncertainties in the assessment. Each step of the risk assessment process contributes uncertainty and uncertainties must be clearly defined and integrated into the conclusions of the assessment.

Risk management was defined in the NAS report as the process of weighing policy alternatives and selecting the most appropriate regulatory actions. It is considered to be separate from the risk assessment process. Risk management decisions are based on the results of the risk assessment and other concerns that are relevant to the situation.

Human chemical exposure is regulated under a number of different laws in the United States, including SDWA, OSHAct, CERCLA, Federal Insecticide, Fungicide, and Rodenticide Act, Toxic Substance Control Act, Hazardous Substance Act (HSA), Clean Air Act, and many state laws. These laws are administered by different agencies (e.g., U.S. FDA, U.S. EPA, OSHA, Consumer Product Safety Commission, and U.S. Department of Agriculture). The approaches and problems associated with estimating exposure and risk, defining acceptable risk, and developing exposure criteria differ both under different regulatory mandates and the circumstances in which exposure is likely to occur. The risk assessment process as described in the NAS paradigm identifies the critical information needed to evaluate risk in the broad sense; however, the resources expended to obtain the information defined in each step may vary in different circumstances. For example, for a new drug, a dose that provides therapeutic benefits and has an acceptably low risk of significant side effects must be identified prior to acceptance for use on the general public. The hazard identification and dose–response assessment steps are likely to be resource intensive. Drugs must be tested in a series of animal tests and clinical trials. The risks of side effects occurring (severity and likelihood of occurrence) have to be well characterized and balanced against the benefit derived from use of the drug. The exposure assessment portion of a risk assessment conducted to evaluate risks associated with environmental contamination regulated under certain state or federal laws (e.g., CERCLA) may be much more resource intensive than the hazard identification and dose–response assessments. The nature and extent of contamination in each environmental media (e.g., soil, groundwater, and water) and the likelihood of human exposure has to be evaluated. Characterization of chemical concentrations in environmental

media may require extensive environmental sampling and complex fate and transport modeling.

Risk Assessment at Sites of Environmental Concern

U.S. EPA's Approach to Assessing Human Health Risks Associated with Environmental Contamination under Superfund

Chemicals may be introduced into the environment (e.g., air, soil, surface water, and groundwater) during ongoing industrial and commercial activities, improper chemical handling, accidental spills and releases, as well as the routine disposal generation of chemical wastes by communities (e.g., household cleaners and solvents).

The risk assessment process has been used to identify when resulting conditions present unacceptable risks. CERCLA (Public L. No. 96-510, 40 CFR 300) was enacted on December 11, 1980, and gave the federal government the authority to act when hazardous substances were released (or could be released) into the environment in quantities that had the potential to endanger the public health. It provided for a system to clean up hazardous waste (e.g., spills, leaks, and abandoned dumpsites), for immediate reporting of chemical releases over specific amounts (reportable quantities), for fines and penalties, for a trust fund to pay for remediation at contaminated sites, and for recovering costs of remediation when the responsible party or parties can be identified. The Superfund Amendments and Reauthorization Act (SARA) to CERCLA was passed in 1986 and provided for more stringent and permanent remedies at sites regulated under the law.

Under CERCLA (sometimes referred to as the "Superfund Law"), actions taken to remediate environmental contamination must protect both human health and the environment. U.S. EPA administers CERCLA and has used the risk assessment process to characterize potential risks at sites of environmental concern. U.S. EPA has developed guidance for conducting quantitative human health risk assessments at Superfund sites. Documents include *Risk Assessment Guidance for Superfund* (RAGS; 1989), *Human Health Evaluation Manual and Supplemental Guidance,* and *Exposure Factors Handbook* (1990). This guidance was intended for use during the remedial investigation/feasibility study process at Superfund sites but has been widely used to address sites regulated under other environmental laws.

U.S. EPA's approach, as defined in RAGs, incorporates the principles defined in 1983 by NAS, U.S. EPA RAGs identifies four steps in an environmental risk assessment: data collection and evaluation, exposure assessment, toxicity assessment, and risk characterization. Tasks involved in characterizing the environmental media have greater emphasis because they often require tremendous resources and time.

- Data collection and evaluation involves characterization of the concentration of contaminants in the media (e.g., soil, groundwater, and air) at the site in question. It includes the collection of samples to characterize soil and groundwater at contaminated property. This phase of a risk assessment may be complex and require significant resources when dealing with environmental contamination. The data collection and evaluation step provides important data to support conducting the exposure assessment.

- Exposure assessment includes both a qualitative and quantitative evaluation of the potential for exposure to site-related chemicals to occur. Assessments commonly address both current and likely future uses of the property (e.g., residential, commercial, industrial, and agricultural). Typically, a conceptual model is developed that summarizes how site-related chemicals may contact receptors (e.g., humans, wildlife, and ecological). The model includes identification of chemical sources, impacted media, potential movement through the environment, identification of the appropriate exposure scenarios, and identification of the points at which contact between receptors and site-related chemicals are likely to occur. Chemical concentrations in environmental media may be estimated based on site data and using statistical analyses and/or fate and transport modeling. An estimate of the dose (intake) attributable to contact with environmental media through significant and completed pathways is made for chemicals of concern at the site. This estimate is based on an estimation of the amount of time over which contact will occur, characterization

of potential routes of exposure (ingestion, inhalation, and skin contact), and the likelihood for the chemical to be absorbed from the contaminated media through those routes.

- Toxicity assessment includes characterization of the toxicity of a chemical, development of a dose–response relationship, and development of exposure criteria. Exposure criteria within this context are often referred to as toxicity values (e.g., RfDs, RfCs, and slope factors). U.S. EPA has developed criteria for many chemicals commonly associated with environmental contamination. [Verified U.S. EPA criteria are available in the Integrated Risk Information System (IRIS) and provisional values are available in the Health Effects Assessment Summary Tables (HEAST). IRIS can be accessed on-line through the National Library of Medicine (sis.nlm.nih.gov) by registered users or through U.S. EPA (www.epa.gov/ngis pgm3/iris/index.html). HEAST can be purchased through the National Technical Information System in Springfield, Virginia.] Typically, criteria is developed based on toxicity data gathered from the literature. (NTP conducts cancer bioassays to evaluate the potential for selected chemicals to cause cancer. Chemicals are included on the list to be evaluated based on a number of different criteria.) Data-bases on individual chemicals may be large or limited. Criteria is not provided by U.S. EPA for chemicals lacking sufficient toxicity data. The toxicity values and criteria developed in the toxicity assessment combine the approaches and procedures described in the hazard identification and dose–response steps described in the NAS paradigms. Because criteria are typically provided by regulatory agencies, for chemicals commonly identified as a concern the effort associated with this step for a given site may be limited. However, if conditions on the site appear to be unacceptable, these may be refined as part of the process.

- Risk characterization includes a comparison between toxicity values (exposure criteria) and exposure (dose or media concentration) to determine whether the exposure is acceptable. U.S. EPA developed a formalized system that is commonly used to determine whether chemicals are likely to present an unacceptable risk based on current and likely future use of the property. The estimated dose is used to calculate an additional lifetime cancer risk for each chemical regulated as a carcinogen. Typically, a total site risk (sum of the risk associated with all carcinogens identified at the site) is presented. Acceptable risk is defined by the agency, in the appropriate laws, or by regulations that govern the site. Acceptable risk is a function of policy or law but is supposed to be rooted in science. The likelihood of noncarcinogenic effects occurring is also evaluated. Total dose is compared to a dose that is considered likely to be safe (exposure criteria; e.g., RfD). The site-related dose is divided by the criteria and the resulting fraction is defined as the hazard quotient. This simply indicates whether the hypothetical dose exceeds the threshold criteria identified as likely to be safe. The hazard quotients for all chemicals at the site are summed and presented as a hazard index for the site. If the total does not exceed 1, it is assumed that the dose attributed to the site will not exceed the criteria and the potential exposure is deemed acceptable. If the hazard index for a site exceeds 1 (hypothetical dose exceeds criteria), a more refined analysis can be completed. Chemicals can be grouped according to target organ and a refined set of indexes can be developed for the site. The sum of the fractions for each target should not exceed 1. Uncertainties associated with each step in the process must be clearly defined and integrated into the conclusions of the assessment.

U.S. EPA's approach is essentially an application of NAS's, tailored to provide guidance for assessing risk associated with contamination in environmental media. Adverse health impacts associated with exposure to the chemical of concern are identified through the hazard identification, dose response, and toxicity assessment. Exposure is evaluated during the exposure assessment and data collection and evaluation steps. Risk is characterized and can then be more efficiently managed (Table R-13).

Risk management is the decision-making process that follows the completion of a risk assessment. The risk assessment provides important information that supports decision making and is integrated with other

TABLE R-13
Comparison of NAS Paradigm and U.S. EPA Superfund Approach

NAS	U.S. EPA
Hazard identification dose response	Toxicity assessment
Exposure assessment	Data collection and evaluation exposure assessment
Risk characterization	Risk characterization

factors, including economic, feasibility, and cost–benefit analysis, in the risk management process.

Risk-Based Corrective Action

The risk assessment process has been used for years to characterize human health risks at sites of environmental concern as well as to define media-specific risk-based criteria (e.g., acceptable concentrations of chemicals in water or soil). However, risk assessment is playing a greater role in driving remedial actions as it provides information that allows actions and resources to be focused on areas in which current or future risks to human health and/or the environment are likely to be unacceptable. The American Society for Testing and Materials (ASTM) *Risk-Based Corrective Action (RBCA) at Petroleum Release Sites Standard* (ASTM, 1995) outlines a systematic approach that integrates risk assessment principles and the decision-making process. It integrates many of the methodologies and approaches defined by U.S. EPA under the Superfund program into a protocol for approaching chemicals in the environment. ASTM provides for the identification of conditions that warrant immediate attention and the development of a time frame for action, if warranted. Risks are evaluated based on a "tiered" approach. Media-specific risk-based screening criteria are developed using very conservative assumptions that are intended to provide a wide margin of safety. If chemical concentrations do not exceed screening criteria, then risks are deemed acceptable and no further action is recommended. Should levels exceed criteria, a more in-depth evaluation of conditions on the site is undertaken. If the assumptions and conditions on which the screening criteria are based do not truly characterize conditions on the site, then site-specific criteria are developed. Tier 2 criteria incorporate limited site data and considerations into criteria development. Tier 3 criteria may be based on sophisticated models and methodology and additional site-specific information. Development of tier 2 and 3 criteria may include a more in-depth evaluation of all components of the risk assessment, including consideration of the appropriateness of particular toxicity values, the likelihood of contact with the media of concern, the presence of populations, and site-specific activities (e.g., home gardens, farms, and fishing). Although the standard specifically addresses sites where there has been a petroleum release, it defines a process that is applicable to environmental problems in general. Versions of the approach have been adopted by many states, and certain state laws actually refer to the ASTM standard. ASTM is currently involved in the development of a standard that provides guidance for the application of this type of approach to sites of environmental concern.

Further Reading

Cohrssen, J. J., and Covello, V. T. (1989). *Risk Analysis: A Guide to Principles and Method for Analyzing Health and Environmental Risks*. National Technical Information Service, U.S. Department of Commerce, Washington, DC.

Graham, J. D., and Rhomberg, L. (1996). How risks are identified and assessed. *Ann. AAPSS, 545.*

National Research Council (1983). *Risk Assessment in the Federal Government: Managing the Process.* National Academy Press, Washington, DC.

National Research Council (1994). *Science and Judgment in Risk Assessment.* National Academy Press, Washington, DC.

National Research Council (1996). *Understanding Risk, Informing Decisions in a Democratic Society.* National Academy Press, Washington, DC.

U.S. EPA (1989). *Risk Assessment Guidance for Superfund Volume I, Human Health Evaluation Manual (Part A),* EPA/540/1-89-002. Office of Emergency and Remedial Response, Washington, DC.

—*Betty J. Locey*

Related Topics

Risk Characterization

Risk characterization represents the fourth component of the risk assessment process and represents an integration of the three other components, i.e., hazard identification, dose–response assessment, and exposure assessment. Procedurally, risk characterization, which includes either qualitative or quantitative estimation of potential health risks, is accomplished by using information and estimates of toxicological thresholds and potency identified in the dose–response assessment and estimates of exposure derived in the exposure assessment. Risk characterization must also include, however, consideration of the uncertainties associated with each component of the risk assessment process so that risk estimates can be appropriately qualified.

In the case of noncancer risk estimation, the no-observed-effect level or no-observed-adverse-effect level from relevant human and animal studies are compared to the estimated exposure levels to determine the margin of safety. In the case of lifetime excess cancer risk estimation, the cancer potency factor [often expressed as $(mg/kg/day)^{-1}$] derived from relevant human studies or animal studies, or both is multiplied by the appropriate reciprocal dose metric (exposure or absorbed dose is usually expressed as mg/kg-day) to yield a risk probability (e.g., 1×10^{-6} or 1 lifetime excess cancer case per 1 million exposed individuals).

As noted previously, in addition to including estimation of health risks, risk characterization should include a qualitative or quantitative analysis of the uncertainties associated with each component of the risk assessment (see Uncertainty Analysis). Uncertainties arise due to the use of assumptions, scientific judgment, and inherent biological variability (e.g., inter- and intraspecies variation in dose response).

Characterizing risk has consistently been considered the weakest and most challenging component of the risk assessment process because it requires the assessor to consider the overall scientific weight-of-evidence to properly describe whether a significant health risk exists in a specific setting. A thorough characterization, for example, should discuss background concentrations of a chemical in the environment and in human tissue, pharmacokinetic differences between animal test species and humans, the impact of uncertainties associated with specific variables in the exposure and dose–response assessment (a sensitivity analysis), as well as other factors that can influence the magnitude of the estimated risks (e.g., the relevance to humans of a given toxicological endpoint observed in animal studies).

Finally, it is important to distinguish risk characterization and the overall risk assessment process from risk management. Risk management combines the results of a given risk assessment, which is a scientific endeavor, with the directives of the pertinent enabling regulatory legislation, together with socioeconomic, technical, political, and other considerations, to reach a decision regarding the acceptability of the health risks associated with a given substance.

—*Jeffrey H. Driver*

Related Topics

Risk Communication

Although communication about risk has a long history, risk communication has been formalized as an area of study for only about 10 years. It is an area that draws from a number of disciplines in the natural and social sciences. Research in this field includes studies of human perception and decision making as well as investigations of a socioeconomic nature. Risk communication practitioners include government officials, business and industry representatives, public interest group members, academics, media professionals, and the general public.

Because of its short history, the concept of risk communication is still evolving. In its landmark report in 1989, *Improving Risk Communication,* a National Academy of Sciences panel defined risk communication in part as "An interactive process of exchange of information and opinion among individuals, groups and institutions." However, this definition describes one group's consensus as to what risk communication ought to be and does not reflect either the diversity of opinion on this topic or the reality of risk communication as it is currently practiced.

Reduced to its essentials, risk communication is just one example of general communication, a process by which information is transferred by one party to another through a variety of channels. The components of the process are the risk communicator or source, the message, the message channel, and the recipient of the message. While this description suggests a linear series of events, communication is often, but not always, an interactive process with feedback among the components. For example, the responses of the recipient may lead to alterations in the source, message, and/or channel and may result in a change in outcome.

Based on this conceptual framework, and using a variety of experimental approaches and case studies, risk communication professionals have been able to identify and, in some cases, roughly quantify many of the factors that determine the effectiveness or success of risk communication. One fundamental problem is, however, that effectiveness can be defined in a number of ways.

From one point of view, success can be measured as increased understanding of risk on the part of the message recipient, whether or not this results in changes in behavior. However, from another perspective, effectiveness is determined by the degree of behavioral change brought about by the risk message being transmitted. There are some who have argued that increased understanding is directly linked to changes in behavior but research suggests that this is not necessarily the case; for example, smokers who become aware of the risk do not always quit.

Many recent communication efforts about public health risks reflect the latter view. For instance, risk communication programs about radon are judged by the number of people who test for radon or take remedial action to reduce radon levels. Similarly, campaigns on smoking are assessed by the number of individuals who stop smoking. When behavior change is the criterion of success, the accuracy of the risk communication is not as important as its impact.

However, many other risk communication efforts reflect more of an educational approach than a persuasive one. Understanding is considered the most important goal and a lack of change in behavior is not a failure in communication; rather it is the result of judgments by the information recipient which may involve consideration of other factors, such as offsetting benefits. Accuracy is considered critical so that the recipient can make an informed decision based on the best information.

The factors that affect the success of risk communication efforts, irrespective of the criteria for judgment, can be considered in the context of the components of the communication process. Studies of the risk messenger (source) have focused on the issue of trust and, not surprisingly, reveal that the effectiveness of the communication increases as trust in the communicator increases. This trust may be related to the degree of expertise of the communicator but more often is determined by other factors such as the messenger's perceived objectivity or social class and accuracy of previous communications by this messenger or other messengers representing the same organization. For example, state agencies may have difficulties in effective risk communication for decades after taking actions or making statements seen as untrustworthy.

The content and form of the message are also critical and it has been demonstrated that the same risk expressed or framed in different ways can have different impacts. For example, describing the risk from a disease or surgery as 60% chance of survival will lead some people to different decisions than if it is stated as 40% chance of death. Similarly, a risk presented in terms of the total number of fatalities may be perceived differently than if it is presented as a probability.

Furthermore, there are a number of factors that come into play if the risk is presented in comparison to other risks—a communication strategy that is often employed. Research suggests that audiences are most receptive to comparisons of the same risk at one time or place to another time or place. For example, an effective message may describe the risk as high 2 years ago but having decreased continuously and now less than a tenth of what it was. Risk comparisons that are not as successful involve comparisons between hazards that are thought of as having incomparable qualities (e.g., smoking and water contaminated with bacteria). Smoking is controllable and the effects delayed, while bacterial contamination of water is involuntary and the effects immediate.

The channel of communication is also an important variable in the communication process. Survey researchers have examined the relative credibility of various channels, including print media, radio and television, magazines, and advertising, and have found differences in the degree of trust people have in each. These differences depend not only on the class of channel but also on the specific representative of that class. For example, coverage of a risk issue in a local newspaper may be viewed differently from stories on the same issue in the national press.

Last, the perception of the recipient is critical to the process and has been the aspect of risk communication that has been studied most intensively. This research has clearly shown that this perception depends both on the way that humans tend to internalize knowledge and on the background of the information recipient. For example, a scientist may perceive risk as a specific quantity and compare risks based on a quantitative approach (e.g., a risk of one in a million is much lower than a risk of one in a thousand).

However, potentially affected citizens may look at these quantitative descriptors as only partially describing the risk. They may consider a number of other factors such as the voluntariness, catastrophic potential, familiarity, and controllability of the hazard. Thus, contrary to the scientist, they may consider the one in a million risk to be more serious than the one in a thousand risk. Government officials are often reminded of this at public meetings when people who smoke express great concern about very low levels of environmental contaminants which pose much lower health risks than smoking.

The results of research on the factors that influence successful risk communication are the bases for sets of risk communication principles that have been developed. One well-known set of principles is titled *Seven Cardinal Rules of Risk Communication*. This guide includes steps to increase trust such as to "coordinate and collaborate with other credible sources," steps to increase the recipients' control such as to "accept and involve the public as a legitimate partner," and steps to increase the interactive nature of the process such as to "listen to the public's specific concerns."

In the main, research to date has been focused on understanding individual differences in risk perception and on the interactions between sources and recipients. However, recently, increasing attention has been paid to the social context in which risk communication is performed. Research in this area has led to a greater awareness of the sociocultural factors that affect the transmission of risk information.

These factors include the ethnic and socioeconomic characteristics of populations as well as the structure and history of communities that must deal with risk concerns. For example, a "company" town may react quite differently than a rural agricultural village or an inner-city neighborhood to information about risk from an environmental chemical.

Another important social issue that influences risk communication is environmental justice and the perceived fairness of the risk to a community or subculture. For example, the perceived risk of adverse health effects from a landfill may depend on whether the waste deposited there is generated locally or transported from somewhere else. Similarly, the perception of risks associated with locating new facilities in an area may be colored by environmental equity concerns.

These sociocultural factors can affect all aspects of the communication process. The credibility of a particular source may vary greatly depending on the cultural experience of the recipient with the organization the source represents, e.g., industry or state government. In addition, the way that the message is framed may have quite different impacts in different cultural settings; an effective comparison in one setting may be an ineffective one in another. Communication channels may be seen as more or less reliable in different socioeconomic classes, e.g., certain media may be seen as more trustworthy by different social groups. Furthermore, the relative importance of the factors that influence perception, e.g., controllability and immediacy, may be different in various cultural settings.

Research in this area, while in the formative stage, has been increasing. One aspect of this expanded effort involves studies aimed at understanding why certain communities have responded to risk, e.g. from Super-

fund sites, quite differently than other communities, or why media focus on some risks has led to increased awareness and concern while attention to other risks has been met with little reaction. It is hoped that research into the reasons behind these differences will help in identifying the roles of various sociocultural factors in risk communication.

It is clear that our current understanding of the risk communication process leaves many important questions unanswered. For example, how do age and gender affect the way risks are perceived? Other questions that are yet to be answered include exactly how the framing of the message affects its impact; how uncertainty in the risk message affects the perception of the risk and risk communicator; and why some risks that appear to share the same characteristics are perceived differently by individuals and/or communities.

Future research will address these questions and our understanding of the influence of various factors on the risk communication process will undoubtedly increase. In addition, innovative ways to present risk messages will certainly be developed. However, some fundamental issues with respect to the goals of risk communication will likely remain.

Further Reading

Freudenburg, W. R. (1988). Perceived risk, real risk: Social science and the art of probabilistic risk assessment. *Science* **242**, 208–213.

Krimsky, S., and Plough, A. (1988). *Environmental Hazards: Communicating Risks as Social Process.* Auburn House, Dover, MA.

National Research Council (1989). *Improving Risk Communication.* National Academy Press, Washington, DC.

Slovic, P. (1987). Perception of risk. *Science* **236**, 280–285.

U.S. Environmental Protection Agency (1988). *Seven Cardinal Rules of Risk Communication,* OPA-87-020. U.S. EPA, Washington, DC.

                                                                            Michael A. Kamrin

Related Topics

Risk Assessment, Ecological

Risk Assessment, Human Health

Risk Characterization

Risk Management

Risk Management

R isk management, in simplified terms, is the decision-making process that follows the assembling and consideration of technical risk assessment data. This definition covers many levels of decision making, however. At one level, it deals with the question of what programs should be undertaken to reduce risk to the population of the country. At another level, it is the allocation of tax dollars to improve the quality of life in a city. At yet another level, it is the decision-making process for mitigating risk from a specific chemical in a specific use.

This latter level of risk management is discussed in some detail in the 1983 National Academy of Sciences (NAS) report on risk assessment (the so-called Red Book). In that report, NAS defined four steps in the risk assessment process: (1) hazard assessment—does the chemical have the intrinsic potential to cause, for example, cancer?; (2) dose–response assessment—what is the risk at low doses in humans (generally estimated from high-dose studies in laboratory rodents)?; (3) exposure assessment—what is the exposure?; and (4) risk management—how do we mitigate this risk?

In this paradigm, the NAS study emphasized the need to separate the technical components of the risk assessment process from the risk management process. In practice, what this translates to is the need to separate the scientific components of the process from the societal decisions that have to be made. To have a credible assessment process, for example, it is necessary not to bury undue conservativeness in the technical component under the guise of public health protection but to deal with the issue of conservativeness or prudence in the societal decision-making process.

Risk management decisions are based on the technical risk assessment information, but they also incorporate other information. Specifically, risk managers consider costs, benefits, risk of alternative actions, and risk of no actions. For example, some drugs have a very small safety margin between the level that is effective and the level that is toxic, but the benefits of the drug are judged to outweigh the risks; therefore, it is decided that these drugs can be used. At the other end of the spectrum, some food-coloring additives have very large safety margins between what is toxic and what is in the diet, but the benefits of food coloring are minimal and alternatives exist, so this risk would generally face a higher level of scrutiny in a risk management decision.

These examples emphasize that risk management decisions are based in part on level of risk, but other factors are also considered to most effectively protect the public. In making risk management decisions, it is also necessary to understand the level of uncertainty in the technical data. For example, there is considerable uncertainty in predictions of potential impacts of global warming, but the consequences are potentially very large. It is often the case in risk assessment that there is considerable uncertainty in the predictions of risk. These uncertainties need to be carried forward into the hands of the risk manager so that they are weighed in the final decision. The necessary level of public protection and the magnitude of margins of safety properly belong in the domain of risk management.

At a more global level, risk management refers to the activity of allocating resources (time and money) to reduce risk in general to the public. For example, the U.S. Congress does this when they allocate money to U.S. federal agencies. The trade-off in agency funding addresses the question of whether more resources should be allocated to improve air and water quality (U.S. EPA), ensure that our food is more strictly inspected and regulated (U.S. FDA and USDA), improve highway safety (NHSTA), improve air safety (FTA), or bolster national defense systems (armed forces). We do not have unlimited resources to expend, so Congress makes these allocations roughly proportional to where it thinks the money is most effectively spent. These allocations reflect risk management on a global level.

The level of risk management closest to what most individuals experience is risk reduction at a city management level. Cities are responsible for ensuring that their drinking water is tested and is safe relative to bacteria, disease, and chemicals. They are also responsible for disposing of garbage in either landfills or incinerators, for keeping beaches and swimming places safe, and for minimizing vehicle risk to pedestrians and vehicle occupants.

A number of cities have a significant problem with keeping their drinking water and beaches clean because of storm water overflowing into sanitary water systems. Risk management at the local level involves deciding whether solving this storm water issue should be the highest priority for the next tax dollars being spent or whether road repair or expanding the capabilities of the local ambulance and hospital system should be highest priority.

—*Colin Park*

Related Topics

Risk Assessment, Ecological
Risk Assessment, Human Health
Risk Characterization
Risk Communication
Sensitivity Analysis
Uncertainty Analysis

Rotenone

♦ CAS: 83-79-4

♦ SYNONYMS: Barbasco; Mexide; Ro-Ko; Fish-Tox; [1]Benzopyrano[3,4-*b*]furo[2,3-*h*][1]benzopyran-6(6a*H*)-one, 1,2,12,12a-tetrahydro-2-α-isopropenyl-8,9-dimethoxy

♦ DESCRIPTION: Rotenone is a naturally occurring alkaloid (rotenoid) extracted from the roots, leaves, seeds, and barks of certain plants (e.g., *Derris elliptica*).

♦ CHEMICAL STRUCTURE:

Uses

Rotenone has been used for centuries as a fish poison. Rotenone is used as an insecticide on livestock, vegetables, fruits, and forage crops. Rotenone is also used in formulations for scabies, chiggers, fleas, ticks, lice, and mange in dogs.

Exposure Pathways

Dermal and ocular exposures are most common, but rotenone may also be ingested or inhaled.

Toxicokinetics

Parenteral exposure is more hazardous than oral exposure; gastrointestinal absorption is slow and incomplete. Fats and oils increase rotenone absorption from the gastrointestinal tract. Rotenone exhibits a significant first-pass effect following oral exposure. Biotransformation of rotenone in rats leads to hydroxylated metabolites (rotenolones). In addition, O-demethylation inactivates rotenone. Rotenone distributes to lipid-rich tissues, including the nervous system. Elimination of rotenone from the body is primarily through the fecal route.

Mechanism of Toxicity

Rotenone inhibits the electron transport chain by blocking transport between the flavoprotein and ubiquinone. The oxidation of pyruvate in rat mitochondria is virtually completely blocked by rotenone *in vitro* (<1 μM concentration).

Human Toxicity: Acute

Ocular exposure to rotenone dusts can cause severe irritation. Inhalation exposure can cause irritation of the nose and throat, and a temporary anesthetic effect may occur. Significant exposures may cause nausea, vomiting, cramps, muscle tremors, loss of coordination, dyspnea, and seizures.

Human Toxicity: Chronic

Long-term exposure to rotenone may cause fatty liver and kidney damage.

The ACGIH TLV-TWA for rotenone is 5 mg/m^3; the suggested no-adverse-response level for chronic exposure is 0.014 mg/liter. The estimated LD_{Lo} (oral) for humans is 143 mg/kg.

Clinical Management

Respiratory and cardiovascular functions should be supported with oxygen, assisted ventilation, and parenteral fluids. If eyes or skin are contaminated, they should be washed immediately. Gastrointestinal decontamination procedures should be used appropriately depending on the patient's level of consciousness and the amount of rotenone ingested. Oils or fats should not be administered because they can promote rotenone absorption. Activated charcoal should be used to block absorption. In animals, 10 mg of menadione (intravenously) reversed rotenone's blocking of mitochondrial oxidative phosphorylation; however, it is not known if this has been tried in humans.

Animal Toxicity

Rotenone has high mammalian hazard potential. Depression of the respiratory center appears to be the primary cause of death. The acute oral LD_{50} values in laboratory rodents range from about 10 to 130 mg/kg. Death following high oral doses in animals can occur within 2 days or as long as 2 weeks after exposure. Intraperitoneal LD_{50} values are considerably lower (about 2 mg/kg). Rabbits appear markedly less sensitive than rodents to oral rotenone exposures ($LD_{50} \gtrsim$ 1 g/kg).

—Janice Reeves and Carey Pope

Saccharin

- CAS: 81-07-2
- PREFERRED NAME: Saccharin (USAN, CTFA)
- SYNONYMS: 1,2-Benzisolthiazol-3(2H)-oal 1,1-dioxide; benzosufamide; glucid; garantose
- CHEMICAL CLASS: Organic acid
- MOLECULAR FORMULA: $C_7H_5NO_3S$
- CHEMICAL STRUCTURE:

Uses

Saccharin is used as a nonnutritive sweetener (500 times sweeter than sucrose). It is also used as a flavor aid in pharmaceuticals.

Exposure Pathway

Oral ingestion is the route of exposure.

Toxicokinetics

Saccharin is excreted primarily unchanged (85–92%) in urine. It has rapid renal clearance via tubular secretion without glomerular filtration. The terminal half-life is 30 min.

Mechanism of Toxicity

Saccharin is a promoting agent in the bladder carcinoma. Slow absorption of high doses may alter metabolism of essential nutrients in the gastrointestinal tract, contributing to changes in biochemistry.

Human Toxicity

Saccharin is a possible carcinogen in humans (but not so listed by IARC and in 1997 considered for delisting as such by NTP). A large daily dose may cause gastric hyperacidity.

Clinical Management

Saccharin is not dangerous unless it is taken in extreme doses. Overingestion should be treated as a nontoxic ingestion.

Animal Toxicity

Saccharin is a weak bladder carcinogen when administered orally in rats and mice. Chronic overingestion alters the physiology and biochemistry of male rats. It is weakly mutagenic in salmonella. Saccharin increases the incidence of chromosome aberrations in Chinese hamster ovary cells.

—*Shayne C. Gad and Jayne E. Ash*

Related Topic

Carcinogenesis

Safe Drinking Water Act

♦ Title: SDWA
♦ Agency: U.S. EPA
♦ Year Passed: 1974; amended 1986
♦ Groups Regulated: Water suppliers

Synopsis of Law

The 1974 Safe Drinking Water Act (SDWA) was enacted to ensure that public water supply systems "meet minimum national standards for the protection of public health." Under the SDWA, U.S. EPA is required to regulate any contaminants "which may have an adverse effect on human health." U.S. EPA was to establish national primary drinking water regulations for public water systems. For each contaminant of concern, the agency was to prescribe a maximum contaminant level (MCL) or a treatment technique for its control.

The 1974 act prescribed a two-stage process. The EPA first was required to promulgate interim national primary drinking water regulations, uniform minimum standards that would "protect health to the extent feasible.(taking costs into consideration)." These interim regulations were later supplanted by regulations formulated on the basis of a series of reports by the National Academy of Sciences (NAS). The charge to the NAS Safe Drinking Water Committee was to recommend the MCLs necessary to protect humans from any known or anticipated adverse health effects. In turn, U.S. EPA was to specify MCLs as close as feasible to the levels recommended by NAS. By 1986, U.S. EPA had established MCLs for 23 contaminants but treatment techniques for none.

In that year, Congress amended the SDWA to cover more contaminants, to apply more pressure to states and localities to clean up their drinking water supplies, and to strengthen U.S. EPA's enforcement role. U.S. EPA was required to adopt regulations for a total of 83 contaminants within 3 years (including all but one of those originally regulated). It was directed to prescribe regulations for two treatment techniques for public water systems: filtration and disinfection. In translating recommended MCLs, now maximum contaminant level goals, into feasible and enforceable regulations, U.S. EPA was directed to assume installation of the best available technology.

Over the past decade, continuing criticism of the SDWA has prompted yet another attempt at reform. Congress failed to pass amendments at the end of the 1994 session, but further revisions to the SDWA are expected in the near future. Congress will likely abandon the current requirement that U.S. EPA issue standards for 25 additional contaminants every 3 years in order to permit U.S. EPA to choose contaminants for regulation based on scientific criteria. There is also support for a cost–benefit approach to standard setting that would release the current feasibility requirement for standards. The financial burden on smaller municipal water systems for achieving compliance is a major political issue.

—*Shayne C. Gad*

Related Topic

Pollution, Water

Salicylates

♦ CAS: 50-78-2
♦ Synonyms: Aspirin; acetylsalicylic acid; acetaminosalol; aluminum aspirin; bismuth subsalicylate; methyl salicylate; phenyl salicylate; potassium salicylate; sodium salicylate; sodium thiosalicylate
♦ Pharmaceutical Class: Nonsteroidal synthetic derivatives of salicylic acid

◆ CHEMICAL STRUCTURE:

Salicylic acid   Aspirin   Methyl salicylate

Uses

Salicylates mainly exhibit analgesic, antiinflammatory, and antipyretic activity.

Exposure Pathway

Ingestion is the most common route of both accidental and intentional exposure to salicylates. Salicylates are available in oral, topical, and rectal dosage forms.

Toxicokinetics

Salicylates are rapidly absorbed from the gastrointestinal tract following oral administration. Absorption occurs primarily from the upper small intestine via passive diffusion of un-ionized molecules. The rate of absorption may be slower with large, potentially lethal salicylate doses due to an inhibitory effect on gastric emptying and impaired dispersion of the drug in gastrointestinal fluids. Formulation characteristics (enteric coating and sustained release) of tablets may have varied effects on the rate of absorption. In general, the smaller the particle size, the faster the absorption rate since the resultant increase in surface area enhances the rate of dissolution. Salicylates are detected in serum within 5–30 min after oral administration of rapidly absorbed dosage forms (aqueous solutions and uncoated or film-coated tablets). Topical application of salicylic acid may result in systemic toxicity. Rectal absorption is slow and unreliable.

Salicylates are metabolized principally in the liver by the microsomal enzyme system and are predominately conjugated with glycine to form salicyluric acid. Salicylates are also conjugated with glucuronic acid to form salicylphenolic glucuronide and salicylacyl glucuronide. In addition, small amounts of salicylates are hydrolyzed to form gentisic acid, which is an active metabolite and a potent inhibitor of prostaglandin synthesis. Salicylates rapidly distribute throughout extracellular fluid and into body tissues and fluids with high concentrations in the liver and kidneys. Under normal homeostatic acid–base conditions, salicylates cross the blood–brain barrier slowly because the ionized (unabsorbable) form predominates. However, systemic acidosis facilitates the penetration of salicylate into the central nervous system (CNS) in the un-ionized form. The volume of distribution in therapeutic amounts is 0.15–0.2 liters/kg, and it is 0.6 liters/kg in toxic amounts. Protein binding is 50–80%.

Overall salicylate elimination involves nonlinear kinetics with elimination half-life increasing with increased dose. Salicylate and its metabolites (salicyluric acid and salicylic acid) are rapidly and almost completely excreted in the urine via glomerular filtration and renal tubular secretion.

Mechanism of Toxicity

Salicylates increase medullary sensitivity to carbon dioxide, directly stimulating the CNS respiratory center (hyperventilation) in the medulla oblongata and resulting in a decreased PCO_2, an increased pH, and respiratory alkalosis. A compensatory increase in the renal excretion of bicarbonate leads to the loss of potassium and sodium in the urine. A metabolic acidosis may follow due to the accumulation of organic acids. Therefore, salicylate poisoning may produce a mixed acid-base abnormality of respiratory alkalosis and metabolic acidosis. Salicylates uncouple oxidative phosphorylation, resulting in a failure to produce high-energy phosphates such as adenosine triphosphate, while at the same time increasing oxygen utilization and carbon dioxide production, thereby increasing heat production. Uncoupling of oxidative phosphorylation also increases tissue glycolysis, cerebral glycolysis, and peripheral demand for glucose.

Human Toxicity: Acute

Ingestions of ≥ 150 mg/kg can produce toxic symptoms such as tinnitus, nausea, and vomiting. Serious toxicity is manifest by severe vomiting, hyperventilation, hyperthermia, confusion, coma, convulsions, hyperglycemia or hypoglycemia, and acid-base disturbances such as respiratory alkalosis or metabolic acidosis. Severe toxicity can also include pulmonary edema, hemorrhage, acute renal failure, or death. Toxic salicylate blood levels appear over 25 mg/dl. Serious symptoms may develop at lower levels due to dehydration, fever, renal dysfunction, or delayed salicylate metabolism. In overdose, the formation of concretions, slow absorption of enteric coated tablets, and delayed gastric emptying may delay toxic reactions and cause salicylate levels to rise over the first 12–24 hr. The Done nomogram is of little value in the assessment of acute ingestions. In patients who have significant acidosis and/or who have ingested multiple doses, the Done nomogram underestimates toxicity.

Human Toxicity: Chronic

Chronic salicylism presents clinically in a similar fashion to the acute situation, although it is often associated with a higher morbidity and mortality as well as more pronounced hyperventilation, dehydration, coma, seizures, and acidosis. Chronic salicylism can result in death at serum salicylate levels as low as 10–15 mg/dl.

Clinical Management

Basic and advanced life-support measures should be utilized as necessary. Gastrointestinal decontamination procedures should be used as deemed appropriate to the patient's level of consciousness and the history of the ingestion. Activated charcoal may be used to adsorb salicylates or concomitant ingestants. Careful correction of fluid and electrolyte abnormalities is essential. Hemodialysis effectively increases clearance and improves fluid/electrolyte balance. This extracorporeal method of elimination should be considered in patients with cardiac or renal failure, intractable acidosis or severe fluid imbalance, or serum salicylate levels over 100–120 mg/dl following acute ingestions. Chronic salicylate intoxications may necessitate dialysis at lower levels of 50–80 mg/dl. Seizures indicate a poor prognosis and may indicate the need to dialyze at lower acute or chronic levels. Multiple-dose activated charcoal may be effective if the gastrointestinal tract contains unabsorbed salicylates as manifest by sustained or increasing salicylate levels.

Animal Toxicity

Cats are sensitive to salicylates. Toxicity may include fever, hyperpnea, seizures, respiratory alkalosis, metabolic acidosis, methemoglobinemia, gastric hemorrhage, and kidney damage. Activated charcoal may be used in cats.

—*Bonnie S. Dean*

Related Topics

Gastrointestinal System
Noise: Ototraumatic Effects

Salmonella

- ◆ SPECIES NAMES: There are 1700 serotypes and variants of Salmonella; the genus has three primary species: *Salmonella typhi, Salmonella choleraesuis,* and *Salmonella enteritidis.*

- ◆ DESCRIPTION: Salmonellae are motile, gram-negative, rod-shaped bacteria which grow aerobically and anaerobically.

Exposure Pathway

Ingestion of contaminated food is the most common route of exposure.

Mechanism of Toxicity

Salmonella produces toxicity as an invasive pathogen. It penetrates the gastrointestinal mucosa, resulting in profound diarrhea. Large numbers of organisms are usually required to cause infection. These organisms produce an inflammatory diarrhea by invading the mucosa and may penetrate the bowel wall, causing septicemia.

Salmonella can contaminate many types of food. Fecal contamination of dairy products and poultry is very common. The incubation period is 8–48 hr. Salmonella food poisoning is an infectious process that may persist for 2–5 days. Salmonella can be destroyed by heat but is resistant to freezing.

Human Toxicity

Salmonella infections are characterized by three clinical syndromes: gastroenteritis, bacteremia with focal extraintestinal infections, and enteric fever.

Gastroenteritis

The incubation period is 8–48 hr. Loose, large, watery diarrhea with or without blood may last 2–5 days.

Symptoms may also include nausea, vomiting, abdominal cramping, weakness, fever, chills, and dehydration. Severity of the symptoms is variable.

Bacteremia with Foca Extraintestinal Infections

This is most commonly observed in infants, sickle cell patients, elderly persons, and immunosuppressed patients. The incidence of bacteremia is highest in children under 2 years of age. The bacteremia usually is transient but both septicemia and meningitis may develop.

Enteric Fever

This is a subacute disease with a 1-week incubation period. Gut invasion occurs initially followed by the establishment of foci of bacteremia in the gut over 1–3 weeks. Symptoms include fever, headache, anorexia, nausea, cough, and sore throat. Fulminant sepsis or local infection of the gastrointestinal tract, urinary tract, respiratory tract, or musculoskeletal system may occur.

Clinical Management

Fluid and electrolyte replacement is the cornerstone of therapy. Antibiotics are not indicated unless a patient is septicemic or at risk from an underlying disease such as HIV.

Most reported cases of Salmonellosis occur through contamination of food. Ways to prevent contamination from Salmonella include ensuring adequate sewage disposal, monitoring water supply, and practicing good hand-washing. Eating raw eggs should be avoided. Effective pasteurization kills salmonella in milk. Meat, eggs, and dairy products should be stored at temperatures less than 40°F. Frozen meat should be thawed in a refrigerator.

—*Vittoria Werth*

Sarin

- CAS: 107-44-8
- SYNONYMS: GB; *o*-isopropyl methyl phosphonofluoridate; G agent; nerve gas

- CHEMICAL CLASS: Nonpersistent anticholinesterase compound organophosphate nerve agent
- MOLECULAR FORMULA: $C_4H_{10}FO_2P$
- CHEMICAL STRUCTURE:

$$(CH_3)_2 CHO \diagdown \quad P \diagup O$$
$$H_3C \diagup \quad \diagdown F$$

Uses

Sarin is an nerve (gas) agent used in chemical warfare. It is an irreversible cholinesterase inhibitor.

Exposure Pathways

Casualties are caused primarily by inhalation but can occur following percutaneous and ocular exposure, as well as by ingestion and injection.

Toxicokinetics

Sarin is absorbed both through the skin and via respiration. It is more soluble in water than the other nerve agents [soman (GD) and VX]; its solubility is directly related to temperature. The half-life of sarin, however, is inversely related to temperature and pH. In water the half-life of sarin is 15 min at 30°C and at pH 7.6. Nerve agents inhaled as vapors or aerosols enter the systemic circulation, resulting in toxic manifestations within seconds to 5 min of inhalation.

The enzyme organophosphate hydrolase hydrolyzes sarin (GB) as well as soman (GD), tabun, and diisopropyl fluorophosphate at approximately the same rate. Sarin consists of two stereoisomers P(−) and P(+) of which the P(−) is more toxic than P(+). The isomers of the asymmetric organophosphates may differ in (1) overall toxicity, (2) rate of aging, (3) rate of cholinesterase inhibitions, and (4) rate of detoxification. The rate of detoxification differs for different animal species and routes of administration. For sarin administered subcutaneously in guinea pigs, the rate of detoxification was reported to be 0.013 mg/kg/min.

Mechanism of Toxicity

Sarin and the other nerve agents are organophosphorus cholinesterase inhibitors. They inhibit the enzymes butyrylcholinesterase in the plasma, acetylcholinesterase on the red blood cell, and acetylcholinesterase at cholinergic receptor sites in tissues. These three enzymes are not identical. Even the two acetylcholinesterases have slightly different properties, although they have a high affinity for acetylcholine. The blood enzymes reflect tissue enzyme activity. Following acute nerve agent exposure, the red blood cell enzyme activity most closely reflects tissue enzyme activity. During recovery, however, the plasma enzyme activity more closely parallels tissue enzyme activity.

Following nerve agent exposure, inhibition of the tissue enzyme blocks its ability to hydrolyze the neurotransmitter acetylcholine at the cholinergic receptor sites. Thus, acetylcholine accumulates and continues to stimulate the affected organ. The clinical effects of nerve agent exposure are caused by excess acetylcholine.

The binding of nerve agent to the enzymes is considered irreversible unless removed by therapy. The accumulation of acetylcholine in the peripheral and central nervous systems leads to depression of the respiratory center in the brain, followed by peripheral neuromuscular blockade causing respiratory depression and death.

The pharmacologic and toxicologic effects of the nerve agents are dependent on their stability, rates of absorption by the various routes of exposure, distribution, ability to cross the blood–brain barrier, rate of reaction and selectivity with the enzyme at specific foci, and their behavior at the active site on the enzyme.

Red blood cell enzyme activity returns at the rate of red blood cell turnover, which is about 1% per day. Tissue and plasma activities return with synthesis of new enzymes. The rate of return of these enzymes are not identical. However, the nerve agent can be removed from the enzymes. This removal is called reactivation, which can be accomplished therapeutically by the use of oximes prior to aging. Aging is the biochemical process by which the agent–enzyme complex becomes refractory to oxime reactivation. The toxicity of nerve agents may include direct action on nicotinic acetylcholine receptors (skeletal muscle and ganglia) as well as on muscarinic acetylcholine receptors and the central nervous system (CNS).

Recently, investigations have focused on organophosphate nerve agent poisoning secondary to acetylcholine effects. These include the effects of nerve agents on γ-aminobutyric acid neurons and cyclic nucleotides. In addition, changes in brain neurotransmitters such as dopamine, serotonin, noradrenaline, and acetylcholine following inhibition of brain cholinesterase activity have been reported. These changes may be due in part to a compensatory mechanism in response to overstimulation of the cholinergic system or could result from direct action of nerve agent on the enzymes responsible for noncholinergic neurotransmission.

Human Toxicity

Toxic effects occur within seconds to 5 min of nerve agent vapor or aerosol inhalation. The muscarinic effects include ocular (miosis, conjunctival congestion, ciliary spasm), nasal discharge,, respiratory (bronchoconstriction and increased bronchial secretion), gastrointestinal (anorexia, vomiting, abdominal cramps, and diarrhea), sweating, salivation, and cardiovascular (bradycardia and hypotension) effects. The nicotinic effects include muscular fasciculation and paralysis. CNS effects can include ataxia, confusion, loss of reflexes, slurred speech, coma, and paralysis.

Following inhalation of sarin, the median lethal dose (LCt_{50}) in man has been estimated to be 70 mg-min/m^3 at a respiratory minute value (RMV) of 15 liters/min and 100 mg-min/m^3 at a RMV of 10 liters/min (resting) for duration of 0.5–2 min.

Following percutaneous exposure of bare skin to sarin vapor, the LCt_{50} has been estimated at 12,000 mg-min/m^3 for a 70-kg man. For liquid percutaneous exposure, the LD_{50} has been estimated as 1.7 g/70-kg man, and for intravenous injection the LD_{50} has been estimated as 1 mg/70-kg man.

Median incapacitation doses estimated for man following inhalation of sarin for a respiratory minute volume of 15 liters/min for a 10-min exposure are as follows: 40 mg-min/m^3 for moderate incapacitation, 56 mg-min/m^3 for severe incapacitation, and 72 mg-min/m^3 for very severe incapacitation. The symptoms for moderate incapacitation include maximal miosis, eye pain, headache, twitching eyelids, difficulty in ocular accommodation, tightness of chest, runny nose, salivation, sneezing and coughing, anorexia, nausea, heartburn, fatigue, weakness, muscle fasciculation, anxiety, and insomnia. Severe incapacitation includes all of the above plus diarrhea, frequent urination, dysphoria, and ataxia. For very severe incapacitation, the principal effects are convulsions, collapse, and paralysis.

The minimum effective dosage for miosis in man has been estimated between 2 and 4 mg-min/m^3. The permissible airborne exposure concentration of sarin for an 8-hr workday or a 40-hr workweek is an 8-hr TWA of 0.00003 mg/m^3.

Sarin is the nerve agent studied most thoroughly in humans. At an estimated concentration of 3–5 mg-min/m^3 in man, it will produce miosis, rhinorrhea, and a feeling of tightness in the throat or chest. Exposure to small amounts of nerve agent vapor causes effects in the eyes, nose, and airways. These effects are from local contact and are not indicative of systemic absorption. Small amounts of liquid agent on the skin cause systemic effects initially in the gastrointestinal tract. Lethal amounts of vapor or liquid cause a rapid cascade of events resulting, within 1 or 2 min, in loss of consciousness and convulsive activity followed by apnea and muscular flaccidity.

Although miosis is a characteristic sign of exposure to the nerve agent, rhinorrhea may be the first indication. Its severity is dose dependent.

Miosis occurs from direct contact of vapor with the eyes. It may also occur from moderate to severe exposure of skin to liquid agent or from a liquid droplet near the eye. Miosis will begin within seconds or minutes following vapor exposure and may not be complete for many minutes if the exposure concentration is low. In unprotected individuals, miosis is bilateral and is often accompanied by complaints of pain, dim and blurred vision, conjunctival injection, nausea, and occasionally vomiting. On occasion, subconjunctival hemorrhage is also present.

Inhalation of nerve agent vapor causes bronchoconstriction and increased secretions of the glands in the airways, which is dose related. Small amounts of the nerve agent will produce a feeling of slight tightness in the chest to severe respiratory distress following large amounts. Large amounts will cause cessation of respiration (apnea) within minutes after the onset. Both CNS effects and peripheral effects (skeletal muscle weakness and bronchoconstriction) may contribute to the apnea.

Systemic absorption of the nerve agent will cause increased motility of the gastrointestinal tract and an increase in glandular secretions. Nausea and vomiting are early signs of liquid exposure on the skin and diarrhea may occur following large amounts of agent.

Nerve agent exposure to glands increases their secretions. These glands include lacrimal, nasal, salivary, and bronchial. Localized sweating will occur at the site of liquid agent on the skin, and after large liquid or vapor exposure generalized sweating is common.

Stimulation of skeletal muscles by nerve agents will produce muscular fasciculation and twitching. Large amounts of the agent will cause fatigue and muscle weakness followed by muscular flaccidity.

Large amounts of the nerve agent in the CNS will cause loss of consciousness, seizure activity, and apnea. CNS effects of smaller amounts of the agent vary and are nonspecific. However, they may include forgetfulness, inability to concentrate, insomnia, bad dreams, irritability, impaired judgment, and depression. These effects may persist up to 6 weeks.

Nerve agent exposure may cause bradycardia due to vagal stimulation or it may often cause the reverse—tachycardia due to fright and hypoxia and adrenergic stimulation secondary to ganglionic stimulation. Bradyarrhythmias such as first-, second-, or third-degree heart block may also occur. Blood pressure may also be elevated because of adrenergic stimulation, but it is usually normal until the terminal decline.

Clinical Management

Management of nerve agent intoxication consists of decontamination, ventilation, administration of antidotes, and supportive therapy.

The three therapeutic drugs for treatment of nerve agent intoxication are atropine, pralidoxime chloride, and diazepam. Atropine, a cholinergic blocking or anticholinergic drug, is effective in blocking the effects of excess acetylcholine at peripheral muscarinic sites. The usual dose is 2 mg, which may be repeated at 3- to 5-min intervals. Pralidoxime chloride (Protopam chloride; 2-PAM CL) is an oxime used to break the agent–enzyme bond and restore the normal activity of the enzyme. This is most apparent in organs with nicotinic receptors. Abnormal activity decreases and normal strength returns to skeletal muscles, but no decrease in secretions is seen following oxime treatment. The usual dose is 1000 mg (iv or im). This may be repeated two or three times at hourly intervals, intravenously or intramuscularly. Diazepam, an anticonvulsant drug, is used to decrease convulsive activity and reduce brain damage that may occur from prolonged seizure activity. It is suggested that all three of these drugs be administered at the onset of severe effects from nerve agent exposure, whether or not seizures occur. The usual dose of diazepam is 10 mg (im).

Miosis, pain, dim vision, and nausea can be relieved by topical atropine in the eye. Pretreatment with carba-

TABLE S-1
Inhalation LCt$_{50}$s of Sarin in Various Species

Species	LCt$_{50}$ (mg-min/m^3)	Exposure duration (min)
Mouse	150	30
Rat	1500	10
Guinea pig	256	2
Rabbit	1200	10
Cat	1000	10
Dog	1000	10
Monkey	1000	10

mates may protect the cholinesterase enzymes before nerve agent exposure.

Supportive therapy may include ventilation via an endotracheal airway if possible and suctioning of excess secretions in the airways.

Animal Toxicity

Small doses of nerve agents in animals can produce tolerance in addition to their classical cholinergic ef-

TABLE S-2
Acute Toxicities of Sarin in Various Species by Various Routes of Exposure

Route of exposure/species	LD$_{50}$ (μg/kg)
Percutaneous liquid	
Mouse	1,080
Rabbit	925
Intravenous	
Mouse	109
Rat	39
Rabbit	15
Cat	22
Dog	19
Monkey	22,300
Intramuscular	
Rat	108
Mouse	164
Intraperitoneal	
Mouse	283
Rat	218
Oral	
Rat	550
Subcutaneous	
Rat	103
Mouse	60
Rabbit	30
Guinea pig	30
Hamster	95

fects. In rats, acute administration of nerve agents in subconvulsive doses produced tumors and hindlimb abduction. In animals, nerve agents can also cause effects on behavior as well as cardiac effects.

The cause of death is attributed to anoxia resulting from a combination of central respiratory paralysis, severe bronchoconstriction, and weakness or paralysis of the accessory muscles for respiration.

Signs of nerve agent toxicity vary in rapidity of onset, severity, and duration of exposure. These are dependent on specific agent, route of exposure, and dose. At the higher doses, convulsions and seizures indicate CNS toxicity.

Following nerve agent exposure, animals exhibit hypothermia resulting from the cholinergic activation of the hypothalamic thermoregulatory center. In addition, plasma concentrations of pituitary, gonadal, thyroid, and adrenal hormones are increased during organophosphate intoxication.

The LCt$_{50}$s (mg-min/m^3) reported following the inhalation of Sarin are presented in Table S-1. The acute toxicities by other routes of exposure and in various animal species are presented in Table S-2.

—Harry Salem and Frederick R. Sidell

(The views of the authors do not purport to reflect the position of the U.S. Department of Defense. The use of trade names does not constitute official endorsement or approval of the use of such commercial products.)

Related Topics

Behavioral Toxicology
Cholinesterase Inhibition
Nerve Agents
Neurotoxicity, Delayed
Organophosphates
Psychological Indices of Toxicity

Saxitoxin

- CAS: 35523-89-8
- SYNONYMS: Mussel poison; clam poison; paralytic shellfish poison

- CHEMICAL CLASS: Complex of amino acids
- MOLECULAR FORMULA: $C_{10}H_{17}N_7O_4$
- CHEMICAL STRUCTURE:

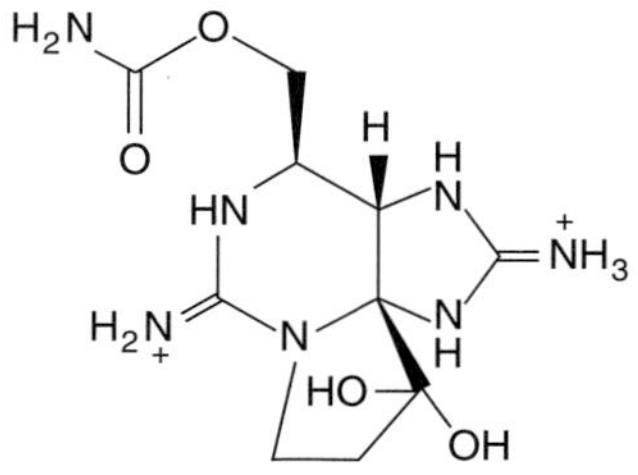

Uses

Saxitoxin is a naturally occurring toxin that is synthesized by various marine dinoflagellates. It is used in neurochemical and molecular biology research.

Exposure Pathway

Ingestion of shellfish containing saxitoxin is the route of exposure (see Shellfish Poisoning, Paralytic).

Toxicokinetics

Saxitoxin is readily absorbed from the gastrointestinal tract and through mucous membranes.

Mechanism of Toxicity

Saxitoxin binds to the sodium channels in the membranes of excitable cells (neurons and muscle cells) blocking synaptic transmission. Saxiton is connected to red tides (see Red Tide).

Human Toxicity

Saxitoxin paralyzes the peripheral nervous system and alters cardiac chronotrophy. Symptoms include gastrointestinal complaints and paresthesias of the face, followed by (in severe cases) muscle paralysis. The estimated lethal dose in humans is 0.3–1 mg. A single contaminated shellfish may contain 50 lethal doses.

Clinical Management

The gut should be decontaminated and the patient observed carefully for signs of respiratory depression.

Animal Toxicity

The LD_{50}s in mice include 10 μg/kg (intraperitoneal), 263 μg/kg (oral), and 3.4 μg/kg (intravenous).

—*Shayne C. Gad and Jayne E. Ash*

Related Topic

Neurotoxicology: Central and Peripheral

Scombroid

- SYNONYMS: Scombroidtoxicosis; form of Ichthyosarcotoxicosi
- IMPLICATED SOURCES (fish): Scombroidae; mahi mahi; bluefish; Bombay duck; kahawai; kingfish; swordfish; pacific amberjack; salmon; tuna

Exposure Pathway

The toxin is contained within the flesh of certain fish. Ingestion of the flesh of these fish causes poisoning. The toxin is not destroyed or inactivated by heating or cooking. Proper refrigeration of fish is the best preventative measure one can take to decrease the likelihood of poisoning.

Toxicokinetics

Histamine, when present in large amounts, is absorbed resulting in histaminic effects. Other compounds, such as saurine, may potentiate histamine absorption. The diagnosis of histamine toxicity may be confirmed (where such confirmation is necessary) by quantitation of histamine in fish flesh.

Mechanism of Toxicity

Contaminated fish contain free histidine in their musculature. Histamine and saurine are produced during spoilage, and these agents are responsible for the symptoms. Generally, symptoms are not seen unless 100 mg of histamine per 100 grams of fish flesh is ingested.

Human Toxicity

Initial symptoms are those of a histamine reaction and typically occur within 30–60 min of ingestion. Common symptoms include dermal flushing, headache, nausea, vomiting, and diarrhea. Facial edema, burning of the throat, palpitations, dizziness, and rash have also been noted. Bronchospasm, urticaria, shock, and death are rare. Symptoms usually resolve within 3–36 hr.

Clinical Management

Basic and advanced life-support measures should be utilized as necessary. Treatment is generally symptomatic and supportive. Gastrointestinal decontamination procedures should be used when appropriate. Other therapy is directed at limiting histaminic symptoms. Use of over-the-counter antihistamines such as diphenhydramine would seem appropriate. Steroids, epinephrine, and H2 antagonists are used in patients refractory to antihistamines.

—*Gaylord P. Lopez*

Scorpions

◆ SYNONYMS: *Centruroides; Vejovis; Hadrurus*

Exposure Pathway

Scorpions inflict a sting and inject their venom subcutaneously with a stinger located at the end (telson) of their multisegmented abdomen.

Toxicokinetics

Scorpion venom may reach systemic circulation through lymphatic transport following a sting. Of those scorpions located in the United States, *Centruroides exilicauda* is an example of a scorpion that can produce systemic symptoms following venom absorption. Onset of symptoms typically occur within 4 hr of the sting. The metabolism of venom components is not well understood. Tissue distribution of venom is complex. Venom components differ among the multitude of scorpion species and thus venom distributes to different tissue sites.

Mechanism of Toxicity

Scorpion venom is composed of many different fractions which vary among the different scorpion species.

These venom fractions act at different tissue receptor sites. The typical local tissue reaction is a result of the inflammatory response to injected foreign proteins and enzymes making up the venom. The venom of poisonous *Centruroides* species contains several different neurotoxins. These toxins block the transmission of nerve impulses in the central nervous system and in muscles by blocking the transport of ions through sodium and potassium channels at the cellular level. Other venom components may decrease heart rate by causing the release of acetylcholine.

Human Toxicity

Most scorpion stings produce some local tissue reaction which is characterized by mild to moderate burning pain. Usually there is minimal swelling and redness. In the United States, this is the limit of the reaction following stings of *Vejovis, Hadrurus,* and several other common species. Wound infection is also possible following the sting. The more poisonous *Centruroides* scorpions, represented by *C. exilicauda,* may also produce systemic symptoms following significant envenomation. However, even these scorpions often produce only pain at the sting site. When systemic symptoms develop, they include increased heart rate, hypertension, dilated pupils, sweating, and increased blood glucose. Also, salivation, tearing, diarrhea, and bradycardia may develop when parasympathetic nerve stimulation predominates from acetylcholine release. Other clinical effects may include blurred vision, nystagmus, opisthotonus, muscle fasciculations, convulsions, breathing difficulty, respiratory failure, and cardiac arrhythmias. Young children, the elderly, and those with preexisting cardiovascular disease are at greater risk for severe systemic symptoms. Occasionally, poisonous exotic species make their way into the United States either by illegal importation or along with agricultural product shipments from abroad. Signs and symptoms following envenomation vary depending on the species of scorpion involved.

Clinical Management

Basic and advanced clinical life support may be required following severe envenomation by several *Centruroides* species. This is especially true in those cases involving young children and the elderly. Most scorpion stings require only local wound care. Cleaning the sting site thoroughly and applying a topical antiseptic such as iodine or isopropyl alcohol (rubbing alcohol) may help prevent infection. Ice may be applied to the sting site for 10–15 min to help decrease pain. Acetaminophen, aspirin, or ibuprofen may be helpful but are often ineffective for moderate to severe pain. Applying ice for long periods of time or immersing the sting site in an ice bath (cryotherapy) is not recommended since this procedure decreases blood flow at the site causing tissue damage. The majority of patients presenting with systemic symptoms can be managed at the hospital with supportive care, pain management, and observation. Careful monitoring of heart rate, blood pressure, and respiratory function are essential. Muscle spasms may respond to diazepam or calcium gluconate. Occasionally hypertension is sufficiently severe or prolonged to require treatment. Depending on the severity of hypertension, nitroprusside, labetalol, or nifedipine may be indicated. Respiratory failure due to neuromuscular blockade is a rare but possible complication. Mechanical ventilation may be required. A polyvalent antivenin and a *C. exilicauda*-specific antivenin are available on a regional basis only for treating severe stings of *Centruroides* scorpions. These preparations are not approved by the U.S. FDA and scientific data regarding their efficacy are lacking. However, anecdotal reports indicate that the antivenin has reversed symptoms associated with neuromuscular blockade within 30 min to 1 hr. The local regional poison information center should be contacted to locate antivenin and assist in determining whether clinical indications exist for its use.

—*Gary W. Everson*

Related Topic

Neurotoxicology: Central and Peripheral

Selenium (Se)

♦ CAS: 7782-49-2
♦ SELECTED COMPOUNDS: Hydrogen selenide, H_2Se (CAS: 7783-07-5); sodium selenate, Na_2SeO_4 (CAS: 13410-01-0); sodium selenite, Na_2SeO_3 (CAS: 10102-18-8)
♦ CHEMICAL CLASS: Metals

Uses

Selenium is used in a wide variety of industries, including electronics, glass, ceramics, steel, pigment manufacturing, and rubber production. Medicinally, selenium is used in antidandruff shampoos and as a dietary supplement.

Exposure Pathways

For the general population, ingestion is the primary exposure pathway; sources include dietary supplements and various foods including seafood, meats, milk products, and grains. Trace amounts are found in drinking water. Selenium is not absorbed from shampoos.

In industrial settings, inhalation may be a significant exposure pathway. Airborne concentrations of selenium are higher in the vicinity of metallurgical industries. Selenium is present in most sulfide ores and is generally a by-product of the roasting of copper pyrite.

Toxicokinetics

Selenium and most of its compounds are rather insoluble and thus not absorbed orally. Soluble selenium compounds (e.g., sodium selenate and sodium selenite) are readily absorbed (up to 90%). Blood concentrations depend on the amount of selenium ingested. After blood levels of 200–240 μg/ml are obtained, homeostatic controls take over. The greatest amount of absorbed selenium concentrates in the liver and kidneys, a lesser amount in the heart and lungs, and the very least in the muscles.

Selenium is an essential trace element and an integral component of heme oxidase. It appears to augment the antioxidant action of vitamin E to protect membrane lipids from oxidation. The exact mechanism of this interaction is not known; however, selenium compounds are found in the selenium analogs of the sulfur-containing amino acids, such as cysteine and methionine. Se-cysteine is found in the active sites of the enzyme, glutathione peroxidase, which acts to use glutathione to reduce organic hydroperoxides.

Selenium is rapidly excreted in the urine; some is incorporated into proteins. Elemental selenium and its oxides can be methylated. Trimethyl selenium is excreted rapidly in the urine; some is exhaled.

Mechanism of Toxicity

Excess selenium results in liver atrophy, necrosis, and hemorrhages. The mechanism of toxicity is unknown but may involve the redox cycling. Sulfhydryl enzymes are attacked by soluble selenium compounds.

Human Toxicity

Toxic manifestations of selenium poisoning include decaying and discoloring of teeth, gastrointestinal tract distress, skin lesions, and loss of hair and nails. In some cases, the skin on the fingertips and toes peels constantly. Excess selenium is metabolized to the dimethyl derivative, which is volatile and produces the "garlic" or "rotten" breath characteristic of selenium toxicity.

The difference between an essential dose and a toxic dose for selenium is quite narrow. Normal intake can range from 50 to 200 μg; in the milligram range, toxicity is noted. Acute selenium poisoning results in nonspecific symptoms (e.g., eye irritation and coughing) and can affect the central nervous system and lead to convulsions. Liver and spleen damage has also been noted.

Inhalation of hydrogen selenide, a gas, may produce irritation of the upper respiratory tract and reduced respiratory flow rates, which can persist for a few years. The ACGIH TLV-TWA for hydrogen selenide is 0.05 mg/m^3. The ACGIH TLV-TWA for selenium and its compounds is 0.2 mg/m^3.

Clinical Management

Currently, there are no antidotes of choice for selenium toxicity. Ethylenediaminetetraacetic acid and BAL (British Anti-Lewisite; 2,3-dimercaptopropanol) should not be used because they may enhance selenium toxicity. Treatment is symptomatic (e.g., cardiopulmonary). Often supplemental oxygen is needed. Corrosive selenious acid (in gun bluing solution) should be treated similar to other agents that cause esophageal burns.

Animal Toxicity

Livestock and other animals are particularly affected by either selenium deficiency or excess selenium. In animals with selenium-deficient diets, liver necrosis arises. In areas with deficient selenium concentrations in soil, calves and lambs develop muscle atrophy, which is referred to as either "white" muscle disease or "stiff" muscle disease. Selenium supplementation (often injections) prevents these symptoms.

In areas with unusually high levels of selenium in the soil, livestock develop "blind-stagger" disease, which is characterized by loss of vision, weakness of the limbs, and possible respiratory failure. Runoff from heavily fertilized farms causes excess selenium in ponds, which results in malformation of birds.

There are two interesting paradoxes concerning selenium. The first is that excess selenium is toxic; however, at lower levels it is a protective agent against the toxicity of cadmium, methylmercury, arsenic, copper, and thallium. The second paradox involves carcinogenicity. The U.S. National Cancer Institute found selenium monosulfide (administered orally) to be carcinogenic in rodents; however, many epidemiological studies associate selenium intake with lower cancer rates in humans. Moreover, in the laboratory, selenium somewhat negates the carcinogenic action of carcinogenic aromatic hydrocarbons, acetylaminofluorene, and azo dyes, and it protects against spontaneous mammary tumors in various species of rodents.

Selenium is teratogenic in chicks and sheep; the evidence for humans is equivocal.

—Arthur Furst and Shirley B. Radding

Related Topics

Metals
Veterinary Toxicology

Sensitivity Analysis

Sensitivity analysis is a method used to evaluate the impact of a single variable or a group of variables on the results from a model calculation. Sensitivity analysis may be used to determine which parameters in a calculation have the greatest influence on the results such that greater emphasis is placed on characterizing these parameters. Moreover, the results from these analyses may be used to identify ways to improve the overall predictive capability of the model by reducing the uncertainty in the parameters to which the model predictions are sensitive. Sensitivity analysis may be applied to the risk assessment process in order to identify those variables that dominate risk estimates as well as those that are relatively unimportant.

Sensitivity analyses are typically conducted at two levels: local and global. The local analysis generally assesses the effect of small perturbations of a single variable to the calculation. The primary assumption for using this technique is that the anticipated response to these slight changes is assumed to be nonlinear for at least one parameter in the calculation. Thus, increasing a specific parameter value by 10% is not predicted to produce a corresponding 10% increase in the result for all the parameters. The model is considered sensitive to a particular variable when small variations in the value produces large changes to the model predictions. There are several methods available for calculating the effect of changes in inputs on model predictions. Perhaps the simplest method is to produce slight changes in a single variable in relative terms based on a fractional change from the base case value using deterministic single-point estimates while keeping the remaining parameters constant. By incrementally changing each parameter value by a set percentage, it essentially normalizes any differences between parameters that may occur due to deviations in units. A general case is to produce a ±10% change in the single variable in order to determine the resulting change in the output. Any arbitrary percentage may be used as long as the value is consistently applied to all the variables in the simulation. The results from the local sensitivity analysis may be used to determine places where additional parameters need to be better defined.

In mathematical terms, the local sensitivity analysis is analogous to determining the partial derivative for the calculation with respect to each parameter. Although the local sensitivity analysis normalizes differences between parameters such that the calculation is not sensitive to unit differences, this method does not address the effect on the results of using the full range of possible values for each parameter. This is an important limitation of conducting a local sensitivity analysis since the analysis is likely to produce misleading results for complicated calculations where parameter values have large uncertainties. For instance, a 10% increase above the default adult body weight in a standard risk calculation will produce a corresponding 10% reduction in risk since the individual has more mass available to equally distribute the pollutant concentration. Thus, the conclusion may be drawn that the risk calculation is not extremely sensitive to changes in body weight; however, this is not completely true since body weights

are known to vary by as much as 20% from the default value. By accounting for this uncertainty, the actual risk may decrease by as much as 20%, which may be significant in certain cases. To resolve this problem, a global sensitivity analysis may be performed that explicitly evaluates the parameters in calculations where the uncertainty is large. It should be noted that the local sensitivity analysis is fairly accurate in cases in which the uncertainties are expected to be small.

The purpose of a global sensitivity analysis is to assess the effect a single parameter has on the results over the full range of possible values for that parameter. The range of values is often based on the frequency distribution assigned to the variable. Global sensitivity analysis identifies the parameters that contribute the most to the overall variance of the result by evaluating the uncertainty for each parameter individually. This approach generally follows the same Monte Carlo method used in uncertainty analysis to quantify uncertainty associated with all parameters defined in a calculation. Monte Carlo involves choosing values from a random selection scheme drawn from probability density functions based on a range of data that characterizes the parameter of interest. A simple method of performing a global sensitivity analysis is to run the Monte Carlo simulation by incorporating the uncertainty for a single parameter and keeping all other values constant. However, this usually leads to a labor-intensive task if the calculation is complex and involves numerous parameters. Instead, the simulation may be performed by entering all the uncertainties for each parameter and somehow assigning the variances in the results to key parameters that are anticipated to be the most sensitive. Statistics such as partial correlations, fractional factorials, or partial rank correlations are often used to estimate which parameter is the most sensitive based on the variance of the result. This approach has the advantage that the same Monte Carlo simulations used to estimate uncertainty can be used to estimate the global sensitivity of the model predictions to the inputs and parameters used in the assessment.

—*Virginia Lau*

Related Topics

Hazard Identification
Risk Assessment, Ecological
Risk Assessment, Human Health
Risk Characterization

Risk Communication
Risk Management
Uncertainty Analysis

Sensory Organs

Due to their complexity, the special sense organs are both protected from and susceptible to toxic insult. Most toxic substances are excluded from the inner milieu of the eye and ear by specialized blood vessel barriers, selective membrane channels, or controlled pH gradients. However, some toxins elude these protective mechanisms and consistently produce toxic effects. Although classification is imperfect, differentiation between agents capable of producing acute toxicity (single exposure) from those only toxic after chronic exposure is important. Although most of these agents are capable of producing multiple toxic effects in humans, the sensory organs effects are often the most unique, identifiable, or disabling.

The Eye

The eye may be our most highly specialized organ. As an outgrowth of the central nervous system (CNS), it is susceptible to many of the same toxins as the brain. However, several toxins stand out in their ability to produce isolated ocular toxicity. In order to understand ocular toxicity, certain ophthalmologic principles need to be reviewed.

Vision has been called the "vital sign" of the eye. Normal visual acuity implies intact light transmission through the optic system (cornea, lens, and vitreous humor) to the retina (the light sensing organ). Abnormalities in any of these components may diminish visual acuity or produce blindness.

Cornea

The cornea is the clear, external layer of the eye. Corneal edema (swelling) or diffuse corneal injury (e.g., a caustic exposure) may result in visual loss. Patients with only mild corneal edema or disruption may experience "halos" around bright objects. Inspection of the external surface of the eye should reveal the corneal swelling or clouding.

Conjunctiva

The remainder of the exposed surface of the eye is covered by a thin mucosal layer known as conjunctiva. Mild irritation results in dilation of small blood vessels in the conjunctiva ("bloodshot"), pain, and a foreign body sensation. Chemosis, or swelling, of the conjunctiva is the typical response to prolonged mild irritant exposure or exposure to a strong irritant.

Pupil

The pupil is the window into the internal eye. The iris, or the colored border of the pupil, opens and closes to vary the amount of light admitted through the pupil into the eye. The pupillary size and reactivity are very useful clinical parameters by which toxic eye exposures are assessed. Control of pupillary size is complex and is dependent on the interaction of the opposing sympathetic and parasympathetic nervous systems (taken together, they make up the autonomic nervous system). The pupil enlarges with sympathetic stimulation and constricts with parasympathetic stimulation. Conversely, sympathetic system blockade results in small pupils, while parasympathetic blockade produces large pupils. Light hitting the retina reduces sympathetic nervous signaling to the pupil and results in pupillary constriction. In patients with retinal dysfunction, the pupils are often enlarged due to the inability of the retina to "see" the light and signal the pupil to constrict (Table S-3).

TABLE S-3
Agents Associated with Pupillary Changes

Miosis
    Organophosphate pesticides
    Opioids
    Pilocarpine
    Clonidine
Mydriasis
    Anticholinergics: atropine, diphenhydramine
    Cocaine, amphetamines
    Withdrawal

Lens

The lens is deep within the eye and is responsible for focusing an image on the retina. Lens opacification (cataracts) may produce similar visual abnormalities as corneal damage but is often undetectable without careful inspection of the inside of the eye.

Vitreous Humor

The remainder of the eye is filled with a viscous, clear substance known as vitreous humor. The vitreous humor is rarely directly altered by toxic exposure, but bleeding into the vitreous may be a secondary effect of a systemic toxin.

Retina

The retina is a true neural structure. It is responsible for converting light and colors into neural signals. Retinal toxicity results in blurry vision, often described as "walking in a snowstorm." Although certain drugs are notable for producing acute retinal toxicity, most agents act indirectly to produce retinal injury. Indirect retinal toxins exert their toxic effect by reducing blood flow or oxygen delivery to the retina, resulting in cell death.

Brain

In addition to a normal eye, vision requires an intact circuit to and from the brain itself. The optic nerve (cranial nerve II), for example, carries retinal information to the occipital cortex in the posterior aspects of the brain. In addition, information concerning pupil size, direction of gaze, simultaneous movement of the eyes (conjugate gaze), and focus clarity is relayed from the brain back to the eye through multiple nerves.

Specific Ocular Toxins

Many agents are capable of producing visual loss in humans. Each component in the visual pathway is capable of succumbing to a toxic exposure. Medical or surgical restoration of sight is routinely available for corneal, lenticular, and vitreal damage. Due to the complexity of the retina, therapeutic interventions are much more limited for patients with retinal toxicity. Unlike most parts of the eye, the retina remains far too complex to be replaced. Table S-4 lists selected agents capable of producing chronic toxicity to various portions of the eye. In the following sections, several common, or important, toxins are discussed (Table S-4).

Caustics

Acid and alkali burns to the eye often result in severe corneal injury and require aggressive intervention. It is often difficult to predict at the outset the amount of damage any given chemical will inflict on the eye. Generally, agents that have extreme pH, particularly alkali, and those which are solid produce the greatest corneal damage. Hydrocarbons and detergents tend to be less problematic but exceptions abound. Immediate treatment should consist of copious irrigation with water or saline. Measurement of pH is useful, but neutralization should never be attempted due to the potential for thermal injury. Even with immaculate initial care, severe caustic injuries may produce corneal scarring and visual loss (see Acids and Corrosives).

Methanol

Of all substances consistently reported to produce direct ocular toxicity in humans, methanol is most frequently responsible. Methanol is locally available as a gasoline additive and as windshield cleaning fluid and is used widely in industry as a solvent. Several well-documented epidemics have occurred in the recent past resulting from the consumption of methanol in place

TABLE S-4
Agents Associated with Chronic
Visual Changes

Corneal
  Metals
  Amiodarone
  Chlorpromazine
Cataracts
  Dinitrophenol
  Steroids
Retinal injury
  Carbon disulfide
  Digitalis
  Vincristine
Neurologic (optic nerve, brain)
  Ethambutol
  Lead
  Methylmercury

of ethanol, and isolated cases are common. Ocular symptoms often take several hours to develop and are actually due to the hepatic metabolite of methanol, formic acid. Formic acid also produces profound systemic acidosis which may be fatal. Management consists of inhibiting the hepatic metabolism of methanol by providing the liver with ethanol, the preferred substrate for metabolism. The remaining methanol, which is poorly excreted without metabolism, must be removed by hemodialysis (see Methanol).

Quinine

The antimalarial agent quinine is derived from the bark of the cinchona tree along with several other alkaloids and salicylate (aspirin). Many of these agents produce similar toxic features (cinchonism) in patients with excessive intake, but only quinine produces blindness. Cinchonism consists of abdominal pain and vomiting, ringing in the ears (tinnitus), and confusion. Visual loss after overdose is due to direct retinal toxicity, although until recently it was believed to be due to spasm of the arterial blood supply to the retina. Treatment is difficult, but limited evidence suggests charcoal hemoperfusion may be beneficial (hemoperfusion is similar to dialysis, except in place of a semipermeable membrane to filter the toxin, charcoal is used to bind the toxin) (see Quinine).

Agents Capable of Indirect Retinal Toxicity

Indirect retinal toxins produce retinal ischemia or reduced oxygen delivery by the blood. Cocaine, amphetamines, and ergot alkaloids (used in the treatment of migraine headaches) may produce retinal ischemia by reducing the caliber of the retinal arteries and thereby reduce blood flow. Foreign bodies, such as talc, introduced by use of impure intravenous drug can result in embolization (mechanical blockade) of the retinal arteries with resultant ischemia. Retinal ischemia can also occur in patients with poor retinal blood flow due to hypotension (low systemic blood pressure). Toxic causes of hypotension reported to induce blindness include calcium channel blockers, β-adrenergic blockers, and nitrates. Additionally, by preventing hemoglobin, the main oxygen transport protein in the blood, from binding and delivering oxygen, carbon monoxide can produced retinal ischemia.

Occupational Exposures

Although occupational inhalation of methanol may produce ocular toxicity, the vast majority of occupational eye toxicity results from exposure to irritant chemicals. Highly water-soluble gases, such as ammonia or hydrogen sulfide, produce immediate pain and tearing upon exposure. Gases that are poorly soluble in the water of the eye only produce irritation after prolonged exposure (e.g., phosgene and ethylene oxide).

The Ear

Nearly as complex as the eye, the ear is also an outgrowth of the CNS. The ear converts sound waves into neural impulses which are transmitted to the brain for processing. Unlike the eye, toxins produce adverse effects only at limited sites in the ear. Like the eye, however, free entry of drugs into the inner ear is prevented by a selective filtering mechanism.

External Ear and Ear Canal

While the external ear and ear canal serve as a pathway for the entrance of sound into the internal ear, they are infrequently affected by toxic exposure. Caustics are the only agent commonly producing toxicity at this level.

Middle Ear

Several tiny bones in the middle ear amplify and convert sound waves from the eardrum into fluid waves in the inner ear. There are no significant toxic exposures affecting the middle ear.

Inner Ear

Two functions are served by the inner ear: hearing (cochlear system) and balance (vestibular system). The cochlea is capable of converting fluid waves into neural impulses. The cochlea is a fluid-filled, snail-shaped organ containing specialized nerve endings known as hair cells. Vibrations transmitted by the middle ear to the cochlear fluid cause movement of the hair cells, triggering signal production which is carried by the auditory nerve (cranial nerve VIII) to the brain. The electrolyte content of the fluid within the inner ear is closely regulated by specialized transport systems. The kidney contains a nearly identical system, which it uses to regulate the electrolyte composition of the blood. This explains why may ototoxic agents are also toxic to the kidney.

Vestibular System

The vestibular system serves to balance the body. Dysfunction results in ataxia (incoordination), nystagmus (abnormal eye movements), and spatial disorientation. Few toxins produce isolated vestibulotoxicity, and most patients demonstrate hearing abnormalities concurrently.

Specific Otic Toxins

Aminoglycoside Antibiotics

The aminoglycoside antibiotics are well-known for their toxic side effects, renal and inner ear toxicity. Destruction of the hair cells of the cochlea produces hearing loss, beginning with high frequencies and progressing toward the lower frequencies. All of the aminoglycosides have the potential for such toxicity, but the relative toxicities differ. Newer techniques of administration such as bolus dosing may prove to reduce the frequency of ototoxicity. Aminoglycosides are also vestibular toxins, and such toxicity often precedes hearing loss (see Aminoglycosides).

Salicylates

Aspirin and other salicylic acid derivatives have a long history of use as analgesics and antiinflammatory agents. Tinnitus, or high-frequency ringing in the ears, is the most common sign of toxicity and is variably accompanied by hearing loss. The ability of aspirin to cause ototoxicity was so widely known that in the early part of this century, tinnitus was used as a clinical marker for therapeutic dosing. Aspirin toxicity, which is almost always reversible, is probably due to interruption of the normal metabolic processes of the sensory hair cells. Several structurally unrelated nonsteroidal antiinflammatory agents are capable of producing toxic symptoms similar to aspirin, suggesting involvement of the prostaglandin system in toxicity (see Salicylates).

Diuretics

The toxic effect of diuretics on the inner ear is related to the alteration of the fluid contained within. Changes in the electrolyte composition of the inner ear fluid causes swelling of the structures of the inner ear and tinnitus. However, this cannot be the only toxic mechanism since symptoms may occur immediately upon large exposure, at which time the fluid composition has not yet been altered.

Occupational Exposures

Surprisingly little research has been performed on the otic effects of chemicals on workers. However, several widely used chemical are known to be ototoxic. However, the combination of toxin exposure and noise may be additive or synergistic in the production of hearing loss. This has made investigation of the isolated toxic effects on exposed workers difficult (Table S-5).

Olfaction

The sense of smell is our most sensitive special sense. We are able to detect the odor of certain chemicals in the parts per million range. However, some chemicals are undetectable altogether, and others cause olfactory fatigue, in which exposure to a substance reduces our ability to detect its odor (e.g., hydrogen sulfide). Even more interesting, certain people are unable to detect certain odors, while others can detect the same odor

TABLE S-5
Agents Associated with Hearing Loss

Occupational
  Bromates
  Carbon disulfide
  Carbon monoxide
  Lead
  Mercury
  Styrene
  Toluene
  Trichloroethylene
  Xylene
Drugs
  Aminoglycosides
  Diuretics
  Chemotherapeutic agents
  Salicylates

at tiny concentrations (e.g., detecting cyanide's almond odor seems to be genetically determined). The most common toxic insult is cigarette smoking, and this must be considered in all patients with olfactory or taste dysfunction.

Olfactory receptors, numbering about 20 million per nares, are receptor ends of neurons that form the olfactory nerve (cranial nerve I). This nerve sends projections to many parts of the brain, which helps explain why smells often elicit profound emotional responses or memories. In addition, irritant receptors exists within the nasal cavity, which are unrelated to smell. This is sometimes called the "common chemical sense" and connects to the brain via the trigeminal nerve (cranial nerve V). This differential recognition of irritants from odors is clinically useful. Patients complaining of dysfunctional odor recognition should have normal recognition of ammonia and other pungents. Failure to recognize the irritant raises the possibility of malingering.

Few agents are acutely toxic to the olfactory receptors. Hydrogen sulfide, as mentioned earlier, causes rapid olfactory fatigue, and the normal "rotten egg" odor quickly vanishes allowing prolonged exposure to this potentially fatal toxin. Occupational exposure to several solvents and metals has been associated with olfactory dysfunction (Table S-6).

Gustation

Disorders of taste, like that of smell, are generally of limited toxicologic interest. The sensation of taste, like smell, is receptor mediated. Most abnormalities of gustation involve detection of abnormal tastes, not reduction in overall function of the sense.

Taste receptors reside within taste buds on the tongue, the larynx, and the palate. There are four primary taste sensations: sour, sweet, bitter, and salty. By mixing these primary taste sensations, the brain can identify many specific tastes (analogous to primary color mixing). Impulses from the taste buds are carried through the facial, glossopharyngeal, and vagus nerves (cranial nerves VII, IX, and X, respectively) to the brain. Taste is modified by the presence of odor, and in the absence of olfactory ability taste is virtually eliminated.

A metallic taste is often noted with exposure to metals or metal-containing compounds, tetracycline, mushrooms (Coprinus), snake venom, and others. Metal fume fever, a febrile reaction to metals volatilized during welding, also produces a metallic taste. An abnormal garlic sensation is experienced after exposure to dimethylsulfoxide, organophosphate insecticides, and arsenic. Such an abnormal taste sensation is likely due to cross-recognition of certain chemical agents by specific taste receptors (Table S-7).

TABLE S-6
Agents Associated with Smell Disorders

Occupational exposures
  Acrylate and derivatives
  Cadmium dust
  Carbon disulfide
  Formaldehyde
  Hydrogen sulfide
  Solvents (volatile hydrocarbons)
Drugs
  Antithyroid medications: methimazole, methylthiouracil
  Antihypertensives: beta blockers, captopril, enalapril
  Levo-dopa
  Opioids: morphine, codeine
Cigarette smoking
Cocaine insufflation

TABLE S-7
Agents Associated with Taste Disorders

ACE inhibitors: captopril, enalapril
Antibiotics
Carbamazepine
Chemotherapeutic agents
Cigarette smoking
Diuretics
Levo-dopa
Phenylbutazone
Metallic taste
  Allopurinol
  Ciguatoxin
  Coprinus mushrooms
  Disulfiram
  Ethambutol
  Heavy metals
  Lithium
  Metronidazole
  Methotrexate
  Penicillamine
  Penicillin
  Tetracycline

Further Reading

Goldfrank, L. R., Flomenbaum, N. E., Lewis, N. A., *et al.* (1994). *Goldfrank's Toxicologic Emergencies,* 5th ed. Appleton & Lange, Norwalk, CT.

Mott, A. E., and Leopold, D. A. (1991). Disorders of taste and smell. *Med. Clin. North Am.* 75, 1321–1353.

Schusterman, D. J., and Sheedy, J. E. (1992). Occupational and environmental disorders of the special senses. *Occup. Med.* 7, 515–542.

—*Lewis Nelson*

Related Topics

Behavioral Toxicology
Eye Irritancy Testing
Metals
Neurotoxicology: Central and Peripheral
Noise: Ototraumatic Effects
Occupational Toxicology
Organophosphates

Sertraline Hydrochloride

◆ CAS: 79559-97-0

◆ SYNONYMS: Zoloft; (1*S-cis*)-4-(3,4-dichlorophenyl)-1,2,3,4-tetrahydro-*N*-methyl-1-naphthalamine hydrochloride

◆ PHARMACEUTICAL CLASS: A naphthalenamine-derivative antidepressant agent; a selective serotonin reuptake inhibitor

◆ MOLECULAR FORMULA: $C_{17}H_{17}NCl_2HCl$

◆ CHEMICAL STRUCTURE:

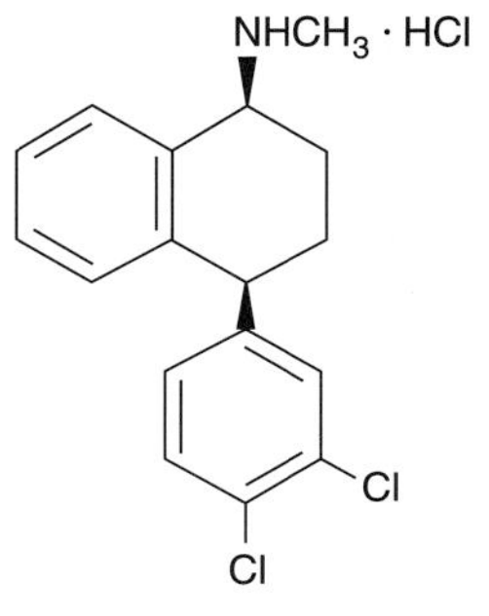

Use

Sertraline hydrochloride is used as an antidepressant.

Exposure Pathway

Ingestion is the most common route of both accidental and intentional exposure to sertraline hydrochloride. It is available only in oral tablet forms.

Toxicokinetics

Sertraline hydrochloride reaches a maximum plasma concentration in approximately 4.5–8.4 hr after the initial dose. Peak absorption may be delayed with large ingestions. When food was administered with the drug, both the peak and plasma levels and total absorption increased by 30–40%. The time to reach peak plasma concentration decreased from 8 to 5.5 hr. With repeated dosing, a steady-state plasma level should be achieved within 7 days.

Sertraline hydrochloride undergoes extensive first-pass metabolism and biotransforms via N-demethylation *in vivo* to form a primary amine. The principal metabolite is *N*-desmethylsertraline, which has a half-life of 2–4 days but is substantially less active than the parent compound.

Sertraline is highly protein bound (98 or 99%). Its volume of distribution is estimated at 20 liters/kg. Sertraline hydrochloride has an elimination half-life of approximately 24–26 hr. The α-hydroxy ketone metabolite is excreted in the urine and feces. The half-life of the metabolite, desmethylsertraline, ranges from 62 to 104 hr.

Mechanism of Toxicity

Sertraline hydrochloride is a potent and highly selective serotonin reuptake inhibitor that increases the availability of this neurotransmitter to the synaptic cleft with minimal effect on the neurotransmitter receptor systems. It shows no significant affinity for adrenergic, cholinergic, γ-aminobutyric acid, dopaminergic, histaminergic, serotonergic, or benzodiazepine receptors *in vitro*.

Human Toxicity: Acute

Sertraline has a low risk of toxicity. It is less sedating and has fewer cardiovascular effects than the tricyclic antidepressants. It has a high therapeutic index that is consistent with other serotonin uptake inhibitors.

Ingestions of up to 4500 mg have been tolerated without profound toxicity. The patient may develop various symptoms including nausea, vomiting, drowsiness, tachycardia, dilated pupils, slurred speech, ataxia, and lightheadedness. Therapeutic plasma concentrations have not been defined.

Human Toxicity: Chronic

Therapeutic chronic use of sertraline has reportedly caused visual defects, cardiac toxicity, gastrointestinal irritation, renal pathology, and nutritional aberrations.

Clinical Management

Basic and advanced life-support measures should be utilized as necessary. Gastric decontamination with syrup of ipecac may be indicated in recent ingestions in accordance with the appropriate level of consciousness and the history of ingestion. Otherwise, treatment should include gastric lavage and administration of activated charcoal. Treatment recommended after decontamination is symptomatic and supportive with cardiac monitoring. There is no antidotal treatment. However, since patients taking sertraline may also have leftover prescriptions of tricyclic antidepressants, the patient should be monitored for cyclic antidepressant poisoning for at least 6 hr.

—*Lanita B. Myers*

Shampoo

- ◆ SYNONYMS: Neutrogena; Head and Shoulders; Prell; Pert; Flex
- ◆ CHEMICAL CLASS: Combination of nonionic, amphoteric, and anionic surfactants (see Surfactants, Anionic and Nonionic)

Uses

Shampoos are used to wash the hair and scalp; they are available in noncoloring and coloring formulations.

Lindane shampoos are available for the treatment of lice; antidandruff formulations are also available.

Exposure Pathways

Ingestion is a common route of exposure. Ocular and dermal exposure occur as well.

Toxicokinetics

There is minimal absorption of anionic, nonionic, and amphoteric surfactants. Antidandruff shampoos may contain zinc pyridinethione and selenium sulfide. The kinetics of zinc pyridinethione have not been studied. Selenium sulfide is poorly absorbed. Peak serum levels of lindane (an ingredient in shampoos used to treat lice infestation) occur approximately 6 hr after a single dermal application. Lindane is highly lipid soluble and is stored in adipose tissue. Lindane is metabolized in the liver to chlorophenols (see Lindane).

Amphoteric, anionic, and nonionic surfactants are eliminated in the urine and feces. Selenium salts are excreted in the urine. Lindane has a half-life of approximately 18–21 hr following dermal application.

Mechanism of Toxicity

The surfactants and other adjuvants in shampoo are primarily irritants, and most dermal, ocular, or gastrointestinal toxicity is a consequence of the irritant properties.

Human Toxicity: Acute

Nonionic and anionic surfactants and selenium and zinc pyrithione shampoos are irritants by nature. Nausea and vomiting can occur following ingestion in large exposures. Spontaneous emesis is common. Persistent vomiting has the potential to cause fluid and electrolyte imbalance. In general, gastrointestinal irritation is self-limiting.

Acute ingestion of lindane shampoo does have the potential to cause central nervous system excitation. Toxicity can occur when children ingest 1 teaspoon or more of 1% lindane shampoo. Ingestion of 1 tablespoon or more of lindane shampoo may result in significant toxicity. Symptoms of lindane toxicity include agitation, tremors, seizures, and respiratory depression.

Human Toxicity: Chronic

Chronic dermal application of 1% lindane shampoo does have the potential to cause lindane toxicity.

Clinical Management

Dilution is generally all that is required in exposures to nonlindane-containing shampoos. If spontaneous emesis does not occur, then it is unlikely that a large ingestion occurred. If persistent vomiting occurs, then fluid and electrolytes should be monitored.

In toxic exposures to lindane shampoos, basic and advanced life-support measures should be utilized as needed. Emesis is not recommended in oral exposures to lindane. Gastric lavage and activated charcoal can be utilized. Milk and fatty foods should not be administered in oral lindane exposures since this may enhance absorption.

Animal Toxicity

In general, nonlindane shampoos do not produce toxicity. Irritant effects are expected. Exposure to lindane shampoos can produce vomiting, tremors, increased salivation, and seizures. Treatment is aimed at appropriate gastrointestinal decontamination and control of seizures.

—*Bridget Flaherty*

Shellfish Poisoning, Paralytic

- SYNONYM: Red tide toxicity (see Red Tide)
- DESCRIPTION: Poisoning caused by a neurotoxin sometimes contained in shellfish found on the east and west coasts of the United States and Canada, the area around Japan, and the area from southern Norway to Spain
- IMPLICATED SOURCES (shellfish): Mussels, clams, scallops, univalve mollusks, starfish, xanthid crabs, sand crabs, turban shells

Exposure Pathway

Ingestion of toxin-infected bivalve shellfish is the route of exposure. There is no reliable taste, smell, or color to detect contaminated shellfish. The toxin is not destroyed or inactivated by heating or cooking.

Toxicokinetics

The toxin is water soluble and absorbed through the mucosa of the mouth and small intestine.

Mechanism of Toxicity

Neosaxitoxin, saxitoxin, and gongantoxin I–IV block transmission of impulses between nerve and muscle. They also block sodium channels in nerve and skeletal muscle, inhibiting the nerve and muscle action potential, thereby blocking nerve conduction and muscle contraction (see Saxitoxin).

Human Toxicity

Initial effects include numbness in the fingertips, lips, and tongue within a few minutes of ingestion. This numbness may then spread to the legs, arms, and the remainder of the body and can lead to muscle paralysis. A brief gastrointestinal episode of nausea and vomiting may also be noted initially. Other symptoms include nystagmus, temporary blindness, irregular heartbeats, drops in blood pressure, headache, dizziness, difficulty in swallowing, and loss of gag reflex. These symptoms may persist for a period lasting a few hours to several days.

Clinical Management

Basic and advanced life-support measures should be utilized as necessary. Treatment is generally symptomatic and supportive. Gastrointestinal decontamination procedures may be used as appropriate to the patient's clinical status and the history of the ingestion. Fluids should be given as necessary for vomiting or for decreased blood pressure.

Animal Toxicity

Shags, terns, and cormorants may develop inflammation of the gastrointestinal tract, hemorrhages in the base of the brain, and other hemorrhages.

—*Gaylord P. Lopez*

Related Topics

Cardiovascular System
Neurotoxicology: Central and Peripheral

Shigella

♦ DESCRIPTION: Shigellae are nonsporing, noncapsulated, gram-negative, rod-shaped bacteria.

Exposure Pathway

Ingestion of food contaminated with shigella organisms is the most common route of exposure.

Mechanism of Toxicity

The shigellae do not produce an enterotoxin outside the intestinal tract. The ingestion of live bacilli is required to produce disease. The pathogenesis of these bacilli requires that they invade and penetrate the epithelial cells of the intestines in which the toxins are produced. The multiplying shigella toxin can cause destruction in gastrointestinal mucosa.

Shigella is usually transmitted via several types of food and through fecal contamination due to poor sanitary habits by food handlers. The organism can be found in fruits, vegetables, and milk. Person-to-person transmission through oral–anal routes is also common. Both an enterotoxin and neurotoxin evolve from the organism. Insects such as flies may also be a source of shigellae contamination of food. Shigellosis is more common than realized.

Human Toxicity

Patients present with rapid onset of fever, severe abdominal cramping, and diarrhea. The incubation period is 24–72 hr before the onset of bloody diarrhea and the influenza-like symptoms. The disease process may persist for 7 days. Children 6 months to 5 years are at highest risk. Vomiting is uncommon. Stools are watery, odorless, yellow-green, commonly mucoid, and bloody. In children febrile seizures are common and are usually generalized and self-limited. Dehydration often occurs with dry mucous membranes, decreased skin turgor, increased heart rate, and decreased blood pressure. Cough and rhinorrhea can also be associated with shigella food poisoning.

Clinical Management

Patients with mild fluid deficits can often be managed with oral fluid therapy. Patients with moderate to severe dehydration are generally treated with intravenous fluids. Documented cases should be aggressively treated with the appropriate antibiotics.

The mode of transmission is the means of prevention. Strict personal hygiene in the preparation of food, use of a sanitary water supply, proper disposal of sewage and human fecal matter, and prompt refrigeration of foods that must be eaten raw or cold following preparation is the key to preventing this type of food poisoning. Some ways to prevent contamination are to remove infected individuals from handling food, properly dispose of human feces, purify the water supply, ensure pasteurization of milk and milk products, maintain cleanliness in preparing food, and properly refrigerate food and milk.

—*Vittoria Werth*

Short-Term Exposure Limit

A short-term exposure limit (STEL) is a concentration of an airborne chemical substance to which workers can be exposed for brief periods without adverse effects. STELs are intended to prevent effects of short-term exposure such as eye and respiratory track irritation, irreversible toxicity to internal organs, or narcosis.

The U.S. Occupational Safety and Health Administration (OSHA) enforces various STELs for airborne chemicals in workplaces (e.g., benzene, ethylene oxide, formaldehyde, and methylenediamine). These standards are worker exposure levels which may not be exceeded over a specified averaging time, usually 15 min. Other OSHA standards use the term "excursion limit" synonymously with STEL (ethylene oxide standard) or simply define a concentration averaged over a short time period which may not be exceeded without assigning it a name (vinyl chloride standard). The definition of a STEL is not consistent throughout OSHA regulations and the actual standard for a compound of interest should be consulted.

The ACGIH Threshold Limit Values (TLVs) are guidelines for the protection of workers, not legally enforceable standards, and include STELs for airbornecontaminants. The TLV-STELs are 15-min time-weighted average concentrations which should not be exceeded at any time during the workday.

Further Reading

American Conference of Governmental Industrial Hygienists (ACGIH). *Threshold Limit Values for Chemical Substances and Physical Agents and Biological Exposure Indices.* ACGIH, Cincinnati, OH.
Occupational Safety and Health Standards for General Industry (1993). *Code of Federal Regulations,* Title 29, Part 1910.1000-1050.

—*Charles Feigley*

Related Topics

American Conference of Governmental Industrial Hygienists
Exposure
Exposure Assessment
Exposure Criteria
Levels of Effect in Toxicological Assessment
Occupational Toxicology
Permissible Exposure Limit
Threshold Limit Value

Sick Building Syndrome

Sick building syndrome (SBS) is a term used to circumscribe office worker discomfort and medical symptoms related to buildings and pollutant exposures, work organization, and personal risk factors. A wide range of definitions exist. Symptoms commonly considered integral parts of the syndrome are listed in Table S-8. In recent years, with increased understanding, odors have generally been dropped from the list and chest symptoms have been included under mucous membrane irritation.

TABLE S-8
Sick Building Syndrome

Mucous membrane irritation	Eye, nose, and throat itching and irritation
Central nervous system symptoms	Headaches, fatigue, difficulty concentrating, lethargy
Chest tightness and asthma-like symptoms (without true wheezing)	
Skin itching and irritation	
Odors	
Diarrhea	

The problem may be viewed from the perspectives of (1) medicine and health sciences, to define symptoms related to work indoors and their associated pathophysiologic mechanisms; (2) engineering, based on design, commissioning, operations, and maintenance strategies and difficulties; and (3) exposure assessment, the formal measurement of specific pollutants.

Health and People

Since the mid-1970s, increasingly voiced office worker discomfort has been studied in formal ways including field epidemiologic studies using buildings or workstations as the sampling unit to identify risk factors and causes, population-based surveys to define prevalence, chamber studies of humans to define effects and mechanisms, and field intervention studies.

Cross-Sectional and Case–Control Studies

Approximately 30 cross-sectional surveys have been published (Mendell, 1993; Sundell *et al.,* 1994). Many of these have included primarily "nonproblem" buildings, selected at random. These consistently demonstrate an association between mechanical ventilation and increasing levels of symptoms. Additional risk factors have been defined in several case–control studies. Table S-9 presents a grouping of widely recognized factors.

Many of these factors overlap. For some, pathophysiologic explanations exist. Women are considered more likely to voice discomfort at any given level of exposure and are exposed, on average, to higher levels of pollut-

TABLE S-9
Risk Factors for and Causes of the Sick Building Syndrome

Personal	Atopy (allergies, asthma, eczema)
	Seborrheic dermatitis
	Work stress
	Gender
	Low job status and pay
	Increased tear-film break-up time
Work activities	More time spent at photoduplication
	Carbonless copy paper
	More time at video display terminals
	Increasing amounts of time spent at workstation
Building factors	Mechanical ventilation
	Inadequate maintenance
	High-fleecing surfaces (high surface area surfaces such as carpets and drapes)
	Carpets
	Recent renovation
	Inadequate operations strategies

ants associated with symptoms such as volatile organic compounds (VOCs) and particulates.

Factor and principal components analyses of questionnaire responses in cross-sectional surveys have explored the interrelationship of various symptoms. Consistently, symptoms related to a single organ system have clustered together more strongly than symptoms relating different organ systems. That is, eye irritation, eye tearing, eye dryness, and eye itching all appear to correlate very strongly, and little benefit is obtained from looking at multiple symptoms.

Controlled Exposure Studies

Animal testing to determine irritant properties and thresholds has become standard. A consensus method, American Society for Testing and Materials, is widely regarded as the basis. This method has been used to develop structure–activity relationships, to demonstrate that more than one irritant receptor may exist in the trigeminal nerve, and to explore interaction of multiple exposure. Recently, it has been used to demonstrate the irritating properties of office equipment offgassing.

Analogous to this method, several approaches have been defined to document methods and dose–response relationships for irritation in humans. This work meanwhile suggests that, at least for "nonreactive" compounds such as saturated aliphatic hydrocarbons, the percentage of vapor pressure saturation of a compound

is a reasonable predictor of its irritant potency. Some evidence also supports that increasing the number of compounds in complex mixtures decreases the irritant thresholds. That is, the more agents present, even at a single mass, the greater the irritation.

Controlled exposure studies have been performed of volunteers in stainless-steel chambers. Most have been performed with one constant mixture of VOCs (Molhave and Nielsen, 1992). These studies consistently document relationships between symptoms and increasing exposure levels. Office workers who perceived themselves as "susceptible" to the effects of usual levels of VOCs indoors demonstrated some impairment on standard tests of neuropsychological performance (Molhave and Nielsen, 1992). Healthy volunteers, on the other hand, demonstrated mucous membrane irritation and headaches at exposures in the range of 10–25 mg/m^3 but no changes on neuropsychological performance. Recently, office workers demonstrated similar symptoms after simulated work in environments where pollutants were generated from commonly used office equipment. Animals, using a standardized test of irritant potency, reacted similarly.

Population-Based Studies

At least three population-based studies have been published in Sweden, Germany, and the United States. The questionnaires differed considerably, and the studies do not allow prevalence estimate comparisons. Nevertheless, between 20 and 35% of respondents were thought to have complaints.

Mechanisms

A number of potential mechanisms and objective measures to explain and examine symptoms within specific organ systems have been identified. None of these have a high predictive value for the presence of disease and are not suitable for clinical diagnostic use. They are useful in field and epidemiologic investigations.

Eyes

Both allergic and irritant mechanisms have been proposed as explanations for eye symptoms. More rapid tear-film break-up time, a measure of tear film instability, is associated with increased levels of symptoms. "Fat-foam thickness" measurement and photography for documentation of ocular erythema have also been used. Some authors attribute eye symptoms at least in part to increased individual susceptibility based on those factors. In addition, office workers with ocular symptoms have been demonstrated to blink less frequently when working at video display terminals.

Nose

Both allergic and irritant mechanisms have been proposed as explanations for nasal symptoms. Measures that have successfully been used include nasal swabs (eosinophils), nasal lavage or biopsy, acoustic rhinometry (nasal volume), anterior and posterior rhinomanometry (plethysmography), and measures of nasal hyperreactivity (visual, using a dental prosthesis as a head fixative, and using an ear surgery microscope to measure distances and swelling).

Central Nervous System

Neuropsychological tests have been used to document decreased performance on standardized tests both as a function of controlled exposure (Molhave and Nielsen, 1992) and as a function of symptom presence (Middaugh *et al.,* 1992).

Engineering and Sources

Beginning in the late 1970s, NIOSH responded to requests for help in identifying causes of occupant discomfort in buildings. Although no standard investigative protocol was used, the primary cause of problems was attributed to ventilation systems (approximately 50%), microbiological contamination (3–5%), strong indoor pollution sources (tobacco, 3%; 14% others), pollutants entrained from the outside (15%), and the remainder unknown. On the other hand, Woods and Robertson published two well-known series of engineering analyses of problem buildings, documenting on average three problems that could be the source (Table S-10).

The current professional ventilation standard (ASHRAE 62-89) suggests two approaches to ventilation: a ventilation rate procedure and an air quality

TABLE S-10
Defined Engineering Problems in a Series of Problem Buildings

Problem category	Physical cause	Frequency Woods	Robertson
System design	Inadequate outdoor air	75	64
	Inadequate distribution	75	46
Equipment	Inadequate filtration	65	57
	Inadequate drain lines and pans	60	63
	Contaminated ducts and liners	45	38
	Humidifier malfunction	20	16
Operations	Inappropriate control strategies	90	—
	Inadequate maintenance	75	—
	Thermal and contaminant load charges	60	—

procedure. The former provides a tabular approach to ventilation requirements: office buildings require 20 cubic feet of outside air per occupant per minute to maintain occupant complaint rates of environmental discomfort at below 20%. This assumes relatively weak pollution sources. When stronger sources are present, that same rate will provide less satisfaction. For example, when smoking is permitted at usual rates (according to data from the early 1980s), approximately 30% of occupants will complain of environmental discomfort. The second approach requires the selection of a target concentration in air (e.g., particulates, VOCs, and formaldehyde), information on emission rates (pollutant per time per mass or surface), and derives the ventilation requirements. Although this is an intellectually much more satisfying procedure, it remains elusive because of inadequate emissions data and disagreement on target concentrations.

In the past, odors were included under the etiologic list of SBS. A recent publication provides at least an overview of common odor sources (Boswell *et al.,* 1994). The single largest source was plumbing, such as dried-up traps (16%), followed by maintenance supplies (14%), renovations (11%), and ventilation (8%).

Pollutants

Environmental scientists have generally defined exposure and health effects on a pollutant-by-pollutant basis. In indoor environments these include multiple air pollutants (i.e., 20–50 different VOCs, including formaldehyde and other aldehydes), microbial products (including spores, cell fragments, viable organisms, and secretion products), and reactive agents such as ozone, fibers, and others. The American Thoracic Society defined six important categories listed in Table S-11.

Environmental criteria have been established for many of these, but the utility and applicability of such criteria for indoor environments is controversial for at least four reasons. For example, the goals of the TLVs often do not include preventing irritation, a primary concern in indoor environments with requirements for close eye work at video display terminals. For most of the pollutant categories, the problem of interactions, commonly termed the "multiple contaminants problem," remains inadequately defined. Even for agents that are thought to affect the same receptor, such as aldehydes, alcohols, and ketones, no prediction models are well established. Finally, the definition of "representative compounds" for measurement is unclear. That is, pollutants must be measurable, but complex mixtures vary in their composition. It is unclear whether the chronic residual odor annoyance from environmental tobacco smoke correlates better with nicotine, particulates, carbon monoxide, or other pollutants. The measure "total volatile organic compounds" is meanwhile considered an interesting concept but is not considered for practical purposes because the various components have such radically different effects (Molhave and Nielsen, 1992; Brown *et al.,* 1994). Particulates found indoors may differ in composition from those found outdoors because filter sizes affect entrained concentrations and indoor sources may differ from outdoor sources.

Finally, emerging data suggest that reactive indoor pollutants may interact with other pollutants and lead to new compounds. For example, ozone, either from office machines or entrained from outdoors, may interact with 4-phenylcyclohexene and generate adlehydes (Wechsler, 1992).

Primary Etiologic Theories

Organic Solvents

Buildings have always relied on general dilution strategies for pollutant removal, but designers have assumed that humans were the primary source of pollutants. Recently, emissions from "solid materials" (e.g., particle board desks, carpeting, and other furniture), from wet products (e.g., glues, wall paints, and office machine toners), and personal products (perfumes) have been recognized as contributors to a complex mixture of very low levels of individual pollutants (summarized in Hodgson, 1994).

Several studies suggest that the presence of reactive VOCs, such as aldehydes and halogenated hydrocarbons, is associated with increasing levels of symptoms. Offices with higher complaint rates have had greater "loss" of VOCs between incoming and outgoing air than did offices with lower complaints. In a prospective study of schools, short-chain VOCs were associated with symptom development. In another survey, higher personal samples for VOCs using a screening sampler that "overreacts" to reactive VOCs such as aldehydes and halogenated hydrocarbons were associated with higher symptom levels. In that study, women had higher levels of VOCs in their breathing zone, suggesting another potential explanation for the increased rate of complaints among women. VOCs might adsorb onto sinks, such as fleecy surfaces, and be reemitted from secondary sources. The interaction of ozone and relatively nonirritant VOCs to form aldehydes is also consistent with this hypothesis.

The presence of multiple potential sources, the consistency of VOC health effects and SBS symptoms, and the widely recognized problems with ventilation systems make VOCs an attractive etiologic agent. Solutions beyond better design and operation of ventilation systems include the selection of low-emitting pollutants, better housekeeping, and prevention of "indoor chemistry."

Bioaerosols

Several studies have suggested that bioaerosols have the potential to contribute to occupant discomfort.

TABLE S-11
Principal Pollutant Categories (American Thoracic Society)

Bioaerosols
Combustion
Environmental tobacco smoke
Radon
Volatile organic compounds
Fibers

They may do this through several different mechanisms: irritant emissions; release of fragments, spores, or viable organisms leading to allergy; and secretion of complex toxins. The proceedings of a recent conference (Johanning, 1994) will summarize this topic. Fewer data exist to support this theory than the others. Nevertheless, it is clear that heating, ventilating, and air-conditioning systems may be sources for microorganisms. They have also been described in building construction materials (as a result of improper curing), as a result of unwanted water incursion, and in office dust. The presence of sensitizers in the office environment, such as dust mites or cat danders brought in from home on clothing, presents another interesting exposure.

If biological agents contribute, dirt and water management become primary control strategies.

Psychosocial Aspects of Work

In all studies in which it has been examined, "work stress" was clearly associated with SBS symptoms. Workers' perceptions of job pressures, task conflicts, and nonwork aspects such as spousal or parental demands may clearly lead to the subjective experience of "stronger" irritation as a function of illness behavior. At times, such perceptions may in fact result from poor supervisory practices. In addition, the persistence of irritants leading to subjective irritation is thought to lead to work stress.

Conclusion

The SBS is a phenomenon experienced by individuals, usually seen in groups, associated with engineering deficiencies, and likely caused by a series of pollutants and pollutant categories. As with all "dis-ease," a component of personal psychology serves as an effect modifier to lead to varying degrees of symptom intensity at any given level of distress.

Further Reading

Boswell, R. T., DiBerardinis, L., and Ducatman, A. (1994). Descriptive epidemiology of indoor odor complaints at a large teaching institution. *Appl. Occup. Environ. Hyg.* **9**, 281–286.

Brown, S. K., Sim, M. R., Abramson, M. J., and Gray, C. N. (1994). Concentrations of VOC in indoor air. *Indoor Air* **2**, 123–134.

Cone, J., and Hodgson, M. J. (Eds.) (1989). Building-associated illness and problem buildings. *State-of-the-Art Rev. Occup. Med.* **4**, 575–802.

Hodgson, M. J., Levin, H., and Wolkoff, P. (1994). Volatile organic compounds and the sick-building syndrome. *J. Allergy Clin. Immunol.* **94**, 296–303.

Johanning, E., and Yang, Chris (1994). Bacteria and fungi in indoor air environments. Proceedings, Eastern New York Occupational Health Program. Boyd Printers, Albany, New York.

Knoeppel, H., and Wolkoff, P. (Eds.) (1991). Chemical, microbiological, health, and comfort aspects of indoor air. In *State of the Art in SBS*. ECSC, EEC, EAEC, Brussels.

Lockey, R., and Ledford, D. (Eds.) (1994, September). Symposium on the sick-building syndrome. *J. Allergy Clin. Immunol.* Supp. **2**, pt. 2, 214–365.

Mendell, M. J. (1993). Non-specific symptoms in office workers: A review and summary of the literature. *Indoor Air* **4**, 227–236.

Middaugh, D. A., Pinney, S. M., and Linz, D. H. (1992). Sick building syndrome: Medical evaluation of two work forces. *J. Occup. Med.* **34**, 1197–1204.

Molhave, L., and Nielsen, G. D. (1992). Interpretation and limitations of the concept "Total volatile organic compounds" (TVOC) as an indicator of human responses to exposures of volatile organic compounds (VOC) in indoor air. *Indoor Air* **2**, 65–77.

Selzer, J. M. (Ed.) (1995). Effects of the indoor environment on health. *State-of-the-Art Rev. Occup. Med.* **10**, 1–245.

Sundell, J., Lindvall, T., Stenberg, B., and Wall, S. (1994). SBS in office workers and facial skin symptoms among VDT workers in relation to building and room characteristics: Two case-referent studies. *Indoor Air* **2**, 83–94.

Walsh, C. S., Dudney, P. J., and Copenhagen, E. (Eds.) (1984). *Indoor Air Quality* pp. 87–106. CRC Press, Boca Raton, FL.

Wechsler, C. J. (1992). Indoor chemistry: Ozone, volatile organic compounds, and carpets. *Environ. Sci. Technol.* **26**, 2371–2377.

World Health Organization (WHO) (1987). *Air Quality Guidelines for Europe.* WHO Regional Office for Europe, Copenhagen.

—Michael Hodgson

Related Topics

Behavioral Toxicology
Dose–Response Relationship
Exposure Assessment
Indoor Air Pollution
Mixtures
Multiple Chemical Sensitivities
Neurotoxicology: Central and Peripheral

Psychological Indices of Toxicology
Respiratory Tract
Sensory Organs

Silver (Ag)

- ◆ CAS: 7440-22-4
- ◆ SELECTED COMPOUNDS: Silver chloride, AgCl (CAS: 7783-90-6); silver nitrate, $AgNO_3$ (CAS: 7761-88-8)
- ◆ CHEMICAL CLASS: Metals

Uses

Silver is used extensively in jewelry, eating utensils, coins, batteries, and dental amalgams. Silver solutions are used as antiseptics, astringents, and germicides. In some domestic water purifiers, silver is used to remove chlorine and kill bacteria. It has also been used in hair dyes. Medicinal use includes silver nitrate eyedrops for newborns (a legal requirement in some states). The main industrial use of silver is in the form of silver halide for the photographic industry. Silver halide is photosensitive, making it an ideal coating for photographic plates.

Exposure Pathways

Ingestion and inhalation are possible routes of exposure; dermal absorption of silver is unlikely. Silver is not a normal constituent of foodstuff. Very little, if any, silver is detected in domestic drinking water; however, some domestic water-purifying systems contain silver.

Toxicokinetics

Approximately 10% of ingested silver is absorbed. Inhaled silver can be absorbed from the lungs. Once absorbed, silver tends to precipitate in various tissues, as the affinity for sulfide by silver is immense. Silver tends to complex with the sulfhydryl groups. It is carried by globulins in the serum and forms complexes with the serum proteins, mainly albumin, which accumulate in the liver. Silver is not excreted in the urine; small amounts are excreted in the feces.

Human Toxicity

Workers chronically exposed to silver have experienced industrial argyria, an occupational disease characterized by discoloring of the skin. Blue-gray patches are noted on the skin and possibly the conjunctiva of the eye or the mucous membranes. Long-term exposure can result in extensive skin discoloration, mainly on parts of the body that are exposed to light (e.g., the face). Light may decompose the silver complex, resulting in extremely fine silver that gives the skin a metallic sheen. In some cases, the dark patches turn black.

Insoluble silver compounds (e.g., silver chloride, silver iodide, and silver oxide) are relatively benign. The soluble silver nitrate is corrosive; ingestion results in irritation of the gastrointestinal tract and possible damage to the lungs and kidneys. Chronic bronchitis has been reported following medicinal use of colloidal silver.

The ACGIH TLV-TWA is 0.1 mg/m³ for silver metal; the TLV is 0.01 mg/m³ for soluble silver compounds.

Clinical Management

Administration of table salt will help precipitate soluble silver as the insoluble silver chloride. BAL (British Anti-Lewisite; 2,3-dimercaptopropanol) has not proven useful.

—Arthur Furst and Shirley B. Radding

Related Topic

Metals

Sister Chromatid Exchanges

Sister chromatid exchanges (SCEs) are reciprocal exchanges of segments of chromatids; chromatids are the subunits of chromosomes, as visualized in metaphase, that become daughter chromosomes upon completion of cell division. Sister chromatid exchanges were discovered by J. H. Taylor in the late 1950s in experiments using the pulsed uptake of ^{3}H-labeled thymidine (TdR) in growing *Vicia faba* root tips, followed by autoradiography, to define the pattern of DNA replication in chromosomes.

Before Taylor's experiments, it was thought that one chromatid might be composed of newly synthesized DNA and the other of preexisting DNA. However, Taylor found that the DNA replicated semiconservatively, i.e., that in the first metaphase after [^{3}H]TdR incorporation (M_1) each chromatid was ^{3}H-labeled, which demonstrated that the chromatid was duplex, containing both preexisting and newly synthesized DNA. Furthermore, in the second division after ^{3}H incorporation (M_2) one chromatid was labeled and the other was unlabeled, i.e., both the preexisting and the newly synthesized DNA in each daughter chromosome had served as a template for the next round of DNA replication, again resulting in two sister chromatids. Thus, in M_3, one half of the chromosomes had one labeled and one unlabeled chromatid, and the other half of the chromosomes were entirely unlabeled. However, Taylor also noted that occasionally in M_2 one otherwise labeled chromatid had an unlabeled segment and, when this occurred, the corresponding segment of the otherwise unlabeled chromatid was labeled with [^{3}H]TdR, indicating a reciprocal exchange of segments between the two sister chromatids, i.e., a sister chromatid exchange.

The definition of SCE assays for genetic toxicology research and testing did not occur until the early 1970s, a time when a plethora of approaches were identified for assessing the potential genetic hazards of chemical exposure. Rather than using [^{3}H]TdR and autoradiography to visualize SCEs, the defined approaches are usually based on the more precise and efficient incorporation of bromodeoxyuridine (BrdU), an analog of thymidine, in two rounds of replication followed by Giemsa, or fluorescence-plus-Giemsa, staining of the chromosomes. Because of semiconservative DNA replication, the chromatids are equally stained in M_1 chromosomes, and the M_2 chromosomes possess one chromatid that is half-BrdU-substituted and one fully substituted chromatid, and the fully substituted chromatid is stained more lightly than the other. In M_3, one half of the chromosomes have differentially stained chromatids. SCEs are revealed in M_2 chromosomes by a "harlequin" pattern of darkly and lightly staining chromatid segment.

This approach has also been used to reveal chemical- and concentration-related delays in the progression of cells through the cell cycle as a preliminary test for selecting exposure conditions for chromosomal aberration assays. The objective of such preliminary tests is to define exposure conditions and harvest times that will yield sufficient numbers of first division, M_1, cells, for cytogenetic analysis because a high percentage of chromosomally damaged cells are unable to progress to the second and following metaphases.

In vitro SCE assays are routinely conducted in cultured Chinese hamster ovary (CHO) cells or human lymphocytes, and assessments of SCEs in human lymphocytes have been used for human population monitoring. *In vivo*, SCEs are usually visualized in bone marrow cells from mice implanted with BrdU. Such SCE assays have been used to test several hundred chemicals and have been shown to be highly sensitive and, in comparison to assays for chromosomal aberrations, to be more rapid, less subjective, and capable of detecting effects at lower dose levels.

SCE assays would, therefore, appear to be ideally suited for inclusion in initial batteries of tests to assess genotoxicity. However, although this was initially the case, the utilization of SCE assays has been greatly reduced for several reasons. First, it was found that the use of BrdU (or [^{3}H]TdR) induces SCEs; thus, there was concern that when SCE frequencies were elevated following chemical exposure, synergistic effects were being measured, which might not be as appropriate for risk assessment as the measurement of direct effects. Second, although there is strong evidence that SCEs result from misreplication of a damaged DNA template, probably from recombination at a stalled replication fork, there was uncertainty concerning whether to classify SCE assays as cytogenetic tests, as approaches for measuring the repair of DNA damage, or as an independent category of tests. Third, alarm was expressed when common chemicals such as NaCl (i.e., table salt) were found positive in *in vitro* SCE assays, although it was subsequently shown that in *in vitro* assays, particularly in the presence of exogenous metabolic activation, such false-positive results are absent if exposure conditions are monitored and adjusted to preclude acidic pH shifts and high osmolality.

However, the most significant reason for an absence of regulatory requirements for the routine use of SCE tests and their discontinuation by industry was the outcome of a NTP comparison of the concordance of results in four *in vitro* tests with results from rodent carcinogenicity bioassays, as described by Tennant *et al.* Specifically, NTP found that, although few positive results were obtained for noncarcinogens in the Ames test in *Salmonella typhimurium* or in the test for chromosomal aberrations in CHO cells, unacceptably high number of false-positive results were obtained in the mouse lymphoma cell mutagenesis assay (MLA) and in the *in vitro* SCE assay.

Thus, SCE tests have been largely discontinued by industry and are recommended by regulatory agencies on a very limited basis. Despite obtaining poor concordance for the SCE assay, however, it has been retained by the NTP, in part because SCE techniques are sufficiently similar to those used for *in vitro* chromosomal aberration assays so that the two tests can be used efficiently, in parallel, by cytogenetic testing laboratories. On the other hand, the NTP has essentially discontinued use of the MLA. However, because it has been shown that the poor concordance of the MLA was the result of the NTP's application of acceptability and evaluation criteria that are considered inappropriate by experts in that assay, the MLA has been retained as an initial genotoxicity test for regulatory submissions.

Future utility of SCE assessments will include population monitoring and, if the basis for the lack of concordance obtained by the NTP is found to be an artifact of techniques or methods of analysis, may again include the initial testing of chemicals for genotoxicity. In addition, with the advances in genomic mapping that have occurred during the past decade, an examination of SCEs may be useful for assessing the influence of exposure to specific agents on the transmission of identified genetic traits.

—*Ann D. Mitchell*

Related Topics

Ames Test
Analytical Toxicology
Carcinogen–DNA Adduct Formation and DNA
   Repair
Chromosome Aberrations
Developmental Toxicology
Dominant Lethal Tests

Host-Mediated Assay
Molecular Toxicology
Mouse Lymphoma Assay
Mutagenesis
Toxicity Testing

Skeletal System

This entry describes the structure and function of the musculoskeletal system and provides an overview of the categories of toxic effects that can affect this body system. The entry is divided into two principle parts, which form the main components of the musculoskeletal system: bone and skeletal muscle.

Bone

Bone, a form of connective tissue, composes the skeletal system. The skeletal system provides mechanical support for the body and protects internal organs such as the brain and heart, which are contained in skull and the chest wall cavity, respectively. The human skeleton is composed of 206 bones that vary in size and shape and include flat, trabecular, and cuboid bones. The body size and shape are determined by the skeletal system.

Bone serves other functions. It is a dynamic tissue that plays a vital role in mineral homeostasis and is a reservoir for several essential minerals including calcium, phosphorus, magnesium, and sodium. Bone houses the delicate bone marrow that forms blood from hematopoietic cells (see Blood). Bone is an extremely vascular tissue and receives up to 10% of the cardiac output.

Joints form the sites where bones come together or articulate. Joints are classified by the type of tissue that lies between the bones. Joints with fibrous tissue between the articulating surfaces are called fibrous joints and include the sutures of the skull. Cartilaginous joints are united by hyaline cartilage and are classified into primary and secondary cartilaginous joints. Primary cartilaginous joints do not allow any movement.

Bone is composed of live cells interspersed in an organic matrix. Inorganic elements or minerals (65%) are deposited into this organic matrix (35%), which makes bone one of the few tissues that normally mineralize. The principal inorganic element in bone is calcium hydroxyapatite (Ca$_{10}$[PO$_4$]OH$_2$), which accounts for approximately 99% of the calcium and 80% of the stores of these respective minerals in the body. Calcium hydroxyapatite provides bone with strength and hardness. The organic matrix provides a degree of elasticity to bone.

The cellular elements of bone include osteoprogenitor cells, which are pluripotential cells derived from mesenchymal tissue. Osteoprogenitor cells produce offspring cells that can differentiate into osteoblasts. Osteoblasts are responsible for the formation of the organic matrix of the bone into which the mineral elements can be deposited. Groups of several hundred osteoblastic cells coordinate activities to facilitate the formation of the organic matrix. The organic matrix is principally composed of type I collagen (90%) and several other noncollagenous proteins, including (1) osteocalcin, which serves to translate mechanical stresses or signals into local bone activity; (2) osteonectin, a calcium-binding protein; (3) osteopontin, a protein that facilitates cell adhesion; (4) cytokines; and (5) growth factors, which help control cell proliferation, mineralization, and metabolism. Osteoblasts have several different types of receptors including those for hormones (e.g., parathyroid hormone and estrogen) as well as other receptors for cytokines and growth factors.

Osteoblastic activity initiates the process of mineralization. Unmineralized bone is known as osteoid. Minerals are deposited in specific holes that are located between collagen fibrils produced by the osteoblast. The architecture of the fibrils is designed to withstand external stress. Mineralization begins shortly after the formation of the secreted matrix. This process occurs in osteons, also referred to as Haversian systems, and is completed in several weeks.

Osteoblasts which have become encased in bone are called osteocytes. The bony covering is not complete and osteocytes maintain communication with other cells and the general circulation through a network of tunnels located in the bony matrix which are called canaliculi. Osteocytes play several roles in body homeostasis, including maintaining normal levels of serum calcium and phosphorus.

Bone tissue also contains osteoclasts, which are multinucleated cells that are derived from the hematopoietic (granulocyte–monocyte) cell line located in bone marrow. Osteoclasts are primarily responsible for bone resorption and they secrete enzymes and hydrochloric acid that break down collagen matrix and help dissolve the bone. The area where osteoclast cell membrane lies adjacent to bony tissue is known as a Howship's lacunae. The osteoclast cell membrane that lies in close proximity to bone can contain numerous villous extensions and form a ruffled border. These areas are also known as resorption pits. The plasmalemma border of the osteoclast cell in this region forms a specialized seal with the underlying bone to prevent the release of enzymes and hydrochloric acid. This process also results in the release of growth factors previously deposited in bone by osteoblasts, which are responsible for maintaining the process of regenerating new bone.

Bone is developed by two methods. Membranous development involves bone formation directly from cartilaginous tissue. Osteoblasts directly deposit calcium and other mineral in this mesenchymal-derived tissue. The skull and portions of the clavicles (collar bone) are formed by this method and, at birth, portions of the membrane persist in the skull and are referred to as soft spots.

The other method of bone formation involves endochondral ossification and is not completed until the eighteenth year or later. The long bones of the limbs are formed by endochondral ossification. Mesenchymal-derived cartilaginous tissue, which is formed during early fetal development, contains chondrocytes and provides a model for future bone. By the eighth week of gestation, this cartilaginous tissue undergoes a series of changes which initiates the process of bone formation. The center of this cartilage undergoes degradative changes that involve mineralization and later resorption by osteoclast-type cells. This process moves up and down the cartilaginous tissue and is accompanied by the ingrowth of blood vessels and osteoprogenitor cells, which will become new bone-forming cells. The remnants of the mineralized cartilage, also known as the primary spongiosa, serve as a framework for new bone deposition. Similar changes occur in the epiphyses of the bone. These changes produce an area of cartilage that lies between two centers of bone formation which is known as the growth plate.

The growth plate chondrocytes undergo proliferation, growth, degradation, mineralization, and resorption and provide the support structure for new bone formation. Bones can increase in length and width through this process. Endochondral ossification occurs near the base of the articular cartilage at the joints.

Bone undergoes continual remodeling through bone resorption and formation. The balance between formation and resorption determines the mass of bone during growth of the skeleton. In childhood, bone formation predominates. Peak bone mass is reached in early adulthood. During adulthood, approximately 10–15% of the skeletal mass undergoes remodeling and resorption yearly and this process remains balanced. The amount of resorbed bone starts to exceed the amount of newly deposited bone by the third or fourth decade.

The osteoblast and osteoclast can be considered to be the basic multicellular units of bone. The osteoblast plays an important role in mediating local osteoclast activity through the release of chemical messengers. The principal factors responsible for stimulation of bone resorption, such as parathyroid hormone, interleukin-1, and interleukin-6, have minimal effects on osteoclasts, but osteoblasts have receptors for these substances.

Increased resorption of bone relative to new bone formation leads to osteoporosis. Osteoporosis is characterized by a net reduction in the mass of bone with no significant decrease in the ratio of mineral components to organic matrix. The bone can be thought of as having increased porosity. Osteoporosis starts to occur in both sexes at the age of 46–50. Trabecular bone loss probably occurs earlier. On average 0.7% of bone is lost on an annual basis. Bone loss accompanies aging for several reasons. Aging is associated with decreased activity of osteoprogenitor cells, decreased synthetic capability of osteoblasts, and the lessened biologic activity of growth factors contained in the organic bone matrix. Diminished physical activity associated with aging acts to reduce bone growth since exercise acts to stimulate the new bone formation. Bone loss occurs rapidly in astronauts who are in a weightless environment, with bed rest, or with the immobilization or paralysis of an extremity. Bone growth is stimulated by skeletal loading and muscle contraction associated with resistive exercises such as weight training.

Postmenopausal women are vulnerable to osteoporosis, which largely involves trabecular bones including the spinal vertebrae. Estrogen deficiency plays a major role since estrogen replacement reduces the rate of bone loss. The mechanism for this effect has not been fully characterized but decreased estrogen resulted in increased Interleukin-1 (IL-1) secretion from blood monocytes. IL-1 stimulates osteoclastic activity and bone resorption. Other risk factors include excessive alcohol consumption and smoking.

Bone is a target tissue for several xenobiotics. Several metals can effect the development of bone. Radiographs of bone in children with significant lead exposure can reveal lead lines, which are areas of increased bone density in the metaphyseal bone region. Lead lines are characteristically seen in rapidly growing tubular bones including the distal femur and proximal tibula and fibula (knee joint), but the vertebral bodies and iliac wing can also be affected. This effect has been attributed to the action of lead on the remodeling of calcified cartilage in the zone of provisional calcification of the metaphyseal bone (see Lead). Bismuth and yellow phosphorus can also produce similar metaphyseal bands.

Chronic ingestion of high concentrations of fluoride can produce fluorosis, whose clinical picture can include osteosclerosis. Osteosclerosis is a painful condition characterized by an increased density in the bones. This is thought to occur because hydroxyapatite is replaced with fluorapatite. Fluoride also accumulates in ligaments X-rays can demonstrate increased bone density; mineral deposits in ligaments, tendons, and muscles; and periosteal outgrowths. Osteosclerotic changes have been observed among aluminum workers with fluoride exposures and among persons with prolonged use of water containing high concentrations of fluoride.

Hypervitaminous A and D have also been associated with bone abnormalities. Vitamin D can cause resorption of calcium from bone. Chronic vitamin D intoxication may result in increased mineralization on bone and metastatic calcifications including joints, periarticular, and the kidney. Excessive vitamin D intake can cause demineralization of bone resulting in multiple fractures from very slight trauma (see Vitamin D).

Osteomalacia and likely osteoporosis among Japanese woman has been linked with ingestion of cadmium-contaminated food, including shellfish. This painful condition, known as "itai-itai byo" (ouch-ouch disease), has occurred primarily in postmenopausal multiparous women. Chronic exposure to cadmium has been associated with microfractures, osteomalacia, radiological decreases in bone density, and disturbances in calcium metabolism. One possible mechanism to account for this finding is increased serum parathyroid hormone and decreased serum vitamin D levels from cadmium-induced renal damage (see Cadmium).

Skeletal Muscle

Skeletal muscle is a major component of body tissue and accounts for 40–50% of the body weight. Skeletal tissue is composed of specialized striated cells which function to convert chemical energy to mechanical work. Skeletal muscle plays a central role in body metabolism and serves as a source of body heat and a storage depot for energy-rich compounds, protein, and intracellular ions (e.g., potassium). It also contains up to 80% of the body water content. In contrast to cardiac and smooth visceral muscle tissue, skeletal muscle is under voluntary control.

Skeletal muscle is composed of individual muscle fibers or cells that are contained in connective tissue. The muscle fibers are composed of hundreds or thousands of myoblasts. Muscle fibers are consequently multinucleated cells that have lengths of up to 10 cm and diameters ranging from 10 to 100 μm. They seldom are as long as the length of muscle which they compose and form interlocking irregular polygons. The size of muscle fiber is influenced by several factors. Proximal large muscles have large-diameter fibers, while smaller distal muscles contain more smaller diameter fibers. Physical activity can increase the size of muscle fibers in both sexes, although men of comparable age have larger fibers than woman. Children have smaller fibers.

Striated skeletal muscle fibers are bound together by collagenous connective tissue to form individual muscles. The connective tissue covering a muscle is known as the epimysium. This forms a resilient elastic sheath covering that separates the muscle from surrounding structures such as tendons and bone. The perimysium extends into muscle fibers and separates groups of individual muscle fibers or fasiculi. This connective tissue is known as the perimysium. Each muscle cell is surrounded by connective tissue known as the endomysium. This collagenous membrane combined with the adjacent muscle cell membrane is termed the sarcolemma. This tissue serves to maintain a framework for striated muscle cells. As long as the connective tissue remains intact, skeletal muscle can regenerate following injury and grow in the pattern provided by this connective tissue.

The muscle cell membrane is termed the plasmalemma. The cytoplasm of the muscle cell is filled with myofilaments, which form the myofibrils. Myofibrils are composed of sarcomeres, which consist of longitudinally directed thin and thick filaments and perpendicularly disposed z bands that are α-actin filaments. The myofibrils form the contractile apparatus of the muscle.

The sarcolemmal membrane has invaginations which run parallel to the z bands. These invaginations are also known as the T system and are involved in the release of calcium into the cell. The release of calcium leads to a contraction of the myofibrils. Sarcoplasm accounts for approximately 40% of the volume of the fiber and contains glycogen, mitochondria, and lipid vacuoles.

There are two principal types of skeletal muscle fibers in humans: type 1 and 2 fibers. They can be thought of as corresponding to red and white muscle. Type 1 fibers, or dark fibers, have more myoglobin and have the capacity for maintaining sustained force and weight bearing. They contain large numbers of mitochondria and maintain activity through sustained aerobic glycolysis. Type 2 fibers, or white fibers, are important in performing sudden and rapid movements. They have abundant glycogen but scant mitochondria and are not able to maintain sustained activity because they accumulate lactic acid. Strength training increases the number and size of type 2 fibers. Aerobic training involves hypertrophy of type 1 fibers.

A number of pathological processes can affect skeletal muscle. Since individual muscle fiber is formed by numerous myoblasts, any injury and pathological changes may only affect a small part of a muscle fiber. This has clinical significance since biopsy of a small segment of muscle may provide a nonrepresentative sample of muscle for assessment of a myopathy. Handling of specimens can be difficult and lead to artefactual lesions from fractures rising from the processing of the muscle. In general, the reactions of muscle are not specific to any disease or toxic agent.

Skeletal muscle can undergo atrophy due to several factors. Loss of innervation from anterior horn cell or a peripheral neuropathy can affect type 1 and type 2 fibers. This is characterized by a diminution of synthesis of myosin and actin and a decrease in size and resorption of the myofibrils; the cells, however, remain viable. Type 2 fiber atrophy can occur in several types of situations including inflammatory disorders involving muscle (e.g., polymyositis and polymyalgia rheumatica), metabolic disorders, corticosteriod myopathy, Cushings disease, and general cachectic states. Disuse of the muscle associated with inactivity (such as the placement of a cast) can lead to atrophy of muscle fibers, especially type 2 fibers. The cellular mechanism of the atrophy is poorly understood but could involve an increase in the rate of protein degradation, a reduction in protein synthesis, or a combination of both factors.

Hypertrophy, an increase in size of muscle cells, can occur in response to several situations. This is largely caused by an increase in the number of myofibrils. Hypertrophy is seen in more slowly progressive muscular dystrophies, a heterogeneous group of inherited muscle disorders (e.g., Becker's muscular dystrophy). Clinically they result in muscle wasting and weakness. Becker's muscular dystrophy is an X-linked disorder. Endocrine disorders including hypothyroidism and an increase in growth hormone or acromegaly can also be associated with hypertrophy.

Necrosis is the other principal muscular pathological process that muscle fibers can undergo. Usually the entire muscle fiber does not undergo necrosis. Segmental necrosis is the term used to describe necrosis confined to a segment of variable length of fiber rather then the entire fiber. The clinical spectrum of persons with systemic necrotizing myopathy typically includes proximal muscle limb weakness and elevated serum creatine kinase. Myoglobinuria can sometimes also be observed.

There are several potential causes for segmental necrosis of muscle fiber. The cause of this type of necrosis is not well understood but could be due to effects on the plasma membrane or the outer boundary of the muscle fiber. Aminocaproic acid (an antifibrinolytic medication used in the treatment of a subarachnoid hemorrhage), chlofibrate (used to treat hyperlipedemia), emetine (found in ipecac syrup), cardiac glycosides, heroin, and phencyclidine have been associated with necrotizing myopathies.

Select drugs injected by the intramuscular route can produce focal necrotizing myopathic changes. Paraldehyde, chlorpromazine, and a number of antibiotics have produced this type of reaction.

Interference with homeostasis of mitochondrial DNA has been linked to segmental necrosis. Medications associated with this type of effect include zidovidine, which is used to treat HIV. Electron microscopic findings include marked increases in mitochondrial enlargement with vacuolation. Chemicals that block aerobic metabolism (e.g., 2,4-dinitrophenol) have been used to experimentally produce a mitochondrial myopathy characterized in some instances by segmental necrosis.

Several drugs that share the chemical property of being a large cationic amhiphillic molecule that has a hydrophobic and hydrophilic region with a primary or substituted amine group with a net positive charge have been shown to produce a necrotizing myopathy. The mechanism of action is that these drugs interfere with lysosomal digestion and lead to autophagic degeneration and accumulation of phospholipds. Autophagic membrane-bound vacuoles containing membranous debris and curvilinear bodies with short, curved membrane structures with light and dark areas are seen. Drugs in this class include chloproquiin, vincristine, colchicine, and amiodarone.

Corticosteroids have produced necrotizing muscle changes. Severity is variable and not always associated with the steroid level or therapeutic regimen but is most likely to occur in persons taking over 40 mg of prednisone per day. Pathologically, degeneration of type 2 fibers is often seen.

Myopathies can also occur as a result of secondary effects. Hypokalemia has been associated with myopathy. Myopathy has been observed in persons consuming large quantities of licorice extract, in persons taking diuretic and some other medications, and in persons with purgative abuse.

Myopathies can be associated with immunologically based reactions that have features of polymyositis or dermatomyosotis. The clinical features of eosinophylia–myalgia syndrome include the abrupt onset of muscle pain. This syndrome was thought to be due to a contaminant in l-tryptophan introduced by the manufacturing process (see Eosinophilia–Myalgia Syndrome). Certain medications are associated with necrotizing myositis, which has similar features. Acute rhabdomyolysis is a severe form of necrotizing myopathy.

—*M. Joseph Fedoruk*

Related Topics

Blood
Eosinophilia–Myalgia Syndrome
Radiation
Tissue Repair

Skin

Introduction

S kin is the largest organ in the body. As the primary interface between the body and its external environment, it serves as a living, protective envelope which prevents the entry of foreign chemicals and microbes, as well as preventing the evaporative loss of body fluids and heat. Although skin is an effective barrier, it is not a complete one in that it is becoming increasingly apparent that skin is an important portal of entry into the systemic circulation for certain drugs and chemicals. Indeed, the increasing awareness that significant adverse health effects can result from skin exposure to toxins has highlighted our inadequate understanding of the function of this organ as it relates to percutaneous absorption and cutaneous toxicity.

Not long ago, studies in skin toxicology were primarily concerned with developing methods to produce and evaluate irritation and allergic reaction in both animal and human skin. However, significant recent advances in tissue culture techniques, cellular and molecular biology, and the understanding of toxicokinetic principles have enormously expanded our horizons in studies of skin function and toxicity, and we are just beginning to appreciate some of the more novel, but important, biochemical, physiological, and metabolic capabilities of this organ. The purpose of this entry is to provide a general overview of cutaneous toxicology. Current knowledge of the etiology and mechanisms of skin toxicity will be summarized and some of the more obvious and typical skin responses to toxic insults will be described. Furthermore, current concepts regarding skin absorption and metabolism will be discussed and, together, it is hoped that a review of these topics will provide a better understanding of the toxicology of the skin.

Skin Structure and Function

Mammalian skin can be described as a multilayer heterogeneous organ that forms the external covering of the body. It is the largest organ in the body and is continuous with the lining of orifices that open onto the body surface. In an adult man, the skin has a total surface area of approximately 2 m^2 and in most places it is no more than 2 mm in thickness, yet it can account for 10–20% of total body weight. The basic structure of mammalian skin can be divided into three main components: (1) a superficial lining of epithelial cells, the epidermis, supported by (2) a subepithelial connective tissue stroma and vasculature, the dermis, which in turn is supported on (3) a layer of subcutaneous fat of varying thickness, called the hypodermis. Impregnated within the epidermis and dermis are specialized "adnexa," which include hair follicles, sebaceous glands, sweat glands, and a complex neural network (Fig. S-1).

The epidermis, which develops from the embryonic ectoderm, comprises approximately 5% of full-thickness skin by weight. It is avascular and is composed primarily of keratinocytes. Based mostly on structural criteria, the epidermis can be subdivided into several layers. The basal layer consists of germinative cells, which retain the capability to undergo cell division and are extremely metabolically active. The daughter cells from the dividing basal layers migrate upward and undergo terminal differentiation to form the next two viable cell layers of the epidermis, the spinus and granular cell layers. During this process, the keratinocytes become flatter and lose many of their cytoplasmic organelles. Their nuclei condense and this is accompanied by the appearance of granules which ultimately form keratin filaments. The endproduct of this terminal differentiation process is the strateum corneum, the outermost layer of the epidermis. The cells of this layer are essentially flat, anucleuated, and devoid of any metabolic activity. These cells are eventually sloughed off to be replaced by terminally differentiating cells from the basal layer. This process of differentiation and outward migration in the epidermis is continuous. The average turnover time varies greatly from species to species: In humans it has been estimated to be about 28 days, but there is considerable variation, depending on anatomical site and disease state.

In addition to keratinocytes, the epidermis contains several "dendritic" cell types. Langerhan's cells, which account for approximately 5–10% of all cells found in the epidermis, are bone marrow mesenchyme-derived cells. These cells are involved in antigen recognition and processing during induction of immune responses in skin. Melanocytes are melanin-synthesizing cells of neural crest origin. These cells are found adjacent to the

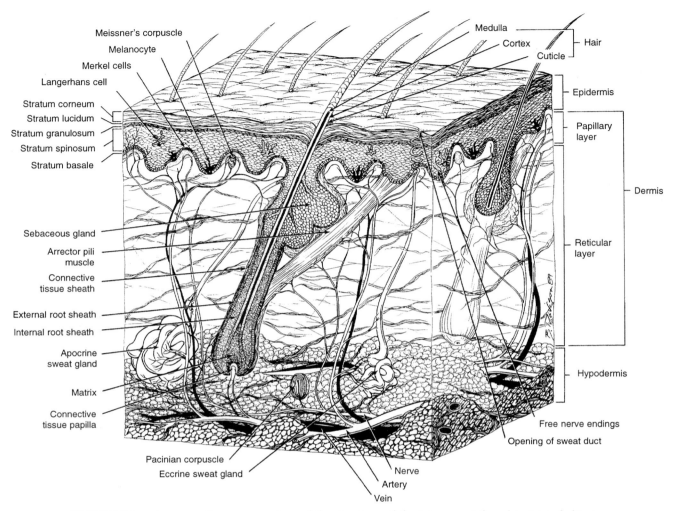

FIGURE S-1. *A composite representation of the structure of the integument found in typical skin in various regions of the body (reproduced with permission from D. W. Hobson (Ed.), Dermal and Ocular Toxicology: Fundamentals and Methods, CRC Press, Boca Raton, FL).*

basal cells and supply them with melanin, the principal pigment in skin, localized in specialized organelles, called melasomes. Merkel cells are the third type of dendritic cell found in the epidermis. They are of neuroectoderm origin and are believed to have a neuroendocrine function in the skin.

The basal lamina, with its characteristic ridge-shaped appearance, forms the epidermal–dermal junction and is the interphase that separates the epidermis from the underlying dermis. The dermis, which originates from the embryonic endoderm, consists of connective tissues and covers the internal organs of the body in a strong, flexible envelope. This envelope can be divided into two anatomical layers—the thin, outer

papillary layer and the thick inner reticular layer. The papillary layer consists of loose connective tissue, whereas dense connective tissue is found in the reticular layer. The elasticity and strength of this envelope are due to the constituent materials forming the dermal matrix. This matrix contains fibrous proteins, such as collagens, elastin, and reticulin, which are embedded in an amorphous material known as the ground substance, consisting of proteins and the glycosaminoglycans chondroitin A sulfate and hyaluronic acid. Fibroblasts are the most important and most numerous cell type found in the dermis. They are motile, capable of mitosis, and are responsible for the synthesis and secretion of the proteins and fibers that make up the bulk of connective tissue. In addition to fibroblasts, the dermis also contains a variety of cells scattered throughout this tissue, including macrophages, lymphocytes, adipocytes, and mast cells. The mast cells are of special interest and are most numerous in areas adjacent to skin appendages (hair follicles), blood vessels, and nerves. Their function in skin homeostatis is unclear; however, it appears mast cells are involved in the pathogenesis of some inflammatory conditions. They are generally indistinguishable from fibroblasts, except they have special intracellular granules containing histamine, heparin, and other vasoactive agents which are released in response to certain chemical irritants.

Within the dermis there are a number of epithelial structures, known collectively as cutaneous adnexa or epidermal appendages. They are all various types of extensions of modified epidermal cell structures into the dermis. The pilosebaceous units are located over the entire surface of the body, consisting of hair follicles and their associated sebacous glands, together with the accompanying arrector pili muscle, capillary plexus, and nerve fibers. The hair follicles are composed of three layers: the inner root sheath of keratinized cells, which form the hair shaft; the outer root sheath, which is a continuation of the epidermis; and the connective tissue layer. This latter layer merges with the papillary layer of the dermis and is continuous with the dermal papilla, located at the base of the hair follicles. The dermal papilla contains the germinative epithelial cells that give rise to hair proper through a complex process of growth, differentiation, and regression. Hair follicles undergo a continuous cycle of growth, called anagen, where new hair shafts are formed, followed by a short period known as catagen, where mitotic activity of the germinative cells in the dermal papilla ceases and the papilla atrophies. At telogen, the resting phase of hair cycle, the hair produced during anagen remains anchored to the skin, and at the next anagen phase, this hair is replaced by the newly synthesized hair. The sebaceous glands develop from the infundibulum of the hair follicle and are found surrounding the hair follicles. They contain differentiating cells that are active in lipid synthesis. The lipids, which form droplets in the fully differentiated cells, are released into the sebaceous ducts and onto the surface of the skin as sebum. This release of sebum is accompanied by the total disintegration of the lipid-laden cells. Sebum, a complex mixture of lipids, has antibacterial and waterproofing functions on the skin's surface.

Eccrine and apocrine glands, the sweat glands, are distributed throughout the skin. They are situated deep within the dermis and are simple coiled tubular structures composed of a long coiled secretory tubule and a long connecting excretory duct that traverses the epidermis and opens directly onto the surface of the skin. The eccrine glands are responsible for producing and secreting aqueous sweat (i.e., water and salts) and participate in thermoregulation. The apocrine glands secrete a viscous material containing proteins, pheromones, sugars, and ammonia. Their function in humans is unclear, but in animals they are believed to be associated with communication, probably acting in a sex-attractant or territorial marker role. The mammary glands are merely enlarged and modified apocrine glands. In man, the apocrine glands are found only in specific body regions, such as the axilla, the areola, the pubis, the perianal region, the eyelids, and the external auditory meatus. They secrete odorless secretions that are decomposed by surface bacteria to form characteristic odiferous products. Due to the combined secretions of the sweat glands and the sebaceous glands, the outer surface of the skin is generally always coated by an acidic film (pH 4–6), composed of various lipids, including triglycerides, phospholipids, and esterified cholesterol, together with salts and water. This film is frequently colonized by certain species of bacteria (e.g., *Micrococciae* and *Corynebacterium*) and the overall composition of this film may vary, depending on such factors as the disease state of the skin or occlusion. Any changes in the composition of the film will have dramatic effects on the makeup of the microflora present on the skin surface.

The dermis is separated from the underlying fiscia of muscle by the subcutis. The subcutis is a layer of adipose tissue of varying thickness, which, in humans, depends on the body region, sex, age, and nutritional status. The extensive dermal vasculature arises from the subcutis and consists of networks of vascular plexuses that are found in the transitional zone of the dermis and subcutis, adjacent to and surrounding the adnexa (eccrine sweat glands, hair follicles, and sebaceous glands). Arterioles branch out from these areas, forming anastomoses generally immediately under the epidermis, in both the reticular dermis and the papillary dermis. The dermal blood supply is usually substantially greater than that required merely to nourish the skin; thus, the dermal vasculature has an additional role in thermoregulation by controlling the dissipation of heat to the body surface. Intertwined with the dermal vasculature is a complex network of nerve plexuses consisting of both encapsulated and free nerve endings. These sensory and sensorimotor nerves ramify throughout the skin and are of extreme importance in the perception sensory stimuli. The skin also has a plexuses of lymphatics which drains into the regional lymph nodes.

This is a generalized and simplified description of mammalian skin. The basic architecture of skin is similar in all mammals. However, there are substantial species differences for aspects such as thickness, blood supply, and types or amounts of the various adnexa, as well as regional differences within each species. The most obvious species difference is hair follicle density in the skin. In lower mammals (rodents, dogs, cats, etc.), hair density is relatively high, whereas hair coat covering in humans and pigs is typically rather sparse. Skin thickness is another parameter that shows extensive species and regional differences, varying from a few microns to several millimeters thick (Table S-12). Furthermore, in humans, the predominant sweat gland

is the eccrine sweat gland which opens directly onto the surface of the skin, whereas in animals the apocrine sweat glands, emptying into hair follicles, predominate. In general, it can be said that the density and distribution of the adnexal structures provide the basis for species differences in skin structure and function while also contributing to the regional differences of skin anatomy noted for each particular species.

Percutaneous Absorption

General Concepts

Mechanisms of Percutaneous Absorption

In order to understand drug and chemical toxicity in the skin, the process whereby the various responsive cell types within skin are exposed to these agents must be examined. Percutaneous, or "via the skin," absorption may be defined as the translocation of surface-applied agents through the various layers of the skin to a location where they can enter systemic circulation via the dermal microvasculature and lymphatics or remain in the deeper layers of the skin. Based on our current knowledge, the important steps involved in skin absorption have been identified as the partitioning of the compound from the delivery vehicle to the stratum corneum, transport through the stratum corneum, partitioning from the lipophilic stratum corneum into the more aqueous viable epidermis, transport across the epidermis, and uptake by the cutaneous microvasculature with subsequent systemic distribution. This process, therefore, is the sum of the penetration and permeation of a chemical into and through the different strata of the skin.

Assessment of this process following topical application of drugs and environmental chemicals is becoming an increasingly important aspect of both toxicological and pharmaceutical investigations. Relative to toxicology, the ultimate aims of skin absorption studies are to identify and quantify the potential cutaneous toxicity, estimate the relative risk, and develop the appropriate strategies to minimize this risk resulting from topical exposure. In contrast, percutaneous absorption studies in transdermal delivery are designed primarily to assess and manipulate the rates of transport of drugs across the skin and ultimately to determine if such rates are sufficient to achieve the desired exposure and provide optimal therapeutic response. The aim is to identify or

TABLE S-12
Comparative Epidermal Thicknesses

Species	Epidermis (mm)	Stratum corneum (mm)
Mouse	9.7 ± 2.3[a]	3.0 ± 0.3
Rat	11.6 ± 1.0	4.6 ± 0.6
Rat	32.1 ± 1.3[b]	18.4 ± 0.5
Rabbit	15.1 ± 1.4	4.9 ± 0.8
Monkey	17.1 ± 2.2	5.3 ± 0.4
Dog	22.5 ± 2.4	8.6 ± 1.9
Cat	23.4 ± 9.9	4.3 ± 1.0
Cow	27.4 ± 2.6	8.1 ± 0.6
Horse	29.1 ± 5.0	7.0 ± 1.1
Pig	46.8 ± 2.0	14.9 ± 1.9
Pig	65.8 ± 1.8[c]	26.4 ± 0.4
Human	46.9 ± 2.3[d]	16.8 ± 0.7
Human	60–120[e]	20–25

[a]Mean + SE ($N = 6$); histologically determined in skin from the ventral abdomen.

[b]Mean ± SE ($N = 9$); histologically determined in skin from the back.

[c]Mean ± SE ($N = 35$); histologically determined in skin from the back.

[d]Mean ± SE ($N = 16$); histologically determined in skin from ventral abdomen.

[e]Range; histologically determined using skin taken from the ankle.

design the therapeutic compound with the appropriate properties for commercial development. Closely associated with these studies are investigations that are designed to assess the cutaneous and systemic bioavailability and bioequivalence of the compounds under development. Thus, depending on one's perspective, the focus of skin absorption studies may be to increase penetration or to reduce absorption.

The rate-limiting barrier to skin absorption is generally considered to be the outermost layer, the nonviable stratum corneum. Consequently, the skin is frequently thought of as a passive, inert barrier and percutaneous absorption of chemicals was thought to be dominated by laws of mass action and physical diffusion. This reduction of percutaneous absorption to diffusion equations and mass transfer coefficients has, unfortunately, overshadowed any considerations of the possible contribution of biochemical factors which may influence the percutaneous fate of topically applied substances. In general, the diffusional theories and the assumption that the skin is merely a physical barrier persist, despite the fact that the skin is an organ active in many essential biochemical and physiological functions. Moreover, for certain lipophilic chemicals, it is clear that the stratum is no barrier at all. The lipid-rich stratum corneum and skin appendages may act as a reservoir for topically applied lipid materials, thus functioning more as a sponge, capable of absorbing a quantity of material which is limited only by the solubility of the substances in the sebaceous and intrinsic epidermal lipids. For such lipophilic chemicals the viable epidermal membrane may be the more important barrier.

Skin appendages, which include sebaceous glands, hair follicles, and sweat glands, are often regarded as channels that bypass the stratum corneum barrier. As such, they are generally thought to facilitate the dermal absorption of topical agents. Because they occupy only a small fraction of the skin's surface area in humans (0.01–0.1%), their overall effect on the extent of percutaneous absorption will be minimal for most compounds. Moreover, it is often overlooked that these appendageal structures are not open pores through the skin, but are usually plugged with hair shafts, dead cells, sebum oils, stratum corneum lipids, and/or aqueous salt solutions (sweat). Thus, this pathway is probably only important immediately after a substance is applied to the skin as a rapid shunt. The bulk of skin absorption takes place via the diffusion processes described previously and later. However, the significance

of this follicular pathway in skin absorption remains to be experimentally assessed. The correlation between permeation and hair density of the different rat skin phenotypes (haired, fuzzy, and hairless) tends to support the hypothesis that the transfollicular pathway may be the more dominant route for the skin absorption of certain highly lipophilic chemicals such as polycyclic aromatic hydrocarbons and coal tars. In addition, these appendages may be important for highly polar molecules, which generally penetrate the stratum corneum very slowly, if at all.

Mathematical Models

Diffusion principles have been traditionally recognized as the most important determinants in skin absorption. Thus, Fick's law of diffusion provided the mathematical basis for early kinetic descriptions of percutaneous absorption. Fick's law simply states that

$$J_s = k_p \, \delta \, C_s$$

where J_s is net flux of substance s (μg/cm^2/hr), k_p is the permeability constant (cm/hr), and C_s is the concentration gradient of s across the diffusion membrane (μg/cm^3). The validity of applying mass diffusion principles to skin absorption rests on at least two main assumptions. First, penetration of chemicals into and permeation within the various layers of skin are passive diffusional processes. Second, the diffusional resistance of the skin layers can be formulated into one or more equations describing diffusion of small particles or molecules across thin layers or barriers, and the layers or barriers must simulate as closely as possible the behavior of dilute polymeric solutions. Because diffusional resistance of the outermost region of skin, i.e., the stratum corneum, is generally far greater than that of other cutaneous substructures, models of skin absorption are frequently simplified to involve diffusion across only one layer, the stratum corneum. This equation forms the simplest mathematical framework describing many percutaneous absorption investigations.

Unfortunately, percutaneous absorption is a complex phenomenon involving a myriad of diffusional and metabolic processess that are proceeding either concurrently or sequentially. Consequently, theoretical models describing the overall process will be approximations and will reflect our current knowledge concerning the most relevant events. Conceptual models of percutaneous absorption which are rigidly adherent to general solutions of Fick's equation are not always applicable to *in vivo* conditions, primarily because such models are not physiologically relevant. Linear kinetic models describing percutaneous absorption in terms of mathematical compartments which have approximate physical or anatomical correlates have been proposed. In these models, the various relevant events, including cutaneous metabolism, considered to be important in the overall process of skin absorption are characterized by first-order rate constants. The rate constants associated with diffusional events in the skin are assumed to be proportional to mass transfer parameters. Constants associated with the systemic distribution and elimination processes are estimated from pharmacokinetic parameters derived from plasma concentration–time profiles obtained following intravenous administration of the penetrant.

As is apparent, these linear kinetic models and diffusion models of skin absorption kinetics have a number of features in common; they are subject to similar constraints and have a similar theoretical basis. The kinetic models, however, are more versatile and are potentially powerful predictive tools used to simulate various aspects of percutaneous absorption. Techniques for simulating multiple-dose behavior; evaporation, cutaneous metabolism, microbial degradation, and other surface-loss processes; dermal risk assessment; transdermal drug delivery; and vehicle effects have all been described. These simulations are only theoretical, but in some cases the predictions compare favorably with the experimental observations. Recently, more sophisticated approaches involving physiologically relevant perfusion-limited models for simulating skin absorption pharmacokinetics have been described. These advanced models provide the conceptual framework from which experiments may be designed to simultaneously assess the role of the cutaneous vasculature and cutaneous metabolism in percutaneous absorption.

Due to the deficiencies of current experimental and analytical methods our ability to appreciate and fully utilize these sophisticated models is limited. As noted previously, the model parameters (e.g., rate constants) are usually based on assumed partitioning phenomena or kinetic behavior. These assumptions are limited by the paucity of kinetic information provided by current experimental methods. For example, little is known about how volatility affects absorption of the applied dose and the concept of mass-balance studies following topical doses has only recently been addressed. From the perspective of absorption, the skin is a portal of entry for a variety of topically applied chemicals, a drug-metabolizing organ, and a target organ for local toxicity. When skin contact with a chemical results in local effects, pathological changes in the skin may be expected to affect its barrier properties and, hence, influence the fate of surface-applied chemicals. The integrity of the stratum corneum is therefore of primary importance. However, biochemical changes in the skin in response to topical exposure to biologically active chemicals may also influence the metabolic capabilities and metabolic status of the skin and thereby modulate the cutaneous disposition of topically applied substances.

Thus, knowledge of the processes involved in the translocation of chemicals through the skin into systemic circulation and the response of the skin to such chemicals are important aspects of skin pharmacology and toxicology. Research in this area is in its infancy and offers many opportunities. Mechanistic and functional approaches to skin absorption need to be developed. Table S-13 presents a fairly comprehensive list of the known factors affecting percutaneous absorption of topically applied drugs and chemicals. It is anticipated that future research will increase our knowledge

TABLE S-13
Factors Affecting Percutaneous Absorption

Factor	Specific examples or other contributing factors
Vehicle	Suspensions, emulsions, lotions, creams, ointments, pastes, PEG and PPG, demulcents, emollients, PH
Solvents	Water, acetone, ethanol, methanol, chloroform, THF
Enhancers	Dimethylformamide (DMF), DMSO, dimethylacetamide, urea, azone, 2-pyrrolidone, surfactants
Species	Skin thickness, hair density, quantity and types of glands, cutaneous vasculature, and blood flow
Application site	Epidermal and stratum corneum thickness, keratinization, blood flow, hair follicles/glands, skin condition (dermatoses, damage, hydration state, occlusion, PH)
Environment	Ambient temperature and humidity, airflow
Dose applied	Surface area, concentration, contact time, vehicle
Physicochemical	Partition coefficients, molecular weight, particle size/shape, dissolution characteristics, absolute aqueous solubility
Miscellaneous	Humans (age, sex, race); metabolic capacity of the skin

of skin absorption, and exploitation of such knowledge would greatly facilitate the continual development of new strategies in reducing the skin absorption of hazardous industrial chemicals. Also, it would provide the basis for improving topical therapy and the transdermal delivery of drugs and prodrugs.

Metabolic Fate of Topically Applied Substances

It is now generally recognized that skin is an organ capable of performing a variety of metabolic functions, including those involved in the metabolism of hormones, carcinogens, drugs, and environmental chemicals. Since skin contains enzymes capable of metabolizing xenobiotics, any chemicals that are applied to the surface of the skin will, during the course of penetration and translocation through this organ, be exposed to available biotransformation systems that are present in the skin. Consequently, the ability of the skin to function as an organ of xenobiotic metabolism is of considerable interest, and questions are raised concerning the functional significance of skin metabolism in the percutaneous fate of chemicals. Is skin metabolism important? Can cutaneous metabolism influence dermal absorption? What, if any, is the functional significance of skin metabolism in the cutaneous and systemic disposition of chemicals, and can it be an important determinant in the development of local and systemic toxicity? What are the modulating factors that may affect skin metabolism and, consequently, the disposition of topically applied chemicals? What are the implications of skin metabolism in dermatotoxicity and dermatopharmaceutics? Indeed, will the ability of the skin to metabolize drug prove to be a desirable advantage or a confounding factor which complicates the development of novel transcutaneous therapeutic devices? Perhaps more important, what conceptual and experimental approaches are readily available for use in evaluating the functional significance of skin metabolism on the percutaneous fate of topically applied chemicals?

Numerous investigations with tissue slices, isolated cell preparations, and subcellular fractions from skin have shown this organ to possess a variety of enzyme activities including those involved in the metabolism of xenobiotics. Although the biotransformation of only two classes of compounds, steroids and polycyclic hydrocarbons, has been studied extensively in the skin, it is evident that a full complement of drug metabolizing

enzyme activities are present in the skin. These drug metabolizing enzymes are generally thought to be associated with the epidermal cells. Recent reports have suggested that high drug metabolizing activities are also localized in the differentiated cells of the hair follicles and adjacent sebaceous glands. It is possible that the resident microorgansims of the skin may also contribute to the metabolism of the topically applied compounds. However, during the development of methods for assessing skin graft viability, it was determined that the metabolic contributions due to skin microorganisms were negligible. Furthermore, experience in our laboratory with metabolism in human skin preparations which have been thoroughly scrubbed with antimicrobial disinfectants support a conclusion that it is the constituent skin cells themselves, rather than surface microbes, which are responsible for the observed biotransformations. Finally, since it is known that the metabolizing activities of the skin readily respond to modulation by inducers and inhibitors, such enzyme modulation could have important implications on cutaneous absorption and disposition experiments.

Experimental Models

In the past, percutaneous absorption investigations have usually concentrated on the physicochemical and biophysical factors that influence skin penetration and permeation of chemicals. There are a wide variety of experimental approaches which have been developed to assess skin absorption; however, it should be noted at the outset that currently there is no generally accepted technique. Debates and conflicting opinions continue to revolve around the various factors that are important in influencing percutaneous absorption. Consequently, the rationales by which experimental models are selected and developed are continually being modified and revised. Fundamentally, skin absorption investigations are concerned with how much, how fast, and what are the modulating factors that may influence the penetration and percutaneous fate of the topically applied agents. To answer these questions the primary methods available are based on *in vivo* and *in vitro* studies. The former is from the Latin phrase for "in life," i.e., using a living organism, while the latter means literally "in glass," or done in the test tube. Both approaches have their advantages and limitations, as will be discussed below.

In Vivo Techniques. It is generally recognized that the most reliable method for learning about skin absorption is to measure it *in vivo* using the appropriate animal model or human volunteers. In principle the *in vivo* approach is simple, but in practice it is often fraught with experimental and ethical difficulties, particularly when studies are conducted in man. Typically, *in vivo* studies are performed by applying the compound of interest, in a suitable vehicle, to the surface of a defined area of skin. To protect the application site, occlusive or nonocclusive covering is often placed over the treated skin area, and absorption is then monitored by various procedures. However, the techniques used to monitor *in vivo* skin absorption often assess absorption indirectly, and frequently measurements are based on nonspecific assays. The validity of the *in vivo* determination will depend, therefore, on the validity of the method used.

When a topically applied compound induces a biological response following skin absorption, the quantitation of that response may provide a basis for assessing skin absorption. Indeed, such physiological or pharmacological responses have been employed as endpoints in assessing skin absorption *in vivo,* and perhaps the most successful example is the vasoconstrictor response to topical corticosteroids. However, while these pharmacodynamic endpoints may be very sensitive and selective for defined classes of compounds, it should be noted that the parameter measured is the product of both the quantity and the potency of the compound under investigation and may not necessarily reflect the extent of skin absorption, cutaneous metabolism, or disposition.

Ideally skin absorption and metabolism should be assessed based on the analysis of the compound and metabolites of interest in the body following topical application, and such analysis should be performed using sensitive, selective, and specific assays. Although it has been possible in some select cases to determine a plasma concentration–time profile of the compound following topical application, such specific analyses in body fluids are not routinely feasible because the low absolute amount normally absorbed via the skin is often too small to quantitate. It is for this reason that radiolabeled compounds are frequently used, and the extent of absorption is typically assessed by monitoring the elimination of radioactivity in excreta over a period of several days. For small laboratory animals, the absorbed radioactivity which may be retained in the animal and not eliminated in the excreta can be determined directly by analysis of the carcass, following removal of the application site and homogenization of the appropriate tissues. However, in larger animals and in humans, such an approach is impractical, and a correction is required to adjust for such pharmacokinetic factors as absorption, distribution, metabolism, and excretion. This correction has often been made by injecting intravenously a single dose of the radiolabeled compound and monitoring radioactivity in the excreta. A correction factor can be obtained which represents the fraction of dose that would be excreted during the time course of the percutaneous absorption study if it were instantly absorbed upon topical application. This has been the standard approach by which the vast majority of *in vivo* skin absorption studies are conducted and has provided invaluable information concerning percutaneous absorption in humans.

When measurements from intravenous dosing are applied as a correction, the validity is dependent on the underlying assumption that metabolism and disposition of the applied compound are not route dependent and that pharmacokinetic behavior of the intravenous and topical doses are similar. Unfortunately, there is little or no experimental basis for substantiating this assumption, and often the pharmacokinetic profile of the compound under investigation has not been fully characterized. Kinetically, skin absorption resembles a slow infusion, but the intravenous dose for correction is often given as a single bolus injection. Subcutaneous injection or a slow intravenous infusion may be the more appropriate delivery method for correction. Moreover, the selection of the size of the intravenous dose is often not rationalized. When differences in the relative amount of radioactivity excreted in the urine and feces following intravenous and topical administration are observed, these differences may be the consequence of route of administration or they may be related to differences in extent of systemic exposure. Furthermore, when metabolites are found in the excreta following topical application, it is difficult, if not impossible, to differentiate between skin metabolism and systemic metabolism. As a result, the significance of cutaneous metabolism in skin absorption cannot be readily established from *in vivo* investigations.

More direct approaches for monitoring skin absorption have been proposed—for example, measuring the rate of disappearance of the chemical at the application site. However, the generally low permeability of the skin means that the rate of disappearance is often very slow, and the accuracy of the measurement will depend on analytical techniques that are capable of accurately quantifying minute differences. Reliable results can only be obtained with chemicals that are rapidly absorbed and/or easily quantitated analytically. The main use of this technique is monitoring the loss of radioactivity from the skin surface, but it should be appreciated that measurements using high-energy emitters whose transmission range may be similar to or greater than the thicknesses of the skin could result in erroneous estimates of skin absorption. Other methods, such as those based on histochemical and fluorescence techniques, are highly specialized and cannot be used with all compounds. The recently described approach in which the extent of percutaneous absorption was correlated to the reservoir function of the stratum corneum, measured by tape-stripping the application site and extracting to determine the amount taken up by the stratum corneum after a short exposure period (30 min), is an interesting alternative. However, its general utility for estimating skin absorption has yet to be established.

In most toxicological and pharmacological investigations, the dose administered is precisely defined and dose–response relationships are usually carefully evaluated. In percutaneous absorption studies, however, this is not always the case. A great deal of absorption information in the literature may be of questionable validity since the dose applied was frequently not clearly defined or reported, even though the extent of skin absorption is usually reported in terms of a percentage of the dose applied. Dose application in skin absorption studies conducted *in vivo* is relatively straightforward. The compound of interest is prepared in an appropriate vehicle which may be liquid or semisolid, and an appropriate amount of this preparation is then applied uniformly onto the surface of the skin. Uniformity of application is important but often difficult to assess and is generally assumed without supporting evidence. Furthermore, very little is known concerning the potential influence of local toxicity on cutaneous metabolism and skin absorption and it is suggested that whenever possible a "no-effect" level of the compound should be used in these types of studies.

Defining the amount of the topical dose applied that is available for absorption is particularly challenging when the compound under investigation is volatile or semivolatile as in the case of solvents and insect repellents. Following topical application, some of the applied dose will penetrate the skin and be absorbed. At the same time, some fraction will evaporate slowly from the surface of the skin and be lost, unavailable for percutaneous absorption. It has been demonstrated that the rate of evaporation, and consequently the relationship between evaporation and skin penetration, can influence the quantity of chemical absorbed dermally. Therefore, the extent of evaporation from the skin surface is a function of the dose applied, airflow, and temperature at the skin surface. The extent to which these variables may be controlled or monitored can have a major impact on the results of *in vivo* skin absorption studies. Furthermore, consideration of the evaporative loss of the applied dose will be particularly important when surface disappearance or stratum corneum concentrations are employed as methods for assessing *in vivo* skin absorption.

Vehicle as a modulating factor that can influence skin absorption has been discussed in great detail, particularly from a standpoint of increasing absorption in the delivery dermatopharmaceutics, and there is much interest in solvents such as dimethylsulfoxide and azone as vehicles because they act as penetrant enhancers. Postapplication loss of volatile components in the vehicles can also alter the permeation characteristics of the applied chemicals. For example, if a highly volatile vehicle is used this may result in the compound under investigation being deposited as a thin film of solid onto the surface of the skin. On the other hand, a nonvolatile vehicle, such as an ointment, may be occlusive and change the diffusional properties of the stratum corneum. Both of these scenarios can greatly influence the extent of percutaneous absorption. Therefore, the rationale used to justify the selection of an appropriate vehicle for dose application will have important bearing on the significance and validity of the *in vivo* observations.

The extent of skin absorption is greatly dependent on the concentration of the applied dose and the surface area of exposure. Increasing the concentration of the applied dose has been shown to result in a decrease in the percentage of the applied dose being absorbed, but total absorption is increased. This effect may be compound specific and may depend on the dose range under investigation. Moreover, increasing the surface area of exposure will also result in increases in the extent of absorption. In defining the dose applied, therefore, one must consider not only the amount of chemical applied per unit area but also the total surface area of application and the total dose applied. The frequency of application and the duration of exposure have also been shown to influence the extent of skin absorption. In the few times that it has been investigated, the results have shown that washing of the application site to remove the applied dose may enhance, reduce, or have no effect on absorption. Studies on the interrelationship and influence of the various parameters pertaining to dose application in skin absorption are in their infancy. How these parameters may influence the extent of skin absorption is being explored, and it is clear that our current knowledge in this area is far from complete.

In Vitro Techniques.

In Vitro Techniques. From a cursory review of the literature on percutaneous absorption it is evident that much of our current understanding of the mechanism of percutaneous absorption was derived from *in vitro* investigations. *In vitro* experiments generally afford the investigator with the ability to manipulate and control the experimental conditions, and the approach provides the unique opportunity to monitor the rate and extent of percutaneous absorption in skin tissues removed from the confounding influences of the rest of the body. *In vitro* methods, primarily those involving excised skin mounted in diffusion chambers, are the most frequently employed techniques used in the assessment of skin absorption.

Generically, these diffusion chambers consist of a donor and a receptor compartment. Skin absorption is then determined based on the assumption that recovery of the compound of interest in the receptor compartment, following application to the skin surface in the donor compartment, will provide an accurate measure of penetration and permeation. The success and popularity of the *in vitro* approach stem from the fact that the techniques are relatively simple. In these experiments the investigator is provided with the ability to monitor the rate and extent of absorption through skin removed from the influence of other bodily organs. Experimental conditions can be readily manipulated and controlled and, compared to *in vivo* studies, *in vitro* results can be obtained relatively quickly. Furthermore, it is recognized that *in vitro* methodology has contributed significantly in defining the important physiochemical parameters underlying percutaneous penetration and is responsible for much of our current understanding on the mechanisms involved.

Typically, an appropriate fluid is placed into the receptor compartment and an appropriate formulation of the compound under investigation (usually radiolabeled) is placed in contact with the skin surface in the donor compartment. The recovery of radioactivity over time in the receptor fluid then provides an estimate of skin absorption. The justification of this methodology centers upon the generally accepted assumption that the stratum corneum is the principal rate-limiting barrier in skin absorption. Since this outermost layer of the skin is composed essentially of nonliving tissues, it is reasoned that biochemical processes are unlikely to influence the diffusional characteristics of the rate-limiting membrane and, hence, *in vitro* diffusion studies will accurately measure skin penetration and absorption.

In vitro diffusion chamber experiments are based on measuring the compound under investigation in the receptor fluid and much of the research activity has naturally focused in this area. Therefore, recovery of material in the skin itself has received only limited attention. Cutaneous distribution, metabolism, and binding of the topically applied agent are integral parts of the percutaneous absorp process, however, implying that assessing the disposition of the applied chemical in the skin tissue should be an important measurement in the evaluations of skin absorption. Indeed, the amount of a chemical that passes through the stratum corneum into the viable epidermis and dermis is an important parameter for assessing local bioavailability, and it also contributes to the overall estimate of *in vitro* percutaneous absorption. Furthermore, analysis of the skin following permeation studies would assist in determining mass balance and dose accountability. Such measurements are often not reported or conducted even though they are important in establishing the validity and the interpretation of *in vitro* observations. Because of obvious advantages, radiolabeled chemicals are routinely used in skin absorption studies, and frequently liquid scintillation spectrometry is the sole method used for detecting the penetrating substances in the receptor fluid. However, skin absorption may be accompanied by cutaneous metabolism; therefore, the radioactivity recovered in the receptor medium reflects not only the permeation of the parent substance but also its metabolites.

Experimental designs of *in vitro* studies utilize one of two main strategies. In the traditional steady-state or "infinite dose" technique, a well-stirred donor solution of the compound of interest, at a defined and constant concentration, is used to deliver the compound across the skin preparation. The absorbed compound is subsequently delivered into a well-stirred receptor compartment. The most important design feature of these studies is that the quantity of compound that penetrates the membrane must be kept small relative to the total amount available—there must be no appreciable reduction in the concentration of the compound in the donor compartment. This is so that steady-state flux conditions are not significantly violated and the studies are performed with rigorous compliance to the laws of diffusion. The conventional approach to presenting data from this type of study is to plot the cumulative amount of drug reaching the receptor as a function of time (Fig. S-2). From the linear portion of this plot, we obtain the most important piece of information, i.e., steady-state flux of the compound across the skin membrane (slope $= J_s$). This value is generally normalized with respect to the area of the skin membrane and is usually expressed as amount of drug per unit area per unit time. The intercept on the X-axis, obtained by extrapolating the linear part of the curve, gives a measure of the time required to establish a linear concentration gradient across the skin membrane and is referred to as the lag time, or τ. From these two parameters it is relatively simple, using the diffusion equations, to calculate the permeability coefficient ($k_p = $ slope/δC_s) and derive the other mass transfer parameters such as the diffusion coefficient of the drug across the skin, the partition coefficient of drug between the skin and the receptor fluid, and the diffusional thickness of the membrane ($\lambda = k_p \cdot \tau$).

While the infinite dose technique has been invaluable in the development of transdermal drug delivery concepts, this methodology may be of limited value as a predictive model for assessing skin absorption *in vivo*. To mimic *in vivo* conditions, the so-called "finite dose" technique was developed. This is essentially a modification of the traditional steady-state method. The important difference is that the skin preparation is supported over the receptor so that the epidermal surface is not covered, and the compound of interest is applied to the surface of the skin in a manner similar to exposure *in vivo*. Recent refinements, such as flowthrough designs for the receptor fluid, to simulate cutaneous blood flow with its concomitant "wash-out" capacity provide additional features that mimic the *in vivo* conditions. This approach has received a great deal of attention since it is potentially a very powerful tool for studying factors that influence percutaneous absorption and may provide a means for assessing the skin absorption of toxic agents.

The techniques described for maintaining viability of the excised skin used in these studies are relatively simple. Basically, the skin preparations are maintained under appropriate conditions as short-term organ culture. They are supported over the culture medium so that their epidermal surfaces are not covered. Material of interest can be applied topically in a manner similar to exposure *in vivo*. This material reaches the epidermal cells by diffusion where it may be metabolized, and recovery of both metabolites and parent compounds in the culture fluid then provides a measure of skin permeation and the extent of cutaneous first-pass metabolism. Two systems have been described. In the "static" system discs of freshly excised skin are maintained, epidermal side up, on filter paper on a stainless-steel ring support within individual culture dishes containing a suitable culture fluid. In the flowthrough system the skin discs, supported on a stainless grid, form the upper seal of tissue wells of a compact, water-jacketed, multisample skin penetration chamber. Fresh, oxygenated culture medium is continuously perfused through the tissue wells and the well effluents may be collected at timed intervals. The skin absorption and metabolism studies described previously utilizing this methodology demonstrated that, by maintaining the metabolic viability of the excised skin under appropriate culture conditions, the *in vitro* approach provided the means whereby the potential influence of skin metabolism may be evaluated in conjunction with the diffusional aspect of percutaneous absorption. This methodology, therefore, offers a possible approach with which an estimate for the contribution by skin to the percutaneous fate of topically applied chemicals may be determined. It has also been suggested that the metabolites found in the perfusion medium result from metabolism of the parent compound after its permeation into the receptor fluid, and this biotransformation is mediated by enzymes that have leaked into the receptor fluid from the cultured tissue. Currently, there is no evidence to support this hypothesis.

The underlying assumptions inherent in the utility of using excised skin for diffusion experiments are (1) skin condition, particularly that of the stratum corneum, is comparable to that found *in situ* (i.e., a key variable is the barrier function of the skin sample once it is removed from the animal or man); (2) recovery of the applied substances in the receptor fluid provides a true reflection of the rates and extent of percutaneous absorption (i.e., tissue binding and partitioning into the receptor fluid are not confounding factors); (3) living processes have little or no effect on percutaneous absorption mechanism or kinetics; and (4) penetration through dermis is not rate limiting. All of these premises are becoming increasingly difficult to justify in light of what we now know about active metabolism of drugs within the skin and the influence of cutaneous blood flow on the clearance of drugs and their metabolites from the skin. Table S-14 lists a number of important design considerations when assessing percutaneous absorption using *in vitro* diffusion chamber experiments. As can be seen, many of these potential sources of error in *in vitro* experiments could be predicted from the list of factors affecting skin absorption *in vivo* (Table S-13). Others, such as skin thickness, barrier integrity/viability, and receptor fluid content, are true artifacts of the *in vitro* system.

In Vivo–In Vitro Correlation. In the skin absorption literature there are only a few instances in which studies were designed specifically to correlate *in vivo* and *in vitro* observations. This is probably because such comparative experiments involve many variables that need to be controlled or monitored and are difficult to perform well. Meaningful comparisons can only be made when experimental parameters of the *in vitro* studies closely resembled those of the *in vivo* study or vice versa. Moreover, ethical and safety concerns often limit the extent to which *in vivo* experiments may be conducted in humans. Thus, *in vivo–in vitro* correlations using human skin are practically nonexistent. In addition, in humans the site of application is frequently the ventral forearm, whereas in animals the back is often used and potentially damaging pretreatments of the animal skin such as shaving, clipping, or chemical depilation (hair removal creams) are frequently necessary before skin absorption experiments can be conducted. Since these are treatment variables that can influence percutaneous absorption, the reported species differences and similarities in skin absorption may reflect the net result of many competing variables, and understanding the significance of these variables would provide additional insights into the mechanism of percutaneous absorption.

A general consensus among investigators in percutaneous absorption is that human skin is preferred and should be used for *in vitro* assessments. On the other hand, it is also recognized that a major liability of human skin as a research tissue *in vitro* is its notoriously high variability in barrier properties. The source of human skin is frequently from cadavers, and since the investigator often has little or no control over the source and characteristics of the donor skin, the high variability observed with human skin preparations is to be expected. Characteristics such as treatment of the cadaver, elapsed time from death to harvest of tissue, skin site, age, health, sex, race, and skin care habits are examples of variables which may bias the *in vitro* penetration studies. Also, when skin samples are derived from elective surgery, the preoperative procedures such as scrubbing with antimicrobial disinfectants, the surgical manipulations, and the manner in which the membrane is prepared from the excised tissue are important details of concern. Again, these variables may influence the *in vitro* penetration observations. It has been recommended that where possible, in an *in vitro* study with human skin, such information should be routinely collected and carefully documented.

Although human skin is the tissue of choice, its limited accessibility to many investigators and the variability experienced with human skin have led many researchers to explore skin from various animals as models for skin absorption. However, species differ considerably in the structure and function of their skin and it is unlikely that animal skin will have barrier properties that are identical to those of human skin. Nevertheless, animal skin is routinely used for evaluating dermal toxicity and percutaneous absorption. Histological evidence and physiochemical studies have concluded that animal skin can provide reasonable per-

TABLE S-14
Experimental Design Considerations with *in Vitro* Assessments of Percutaneous Absorption

Factor	Contributing factors
Skin source	Animal species differences, fresh human vs cadaver skin
Membrane thickness	Full thickness vs split thickness (no dermis) vs stratum corneum alone
Barrier integrity	Preparation and storage conditions, follicles and other holes, viability (metabolic capacity)
Receptor fluid	Solubility, maintenance of skin viability/barrier integrity
Dosing method	Finite vs infinite, dose formulation
Environment	Ambient temperature and humidity, hydration state

cutaneous absorption models that approximate human skin; however, the debate concerning the appropriate animal model continues. Numerous comparative studies, both *in vivo* and *in vitro,* have been conducted to identify the ideal animal model. From the results obtained thus far, it would appear that the choice of animal model will depend on the preference of the investigators and the compound under investigation. The pig, the monkey, hairless mice, and, recently, the fuzzy or hairless rat have been described as species with the potentials to be good candidates as predictive models of skin absorption in humans.

Because physical diffusion is assumed to be the principal determinant in skin absorption, and since the selection of an appropriate animal model remains controversial, artificial barrier systems have been explored as potential models for evaluating absorption in human skin. These systems offer some advantages over biological models in that they are reproducible, easily prepared, and the composition of the membrane can be readily manipulated. Such membranes offer a defined matrix with which basic physical concepts regulating permeation may be examined. Various materials have been used in the construction of artificial membranes, and they include chemicals such as cellulose acetate, isopropyl myristate, mineral oil, and dimethyl polysiloxane. Materials such as collagen and egg shell membrane which are of biological origin have also been used. In general, the construction of these artificial membranes attempts to mimic the stratum corneum barrier, and their use in diffusion studies has provided some useful information on the underlying mechanisms governing the physiochemical properties of chemicals and the relative abilities of the chemicals to diffuse through lipid membranes. However, use of artificial membranes in skin absorption assessment has been limited, although they have been used as models for evaluating potential drug formulations during the development of topical preparations and transdermal delivery systems. Nevertheless, the extent to which artificial membranes may serve as useful surrogates for skin in dermal absorption studies has yet to be established.

While there is no doubt that *in vitro* diffusion chamber experiments have made substantial contributions to our current knowledge concerning the diffusional aspects of skin absorption, there are limitations to their usefulness, particularly for developing predictive pharmacokinetic models of skin absorption. Today, standardized versions of these diffusion chambers are readily available commercially (e.g., Vangard International, Inc., Neptune NJ; Crown Bioscience, Inc., Clinton, NJ; and Laboratory Glass Apparatus, Inc., Berkley, CA). The commercial systems are easy to use and are well designed, having incorporated many of the desirable features of diffusion chambers described by investigators with many years of research experience in the field of skin absorption.

Advanced Models: Isolated Organ Perfusion Methods

A major limitation of all diffusion chamber experiments is the lack of normal vascular uptake mechanisms. One consequence of this inadequacy has been a large effort aimed at better diffusion chamber designs, particularly through the use of flowthrough devices mentioned previously. It should be noted that the recovery of material in the receptor fluid, which provides an overall measure of *in vitro* permeation (i.e., the net penetration through the various layers of the skin into the receptor), does not necessarily provide an accurate measure of percutaneous absorption. Material that permeates the skin and remains in the tissue is absorbed material that would not be included when receptor fluid only is assessed. Furthermore, various tissue slicing techniques or apparatus design considerations have failed to totally resolve the possibility that the thick dermis may represent an artificial and selective barrier limiting the permeation of lipophilic penetrants *in vitro.* There is no consensus on what constitutes an ideal receptor fluid. The selection of the optimum receptor fluid for a particular compound of interest is frequently empirical and often reflects the biases of the investigator. Therefore, caution should be exercised and *in vitro* observations should not always be considered to be true and accurate representations of the *in vivo* situation with respect to cutaneous absorption and metabolism.

Given the obvious physiological limitations of the previously discussed organ culture and diffusion cell approaches, perfused skin preparations, with an intact and functional cutaneous microcirculation, appear to represent an ideal experimental methodology for investigating the pharmacokinetics and mechanisms of percutaneous absorption and metabolism. Following the development of the perfused rabbit ear model in the 1930s and the subsequent demonstration of its potential as a tool for studying skin absorption, surprisingly little progress was made over the next half century. Reports of perfused feline and canine skin flaps, both *in situ* (meaning "on site," or still attached to the animal) and *in vitro*, appeared sporadically in the literature. These models have provided useful information in studies of skin physiology and the basic pathways of cutaneous respiration and energy production. However, they have not been widely used to study skin absorption. In addition, although early attempts to develop human skin perfusion models have been documented and recently an isolated perfused human groin flap was reported, progress has been slow.

Recently, techniques for creating and maintaining isolated arterial sandwich skin flaps *in situ* in rats have been described. This rat skin flap (RSF) is created on athymic nude rats by surgically raising a small area of skin, perfused by the superficial epigastric artery, and grafting a split-thickness skin sample from syngenetic rats (i.e., from the same breeding stock) onto the underside. Although athymic rats reject foreign skin grafts at a relatively high rate (up to 90%), some success has been achieved in creating and maintaining a hybrid rat–human sandwich flap (RHSF) on this animal by repeated low-dose cyclosporine therapy. It has been proposed that the RHSF might be useful for studying percutaneous absorption in human skin if it can be shown that the absorption mechanisms are unaffected by the surgical manipulations and cyclosporine treatments. The experimental advantages afforded by such human-grafted skin flaps, in addition to the fact that they are reusable, are intuitive. Unfortunately, the complicated surgical procedures, costly animal and housing requirements, and expensive cyclosporine therapy and its confounding effects on skin absorption, coupled with the apparently high variability in xenobiotic flux through the xenografts, place severe limitations on their utility as experimental models for studying cutaneous metabolism and skin absorption.

Since it has long been known that there is a high degree of anatomical and physiological similarity between skin obtained from certain pale-skinned porcine species and that of man, it is not surprising that various pig skin flaps have been pursued. The biochemistry and utility of pig buttock flaps, created surgically in several different patterns, for dermatological purposes have been extensively investigated. Proposed advantages for using pig skin flaps include the availability of large surface areas, similar vasculature and anatomic structure, and the ease and similarity in the types of clinical observations which can be made. Perhaps the most promising perfused skin preparation is the isolated perfused porcine skin flap (IPPSF), which has been developed recently and provides a novel *in vitro* approach for examining percutaneous absorption processes in intact, living skin. The biochemistry and morphology of the IPPSF, maintained using an isolated organ perfusion technique for skin which is essentially analogous to methods developed for other organs such as liver, lung, and kidneys, have been examined in great detail and appear to be consistent with that found in porcine integument *in vivo*. The absorption of a wide variety of topically applied xenobiotics has already been demonstrated using the IPPSF, including such diverse chemicals as organic acids and bases, organophosphate insecticides, and steroid hormones and organochlorines. In addition, the effects of applied surface concentration and coadministration of vasoactive drugs (tolazoline and norepinephrine) on lidocaine iontophoresis (electrically driven drug transport across biological membranes), as well as the iontophoretic transport of small peptides and proteins (insulin), have been examined using the IPPSF, demonstrating its potential for testing novel transdermal drug delivery systems. Cutaneous biotransformation of xenobiotics during percutaneous absorption has been demonstrated using the IPPSF with the chlorinated hydrocarbon, chlorbenzilate, and with the organophosphate, parathion.

Preliminary studies using the IPPSF have shown that compounds such as the cancer chemotherapeutic agents cisplatin and carboplatin and the antibiotics tetracycline and doxycycline readily distribute into the skin following intravascular administration. Also, compounds such as parathion, ABT, and 25-hydroxyvitamin D are bioactivated in the skin following intravascular administration in the IPPSF. This demonstrates a role for the IPPSF as an ideal experimental model for studying the disposition of xenobiotics which are distributed to skin from the systemic circulation. Interest in the so-called outward transdermal migration or reverse penetration concept, namely, that skin may function as a clearance organ following delivery of systemically administered substances via the cutaneous vasculature, has been stimulated by the development of noninvasive techniques for measuring and analyzing the pharmacokinetics of the distribution of substances to skin *in vivo*. The absence of confounding, extracutaneous metabolizing organs, such as the liver, lungs, and kidneys, is a distinct advantage in IPPSF investigations of this reverse penetration phenomenon.

In conclusion, there are many fundamental questions concerning skin absorption and metabolism that remain to be addressed. The potential role of the dermal vasculature, the contribution of skin appendages such as hair follicles and sebaceous glands, the influence of skin condition, age, disease state, and anatomic sites are just a few examples of questions that need to be resolved. When topical exposure results in local effects, pathological changes in the skin may be expected to affect its barrier functions. These changes may involve alteration of the physical barrier as well as the biochemical properties, such as the metabolic status of the of the skin. Such local changes may have important implications on the outcome of percutaneous absorption and fate of topically applied xenobiotics. The experimental techniques necessary to address these questions are available, and productive research in these areas will provide means whereby species differences in skin absorption and metabolism may be investigated. These studies should provide not only a better understanding of the mechanisms important in the percutaneous fate of topically applied chemicals but also a rational basis for cross-species extrapolation and, therefore, more predictive estimates for skin absorption and metabolism in man.

Etiology of Skin Toxicity

General Concepts

Because the skin is in direct contact with the external environment, it is constantly being exposed to drugs, chemicals, electromagnetic radiation, and physical materials capable of producing toxic responses in this organ. In addition, many drugs are delivered into the skin via the systemic circulation, which also may result in cutaneous toxicity. It is the purpose of this section to review and categorize the extensive list of agents that exert toxic effects within the skin. Without discussing specific mechanisms, which are described in detail later in this entry, it is necessary here to make a distinction between those agents which produce a direct irritant response and those which act via a systemic, immune-mediated pathway. The former is called irritant contact dermatitis (ICD), while the latter is called allergic contact dermatitis (ACD). Both ICD and ACD involve the participation of many immune cell types found in the skin and are often histologically and biochemically in-

distinguishable from each other. Moreover, a third category of skin reactions has emerged, called contact uticaria. Unfortunately, the mechanistic distinction (discussed later) between this syndrome and ICD is even more blurred than the ACD vs ICD comparison. Because the list of urticariants (Table S-15) appears to be a subset of the contact irritants, representing materials from every chemical class, these agents will not be described separately in the categories developed for this section.

Direct cutaneous irritation, or ICD, is one of the most common maladies in industrialized society. The symptoms of ICD are the classical inflammatory response markers: redness, swelling, pain, and loss of function. Although ICD is not often fatal, this disease does involve significant morbidity and takes a heavy economic toll due to its sheer prevalence. The incidence of ICD in the general populations of the United States and Western Europe has been variously estimated at between 1 and 10%; however, counting undiagnosed cases the true incidence may lie closer to 25%. It is well documented that ICD is the single most common occupational disease seen in the United States, with over 5000 man-made and natural chemicals known to be capable of irritating the skin. A simplistic classification of these irritants includes such agents as dessicants, abrasive materials, organic solvents, acids and alkalis, concentrated metallic salt solutions, oxidizing/reducing agents, enzymes, plant extracts, and surfactants. The latter group of agents represents the various soaps and

TABLE S-15
Contact Urticariants

Category	Examples
Natural agents	Birch bark, butter, cabbage, capsaicin, chicken, cinnamon, cobalt chloride, copper, cotton oils, eggs, fish, fruits (kiwi, strawberry), hawthorn, honey, horse saliva, laboratory animals, mahogany, milk, nickel, papain, prawn crust, seminal fluid, sorbic acid, spices, spider mites
Industrial sources	Alcohols, benzoates, BHT, carbonless copying paper, chloramine, chlorhexidine, diethyl fumarate, "DEETS," DMSO, formaldehydes, *p*-phenylene diamine, phosphorus sesquisulfide, plastics, rouge, rubber, sorbitan monolaureate
Pharmaceuticals	Aminophenazone, benzocaine, benzoyl peroxide, penicillins

detergents used in the form of complex mixtures and marketed extensively as cleansers in personal, fabric, and hard surface care products. As such, surfactants are primarily responsible for ICD of household origin and are considered second only to organic solvents in producing occupational dermatitis.

Equally important from a dermatological viewpoint, although not nearly as prevalent, are the immune-mediated skin reactions, which can be broadly categorized as ACD. Whereas ICD is commonly thought to account for 60–80% of clinically recognized human contact dermatitis, ACD accounts for most of the remainder (20–30%). As will be discussed later in this entry, ACD is clinically and histologically indistinguishable from ICD in most cases. However, the presence of two etiologic factors renders this condition perhaps even more dangerous than ICD. First, once an individual has become sensitized to contact allergens, quite low amounts of the offending agent can subsequently elicit massive skin responses. Second, once induced, this hypersensitivity may persist for a long and varied period of time, possibly even for the rest of one's life.

Like ICD, ACD may also occur from a very large number of chemicals, but not from electromagnetic radiation or physical stimuli alone. Most substances are rarely allergenic and there is a great range in allergenic potency, with a small number of known strong sensitizers having been identified experimentally in man. These strong allergens are often aromatic substances with molecular weights less than 500, highly lipid soluble, and quite reactive with proteins (a mechanistic requirement, as will be detailed later). The simplistic classification of the principal ACD agents includes metallic salts, plant polyunsaturated alcohols and ketones, acrylates, plasticizers, antibiotics, aliphatic amines and phenols, and formaldehyde. The possibilities for human exposure to both contact allergens and contact irritants can be divided among four broad categories: consumer products, occupational or industrial chemicals, environmental agents, and pharmaceuticals.

Root Causes of ICD and ACD

Consumer Products

As mentioned earlier, the soaps and detergents in cleaning products and cosmetics comprise the bulk of the household materials which are irritating to human skin. The molecules responsible for this type of ICD are called surface-active agents, or surfactants. There are four main classes of surfactants, which are listed in order of their irritancy: anionics (used as industrial-strength cleaners and fat-based soaps), cationics (mostly disinfectant cleaners), and nonionics and amphoterics (fabric cleaners, cosmetics, shampoos, and mild cleansers). The irritancy of surfactants is roughly correlated to their cleaning power and their ability to foam when mixed with water and air. Other consumer products likely to cause ICD are wool and fiberglass due to mechanical action of the fibers on the skin surface; formaldehyde residues found in newspaper inks, building materials, and clothing; and dry cleaning fluid residues of polychloroethylene. Diaper dermatitis is also a form of consumer product-induced ICD caused by the combination of enzymes in urine/feces and disinfectant cleansers used on the skin.

The largest group of agents capable of causing ACD in the household are perfumes and dyes used in cosmetics, toiletries, and clothing (Table S-16). Metal salts, such as nickel salts, chromium salts, and cobalt; organomercurials; and formalin are all sometimes used as preservatives in household products and cosmetics and can also become allergenic. In fact, reactions to nickel (jewelry) and nickel salts are typically the most prevalent response in diagnostic patch test studies involving a wide variety of known allergens. Certain pesticides (Kathon biocide) and sunscreens also produce ACD, although the more potent sensitizers, such as p-aminobenzoic acid, are no longer in general use as sunscreens. Finally, the component monomers from certain rubber and plastic materials may also leach out and cause ACD, although humans are more likely to be exposed to these molecules in the workplace.

Industrial Chemicals

With the possible exception of consumer products, this category represents the largest and most widely studied group of irritants and sensitizers. Certainly, it consists of the widest range of chemical classes to which humans are routinely exposed. It has been estimated that occupational skin disease accounts for 40–60% of all lost work days and nearly 95% of the cost, with ICD being more prevalent than ACD. Moreover, 25% or more of the general population is considered to be atopic or predisposed to skin eruptions despite the lack of visual, or even histologic, evidence that the skin is compromised. Chronic exposure to damaging consumer prod-

TABLE S-16
Chemical agents associated with ACD specific examples

Category	Specific examples
Plants	Barley dust, lichens (D-usnic acid), hops (colophony), hetzil, sawdust, sesquiterpene lactones (*Compositae, frullania* spp.), tulips (tulipalin A), poison ivy (urushiol)
Plastics	Cyanoacrylate, epoxy resins, polyacrylates, phenolformaldehyde resins, polyurethane, rubber additives (thiuram, carbamates)
Metals	Nickel, cobalt, mercury, silver, chromates (welding fumes and cement), beryllium
Industrial chemicals	bis-(4-Chlorophenyl) Methylchloride, 3-bromo-3(4-chlorobenzoyl) propionic acid, 4-bromomethyl-(6,8)-dimethyl-2 (1*H*) quinolone, bromomethyl-4-nitrobenzene, bromophthalide, 2-chlror-6-fluorobenzaldehyde chlorooxime, hydrogen sulfide, *N*-hydroxyphthalimide, trimethyl hexamethylene diisocyanate, solvents—formaldehyde, turpentine, persulfate, phosphorus sesquisulfide, thioureas, allylphenoxyacetate, dimethoxane, chloracetamides (paints, wood shavings)
Pharmaceuticals	Chloroquine sulfate, benzocaine, chlorpromazine, cytosine arabinoside, 4,7-dichloroquinoline, 2,6-dichloropurine, streptomycin, neomycin, vincamine tartarate, 2[4(5)methyl-5-(4)imidazolyl methylthio] C_{13} pyritinol hydrochloride
Pesticides	Calcium lignosulfate, captafol, captan, carbamates, dithianone, ethoxyquin, naled, pyrethrum, spiramycin, tetrachloroisophthalonitrile, thiuram, tylosine, virginiamycin
Cosmetics	Perfumes, deodorants, hair sprays, sunscreens, skin lotions/creams, nail polish, dyes, shampoos

ics, manufacturing, and construction industries. They include, in order of their irritancy, chlorinated aliphatics (e.g., trichloroethylene and polychlorinated biphenyls), aromatics (benzene/toluene), aliphatics (*n*-hexanes), ketones (acetone), and alcohols. Surfactants, discussed earlier in the context of consumer products, represent the second most important class of industrial irritants.

Miscellaneous industrial irritants include alkalis, such as caustic soda, NaOH, cement, and lime used in mining, dying, tanning, and construction, as well as strong acids (sulfuric, chromic, nitric, hydrochloric, and hydrofluoric) used in ironworks, glass etching, and masonry. Hydrogen peroxide and organic peroxides in plastic manufacture and reducing agents, such as phenols, hydrazines, aldehydes, and thioglycollates, may also produce ICD in the workplace. Moreover, enzymes released from meats and fish have been known to cause ICD in processing/packing plants. Besides the rubber and plastics industries (monomers), the primary

TABLE S-17
High-Risk Occupations for ICD

Occupation	Specific exposures of interest
Baker	Soaps and detergents, fruit juices, spices, enzymes
Construction worker	Cement, chalk, acids, wood preservatives, glues, detergents, industrial solvents
Canner, food service industry	Soaps and detergents, brine, syrup, fruit and vegetable juices, fish, meat, poultry
Dental technicians	Soaps and detergents, soldering fluxes, adhesives, acrylics, solvents, mercury
Electricians	Soldering fluxes, metal cleaners (solvents), epoxy resins, PCB and PBB
Hairdressers	Soaps and detergents, shampoos, permanent wave liquids, bleaches and dyes
Horticulture	Manure, fertilizers, pesticides, irritating plants
Mechanics	Detergents, degreasers (solvents), lubricants, petroleum products, battery acids, soldering fluxes, cooling system fluids (PEG), metal shavings
Nurses	Soaps and Detergents, alcohols, disinfectants, hand creams
Printer	Solvents, acrylates, formaldehyde, phthalate esters (inks)
Agriculture	Pesticides, fertilizers, disinfectants, detergents, petroleum products, irritating plants, animal secretions

ucts or environmental agents no doubt contributes to occupationally induced skin disease.

Table S-17 lists high-risk occupations for developing ICD and ACD. A common factor in these occupations is the presence of water ("wet work") and exposures to organic solvents and surfactants. While water itself is not considered an irritant, continual wetting and drying of the skin usually produces many of the hallmark symptoms of ICD. Organic solvents are by far the chemical class most responsible for occupationally induced ICD. These chemicals are used as degreasing agents and lubricants in many processes in the electron-source of occupationally induced ACD is the manufacture of consumer products, or the raw materials thereof, containing perfumes, dyes, preservatives, biocides, and other specialty chemicals.

Environmental Agents

Many plants contain rough hairs or large calcium oxalate crystals (*Dieffenbachia, Caladium,* and *Philodendron* spp.), both of which are capable of producing mechanical damage to the skin. In addition, enzymes like bromelin (pineapples) or mucanain (cowhage) and chemicals like capsaicin (nightshade) or polycyclic diterpene alcohols (spurges) are also somewhat irritating. Nettles produce a contact urticaria by direct injection of the inflammatory mediators acetylcholine, histamine, and 5-hydroxytryptamine (serotonin). Anthralin, a synthetic drug, but originally isolated from the araroba tree, is also an important environmental contact irritant. The primary plant allergens are catechols present in the *Toxicodendron* genus which are responsible for the most common form of plant-induced ACD: poison ivy (urushiol)/oak/sumac. ACD is also caused by butryo- and sesquiterpene-lactones found in the *Primula obconica* and *Compositae* (ragweed and Australian bush) plant families. Finally, atmospheric changes can also cause or predispose certain individuals to have ICD since it has long been known that low ambient humidity (more common in winter) can impair the barrier function of skin.

Ultraviolet (UV) light, a principal toxic component of solar radiation, interacts with skin in a variety of different ways which deserve special mention here. Visible light, having wavelengths of 400–760 nm, is relatively harmless, but shorter wavelengths can produce devastating effects alone or in combination with "photoreactive" drugs and chemicals (described below). The three important divisions of UV light are UVA (320–400 nm), UVB (280–320 nm), and UVC (220–280 nm). UVC is of little natural concern because these shorter wavelengths are almost entirely absorbed, or blocked, by the stratospheric ozone layer. UVB is the part of the solar spectrum responsible for the most damaging effects on the skin, although UVA is now felt to play a more prominent role in certain types of skin disorders. The more serious effects of UV exposure are pigmentation defects, actinic elastosis (premature skin aging), selective defects in immune function, actinic keratosis, squamous/basal cell cancers, and malignant melanomas. UVB alone produces a characteristic,

ICD-like inflammatory response (sunburn) or can react with chemical agents in and on the skin to produce photoirritation, or photo-ICD. A list of common phototoxic chemicals is shown in Table S-18. Most of the recognized photoirritants are drugs delivered into the skin from systemic, not topical, administration, although plant-derived phototoxins are also known. For example, a pigment from St. John's wort is delivered to the skin upon ingestion, reacts with sunlight, and causes a massive vascular leakage which may progress to sloughing of large patches of dead skin.

Photo-induced ACD, or photosensitization, is also a consequence of combined exposure to sunlight and certain chemicals. The vast majority of these reactions appear to result from UVA wavelengths acting on topical agents, although isolated and incompletely documented reports of photosensitization resulting from systemic administration have appeared in the literature. A list of selected photoallergens is shown in Table S-19. All are substances that absorb UV light and most have a resonating structure, i.e., aromatic ring(s). An important complication of photo-ACD is the development of persistent light reaction, seen with phenothiazines, wherein a marked sensitivity to light persists long after exposure to the photoallergenic chemical has ended.

Pharmaceuticals

Adverse drug reactions account for 3–5% of hospital admissions and occur in as many as 5% of patients who are already hospitalized. Cutaneous involvement is particularly common in these circumstances, especially in children, in part because these so-called skin rashes are easily identified. Although many of these conditions are relatively harmless, cutaneous adverse

TABLE S-18
List of Common Photoirritants

Class	Chemical agents
Coumarins	8-Methoxypsoralen, 5-methoxypsoralen, trimethoxypsoralen
Polycyclic aromatic hydrocarbons	Anthracene, fluoranthene, acridine, phenanthrene
Pharmaceuticals	Tetracycline, sulfonamides, chlorpromazine, nalidixic acid, NSAID[a] (benoxaprofen)
Dyes	Eosin, acridine orange
Miscellaneous	Porphyrins, amyl-O-dimethylaminobenzoate

[a]Nonsteroidal antiinflammatory drugs.

TABLE S-19
List of Common Photoallergens

Class	Chemical agents
Halogenated salicylanides	Tetrachlorosalicylanide, bithional, dibromosalicylanide, tribromosalicylanide, 4-chloro-2-hydroxybenzoic acid, *N*-butylamide (JADIT)
Columarins	6-Methylcoumarin, 4-methyl-7-ethoxycoumarin, 7-methylcoumarin
Plants	*Compositae* family (Ragweed, australian bush)
Sunscreens	*p*-Aminobenzoate (PABA), Glyceryl-PABA
Miscellaneous	Sulfonamides, phenothiazides, 4,6-dichlorophenylphenol, quinoxaline-1,4-di-*N*-oxide, musk ambrette

drug reactions (CADRs) may be only one symptom of a much larger, and potentially life-threatening, immune response to a drug, or CADRs may be severe in and of themselves. Moreover, except for the occasional irritancies produced by topical ointments or transdermal drug delivery devices (patches), which are often not drug related but are due to other chemicals/materials present in the formulation/device, CADRs are almost always a form of ACD.

The most common CADRs are the less severe exanthem-like (characterized by a small papular rash which can cover large surface areas) and urticarial reactions, together accounting for over two-thirds of drug-induced skin rashes. Table S-20 presents a list of drugs which are often associated with CADRs. Antibiotics are the drug class most likely to produce skin reactions, particularly in children, in which they account for more than 50% of all prescriptions. In addition, this drug class is mostly responsible for the non-life-threatening

TABLE S-20
List of Drugs often Associated with CADRs

Class	Specific drugs involved
Antibiotics	Penicillins, cephalosporins, sulfatrimethoprim, sulfonamides, nitrofurantoin, isoniazid, rifampin
Anticonvulsants	Phenytoin, carbamazepine, barbiturates
Antiinflammatories	Corticosteroids, gold, NSAIDs
Others	Antineoplastics, allopurinol, diuretics (sulfa derivatives)

skin rashes. However, as can be seen from Table S-20, the situation is complex in that most of the drug-induced skin diseases have multiple causes, and many of the drugs are capable of causing more than one type of skin lesion. The probability that an individual drug will cause a particular CADR is under the control of many "host" factors, such as genetics, age, sex, the presence of other drugs (interactions), and concurrent diseases (liver or kidney failure, for example). Thus, generalizations are not very useful in the case of CADRs.

Other types of skin reactions to drugs (Table S-21) are less frequent, but some are much more severe and deserve special mention. These CADRs fall into three major classes: severe and life-threatening dermatoses,

TABLE S-21
Types of CADRs

Category	Subclass	Examples of drugs involved
Exanthem-like erythemas (46%)[a]		Antibiotics, anticonvulsants
Urticarias (23%)		Antibiotics, antiinflammatories, opiate analgesics
Other erythemas (<20%)	SLE[b]	Antibiotics, anticonvulsants, oral contraceptives
	Erythrodermas	sulfonamides, gold, isoniazid, streptomycin
	Lichenoid photosensitivity	quinicrine antimalarials
Blistering diseases (<10%)	Erythema multiforme	Sulfonamides, penicillins, diclofenac, oxyphenbutazone, piroxicam, phenytoin, carbamazepine same as for erythema multiforme
	Ten[c] Pemphigus	Penicillamine, captopril, piroxicam, penicillins, rifampicin
	Bullous pemphigoid	Frusemide, penicillamine, penicillin, PUVA therapy
Skin cancer (<1%)		Immunosuppressants, mexiletine, thioridazine, penicillamine, moduretic, atenolol, quinacrine

[a]Where percentages are noted, this is the approximate frequency among all patients experiencing CADRs.
[b]Systemic lupus erythematosis.
[c]Toxic epidermal necrolysis.

skin malignancies, and other skin reactions. The severe dermatoses are the erythrodermas, erythema multiforme, toxic epidermal necrolysis, and bullous or blistering diseases. The dangerous symptom common to most of these CADRs is the sloughing off of large areas of epidermis, leaving the underlying dermis unprotected from bacterial infection. Although exceedingly rare, the mortality of toxic epidermal necrolysis has been estimated at 34%. The other dermatoses generally respond better to withdrawal of therapy. Phototoxicity and photoallergic reactions to common drugs were described previously. Drug-induced skin tumors have provided increasing evidence for the role of the immune system in the inhibition of malignancy due to the observed higher frequency of skin tumors of patients receiving immunosuppressants. It is also possible that certain drug-induced dermatoses result in greater propensity toward skin malignancies. Although ideopathic lichen planus does not appear to result in greater incidence of skin tumors; lichenoid eruptions due to quinicrine appear to have predisposed some individuals to a subsequent squamous cell epithelioma. Since the latency period for skin cancers can be many years to decades, more examples of drug rashes leading to skin malignancies may be forthcoming. Finally, although not generally life-threatening, but often severely and socially debilitating, are pigmentation, hair, and nail changes, acne, and vascular inflammation, all of which are listed in Table S-21 under "other lesions."

Skin's Response to Toxic Insult

General Considerations

It is axiomatic that the body's reaction to injury is limited and it is often impossible to identify the causal agent based solely on the observed responses. Toxic insults on the skin can result in a combination of functional, biochemical, and morphological changes. These alterations induced by toxicants do not differ, in general terms, from changes caused by physical or biological agents, but the magnitude of the changes that are observed at any point in time depends on the nature, rate, extent, depth, and duration of the insult. From a mechanistic viewpoint, toxic insults to the skin can be classified into two main categories, namely, direct injury (i.e., ICD or contact urticaria) and immune injury (i.e., ACD). However, as mentioned earlier, the basic pathological lesions and clinical features that are en-

countered in all inflammatory skin responses are essentially indistinguishable. Thus, irrespective of the mechanism, the manifestations of toxic responses of the skin to an insult are basically the same and are similar to those following any other cause of cell injury in other organs and tissues: degeneration, proliferation and repair, or any combination of these basic dynamic responses.

Degenerations are regressive changes within a cell or cell population in response to injury. They range from reversible changes such as atrophy, which may be considered an adaptive homeostatic response to an adverse environment, to irreversible changes such as necrosis or cell death, while still forming part of the living organ. In between are a range of cellular alterations, including hydropic changes, fatty changes, and other inclusions, resulting from cytoplasmic accumulation of water, lipids, and granular materials, respectively, all of which are derived from breakdown of intracellular components.

Proliferation, in contrast to degeneration, involves increased growth in response to an injurious stress. The hypertrophy and hyperplasia experienced may range from adaptive homeostatic responses to irreversible proliferation of a cell population, leading to cancer. Inflammation and repair are extracelluar responses that often accompany degeneration and proliferation. They represent tissue responses that attempt to contain or remove the injurious agent and revitalize the damage tissue. The extent and nature of the inflammatory response varies according to the nature, extent, and duration of the injury and include vascular, neurological, humoral, and cellular responses at the site of injury. Acute inflammation is typically an immediate and early response to an injurious agent. The vascular and connective tissues adjacent to the injured cells are usually involved and may include local vasodilation with transient increased blood flow and increased vascular permeability, with egress of white blood cells into the injured tissue. These processes are coordinated and integrated by numerous inflammatory mediators (e.g., histamine and bradykinin) that are produced or released at the site of injury. Where injury persists, chronic inflammation ensues and is characterized by the accumulation or proliferation of macrophages, lymphocytes vascular endothelium, and fibroblasts at the damage site.

The goal of the inflammatory process is to rapidly effect the elimination of the causal agent and removal of debris from damaged cells by dilution and phagocytosis, as well as to initiate the repair process. Repair of the damage tissue may be achieved by a process of regeneration, which involves the replacement of damaged cells with viable cells of the same type through proliferation of adjacent healthy cells. Where the intrinsic regenerative capacity of cells of the damaged tissue is limited or tissue damage is severe, repair will involve fibrosis, a process in which fibroblasts from adjacent connective tissue mediate the replacement of damaged cells, with a characteristic scar tissue formation as the inevitable consequence.

Nonneoplastic Lesions

Epidermal Lesions

Since the skin is composed of various structures, the extent and degree of involvement of each component will depend on the agent itself and on the severity of the exposure. However, because of its location, the epidermis is always first exposed to externally applied toxicants. Consequently, many skin responses to adverse reactions are epidermal in nature and usually involve inflammation. In the mildest form of superficial skin injury, where damage is restricted solely to the epidermis and there is some degree of epidermal destruction, hyperplasia is generally the dominant response. The epidermal destruction ranges from focal keratinocyte swelling (e.g., spongiosis) to hydropic degeneration of the basal layers and focal cellular necrosis. Under these conditions, the basal cells typically respond by increasing cell division and the epidermis quickly regenerates to normal. However, when the insult is sustained the proliferative response continues and ultimately results in a thickening of the epidermis. A good example of such proliferative response is that observed in the thickened skin on the palms of manual workers and it is the result of a continued low-level abrasive injury. Depending on the particular cell layers of the epidermis that are affected, these hyperplasias are described as hyperkeratosis, hypergranulosis, and acanthosis for thickening of the stratum corneum, stratum granulosum, and stratum spinosum, respectively.

In severe injuries (e.g., corrosions), extensive epidermal necrosis, with accompanying damage to the cells of the basement membrane as well the superficial dermis, is frequently encountered. In this case, the extensive epidermal necrosis may lead to various degrees of ulcerations and be seen as devitalized epithelial layers with pyknotic nuclei that loosely line the dermis. Alternatively, the epidermis itself has sloughed off leaving a denuded dermal surface exposed to the external environment. These ulcerations are frequently accompanied by inflammatory changes, with migration of inflammatory cells such as polymorphonuclear leukocytes to the site of ulceration at the junction of the necrotic and viable tissues. This is followed by regenerative and proliferative changes involving the surrounding viable epithelial and connective tissue elements in an attempt to repair the damage. The undamaged adnexal components (e.g., hair follicles) are generally the source of precursor cells involved in the regeneration of the epithelial layers, and fibroblasts from the surrounding dermis are responsible for repair by fibrosis. As the damaged epidermis is repaired, the dead layers are sloughed, eventually leaving a scar.

These scenerios represent the two extremes, with most forms of dermatitis falling somewhere in between. Mild to moderate injuries usually produce clinical conditions described as eczema and they represent a wide range of responses essentially involving various combinations of degeneration, proliferation, and inflammation. Inflammatory responses often dominate during the early stages and are characterized by erythema, exudation, and leukocyte migration. These responses are sometimes accompanied by bullae, or blisters, and abscesses, or pustule formation, resulting from epidermal accumulation of fluids and cellular debris, respectively. With chronic or protracted exposure to mild irritants, proliferation of the epithelium increases. The skin becomes thickened, fissures may develop, and the proliferating keratinocytes begin to differentiate abnormally in a process known as parakeratosis, where the nuclei are retained in the stratum corneum. Although proliferation, involving hyperplasia and/or hypertrophy, is the usual pattern of epidermal response to toxicant exposure, on rare occasions epidermal atrophy is observed wherein the epidermis responds with decreases in cell size or decreases in number of epidermal layers.

Dermal Lesions

As eluded to previously, dermal responses to toxic insults can be elicited by direct penetration of the toxicant through the epidermis to the dermis and this may occur with or without the destruction of the epidermis. Furthermore, reactive processes, initiated in the epidermis as a consequence of epidermal exposure, while not injuring the dermis directly can also elicit dermal responses. In addition, dermal exposure to toxicants of systemic origin via diffusion through dermal capillaries may produce toxic responses in the dermis in the absence of associated injuries to the epidermis. As previously described, the extent and nature of the toxic response will depend largely on the severity of the insult and will likely involve a combination of mechanisms. Mild acute injuries can produce focal necrosis which may be accompanied by localized inflammatory infiltrations and possibly abscesses. On the other hand, severe injuries resulting from exposure to corrosive substances can produce dermal and eventual subcutaneous coagulative necrosis which may be very painful. Edema and congestion in both the dermis and the epidermis, with eventual formation of vesicles, often accompany allergic reactions in the dermis that result from either systemic or local exposure to toxicants. Prolonged dermal exposure to mild toxicants can result in chronic dermatitis and this is often associated with extensive subepidermal mononuclear infiltrates or with perivascular infiltrates. The presence of secondary infections often complicates the overall picture of the toxic response. Finally, profiltration of dermal fibroblasts accompanied by angioblastic activity completes the repair process and this frequently culminates in fibrosis, or dermal scarring.

Adnexal Lesions

In response to toxicologic insults, the cutaneous adnexa will also undergo the dynamic changes of degeneration, proliferation, inflammation, and repair in a manner similar to that described. Thus, during toxicant exposure, typical destructive and involutional changes (e.g., focal necrosis, edema, hyertrophy, and hyperplasia) are evident. However, severe acute or chronic injuries can result in the partial or, in certain instances, complete loss of skin appendages from the exposed area. This is due to the fact that although the epidermis can regenerate completely by cell migration from unaffected sites, the newly formed epidermis is unable to reconstitute the adnexal elements. When hair is the target of the toxic insult, alopecia (hair loss) is the main consequence. Hair is susceptible to damage by both external agents and agents reaching the hair matrix through the dermis. Two major types of injury are experienced,

namely, matrix cell damage and keratolytic damage. Keratolysis, the dissolution of hair keratin, is generally associated with local or surface contact of the toxic agent with hair. The resulting hair loss, due to the increased fragility of the hair shaft, may involve local patches or extensive areas, depending on the extent of the exposure. Regrowth of hair generally occurs following removal of the toxic agent as the hair matrix cells are not damaged.

Agents that damage the hair matrix cells may affect hair follicles at a specific phase of hair cycle, i.e., during anagen or telogen. The effect of anagen toxicity is typically hair loss (anagen effluvium). The mechanism of toxicity involves interference of the rapid mitotic activity of the follicular cells, leading to either a cessation of growth and the loss of the hair or the later loss of excessively brittle hair at the site of a weak, constricted area in the hair shaft. Anagen effluvium can occur within 1 or 2 weeks of exposure to the toxic agent and a number of commonly used cancer chemotherapeutic agents are known to be anagen toxicants. Hair loss is also a consequence of telogen toxicity. The onset of telogen toxicity is slower and occurs over months of exposure and may involve a variety of mechanisms. Anagen and telogen toxicity can occur simultaneously and typical early histological signs of toxicity may include the vacuolization, disappearance of mitosis, pyknosis of the nuclei in the follicular matrix, or the presence of nuclear and other debris in the hair shaft. When damage to the hair follicles is severe, there is the potential for complete and irreversible loss of hair follicles resulting in permanent alopecia. As indicated previously, although the epidermis has full regenerative capacity, the newly formed epithelium usually cannot regenerate the skin adnexa.

Another class of lesions of adnexal origin which is frequently seen as a result to exposure to a variety of agents, including grease, oils, coal tar, and cosmetic preparations, is acne. These acneiform lesions originate from the sebaceous glands and typically start with comedones and inflammatory folliculitis on the skin surface that is in direct contact with the causal agents. The resultant proliferation of the sebaceous gland follicular epithelium leads to the formation of lipid-filled keratin cysts, similar to those observed in acne vulgaris. Chloracne is a somewhat specific type of acneiform eruption which occurs after exposure to a group of halogenated aromatic hydrocarbons (e.g., polyhalogenated dibenzofurans, polychlorinated dioxins, polychlorinated naphthalenes, and polychlorinated biphenyls). Chloracne is characterized by small, straw-colored cysts, comedones, and, in severe cases, inflammatory pustules or abscesses may be seen. Histologically the changes that are seen during the development of chloracne begin with keratinization of the sebaceous gland epithelial duct and the outer root sheath of the hair follicle. The sebaceous gland is eventually replaced by a keratinous cyst and the typical fully developed lesion consists of a dilation of the upper third of the hair follicle, which is usually bottle shaped. No differentiation can be seen between the epithelia of the infundibulum and the sebaceous glands. Edema and mononuclear perivascular infiltrates are sometimes seen in the papillary dermis and late manifestations of chloracne often include mild fibrosis of the dermis, hypotrichosis, and hyperpigmentation. The affected areas are usually those located in the malar crescent of the face and behind the ears. The external genitalia, axillae, shoulders, chest, back, abdomen, and buttocks are sometime involved, but leisons are rarely seen in the extremities. Chloracne often continues to appear even after exposure to the chemical agent responsible has ceased, possibly as a consequence of release from tissue depots since most chloracnegens tend to be highly lipophilic. Experimental chloracne has been produced in rabbits, monkeys, and hairless mice. This latter species is thought to be the most useful animal model for the disease, but the occasional presence of degenerative cystic hair follicles in normal hairless mice is a confounding factor with this model.

Selective local damage to other skin adnexa, such as the sweat glands, can occur with exposure to a number of cytostatic agents, such as cytarabine and bleomycin, which are used in human cancer therapy. The condition is characterized by necrosis of the epithelium lining the eccrine sweat duct, accompanied by acute inflammation and squamous metaplasia of the remaining cells of the eccrine apparatus. The mechanism for the selective toxicity is unknown, although high concentrations of these compounds in sweat may provide an explanation. Other chemicals that are toxic to the sweat gland include formaldehyde, arsenic, lead, fluorine, and thallium, all of which produce generalized anhidrosis (loss of the sweating mechanism) due to partial or total destruction of the eccrine system.

Neoplastic Lesions

Cellular proliferation is one of the ways in which cells and tissues respond to an injurious insult and when these proliferations show partial or complete loss of responsiveness to normal growth controls the result is neoplasia, or cancerous growth. Neoplastic lesions induced in the skin of experimental animals have played an important role in our understanding of the multistage process of chemical carcinogenesis. Tumors produced in this multistage process are initially benign exophytic lesions (e.g., papillomas), some of which may regress while others gradually convert into fully invasive, malignant, endophytic tumors (i.e., carcinomas). The mechanisms by which chemicals may lead to uncontrolled cell proliferation are outside the scope of this entry, but suffice it to say that chemical carcinogens may be divided into two categories based on their proposed mechanims of action: (1) genototoxic, or those acting intracellularly, usually directly damaging to DNA, and (2) nongenotoxic, or those which act via regulatory factors in the extracellular environment.

Papillomas are the most common neoplastic lesions occurring in rodent skin after exposure to chemical carcinogens. They generally arise from the infundibular region of metaplastic or hyperplastic hair follicles. They are composed of a series of folds, united by common stalks to the underlying skin, and have a cauliflower-like structure and appearance. The folds of a papilloma consist of a central connective tissue core covered by a thick layer of epidermis-like epithelium. The germinative layers of the epithelium contain numerous mitoses and there are distinct spinous and granular layers as well as a thick, fully keratinized stratum corneum. Papillomas may regress or continue their progression toward carcinomas, and confluency into larger malignant tumors can also occur.

Keratocanthomas are benign neoplastic skin lesions often found after exposure to UV radiation or complete carcinogens in various species, including humans. They originate in the hair follicles as an intradermal growth of epithelial prolongations. They have a cup-shaped architecture with a central horny crater that has a papillomatous exophytic component and an endophytic component of deeply penetrating epithelial cords, which appear not to invade the subcutaneous tissues. In mice, keratocanthomas generally progress to squamous carcinomas and regression is uncommon. In humans, however, they are generally considered to be abortive neoplasias that usually regress. Preneoplastic, intraepithelial lesions are commonly found in humans as the result of exposure to sunlight or arsenicals, but such lesions are not frequently inducible in animal models of chemical carcinogenesis. These preneoplastic lesions have the potential to progress to carcinoma.

Carcinomas of various types, e.g., squamous cell and basal cell carcinomas, have been induced in many different laboratory species using UV light, other forms of ionizing radiation, and chemical carcinogens. Generally these tumors arise from existing papillomas, keratocanthomas, or intraepidermal preneoplastic lesions (in humans), as well as from otherwise normal or hyperplastic epidermis. In humans, cutaneous squamous cell and basal cell carcinomas are extremely common clinical problems and the major etiologic agent is generally considered to be chronic sun exposure. Fortunately, these tumors rarely metastasize and thus have low mortality, but they are locally destructive and can be associated with considerable morbidity. Melanomas, arising from the pigment-producing melanocytes in the epidermis, have been produced using chemical carcinogens in experimental animals. These melanotic tumors, which include both benign and malignant types, have generated considerable concern in recent years, particularly in relation to skin cancer in man. This is because the rate of skin cancer in humans has been increasing at an alarming rate for the past several years and melanomas, which metastasize widely, are responsible for more deaths than any other type of skin cancer. Chronic sun exposure is believed to be a major risk factor and the implication that UV radiation is a major causative agent in the pathogenesis of melanoma remains controversial. In experimental species, chemically induced melanotic tumors are less aggressive than the human malignant melanomas, thus they tend not to metastasize readily.

Other Responses

Urticarias, or "wheal and flare" reactions, are common skin responses produced by topical exposure to a variety of topical agents (Table S-15), especially biogenic polymers released from plants and insects. The response generally occurs within 1 hr of exposure and involves the local release of vasoactive substances including histamine. Frequently, urticaria is associated with immunologic responses and is often an integral part of immediate hypersensitivity reactions to ingested agents (e.g., drugs involved in CADRs). Undesirable color or pigmentary changes are also encountered as adverse cutaneous responses to topical agents. Chemicals which show structural similarities to tyrosine, the major building block of melanin, are known to cause

local loss of pigmentation, whereas increased pigmentation may result as a secondary consequence to a phototoxic response. Color changes in the skin may also occur as the result of cutaneous accumulation of endogenous (e.g., carotenemia from eating too many carrots) as well as exogenous (argyria from contacting silver) pigments. Subjective reactions such as itching, burning, or stinging sensations are often encountered by sensitive individuals following exposure to a variety of topical agents, primarily cosmetics and detergents. These reactions are entirely subjective and do not have any obvious manifestations that can be perceived by the outside observer. Nevertheless, they are considered by the affected individuals to be completely undesirable.

Mechanisms and Methods for Assessing Skin Toxicity

General Considerations

The classic signs of the inflammatory response in skin were recognized long ago in ancient Rome by the physician Celsus, who coined the Latin phrase, "*Rubor et tumor, cum calore et dolor,*" roughly meaning redness and swelling, resulting in heat and pain. The underlying mechanisms whereby these processes take place in ICD, ACD, and contact urticaria were, of course, unknown at the time. Much work in the past 50 years has helped clarify this mystery, however, and inflammation is best described within the paradigm of two major phases: the vascular phase and the cellular phase. Although a third, more immediate, "neurologic" phase has been identified recently, it is such a transient and poorly understood component of the inflammatory response that it bears little mention here.

The vascular phase represents the most acute response of the skin to the presence of an irritating chemical or to a potential allergen, taking place within minutes and generally lasting only a few hours. This phase is induced by several systems, first and foremost of which is the nonspecific release of inflammatory mediators by epidermal keratinocytes and dermal fibroblasts. Such vasoactive materials as IL-1β, other cytokines, and the arachidonic acid metabolites prostaglandin E_2, leukotriene D_4, and prostacyclin initiate a cascade of events resulting in vasodilation, increased vascular permeability, and the influx of blood cell constituents. A good analogy would be the situation presented by an overturned fuel tanker on a major highway. The roads would become swelled with traffic and the influx of police, fire trucks, ambulances, and onlookers would spill over into the surrounding countryside. Other systems involved in this early phase are the complement pathways, primarily C3a and C5a; the coagulation system (fibrin split-products, Factor XIIa, and thrombin); plasma bradykinin; and an immunological reaction mediated by mast cells, which are abundant in the dermis (7,000–10,000 cells/mm³). This latter component is the principal mechanism in contact urticaria (mentioned earlier) and the release of histamine, serotonin, heparin, and chemotactic factors from these mast cells is also important in initiating the cellular phase of the inflammatory response.

The cellular phase takes place over a period of several days and begins with leukocyte margination (contact with vascular walls) and the release of chemotactic factors causing the migration of neutrophils into the injured tissue. Neutrophils contain granules which provide microbicidal enzymes (myeloperoxidase and lysozyme), neutral serine proteinases (e.g., elastase and cathepsin G), β-glucuronidase, α-mannosidase, vitamin B_{12}-binding proteins, and collagenase. The net effect of these mediators is increased tissue oxygen consumption and the generation of reactive oxygen species, or free radicals (e.g., superoxide anions, peroxide radicals, and halide acids), all of which are lethal to invading pathogens and are somewhat responsible for the heat and pain which accompany the inflammatory response.

Basophils and eosinophils may also be involved, especially in ACD. Basophils are similar to mast cells and play a role in delayed-type hypersensitivity, whereas eosinophilic migration is dependent on complement and chemotactic factors released early upon exposure to a contact allergen. Langerhan's cells and other macrophagic monocytes release IL-1 and other cytokines early and are important in antigen presentation to lymphocytes. The latter white blood cell type then proceeds to influence a number of other cellular responses (Table S-22). Overall, there are three types of allergic reactions in skin: type I (anaphylaxis), typified by the "wheal and flare" produced by IgA- and IgE-responsive mast cells; type III (immune complex), which is an anitgen–antibody response involving complement; and type IV (delayed-type hypersensitivity). The latter is by far the most prominent type of chemically induced ACD and begins with Langerhan's cell and lymphocyte presentation of antigen to regional lymph nodes, followed by a

TABLE S-22
Lymphocyte Products Acting on Other Cell Types

Cell types affected	Lymphocyte products involved
Macrophage	Migration inhibitory factor (MAF), macrophage activating factor, macrophage aggregating factor, chemotactic factor, AG-dependent MIF
Neutrophil	Chemotactic factor, leukocyte inhibitory factor (LIF)
Lymphocyte	Interleukins IL-2, -3, -4, and -5; chemotactic factors
Eosinophil	AG-AB-dependent chemotactic factor, IL-5, migration stimulation factor
Basophil	Histamine releasing factor, IL-3
Other cells	Lymphotoxin, growth inhibitory factors, osteoclast activating factor (OAF)

vascular phase 24–48 hr later. As mentioned earlier, it is prerequisite for a molecule to produce ACD that it react chemically with proteins in the antigen presenting cells.

Experimental Models

In Vivo Techniques

Determination of eye and skin irritation potential is mandated for proper labeling of all consumer products and for any chemicals or products which are to be transported across state lines by the U.S. FDA and DOT, respectively. Animal testing for skin irritation (ICD) is almost exclusively restricted to modifications of the test first proposed by John Draize at the FDA in 1944. The rabbit primary dermal irritation (PDI) bioassay, as recommended by the CPSC in 1981 (in the Federal Hazard Substances Act), provides the basis for the most modern version of this model. Slight modifications proposed by the OECD (1981) and the U.S. EPA (1983) have recently been incorporated to reduce total animal use and eliminate the unnecessary discomfort of abraded test sites and overly long exposure periods. Briefly, 0.5 ml (liquids) or 0.5 g (solids) of each test substance are applied to unabraded sites only on three New Zealand White rabbits, under a 1 × 1-in. gauze pad, and the site is occluded with gauze and tape wrappings. Following a 4-hr exposure period, the wrappings and patches are removed and the sites are gently swabbed free of residual test material. Scores ranging from 1 to 4 for both erythema and edema are based on visual observation of the test site immediately after patch removal and at 24- and 48-hr postexposure. These scores are summed and divided by the total number of scores to calculate the PDI index (PDII), which serves as the *in vivo* response variable for each test substance. Occasionally, the guinea pig is used in place of rabbits in this assay or in full immersion studies and cumulative irritation (multiple doses) tests, while other species are used very infrequently. An exception to this rule may be the mouse ear swelling test (MEST), which has been undergoing extensive evaluation and validation in the past few years. Nevertheless, despite clear evidence that these animal models may not be relevant to the human condition, Draize-type testing is still standard practice in the consumer products and cosmetics industries.

Human testing for ICD involves either single application patches or cumulative patching, usually fresh doses of the chemical every 48 hr over a 21-day period, for most compounds. For soaps and detergents, specialized assays, called soap chamber tests and arm wash tests, are utilized. In the former, a modified Franz diffusion cell-type donor chamber is affixed to the forearm and the soap solution is left in contact with the skin surface for a few hours. Arm wash tests were devised to mimic actual use conditions. This test has been further modified to include multiple washes over a short time period, or an "exaggerated" arm wash test, to help discriminate among milder irritants, which produce little or no response in the standard soap chamber, arm wash, or patch tests. The major limitation to all human tests is the large intersubject variability coupled with the heavy influence of environmental conditions on the skin's initial condition, which ultimately affects its ability to respond to irritant challenge. It is for this latter reason that most of these clinical assessments of ICD are performed in the summer months because cold, dry air alone can be very damaging to skin.

The situation with animal testing for ACD is somewhat more complicated than that for ICD tests, probably because the disease process and underlying mechanisms are more complex. The guinea pig is the standard animal model for ACD, based on the original intradermal injection studies of the nitro -and chlorobenzene classes of sensitizers in 1935. The following modifications of the original protocol are now in routine use: occluded patch test, ear-flank test, guinea pig maximization test, split-adjuvant test, guinea pig optimization or Freund's complete adjuvant test, and open epicutaneous test. The

common feature to all these tests is that they are biphasic, employing an induction phase followed by a challenge phase. Their major limitations are the subjective nature of the visual scoring system and the fact that these are rather costly, time-consuming bioassays compared to the ICD counterparts. In addition, there are ethical concerns with the use of adjuvants, which are basically allergenic materials added to the assay to increase the response. Adjuvants alone, when injected intradermally, can cause considerable redness, swelling, and intense pain. Finally, there are two newer models under evaluation, neither of which has been validated for regulatory purposes: a variation of the MEST and the local lymph node assay (LLNA). The LLNA is very promising and is based on measurement of cellular proliferation and other parameters in white blood cells (lymphocytes) collected from the lymph node draining the site of exposure.

Human ACD assays are of two basic types: the so-called prophetic patch test or single-induction dose, which is insensitive and rarely used, and the repeat insult patch test (RIPT). The latter involves multiple applications (every other day for 2 or 3 weeks) of low concentrations of the test article during the induction phase, followed by a single 24-hr exposure to a higher dose and visual scoring over a 3- to 7-day period during the challenge phase. A modification of the RIPT is Kligman's maximization procedure, which utilizes the irritating surfactant sodium lauryl sulfate to increase the skin's responsiveness to the test material. Besides the interfering factors cited previously for human ICD tests, a major limitation of the RIPT is the selection of nonirritating induction doses, vehicle effects, and the inability to properly evaluate the skin reactions. In addition, the results of RIPTs are the least quantitative of all the *in vivo* irritation and sensitization tests. This issue of quantitation is particularly important in human tests for both ICD and ACD, which normally depend entirely on subjective, visual assessments of erythema and edema. This need has led, in turn, to a large effort to develop instrumental methods for measuring the vast array of skin responses to toxic compounds (Table S-23). Besides providing quantitative data for such diverse responses to cutaneous toxins as inflammation, altered hydration state (e.g., dryness and "tight feel"), changes in elastic or mechanical properties, or altered surface morphology (e.g., roughness, scaliness, and flaking), these biophysical methods are much more sensitive than visual techniques. Moreover, some of these

TABLE S-23
Instrumental Methods for Assessing Cutaneous Toxicity

Category	Instrument used (measured response)
Spectraphotometry	Dia-Stron erythema meter, Minolta Chromameter, Cortex Dermaspectrometer (all measure a "redness" index of erythema); laser Doppler Velocimeter (blood flow)
Evaporimetry	Servo-Med Evaporimeter (transepidermal or skin surface water loss)
Electrical properties	Skicon, Corneometer, Nova Dermal Phase Meter (all measure conductance/capacitance to assess hydration state)
Calorimetry	Skin surface temperature, thermography
Mechanical properties	Dia-Stron Dermal Torque Meter, Rheometers, SEM 474 Cutometer, Gas-Bearing Electrodynamometer (all measure elasticity); Newcastle Friction Meter (roughness); Cortex Dermascan and other ultrasound equipment (epidermal thickness)
Surface features	Anjinomoto scopeman or Microwatcher image analyzers, ultrasound equipment, profilometers (roughness, flakiness, scaliness, etc.)
Miscellaneous	Differential scanning calorimetry, Fourier transform infrared spectrometry (changes in stratum corneum lipid structure/function)

instruments have demonstrated utility in animal or *in vitro* studies of cutaneous toxicity.

In Vitro Techniques

During the past two decades, public pressure to reduce the use of animals in all areas of biomedical research has resulted in animal experimentation coming under close scrutiny and increased governmental regulation. One area in which alternatives to animal systems seem both feasible and justified is that of early screens in premarket safety evaluations. In fact, there are *in vitro* alternatives which have undergone large multiinstitutional validation studies and which are being extensively utilized for mutagenicity and ocular irritancy testing in the industrial setting. Furthermore, a number of alternative assays have been proposed as screens for cutaneous irritation, although the validation process has been much slower. Nevertheless, assays based on disruption of cell membrane integrity, metabolic activity, or growth; incorporation of radiolabeled nucleotides and amino acids; cellular release of inflammatory

mediators; or induction of morphological alterations at the cellular level are all currently under evaluation (Table S-24). These types of assays may be performed using human and nonhuman fibroblast and keratinocyte cell cultures or using the more complex, organotypic skin tissue and organ culture models. In addition, there are a number of techniques which do not involve tissue cultures, operate via unknown mechanisms or mechanisms that are unrelated to the ICD response *in vivo,* or which are known to be entirely correlative in nature. Many of the commonly used biochemical markers or endpoints associated with these alternative methods share significant limitations: (1) They often require high test substance concentrations, effectively killing a large fraction of the exposed cells, and there is no clear evidence that this degree of cytotoxicity is mechanistically relevant in ICD; (2) they produce extremely variable and unreliable results for diverse sets of test materials and are sometimes more costly than animal or human patch tests; and (3) they were primarily validated against *in vivo* ocular irritation

TABLE S-24
In Vitro Endpoints for Predicting ICD

Class	Specific examples of proposed markers
Membrane integrity	Vital dyes (trypan blue, eosin); fluorescence (Hoechst, fluoresceins, rhodamine, ethidium bromide, propidium iodide); exogenous (^{51}Cr release); endogenous (LDH or alkaline phosphatase leakage, intracellular K^+, lipid peroxidation)
Subcellular function	Mitochondrial (MTT, XTT, Alamar blue, ATP); ribosomal ([^{14}C]-LEU or [^{14}C]URI Incorporation); lysosomal (neutral red uptake/release); Nuclear ([^{3}H]THY Incorporation, DNA binding)
Cellular metabolism	Glucose utilization, O_2 consumption, growth inhibition, lactate/pyruvate ratios, glutathione/redox status
Inflammatory mediators	Arachidonic acid cascade ([^{3}H]AA release PGE_2 release, leukotrienes and HETEs); cytokine release (IL-1$_B$, TNF-α)
Morphology	Light and electron microscopic changes
Unknown mechanisms	Collagen swelling, skintex, Coumassie blue dye extraction from gelatin, quantitative structure–activity relationships (QSAR computer models)

data. Since it is well documented that the potential for a chemical to produce eye irritancy is not well correlated with its irritability to the skin, the latter point is an important distinction to make in the validation of alternative models for predicting ICD.

The situation with *in vitro* models for predicting ACD, unlike its *in vivo* counterpart, is less complicated than that for ICD because there are very few *in vitro* systems which have even been proposed for ACD testing. This is also a consequence of the complexity of this disease since ACD involves the interaction of many organ systems, which cannot be properly simulated in any currently available cell or tissue culture model. Nevertheless, two assays that have shown some promise for predicting ACD with certain classes of allergens are the lymphocyte transformation test and the macrophage migratory inhibition test.

Conclusions

It is clear that skin is not just an inert, protective barrier which surrounds the body's internal organs, but rather is an active participant in the overall outcome of exposure to potentially injurious materials in the external environment. The significance of cutaneous reactions to topical agents, particularly the inflammatory response and carcinogenesis, is the subject of an increasing number of scientific investigations. From the perspective of the skin absorption process, this organ is at once a portal of entry for a variety of topically applied chemicals, a drug-metabolizing organ, and a target organ for local toxicity. Thus, knowledge of the mechanisms involved in translocating chemicals into and through the skin, coupled with its effect on the physiological disposition or availability of topically delivered chemicals to interact with skin and other body organs, is key to understanding cutaneous pharmacology and toxicology.

In this review, some of the theoretical models and experimental methodologies employed in dermatotoxicity studies, both *in vivo* and *in vitro*, have been described. It is suggested that a combination of these techniques may provide the basis for future experimental approaches toward increasing our knowledge of the mechanisms of cutaneous toxicity. It should be emphasized that research in this area is evolving, such techniques are being developed, and the rationale by which *in vivo* or *in vitro* models are selected and utilized is

under continual scrutiny. Further development in this area will necessitate improvements in bioanalytical techniques and a better understanding of the interplay between skin penetration, permeation, and metabolism, as well as the role of modulating factors that may influence the structure, function, and toxicology of the skin.

Further Reading

Barile, F. A. (1994). *Introduction to in Vitro Cytotoxicology. Mechanisms and Methods.* CRC Press, Boca Raton, FL.

Bronaugh, R. L., and Maibach, H. I. (Eds.) (1989). *Percutaneous Absorption. Mechanisms–Methodology–Drug Delivery,* 2nd ed. Dekker, New York.

Drill, V. A., and Lazar, P. (Eds.) (1984). *Cutaneous Toxicity.* Raven Press, New York.

Goldsmith, L. A. (Ed.) (1983). *Biochemistry and Physiology of the Skin,* Vols. I and II. Oxford Univ. Press, London.

Hobson, D. W. (1991). *Dermal and Ocular Toxicology: Fundamentals and Methods.* CRC Press, Boca Raton, FL.

Jackson, E. M., and Goldner, R. (Eds.) (1990). *Irritant Contact Dermatitis.* Dekker, New York.

Kemppainen, B. W., and Reifenrath, W. G. (Eds.) (1990). *Methods for Skin Absorption.* CRC Press, Boca Raton, FL.

Marzulli, F. N., and Maibach, H. I. (Eds.) (1991). *Dermatotoxicology,* 4th ed. Hemisphere, New York.

Mukhtar, H. (Ed.) (1992). *Pharmacology of the Skin.* CRC Press, Boca Raton, FL.

—*Michael P. Carver and John Kao*

Related Topics

Acids
Alkalies
Carcinogenesis
Delayed-Type Hypersensitivity
Ocular and Dermal Studies
Organophosphates
Photoallergens
Poisoning Emergencies in Humans
Radiation
Tissue Repair
Toxicity Testing, Alternatives
Toxicity Testing, Dermal

Snake, Crotalidae

◆ Synonyms: Pit viper; Crotalus (rattlesnake); Agkistrodon (copperhead, cottonmouth); Sistrurus (pigmy rattler, massasauga) (see Snake, Elapidae)

Exposure Pathways

Most frequently, envenomation by Crotalidae species occurs subcutaneously. However, envenomation directly into an artery or vein has been documented and is associated with a rapid progression of life-threatening symptoms such as shock and cardiovascular collapse.

Toxicokinetics

Systemic absorption of Crotalidae venom is dependent on lymphatic transport following subcutaneous envenomation. The onset of local symptoms such as swelling and ecchymosis occurs within several hours following envenomation. Cardiovascular, neurologic, or hematological compromise varies in onset but may occur within as little as 10–15 min following an intravenous envenomation. The metabolism of venom components is not well understood. It is likely that venom components are inactivated by enzymes within tissues where the venom is ultimately distributed. The distribution of venom is variable and complex. Venom components differ among Crotalidae species and will distribute unevenly to different tissue sites. The biological half-life of Crotalidae venom has not been determined. Metabolized venom fractions are primarily eliminated by the kidney.

Mechanism of Toxicity

Crotalidae venom is composed of many different fractions that vary among Crotalidae species and act at different tissue receptor sites. Hyaluronidase, a common venom component, hydrolyzes connective tissue, increasing the spread of the venom. Phospholipases effect nerve conduction and disrupt cell membranes leading to hemolysis. Thus, multiple organ systems may be affected. Historically, Crotalidae venom was classified as neurotoxic, hemotoxic, cardiotoxic, or myo-

toxic. This oversimplifies the complex nature of Crotalidae venom. Clinically, a patient may develop such multisystem disorders as platelet destruction, internal bleeding, hypotension, paresthesias, and rhabdomyolysis.

Human Toxicity

The severity of envenomation varies greatly and is dependent on such things as the species of snake, amount of venom injected, and victim age. Crotalidae venom initially produces local tissue changes that manifest as swelling, ecchymosis, bruising, petechiae, pain, and erythema. Swelling may progress to involve the entire affected limb. These local symptoms commonly develop within minutes to several hours following envenomation. Because of poor tissue perfusion, local skin sloughing and tissue necrosis may occur. Swelling of an envenomated extremity may be severe but is commonly subcutaneous in nature. True compartment syndrome is unlikely since subfascial envenomations are rare. Approximately 25% of Crotalidae bites are "dry" or no venom is injected during the bite. Patients with dry bites may exhibit symptoms of erythema and slight swelling due to the trauma of the inflicted bite. However, these symptoms are limited to the immediate area of the bite and require only wound management and follow-up for infection, which is common in snakebites. Systemic symptoms following envenomation may include paresthesias, coagulation disorders, thrombocytopenia, active bleeding, decreased hemoglobin, disseminated intravascular coagulation, hypotension, EKG changes, decreased level of consciousness, and rhabdomyolysis.

In contrast to most rattlesnakes, the Mojave rattlesnake generally produces less tissue destruction and, therefore, less swelling and pain. In addition, central nervous system depression and respiratory paralysis are more frequently seen following envenomation by this species.

Clinical Management

Basic and advanced clinical life support is essential in the successful management of a crotalid envenomation. This is especially true in those rare cases involving intravenous envenomation in which immediate and aggressive cardiovascular and respiratory support can be life saving.

It is important to evaluate the clinical presentation of the patient as well as laboratory data to determine whether the administration of Crotalidae antivenin is indicated. Antivenin is not necessarily required in every patient who is envenomated. Envenomation by the copperhead (*Agkistrodon contortrix*) often requires little or no antivenin. Antivenin is generally not necessary in patients lacking significant tissue swelling in the absence off any systemic symptoms or laboratory abnormalities. Crotalidae antivenin is derived from horse serum. Anaphylaxis as well as delayed hypersensitivity reactions are not uncommon. Therefore, antivenin should not be used indiscriminately. A small amount of diluted antivenin is administered as a test dose to check for allergic response. This procedure is outlined in the package insert included with the antivenin. Traditionally, the degree of envenomation is graded according to the severity of local tissue changes, systemic symptoms, and laboratory changes. The quantity of antivenin to be administered is determined by the "grade" or severity of symptoms present. Typically, antivenin is administered in multiples of five vials. Patient response to the antivenin must be evaluated at various time intervals following administration to determine if further antivenin is required. Uncomplicated envenomations resulting in minimal swelling of a limb may require only 5 vials of antivenin. In contrast, patients exhibiting life-threatening symptoms may require up to 30 or more vials of antivenin. Antivenin should be administered within 4 hr of the envenomation. Effectiveness decreases with time.

Research is being conducted on a rattlesnake immune globulin (Fab). This highly specific and less antigenic antivenin may improve therapy and reduce or eliminate the adverse effects associated with the current antivenin. However, at this time it is only investigational.

Historically, fasciotomy was frequently employed to relieve pressure due to extensive swelling. However, fascially bound compartments are rarely involved, making this disfiguring procedure unnecessary. Most first-aid measures are of little value and some are dangerous. The use of ice to prevent the spread of the venom has been linked to an increased frequency of limb amputations and should never be employed. Field procedures such as fang mark incisions may result in vein or artery damage, and improperly placed tourniquets may impede blood flow. Electric shock directed at the site of envenomation has not been proven effective and is a dangerous procedure.

—*Gary W. Everson*

Related Topic

Poisoning Emergencies in Humans

Snake, Elapidae

◆ SYNONYMS: Coral snake; Micrurus; Micruroides

Exposure Pathways
Envenomation by North American species of Elapidae occurs subcutaneously. The bite differs from that of the Crotalidae species. Coral snakes possess smaller fangs and the snake tends to grasp and hold on rather than strike and release. The venom is discharged through grooved fangs and worked into the bite site with a chewing motion. Due to these characteristics, envenomation into an artery or vein is not likely [see Snake, Crotalidae (e.g., rattlesnake and pit viper)].

Toxicokinetics
Systemic absorption of Elapidae venom is dependent on lymphatic transport following a subcutaneous envenomation. The onset of neurotoxic symptoms usually occurs within 4 hr but can be delayed up to 4–10 hr following envenomation. The metabolism of venom components is not well understood. It is likely that venom components are inactivated by enzymes within tissues where the venom is ultimately distributed. The distribution of venom is variable and complex. Venom components differ among Elapidae species and will distribute unevenly to different tissue sites. The biological half-life of Elapidae venom has not been determined. It is likely that metabolized venom fractions are eliminated primarily by the kidneys.

Mechanism of Toxicity
Elapidae venom is composed of different fractions which vary among Elapidae species. The venom con-

tains fractions which are primarily neurotoxic in nature. The venom results in a bulbar-type cranial nerve paralysis. In contrast to Crotalidae species, Elapidae venom from North American species lacks most of the enzymes and spreading factors that cause local tissue destruction. The toxicity of exotic species of elapids from other countries possess venom components much different than that of coral snakes.

Human Toxicity

The severity of an envenomation varies greatly and is dependent on such things as the species of snake, amount of venom injected, and victim age. Envenomation may occur despite the absence of identifiable fang marks at the bite site. Coral snake venom causes very little local tissue changes. Mild swelling, pain, and redness at the immediate bite site are generally the limit of the local reaction. There may be some numbness and weakness of the bitten extremity within 1 or 2 hr. The degree of local tissue reaction does not correlate with the degree of systemic symptoms which may appear much later. Typically, lightheadedness, dizziness, or drowsiness mark the onset of systemic toxicity. Generalized muscle weakness, fasciculations, and tremors may develop. Increased salivation, nausea, and vomiting are also common. Neurological symptoms may progress to include slurred speech, ptosis, dysphagia, visual disturbances, muscle paralysis, and respiratory depression. Neurological symptoms may be delayed for up to 12 hr after envenomation. Seizures may occur, especially in children. Death results from respiratory depression, hypotension, and cardiovascular collapse. The bite of the Sonoran (Arizona) coral snake is associated with a much less severe progression of symptoms than that of the eastern coral snake. Headaches, blurred vision, and ataxia may be the limit of neurological symptoms following envenomation by the Sonoran (Arizona) coral snake. Exotic Elapidae species of snakes include cobras, kraits, and others which possess toxins that target the heart, coagulation factors, and other sites in addition to the central nervous system. Although rare, bites may occur to individuals illegally possessing these poisonous snakes. In these cases, contacting the local poison center is essential in determining the nearest location of specific antivenin, if available.

Clinical Management

Basic and advanced clinical life support is essential in the successful management of coral snake envenoma-

tion, especially those involving the eastern coral snake. A coral snake bite is a medical emergency and requires immediate transport to a hospital. Aggressive respiratory and cardiovascular support can be life-saving. Establishing intravenous fluid support should be started soon after the bite. Early administration of antivenin, *Micrurus fulvius* (Equine), is essential following envenomation by the eastern and Texas coral snakes. The antivenin is not effective for bites of the Sonoran (Arizona) coral snake. Since local symptoms do not correlate with the severity of the envenomation, antivenin should be administered as soon as possible following envenomation, despite the absence of neurological symptoms. Three to five vials of antivenin should be diluted in 100–500 cc of 0.9% sodium chloride. This should be infused intravenously over 30 min. Antivenin, *M. fulvius*, is derived from horse serum. Anaphylaxis as well as delayed hypersensitivity reactions are not an uncommon response to the antivenin. A small amount of diluted antivenin should be administered as a test dose to check for allergic response. This procedure is outlined in the package insert included with the antivenin. Patients who exhibit a negative allergic response following the test dose may still develop an anaphylactic reaction. Therefore, one should be prepared at all times to treat an allergic reaction to the antivenin. Epinephrine, intravenous antihistamines, and corticosteroids should be readily available. Close observation of the patient is required to determine the patient's response to the antivenin. Additional doses of antivenin may be required should neurological symptoms progress. Most envenomations require from 3 to 10 vials of antivenin. Further doses of antivenin may be required in those patients exhibiting life-threatening symptoms. As in any snakebite, infection is common and a broad-spectrum antibiotic should be considered. In addition, tetanus prophylaxis should be provided. Serum sickness may occur following the use of the equine-based antivenin. Although serum sickness is usually mild, an outpatient course of corticosteroid may be required in some cases.

Most first-aid measures are of little value and some are dangerous. The use of ice to prevent the spread of the venom has been linked to an increased frequency of limb amputations and should never be employed. Field procedures such as fang mark incisions may result in vein or artery damage and improperly placed tourniquets may impede blood flow. Electric shock directed at the site of envenomation has not been proven effective and is a dangerous procedure.

—*Gary W. Everson*

Related Topic

Poisoning Emergencies in Humans

Society of Environmental Toxicology and Chemistry

In the 1970s, no forum existed for interdisciplinary communication among environmental scientists—biologists, chemists, and toxicologists—and others interested in environmental issues such as managers and engineers. The Society of Environmental Toxicology and Chemistry (SETAC) was founded in 1979 to fill the void. Based on the growth in membership, annual meeting attendance, and publications, the forum was needed.

A unique strength of SETAC is its commitment to balance the interests of academia, business, and government. The society by-laws mandate equal representation from these three sectors for officers, board of directors, and committee members. Although there is no control mechanism, the proportion of members from each of the three sectors has remained nearly equal over the past 15 years.

Like many other professional societies, SETAC publishes an esteemed scientific journal and convenes an annual meeting replete with state-of-the-science poster and platform presentations. Because of its multidisciplinary approach, however, the scope of the science of SETAC is much broader in concept and application than that of many other societies.

SETAC is concerned about global environmental issues. Its members are committed to good science worldwide, to timely and effective communication of research, and to interactions among professionals so that enhanced knowledge and increased personal exchanges occur. A sister organization, SETAC-Europe, was founded in 1989, and the nonprofit SETAC Foundation for Environmental Education was founded in 1990. As evidence of international acceptance of the SETAC model, member groups have been proposed in South America, Russia, South Africa, Japan, India, and Australia/New Zealand.

SETAC membership has increased from 230 charter members in October 1980 to more than 4500 members from 50 U.S. states, 9 Canadian provinces, and more than 50 other countries worldwide. Participants and technical presentations at SETAC annual meetings have increased from 470 attendees and 86 technical presentations in 1980 to more than 2600 participants and nearly 1350 presentations in 1994.

Environmental Toxicology and Chemistry, an internationally acclaimed scientific journal, has grown from a quarterly publication of fewer than 400 pages annually (1980) to a monthly publication of more than 2000 pages in 1994. SETAC Press publishes the journal along with peer-reviewed workshop and symposia proceedings and a variety of technical reports.

SETAC Foundation for Environmental Education

In 1991, SETAC began to reach out to the public and to provide understandable information about the risks and benefits of chemicals in the environment through the activities of the SETAC Foundation for Environmental Education. The major goals of the Foundation are listed below.

Goals

- Enhance the scientific approach to assessing the risks/benefits of chemicals in the environment;

- Foster a science-based, holistic vision for improving environmental quality;

- Expand public awareness of chemical risks/benefits in everyday life;

• Provide training and education grants and other support to environmental educators and education programs; and

• Emphasize an objective and balanced approach to assessing chemical use and misuse.

Accomplishments

Since 1991, when an executive director was named and a permanent office was established, the foundation has accomplished the following:

• Organized and hosted or cohosted 11 weeklong technical workshops on subjects ranging from life cycle assessment to the mechanisms of bioavailability of chemicals in aquatic environments;

• Provided administrative support for three SETAC/SETAC Foundation interactive groups on life cycle assessment, aquatic model ecosystems, and ecological risk assessment;

• Sponsored open forums to communicate workshop results;

• Organized continuing-education courses and training courses on diverse subjects;

• Published numerous technical and informal reports;

• Organized and funded student travel awards (1991–1994) and minority student travel awards (1993–1994) to SETAC annual meetings; and

• Assisted SETAC in establishing science fellows in both the U.S. House and Senate.

For additional information, contact the SETAC Foundation for Environmental Education, 1010 North 12th Avenue, Pensacola, FL 32501. Telephone: 904-469-9777; fax: 904-469-9778. The SETAC Foundation for Environmental Education is a tax-exempt 501(c)(3) organization, and gifts are tax-deductible.

Also for additional information, contact SETAC at SETAC Office, 1010 North 12th Avenue, Pensacola, FL 32501-3370. Telephone: 904-469-1500; fax: 904-469-9778; e-mail: SETAC@SETAC.ORG.

—*Harihara M. Mehendale*

(Adapted from the information supplied by SETAC)

Related Topics

Academy of Toxicological Sciences
American College of Toxicology
Chemical Industry Institute of Toxicology
European Society of Toxicology
International Life Sciences Institute--North America
Society of Environmental Toxicology and Chemistry
Society of Toxicology

Society of Toxicology

Objectives

The Society of Toxicology (SOT) is a professional organization of scientists from academic institutions, government, and industry representing the great variety of scientists who practice toxicology. SOT promotes the acquisition and utilization of knowledge in toxicology, aids in the protection of public health, and facilitates the exchange of information among its members as well as among investigators in other scientific disciplines. The society has a strong commitment to education in toxicology and to the recruitment of students and new members into the profession.

History and Organization

SOT was founded in 1961 as a not-for-profit scientific society. The society is governed by an 11-person elected council and managed by an administrative office in the Washington, DC, area.

The society's activities are highly diverse and assisted by the efforts of nearly 20 committees, such as: Animals in Research, Public Communications, Education, Regulatory Affairs, and Legislative Assistance.

Membership

Currently, SOT has more than 3500 members in 34 countries. The majority of members are practicing toxicologists and scientists from allied disciplines. The society offers three kinds of individual memberships: full, associate, and student. Also offered are institutional memberships for companies and other organizations. In addition, honorary memberships may be awarded by council to persons who are not members of the society in recognition of outstanding and sustained achievement in the field of toxicology.

Specialty Sections

The society has established 12 specialty sections that may propose sessions for the annual meeting, exchange information via newsletters, present awards, and participate in other scientific activities. The specialty sections of the society are Carcinogenesis, Food Safety, Immunotoxicology, Inhalation, Mechanisms, Metals, Molecular Biology, Neurotoxicology, Reproductive and Developmental Toxicology, Regulatory and Safety Evaluation, Risk Assessment, and Veterinary Toxicology.

Regional Chapters

The SOT has 16 regional chapters that sponsor regular local meetings throughout the year. The purpose of the regional chapters is to foster scientific exchange at a local level.

Publications

The society has two official journals: *Toxicology and Applied Pharmacology* and *Fundamental and Applied Toxicology*. *Toxicology and Applied Pharmacology* is published 12 times per year. This journal publishes original scientific research pertaining to effects on tissue structure or functions resulting from administration of chemicals, drugs, or natural products to animals or man. Manuscripts address mechanistic approaches to the physiological, biochemical, cellular, or molecular understanding of toxicologic/pathologic lesions and to methods used to describe these responses. *Fundamental and Applied Toxicology* is published 8 times per year. This journal publishes scientific articles and reports relating to those broad aspects of toxicology that are relevant to assessing the risks or effects of toxic agents on the health of humans and other animals. Examples include statistical and mathematical methods of risk assessment and safety evaluation studies that are structural, biochemical, or functional in nature. Also included are articles on methods, regulatory issues and policy relevant to the practice of toxicology, scientific reviews on specific topics, and summaries of symposia.

Meetings

SOT conducts an annual meeting, the largest of its kind in the world. The meeting occurs in March of each year and nearly 2000 papers are presented on a variety of subjects. Continuing education courses and symposia sponsored by specialty sections of the society are regular features at the annual meeting. The abstracts of all presented papers are published annually in *The Toxicologist*, a special edition of *Fundamental and Applied Toxicology*.

Awards and Grants

The society presents several awards annually which recognize outstanding achievement in the field of toxicology. Special awards may be presented at the discretion of the council. SOT also presents a number of sponsored awards and grants for both graduate and postdoctoral research positions, including the prestigious Burroughs–Wellcome Toxicology Scholar Award.

Related Societies

SOT maintains liaison with 41 affiliated societies and participates in the International Union on Toxicology and intersociety activities.

For additional information, contact the Society of Toxicology, 1767 Business Center Drive, Suite 302, Reston, VA 22090-5332. Telephone: 703-438-3115; fax: 703-438-3113.

—*David M. Krentz and Harihara M. Mehendale*

(Adapted from information supplied by SOT)

Related Topics

Academy of Toxicological Sciences
American Academy of Clinical Toxicology
American College of Toxicology
European Society of Toxicology
International Union of Toxicology
Society of Environmental Toxicology and Chemistry

Sodium (Na)

- CAS: 7440-23-5
- SELECTED COMPOUNDS: Sodium bicarbonate, NaHCO$_3$ (CAS: 144-55-8); sodium chloride, NaCl (CAS: 7647-14-5); sodium hydroxide, NaOH (CAS: 1310-73-2); sodium hypochlorite, NaOCl (CAS: 7681-52-9)
- CHEMICAL CLASS: Metals

Uses

Numerous industries use sodium compounds. They are used in detergents, hair straighteners, glass, paper, textiles, and wood pulp. Sodium metal is used in sodium vapor lamps. Sodium hypochlorite is used in bleaches and sodium hydroxide in drain cleaners. Sodium chloride (table salt) is used in ion exchangers to soften water, and sodium bicarbonate is used in beverages, baking soda, and antacid pills.

Exposure Pathways

Ingestion is the primary route of exposure to sodium. Many foods contain sodium chloride naturally (e.g., milk, cheese, shellfish, and, to a lesser extent, meat and poultry). Nonetheless, most people add extra table salt to their food to the extent of 2000–7000 mg/day. In addition, all water supplies tested and nearly all carbonated beverages contain sodium (the American public consumes more carbonated beverages than drinking water).

Inhalation of sodium is a minor route of exposure. Sodium in the air comes from the oceans. Dermal absorption is not normally considered an important exposure pathway.

Toxicokinetics

Ingested sodium compounds are usually completely absorbed. Once absorbed, sodium is distributed throughout all tissues in the body. Most sodium is found in the plasma. Urine and perspiration are the major routes of excretion. Heat and hard physical labor can contribute to excessive loss of sodium.

Mechanism of Toxicity

Very little is known about sodium's mechanism of toxicity. There is practically no information on the effect of sodium on enzymes. No information is available on metabolic alterations of the sodium ion.

Human Toxicity

Sodium is associated with hypertension. Excess sodium results in an increase of extracellular fluid volume. Under these conditions the plasma protein concentration decreases. Sodium is an emetic; intake of excess sodium leads to nausea and vomiting.

The accidental substitution of table salt for sugar has resulted in sodium poisoning in infants. These infants experienced increased body temperature, muscle twitching, and convulsions; in some cases, their kidneys were damaged.

Sodium compounds with high pH values in solution (e.g., sodium hydroxide) are extremely corrosive to the skin and mucous membranes. The ACGIH STEL-C for sodium hydroxide is 2 mg/m^3.

Clinical Management

For extremely high sodium intake, peritoneal dialysis is the treatment of choice to lower the plasma sodium concentration. For exposure to sodium hydroxide, clinical management of skin corrosion is indicated.

Animal Toxicity

In animals, sodium has proven to be teratogenic but not carcinogenic. Laboratory animals given a high salt diet develop hypertension.

—Arthur Furst and Shirley B. Radding

Related Topic

Metals

Sodium Fluoroacetate

♦ CAS: 62-74-8

♦ SYNONYMS: Fluoroacetate-1080; Fratol; Ten-Eighty

♦ CHEMICAL STRUCTURE:

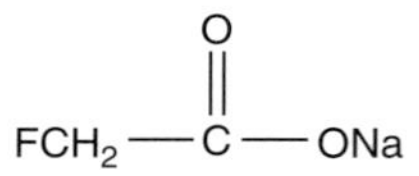

$$FCH_2 - \overset{\overset{\textstyle O}{\|}}{C} - ONa$$

Use

Fluoroacetate is primarily used as a rodenticide and is only available to licensed pesticide applicators.

Exposure Pathways

Ingestion, inhalation, and dermal exposures are all possible, but ingestion is the major route of exposure.

Toxicokinetics

Fluoroacetate is rapidly absorbed by the gastrointestinal tract but not well absorbed dermally. Fluoroacetate is converted to fluorocitrate, the ultimate toxicant. Fluoroacetate is distributed to lipid-rich organs, such as the liver, brain, and kidneys. Fluoroacetate is primarily eliminated through urine.

Mechanism of Toxicity

Fluoroacetate produces its toxic action (after conversion to fluorocitrate) by inhibiting the Kreb's cycle. The compound is incorporated into fluoroacetyl coenzyme A, which condenses with oxaloacetate to form fluorocitrate. This inhibits the enzyme aconitase, which inhibits conversion of citrate to isocitrate. Mitochondrial uptake of acetate may also be affected. The heart and central nervous system (CNS) are the tissues most affected by this inhibition of oxidative energy metabolism. Oxygen consumption is markedly reduced.

Human Toxicity

Acute fluoroacetate poisoning can result in nausea, vomiting, cardiac arrythmia, cyanosis, generalized convulsions, hypotension, and death from ventricular fibrillation or respiratory failure. Long-term effects are not common if the patient survives the acute toxicity.

The LD_{Lo} (oral) for humans is 714 μg/kg. The ACGIH TLV-TWA is 0.05 mg/m^3. The probable lethal dose in humans is less than 5 mg/kg.

Clinical Management

The patient should be moved to fresh air. Decontamination of eyes and skin should be immediate. For oral exposure, gastric lavage is preferable to emesis and should be prompt. Charcoal should be administered as a slurry to block absorption of sodium fluoroacetate. Treatment is symptomatic. Respiratory and cardiovascular support is often necessary with significant exposures. Anticonvulsants (barbiturate) and antiarrythmic (procainamide) agents are useful. Competition with acetate (in the form of acetamide or monoacetin) is recommended. Ethanol appears to be beneficial. Mephentermine is more efficacious than norepinephrine in raising blood pressure.

Animal Toxicity

Fluoroacetate is a compound of very high acute toxicity. Oral LD_{50} values in laboratory rodents range from 0.2 to 2 mg/kg. There is a large variation in toxicity of fluoroacetate which is not due to differences in size of animal, type of digestive system, or basal metabolic rate. The variation may be due to the rate of elimination or rate of condensation of the poison with oxaloacetate. The oral LD_{50} in mammals is 110 μg/kg.

—Janice Reeves and Carey Pope

Related Topic

Pesticides

Sodium Sulfite

- CAS: 7757-83-7
- SYNONYMS: Anhydrous sodium sulfite; disodium sulfite; exsiccated sodium sulfite; sulftech; natriumsulfit (German); sodium sulfite anhydrous; sodium sulphite; sulfurous acid; disodium salt; sodium salt (1:2)
- CHEMICAL CLASS: Inorganic salt
- MOLECULAR FORMULA: Na_2SO_3

Uses

Sodium sulfite is a solid white powder with a salty sulfurous taste that is soluble in water. It is a reducing agent that is used as a food preservative and antioxidant. Its use is prohibited in meats and other sources of vitamin B_1. Sodium sulfite is also used in the treatment of semichemical pulp in the paper industry, in the treatment of water, as a photographic developer, and in textile bleaching (antichlor).

Exposure Pathways

Sodium sulfite is not an "environmental" pollutant per se, but its wide use as a food additive may lead to widespread exposure of the general population through ingestion. This may pose a problem for a small percentage of people who are hypersensitive to this chemical. Exposure to elevated concentrations (i.e., those that might cause abject toxicity) of this compound would only be expected to occur in the workplace, primarily involving sources of production or bulk use as mentioned previously. Because this compound is packaged as a powder, exposure would be expected to occur from airborne dust. Potential exposure routes would thus include skin, inhalation, and possibly involvement of the eye, nose, and throat.

Mechanism of Toxicity

The exact mechanism of toxicity has not been elucidated, although there is a lot of information on how sulfur-based compounds are detoxified by the liver. Sodium sulfite is a mild reducing agent which would most likely cause burning or irritation at the site of exposure or application by altering oxidation-reduction potential and pH.

Sulfites are used widely as antioxidants to keep foods from prematurely spoiling and to keep them looking "fresh" by preventing oxidation and subsequent "browning." Many people, however, are "sulfite sensitive." After ingestion of food or beverages containing sulfite, these people may have allergic-type reactions such as asthmatic wheezing, hypotension, tingling sensations, and flushing of the skin. The mechanism is unclear but probably has to do with an individual-specific chemical stimulation of the immune system, which in turn releases small amounts of vasoactive substances (see Sulfites).

Human Toxicity

At the concentrations used as a food additive, sodium sulfite is not toxic per se in humans; however, as mentioned previously, it will pose a problem for individuals who are sensitive to this chemical following ingestion. Allergic-type responses include asthmatic wheezing, a feeling of increased warmth and flushing of the skin, hypotension, and tingling sensations. Because some food manufacturers may use sulfites sporadically, it may be difficult for sensitive persons to avoid these additives altogether.

Clinical Management

Persons who are sensitive to sulfites should avoid foods containing this additive (e.g., wines), and those exhibiting severe allergic-type reactions (e.g., difficulty breathing) following a meal or beverage should seek immediate medical attention.

Persons exposed to large quantities of the dust in air should vacate the high-exposure area and seek conventional medical treatment if adverse symptoms are seen or if discomfort persists. As with exposure to any potentially irritating dust, eyes should be irrigated with water immediately following exposure and skin should be thoroughly washed with warm soapy water.

Animal Toxicity

The median lethal dose (LD_{50}) measured for a mouse was 820 mg/kg. The LD_{50} for a rabbit (2825 mg/kg) indicated that sodium sulfite was more than three times less toxic to rabbits than to mice. The lowest lethal dose for a cat or dog, administered subcutaneously, was 1300 mg/kg, whereas only half that dose was required to have the same effect on a guinea pig or rabbit. The median lethal dose for a mouse, administered intraperitoneally, was similar to the oral route (950 mg/kg).

—Stephen Clough

Related Topic

Food Additives

Solanum Genus

- ◆ SPECIES NAMES: *Solanum tuberosum* (potatoes); *Solanum dulcamara* (bittersweet); *Solanum nigrum* (nightshade); *Solanum pseudocapsicum* (Jerusalem cherry); *Solanum aculeatissimum* (coachroach berry); *Solanum seaforthianum* (blue flowered "potato vine"); *Solanum sodomeum*; *Solanum americanum* (black nightshade); *Solanum carolinense* (horse nettle, wild tomato); *Solanum lycopersicum* (tomato); *Solanum melongena* (eggplant)
- ◆ CHEMICAL CLASS: Solanine is a glycoalkaloid combination of solanidine (alkamine aglycone) and a glycoside linkage of sugars (galactose, glucose, and rhamose). Solanine is found throughout the plant with highest concentration in the unripe fruit. This decreases as the fruit ripens.

Exposure Pathway

Ingestion of berries and herbage is the route of exposure.

Toxicokinetics

Absorption increases as the amount ingested increases. Serum peak alkaloid levels are attained in 4–8 hr (depending on the species and amount ingested). Solanine is converted to solanidine by hydrolysis in the gastrointestinal tract. The sugar portion is split off and the alkamine solanidine is produced. Solanine is rapidly excreted in urine and feces. The half-life of solanidine is 11 hr.

Mechanism of Toxicity

The concentration of glycoalkaloids is increased with exposure to cold temperatures for prolonged periods of time. Solanine displays weak cardiac activity. It inhibits hepatic microsomal enzymes. Both the α-chaconine and α-solanine inhibit blood and brain cholinesterase. Solanine causes hemolytic and hemorrhagic damage to the gastrointestinal tract. The unhydrolyzed solanine is not rapidly absorbed from gastrointestinal tract. Its saponin-like quality makes it highly irritating.

Human Toxicity

Most plants of this genus primarily contain solanine. Some also contain anticholinergic alkaloids, which may lead to a confusing clinical picture.

Acute

Initially a harsh, scratchy sensation may be noted in the mouth and throat. The most common effects include nausea, vomiting, diarrhea, headache, drowsiness, hyperthermia, and dehydration. Other less common effects include blurred vision, mydriasis (if anticholinergics are involved), positive inotropic effect (due to the similarity of solanine to cardiac glycosides), rapid respiratory rate, dyspnea, confusion, weakness, hallucinations, salivation, diaphoresis, and muscular cramping.

Chronic

Continued exposure may include signs of acute toxicity as well as severe dehydration and coma.

Clinical Management

The onset of most symptoms occurs in 2–24 hr after the exposure. Management is basically symptomatic and supportive. Decontamination of the stomach with activated charcoal may be effective even if several hours have passed since the ingestion. The use of pharmacologic antagonists, such as physostigmine and atropine, is rarely indicated and their inappropriate use may complicate toxic manifestations arising from the ingestion. Dehydration can occur rapidly, so supportive fluid and electrolyte replacement may be needed. Recovery can occur within a few hours or 1 or 2 days later. Significant morbidity is unlikely unless large amounts of the plants are ingested.

Animal Toxicity

Both solanine and solanidine are poorly absorbed. Peak serum levels can be detected 12 hr postexposure. Characteristic symptoms seen in animals include decreased central nervous system, dullness, indifference to surroundings, and stupor. Symptoms can be particular to specific animals. Cows and pigs display ulcerative stomatitis, conjunctivitis, vesticular and scruffy enzyme of the legs, and diarrhea. Postmortem findings include gastroenteritis and congestion of cerebral membranes and kidneys. Treatment is the same as in humans.

—*Regina M. Rogowski*

Soman

- CAS: 96-64-0
- SYNONYMS: GD; phosphonofluoridic acid; zoman; PFMP; G-agent; methyl-1,2,2-trimethylpropyl ester; pinacolyl methylphosphonofluoridate; methylpinacolyloxyfluorophosphine oxide; pinacolyloxymethylphosphonyl fluoride; pinacolyl methanefluoropohosphonate; methylfluoropinacolylphosphonate; fluoromethylpenacolyloxyphosphine oxide; methylpinacolyloxyphosphonyl fluoride; pinacolyl methylfluorophosphinate; 1,2,2-trimethylpropoxyfluoromethylphosphine; nerve gas; nerve agent
- CHEMICAL CLASS: Soman is a nonpersistent anticholinesterase compound or fluorinated organophosphate (OP) nerve agent, irreversible cholinesterase inhibitor, and chemical warfare agent.
- MOLECULAR FORMULA: $C_7H_{16}FO_2P$

- CHEMICAL STRUCTURE:

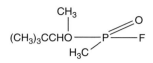

Use

Soman is a nerve agent used in chemical warfare.

Exposure Pathways

Casualties are caused primarily by inhalation but can occur following percutaneous and ocular exposure, as well as by ingestion and injection.

Toxicokinetics

Soman is absorbed both through the skin and via respiration. The half-life of soman in water at 30°C and pH 7.6 was reported to be 577 min compared to sarin at 30°C and pH 7.6 which was 5 min (see Sarin). Soman consists of a mixture of four stereoisomers:

1. C(−) P(−)
2. C(−) P(+) C(−)-soman
3. C(+) P(−)
4. C(+) P(+) C(+)-soman

The enzyme OP hydrolase hydrolyzes soman, tabun, sarin, and diisopropyl fluorophosphate at approximately the same rate.

Mechanism of Toxicity

Soman and the other nerve agents are organophosphorus cholinesterase inhibitors. They inhibit the enzymes butyrycholinesterase in the plasma, the acetylcholinesterase on the red blood cell, and the acetylcholinesterase at cholinergic receptor sites in tissues. These three enzymes are not identical. Even the two acetylcholinesterases have slightly different properties, although they have a high affinity for acetylcholine. The blood enzymes reflect tissue enzyme activity. Following acute nerve agent exposure, the red blood cell enzyme activity most closely reflects tissue enzyme activity. However, during recovery, the plasma enzyme activity more closely parallels tissue enzyme activity.

Following nerve agent exposure, inhibition of the tissue enzyme blocks its ability to hydrolyze the neurotransmitter acetylcholine at the cholinergic receptor sites. Thus, acetylcholine accumulates and continues to stimulate the affected organ. The clinical effects of nerve agent exposure are caused by excess acetylcholine.

The binding of nerve agent to the enzymes is considered irreversible unless removed by therapy. The accumulation of acetylcholine in the peripheral nervous system and central nervous system (CNS) leads to depression of the respiratory center in the brain, followed by peripheral neuromuscular blockade causing respiratory depression and death.

The pharmacologic and toxicologic effects of the nerve agents are dependent on their stability, rates of absorption by the various routes of exposure, distribution, ability to cross the blood–brain barrier, rate of reaction and selectivity with the enzyme at specific foci, and their behavior at the active site on the enzyme.

Red blood cell enzyme activity returns at the rate of red blood cell turnover, which is about 1% per day. Tissue and plasma activities return with synthesis of new enzymes. The rates of return of these enzymes are not identical. However, the nerve agent can be removed from the enzymes. This removal is called reactivation, which can be accomplished therapeutically by the use of oximes prior to aging. Aging is the biochemical process by which the agent–enzyme complex becomes refractory to oxime reactivation. The toxicity of nerve agents may include direct action on nicotinic acetylcholine receptors (skeletal muscle and ganglia) as well as on muscarinic acetylcholine receptors and the CNS.

Recently, investigations have focused on OP nerve agent poisoning secondary to acetylcholine effects. These include the effects of nerve agents on γ-aminobutyric acid neurons and cyclic nucleotides. In addition, changes in brain neurotransmitters, such as dopamine, serotonin, noradrenaline, as well as acetylcholine, following inhibition of brain cholinesterase activity have been reported. These changes may be due in part to a compensatory mechanism in response to overstimulation of the cholinergic system or could result from direct action of nerve agent on the enzymes responsible for noncholinergic neurotransmission.

Human Toxicity

Toxic effects occur within seconds to 5 min of nerve agent vapor or aerosol inhalation. The muscarinic effects include ocular effects (miosis, conjunctival congestion, ciliary spasm, and nasal discharge); respiratory effects (bronchoconstriction and increased bronchial secretion); gastrointestinal effects (anorexia, vomiting, abdominal cramps, and diarrhea); sweating; salivation; and cardiovascular effects (bradycardia and hypotension). The nicotinic effects include muscular fasciculation and paralysis. The effects on the CNS can include ataxia, confusion, loss of reflexes, slurred speech, coma, and paralysis.

Following inhalation exposure of soman, the median lethal dosage in man has been estimated to be 70 mg-min/m^3 at a respiratory minute volume of 15 liters/min for 10 min. For percutaneous liquid the LD_{50} has been estimated to be 350 mg/70-kg man. The permissible airborne exposure concentration of soman for an 8-hr workday or a 40-hr work week is an 8-hr TWA of 0.00003 mg/m^3.

Doses that are potentially life-threatening may be only slightly larger than those producing minimal effects. Vapor exposure to the eyes and nose causes miosis and runny nose at ECt_{50} dosages of less than 2 mg-min/m^3. The median incapacitation dosage (ICt_{50}) of vapor inhalation has been estimated as 35 mg-min/m^3, while the LCt_{50} is 70 mg-min/m^3. These vapor exposure durations are from 2 to 10 min. Individuals intoxicated with soman exhibit miosis, visual disturbances, headache and pressure sensation, runny nose, nasal congestion, salivation, tightness in the chest, nausea, vomiting, giddiness, anxiety, difficulty in thinking, difficulty sleeping, nightmares, muscle twitching, tremors, weakness, abdominal cramps, diarrhea, and involuntary urination and defecation. These effects may progress to convulsions and respiratory failure. Depending on dose, the onset of signs and symptoms may occur within minutes or hours.

Clinical Management

Management of nerve agent intoxication consists of decontamination, ventilation, administration of antidotes, and supportive therapy.

The three therapeutic drugs for treatment of nerve agent intoxication are atropine, pralidoxime chloride, and diazepam.

Atropine, a cholinergic blocking or anticholinergic drug, is effective in blocking the effects of excess acetylcholine at peripheral muscarinic sites. The usual dose is 2 mg, which may be repeated at 3- to 5-min intervals. Pralidoxime chloride (Protopam chloride; 2-PAM CL) is an oxime used to break the agent–enzyme bond and restore the normal activity of the enzyme if given before aging. However, aging occurs in about 2 min after administration. This is most apparent in organs with nicotinic receptors. Abnormal activity and normal strength returns to skeletal muscles, but no decrease in secretions is seen following oxime treatment. The usual dose is 1000 mg (iv or im). This may be repeated two or three times at hourly intervals, intravenously or intramuscularly. Diazepam, an anticonvulsant drug, is used to decrease convulsive activity and reduce brain damage that may occur from prolonged seizure activity. It is suggested that all three of these drugs be administered at the onset of severe effects from nerve agent exposure, whether or not seizures occur. The usual dose of diazepam is 10 mg (im).

Miosis, pain, dim vision, and nausea can be relieved by topical atropine in the eye. Pretreatment with carbamates may protect the cholinesterase enzymes before nerve agent exposure. It is available in 30-mg tablets, and the tablets should be administered every 8 hr. When used prior to exposure, it should be followed by atropine and pralidoxime chloride after exposure. $LD_{50}s$ are increased several fold and survival is increased if the nerve agent is soman (GD).

Supportive therapy may include ventilation via an endotracheal airway if possible and suctioning of excess secretions from the airways.

Animal Toxicity

The stereoisomers of soman have different median lethal doses. The C(+)P(+) soman and C(−)P(+)-soman are the least toxic with subcutaneous LD_{50} values of ≥5000 and ≥2000 μg/kg, respectively. The more toxic stereoisomers, C(−)P(−)-soman and C(+)P(−)-soman, have subcutaneous LD_{50} values of 38 and 99 μg/kg, respectively. The racemic mixture of soman has a subcutaneous LD_{50} of 156 μg/kg in mice.

The cause of death is attributed to anoxia resulting from a combination of central respiratory paralysis, severe bronchoconstriction, and weakness or paralysis of the accessory muscles for respiration.

Signs of nerve agent toxicity vary in rapidity of onset, severity, and duration of exposure. These are dependent on specific agent, route of exposure, and dose. At the higher doses, convulsions and seizures are indication of CNS toxicity. Following nerve agent exposure, animals exhibit hypothermia resulting from the cholinergic activation of the hypothalamic thermoregulatory center. In addition, plasma levels of pituitary, gonadal,

TABLE S-25
Acute Toxicities of Soman in Various Species by Various Routes of Exposure

Route of exposure/species	LD_{50} (μg/kg)
Percutaneous	
Rat	7800
Subcutaneous	
Chicken	50
Dog	12
Guinea pig	24
Monkey	13
Rabbit	20
Mouse	40
Rat	71
Intramuscular	
Monkey	9.5
Mouse	89
Rat	62
Intraperitoneal	
Chicken	71
Frog	251
Mouse	393
Rat	98
Intravenous	
Cat	15
Rat	44.5
Mouse	35

thyroid, and adrenal hormones are increased during organophosphate intoxication.

An LCt_{50} of 30 mg-min/m³ was reported following a 30-min inhalation exposure to soman by rats. The acute toxicities by other routes of exposure in various animal species are presented in Table S-25.

—Harry Salem and Frederick R. Sidell

(The views of the authors do not purport to reflect the position of the U.S. Department of Defense. The use of trade names does not constitute official endorsement or approval of the use of such commercial products.)

Related Topics

Behavioral Toxicology
Cholinesterase Inhibition
Nerve Agents
Neurotoxicity, Delayed
Organophosphates
Psychological Indices of Toxicity

Speed

- SYNONYMS: Street speed; "Look alike" drugs; white crosses; pink hearts; black beauties; 357s; 357 magnums; dexies; robin eggs

- DESCRIPTION: Speed is the nomenclature used for a number of preparations that resemble and are often misrepresented as prescription amphetamines. They are used as substitutes for amphetamines. Speed is commonly composed of ephedrine, phenylpropanolamine, caffeine, or a combination of those agents. Ephedrine is probably the most frequently encountered component of street speed (see Caffeine and Phenylpropanolamine).

- PHARMACEUTICAL CLASS: Speed generally contains one or more agents belonging to the drug class of sympathomimetics.

- CHEMICAL STRUCTURE:

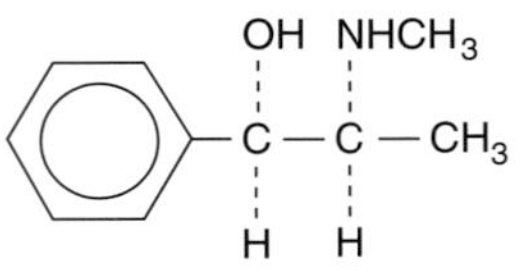

Ephedrine

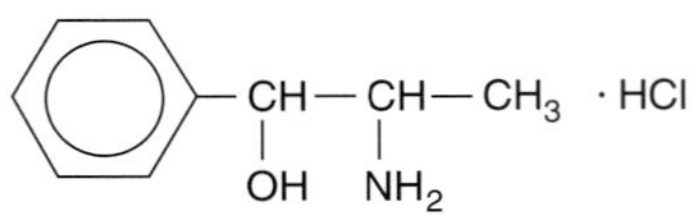

Phenylpropanolamine hydrochloride

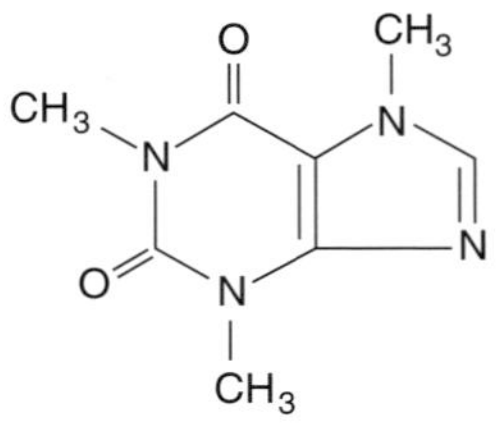

Caffeine

Uses

Ephedrine is labeled for sale as a bronchodilator. Phenylpropanolamine may have a labeled use as a decongestant or diet aid. Caffeine is marketed as a minor stimulant to produce greater alertness or reduce drowsiness.

Exposure Pathways

Oral speed preparations may be in tablet or capsule form. Ingestion is the most common route of intentional or accidental exposure.

Toxicokinetics

Ephedrine, phenylpropanolamine, and caffeine are all well absorbed from the gastrointestinal tract. Following ingestion of an oral dose, clinical effects are seen within 60 min and persist from 2 to 6 hr. With time-released preparations of phenylpropanolamine, the duration of action may be extended to 12 hr or longer.

Ephedrine is metabolized in the liver by oxidative deamination, demethylation, aromatic hydroxylation,

and conjugation. Metabolites of ephedrine include norepinephrine, benzoic acid, and hippuric acid. Ephedrine is resistant to metabolism by monoamine oxidase. Ephedrine, phenylpropanolamine, and caffeine have wide distribution throughout the body following oral administration. Ephedrine is presumed to cross the placenta and to distribute into breast milk.

Ephedrine, phenylpropanolamine, and caffeine are excreted in the urine. The rate of urinary excretion of ephedrine is dependent on urinary pH. The elimination half-life of ephedrine is 3 hr when the urine is acidified to pH of 5.0 and 6 hr when urinary pH is 6.3.

Mechanism of Toxicity

Sympathomimetic agents, frequently found in speed, stimulate α-adrenergic and β-adrenergic receptors and also stimulate the release of neuronal norepinephrine. Sympathomimetic drugs stimulate the sympathetic division of the autonomic nervous system. Stimulation of β-adrenergic receptors in the heart initially produces a positive inotropic effect on the myocardium. However, large or frequent doses produce a negative inotropic effect. With prolonged use, ephedrine, in particular, may deplete norepinephrine stores in sympathetic nerve endings and tachyphylaxis to the cardiac and pressor effects may develop.

Human Toxicity: Acute

The clinical effects following overdose of sympathomimetic agents depend on the particular receptor selectivity and consist of α-adrenergic and/or β-adrenergic stimulation. Hypertension is usually the predominating symptom and may be accompanied by tachycardia or bradycardia, depending on the drug involved. Cardiac arrhythmias, hypertensive crisis, and myocardial ischemia are possible effects of excessive exposure to sympathomimetic agents. Anxiety, muscle tremor, central nervous system (CNS) stimulation, seizures, and cerebral hemorrhage may occur, particularly following the ingestion of large doses of phenylpropanolamine. Hypokalemia is a possible serum electrolyte manifestation.

Human Toxicity: Chronic

Long-term use of large doses of ephedrine (350–2500 milligrams per day for 3–20 years) may produce psychotic episodes characterized by paranoia, hallucinations, depression, and bizarre mentation. Following withdrawal of the drug, aberrant mental effects will resolve but reinstitution of ephedrine use may result in a return of the psychotic symptoms.

Clinical Management

Basic and advanced life-support measures should be instituted as indicated. Gastric decontamination may be performed depending on the specific drug involved, the patient's symptomatology, and the history of the ingestion. Activated charcoal may be used to adsorb phenylpropanolamine, ephedrine, and/or caffeine. Careful monitoring of the heart and hemodynamic status should be performed. Hypertension and symptoms of CNS stimulation usually resolve spontaneously with only supportive measures. Antiarrhythmic and antihypertensive agents may be necessary in severe exposures. Management of concurrently ingested drugs should be appropriate to the agent involved. Laboratory analysis of the serum electrolytes, creatine phosphokinase, and urinalysis should be performed.

—*Carole Wezorek*

Spider, Black Widow

♦ SYNONYMS: *Latrodectus mactans*; "hour glass" spider

Exposure Pathway

Envenomation occurs subcutaneously due to the small biting apparatus of the spider. Bites occur most frequently on the extremities.

Toxicokinetics

In humans, the specific disposition of *Latrodectus* venom is not well understood. Distribution of venom to the central and peripheral nervous system occurs following absorption through the lymphatic system. The onset of muscle cramping and pain ranges from 30 min to several hours. Resolution of symptoms is usually complete within 24 hr. Occasionally a longer clinical course is experienced with symptoms persisting for several days.

Mechanism of Toxicity

Black widow spider venom contains several different protein fractions. However, the high-molecular-weight neurotoxin is the only fraction that is of clinical significance. This neurotoxin acts at the neuromuscular synapse damaging nerve terminals and causing the release and ultimately the depletion of such neurotransmitters as α-aminobutyric acid, norepinephrine, and acetylcholine. Neurotransmitter release is most likely responsible for hypertension, muscle fasciculations, and spasm most frequently experienced by victims of a bite. Later, generalized muscle weakness and labored breathing may develop in severe cases. While the venom of the black widow spider has been characterized as being more potent than that of many poisonous snakes, the small amount of venom injected limits the degree of toxicity.

Human Toxicity

Several species of *Latrodectus* exist and all produce a similar clinical course. The severity of an envenomation is dependent on patient age and the presence of any preexisting cardiovascular disease. Fatalities are extremely rare. Infants and the elderly are at greater risk for developing severe symptoms. Initially, the bite is associated with mild local pain and redness limited to the immediate bite site. The venom causes no tissue damage so little or no swelling occurs. Most frequently, patients present with painful muscle cramping, spasms, and rigidity which commonly occur within a few hours of the bite. The location of the bite determines which muscle groups will be affected. Bites occurring to the upper body commonly affect muscles of the back, shoulders, and chest. Lower extremity bites are associated with abdominal spasms and rigidity. In some cases, the presentation resembles an acute abdomen. Most problematic is the severe pain which commonly accompanies the muscle spasms. Nausea, vomiting, headache, dizziness, diaphoresis, and mild hypertension are other commonly encountered symptoms. Severe clinical manifestations are rare but can include clinically significant hypertension, respiratory insufficiency, and seizures.

Clinical Management

Although life-threatening envenomations are extremely rare, basic and advanced clinical life support should be employed when necessary. Normally healthy adult patients bitten by black widow spiders often do not develop symptoms significant enough to require medical evaluation. However, the primary complaint of most patients evaluated in the emergency department is moderate to severe pain due to muscle spasms. Therapy is directed toward making the patient as comfortable as possible while monitoring for the development of severe symptoms such as hypertension and labored breathing. Agents commonly employed to treat muscle spasms and pain include intravenous calcium, muscle relaxants, and narcotic analgesics. Intravenous calcium salts, as either calcium chloride or calcium gluconate, alone often fail to provide adequate relief. Calcium gluconate, a less concentrated form of calcium than calcium chloride, is often preferred since it is less irritating to blood vessels during administration. Following calcium, the addition of both intravenous diazepam and a narcotic analgesic appears to be the most effective method in reducing muscle spasm and pain. However, readministration of this combination is frequently required. The use of *L. mactans* antivenin, in general, should be limited only to those patients experiencing severe symptoms. Most studies indicate that the routine use of antivenin is unnecessary and therefore should be discouraged. The vast majority of patients will recover

fully with only observation and the use of muscle relaxants and narcotic analgesics to manage pain. "High-risk" patients, such as infants, the elderly, or those with significant cardiovascular disease, are potential candidates for receiving antivenin. *Latrodectus* antivenin is derived from horse serum. Both anaphylactic and delayed hypersensitivity reactions have occurred. If antivenin use is indicated, intravenous administration is the preferred route. One vial of *Latrodectus* antivenin is reconstituted and commonly diluted further in 50 cc of normal saline and administered intravenously over 30 min. A small dose should be administered subcutaneously prior to intravenous infusion to check for hypersensitivity reactions. Wheal formation at the test site indicates the possibility that an allergic reaction may occur and antivenin use should be reconsidered. A negative reaction to the test dose does not necessarily rule out the possibility of an allergic reaction to the antivenin. One should always be prepared for the possibility of anaphylaxis whenever antivenin derived from horse serum is given. The benefits of giving antivenin in a particular patient should be weighed against the potential risks. Elevation in blood pressure is frequent following black widow spider envenomation but rarely requires treatment with an antihypertensive agent. Tetanus prophylaxis should be provided as necessary.

—*Gary W. Everson*

Related Topic

Neurotoxicology: central and peripheral

Spider, Brown Recluse

♦ SYNONYMS: *Loxosceles reclusa;* Fiddle-back spider; violin spider

Exposure Pathway

Envenomation occurs subcutaneously due to the small biting apparatus of the spider. The brown recluse spider, as its name implies, is found in secluded areas. Bites occur most frequently to the hands and arms while reaching into woodpiles or other well-protected areas.

Toxicokinetics

In humans, the specific disposition of venom is not well understood. The local distribution of venom is enhanced by the presence of hyaluronidase and other spreading factors found in the venom. Systemic absorption of venom components is likely dependent on lymphatic transport. The onset of local symptoms such as redness and pain may develop within a few hours of the bite.

Mechanism of Toxicity

Brown recluse spider venom contains many diverse protein fractions including spreading factors and enzymes such as hyaluronidase, collagenase, protease, phospholipase, and others. These venom components cause coagulation of blood and, ultimately, the occlusion of small blood vessels at the bite site. This leads to local skin and tissue necrosis due to ischemia. Hemolysis of red blood cells also occurs. The normal inflammatory processes that follow, such as edema and hemorrhage, contribute to the tissue damage caused by the venom. Occasionally, the local tissue necrosis expands as the tissue ischemia spreads from the initial bite site.

Human Toxicity

The brown recluse spider bite produces symptoms that range from mild local tissue inflammation to widespread systemic toxicity. The extent of toxicity is dependent on the amount of venom injected and the age and general health of the patient. Although life-threatening symptoms are possible, a localized skin and tissue reaction is much more common. It is important to note that brown recluse spider bites often do not progress to a necrotic lesion. However, when present, the tissue necrosis is usually self-limiting and often responds to general wound management.

There are some brown spiders that somewhat resemble the brown recluse to the layperson. Without proper identification, it is often not possible to immediately diagnose a brown recluse spider bite. Complicating the diagnosis further is the fact that there are several other species of spider (e.g., *Chiracantheum*, *Argiope*, and Phidippus) that can cause a necrotic skin lesion, although not as severe.

The bite of the brown recluse is usually painless and often goes unnoticed initially. The spider is seldom seen. Therefore, most patients do not seek treatment until a necrotic lesion develops. Within several hours of envenomation, local symptoms of redness and pain occur. Within 24 hr, a reddish to violet colored blister becomes surrounded by a blanched, ischemic ring that is bordered by a reddish ring. This represents the often described "bull's eye lesion." Over the next several days, the blistered, ischemic area may turn darker and sink below the level of skin due to subcutaneous tissue necrosis. This necrotic reaction may stop or continue to expand, producing a lesion as large as 5–30 cm in diameter. In 7–14 days, the top layer of the blister sloughs off leaving an ulcerative lesion. Depending on the size of the lesion, healing may require several months. The necrosis tends to be more extensive following bites in fatty areas such as the thighs and buttocks. Rarely, systemic involvement may occur. Symptoms can include fever, chills, weakness, vomiting, muscle pain, generalized rash, seizures, disseminated intravascular coagulation, thrombocytopenia, and hemolytic anemia. Renal failure and death may occur due to widespread hemolysis.

Clinical Management

Many controversial techniques have been employed in the management of brown recluse spider bites. Unfortunately, no scientific evidence exists which supports an ideal method or methods of management. However, case reports advocate a variety of therapies as potentially useful. Most agree, however, that good local management of the cutaneous lesion is the most important aspect of care. Tetanus prophylaxis should always be included. In general, antibiotics should be withheld unless there is evidence of infection. Local and systemic injection of steroids have also been employed, but research has shown that neither the extent nor the duration of tissue necrosis is affected. Dapsone has been shown to be effective in research done in the animal model. In addition, several case reports have described some success with Dapsone in decreasing local pain and preventing further induration and necrosis. Doses have ranged from 50 to 200 mg per day. This drug appears to decrease the extent of tissue necrosis by inhibiting polymorphonuclear leukocytes, the mediators of the inflammatory response to the bite. However, side effects of Dapsone are potentially severe. Hemolytic anemia and liver toxicity have also been described.

There appears to be no benefit to early surgical excision of the bite site. Some time is required before a clear boundary is established marking the end of the spread of venom. Excising the necrotic area too soon may leave some venom at the boundary that can produce further tissue necrosis. In cases in which the necrotic lesion expands and is unresponsive to local treatment, "delayed" surgical excision of the wound after 2 or 3 weeks may be indicated.

Although several case reports describe some success in patients following the use of hyperbaric oxygen, no scientific studies have been completed yet to determine the effectiveness of this approach. Management of systemic toxicity is primarily supportive and includes the use of steroids to prevent red blood cell hemolysis. Also, adequate hydration is important to maintain good urine output. Clotting abnormalities and anemia should be managed with appropriate blood products. The brown recluse spider antivenin is still experimental and not commercially available.

—*Gary W. Everson*

Staphylococcus aureus

- ♦ SYNONYMS: Enterotoxin A, B, C, D, E
- ♦ DESCRIPTION: Staphylococci appear as grape-like clusters of cells on solid media. In broth culture, they occur in pairs, short chains, or small clusters of cells. Staphylococci produce heat-stable enterotoxins. These toxins are simple proteins with molecular weights of 30–35 kD.

Exposure Pathway

Ingestion of staphylococcal-contaminated food (containing the enterotoxin) is the most common route of exposure to this foodborne bacteria.

Mechanism of Toxicity

The concentration of enterotoxins in foods depends on the degree of contamination, type of food, time, temperature, and other conditions of incubation. The toxins differ in heat stability. Staphylococcus toxins can be formed within a few hours when foods are kept at room temperature. Most food, especially those high in protein and previously cooked food, will support staphylococcal growth. Some of these foods include cream and custard-filled pastries, mayonnaise, ham, poultry, potato salad, egg salad, dairy products, and meat.

Staphylococcus aureus has a short incubation period of 4 hr with a range of 1–6 hr. The duration of the illness is 8–36 hr with a mean of 20 hr. Staphylococcal toxin forms at temperatures between 68 and 112°F and staphylococcal growth can be limited by keeping foods at temperatures below 40 or above 140°F.

The toxin produces vomiting by acting on the emetic receptor sites in the lower parts of the gastrointestinal tract. The stimulus reaches the vomiting center in the brain by way of the vagus and sympathetic nerves. The diarrheal action appears to be due to enterotoxin's interference of the water absorption in the intestinal lumen.

Human Toxicity

Food poisoning generally occurs with a common food source exposure resulting in several individuals becoming ill within a common time frame. Heat-stable enterotoxin A–E produces gastroenteritis. Staphylococcus is the most common cause of foodborne bacterial illness. *Staphylococcus aureus* proliferates in food and produces heat-stable toxins. When contaminated food is ingested, the toxins, not the bacteria, produce the gastrointestinal symptoms.

Vomiting is most the common symptom, followed by abdominal pain and diarrhea. Headache is also common and can be followed by weakness and dizziness. Fever is not usually present. Fatalities are rare with this type of food poisoning; however, it can lead to serious complications in children and the elderly.

Clinical Management

The symptoms are usually self-limited and resolve over 24–48 hr. Symptomatic and supportive care, including fluid replacement, is normally adequate intervention. This is not an infectious process; therefore, antibiotics are of no value. Antidiarrheal and antiemetics are not indicated since they may slow down the toxin elimination process. Most cases are mild. Only 5% of type A require hospitalization.

Staphylococcal food poisoning can be prevented by maintaining clean kitchens, proper refrigeration, and good hand washing. Moreover, people with skin infections should avoid preparation of foods.

—Vittoria Werth

State Regulation of Consumer Products

There are two unique facts about the safety of consumer products. From a legal/regulatory standpoint, consumer products such as cosmetics are well regulated at the federal level, which is enhanced by both cooperative, voluntary, and government-industry regulations and industry self-regulation. From a toxicological perspective, no class of consumer products has a higher exposure (frequency and dose) for the consumer than cosmetics. The duration of exposure often approaches the lifetime of the consumer (chronic exposure), and multiple exposures from different products (e.g., preservative ingredients in cosmetics from several different cosmetics used daily) are also possible. Despite such chronic exposure to multiple doses, cosmetics remain the safest class of consumer products by far.

It would not seem that current regulations would need any additions. However, after the decentralization of the federal government under the Reagan administration, states have become more aggressive in passing laws regulating all consumer products, including cosmetics.

California

California has exceeded the State of New York (which is now third, behind Texas) in sheer numbers of citizens. This has been coupled with a legislative tendency toward independence from federal regulatory requirements and onerous laws. Other states are beginning to follow California's lead.

In 1986, California passed the Safe Drinking Water and Toxic Enforcement Act requiring the governor to revise and publish a list of chemicals known to the state to cause cancer or reproductive toxicity. This law is more commonly known as Proposition 65. If determinations of carcinogenicity or reproductive toxicity were made within the bounds of the published scientific and medical literature by nationally or internationally renowned experts in these highly specialized fields, the law might seem reasonable; if the determination took into account information supplied by national scientific, medical, and/or trade associations, the law might seem reasonable; or if trace quantities of known carcinogens or teratogens in consumer products were suspect of their true biological risk to consumers, the law might seem reasonable. The administrative record of Proposition 65 in California clearly shows that the previous arguments and suggestions have been and are being repeatedly made, but the California regulation relying on California state toxicologists and physicians retains its sovereignty in such determinations. As in the case of organic colors, some useful and safe consumer products will be lost due to California's Proposition 65, making certain ingredients or trace contaminants unlawful in consumer products.

In June 1991, the California state assembly passed Assembly Bill 1474, which would require specific tests in support of the cosmetic product claims hypoallergenic (including the terms allergy-tested and safe for sensitive skin), dermatologist-tested, and ophthalmologist-tested. In July 1991, this bill failed by one vote but was brought before the California Senate's Health and Human Services Committee. This committee can reconsider the measure at any time, and the originators of the bill in the assembly are attempting to do just that.

Cosmetic product claims are regulated by U.S. FDA, the FTC, the Lanham Act, the NAD, and the research and legal staff of the major commercial television networks if the claim is in a television advertisement. The cosmetics manufacturer generates technical support for such claims, which are reviewed and approved by their own scientists and legal counsel.

—Shayne C. Gad

Related Topics

Consumer Product Safety Commission
Cosmetics

Food and Drug Administration
Food Drug and Cosmetic Act
Proposition 65

Structure–Activity Relationships

Structure–activity relationship (SAR) methods have become a legitimate and useful part of toxicology since the early 1970s. These methods are various forms of mathematical or statistical models which seek to predict the adverse biological effects of chemicals based on their structure. The prediction may be of either a qualitative (carcinogen/noncarcinogen) or quantitative (LD_{50}) nature, with the second group usually being denoted as QSAR (quantitative structure–activity relationship) models.

The concept that the biological activity of a compound is a direct function of its chemical structure is now at least a century old. During most of this century, the development and use of SARs were the domain of pharmacology and medicinal chemistry. These two fields are responsible for the beginnings of all the basic approaches in SAR work, usually with the effort being called drug design. An introductory medicinal chemistry text is strongly recommended as a starting place for understanding SARs. Additionally, Burger's *Medicinal Chemistry,* with its excellent overview of drug structures and activities, should enhance at least the initial stages of identifying the potential biological actions of *de novo* compounds using a pattern recognition approach.

Having already classified SAR methods into qualitative and quantitative, it should also be pointed out that both of these can be approached on two levels. The first is on a local level, where prediction of activity (or lack of activity) is limited to other members of a congeneric series or structural near-neighbors. The accuracy of predictions via this approach is generally greater but is of value only if one has sufficient information on some of the structures within a series of interest.

The second approach is a prediction of activity over a wide range, generally based on the presence or absence of particular structural features (functional groups).

For toxicology, SARs currently have a small but important number of uses. These can all be generalized as identifying potentially toxic effects or restated as three main uses:

1. They can be used for the selection and design of toxicity tests to address endpoints of possible concern.

2. If a comprehensive or large testing program is to be conducted, SAR predictions can be used to prioritize the tests so that outlined questions (the answers to which might preclude the need to do further testing) may be addressed first.

3. They can be used as an alternative to testing. Though in general it is not believed that the state of the art for SAR methods allows such usage, in certain special cases (such as selecting which of several alternative candidate compounds to develop further and then test) this may be valid and valuable.

Basic Assumptions

Starting with the initial assumption that there is a relationship between structure and biological activity, we can proceed to more readily testable assumptions. First, the dose of chemical is subject to a number of modifying factors (such as membrane selectivities and selective metabolic actions) which are each related in some manner to chemical structure. Indeed, absorption, metabolism, pharmacological activity, and excretion are each subject to not just structurally determined actions but also (in many cases) stereo-specific differential handlings.

Given these assumptions, actual elucidation of SARs requires the following:

1. Knowledge of the biological activities of existing structures

2. Knowledge of structural features which serve to predict activity (also called molecular parameters of interest)

3. One of more models which relate 2 to 1 with some degree of reliability

There are now extensive sources of information as to both toxic properties of chemicals and, indeed, biological activities. These include books, journals, and manual and computerized databases.

Molecular Parameters of Interest

Which structural and physicochemical properties of a chemical are important in predicting its toxicologic activity are open to considerable debate. Table S-26 presents a partial list of such parameters. The reader is referred to a biologically oriented physical chemistry text both for explanations of these parameters and for references to sources from which specific values may be obtained.

There are now several systems available to study the three-dimensional structural aspects of molecules and their interactions. The first are the various molecular modeling sets, which can actually be very useful for some simpler problems. The second are the molecular design and analysis packages that are available for mainframe computers. Lastly, molecular graphics software is available for microcomputers. Use of such forms of graphic structural examination as a tool or method in SAR analysis has been widely discussed. Such methods are generally called topological methods.

SAR Modeling Methods

All the current major SAR methods used in toxicology can be classified based on what kinds of compound-related or structural data they use and what method is used to correlate this structural data with the existing biological data.

The more classical approaches use physicochemical data (e.g., molecular weight and free energies) as a starting point. The major approaches to this are by manual pattern recognition methods, cluster analysis, or regression analysis. It is this last approach, in the form of Hansch or linear-free energy relationships (LFERs), which actually launched all SAR work (other than that on limited congeneric cases) into the realm of being a useful approach. Indeed, still foremost among the QSAR methods is the model proposed by

TABLE S-26
Molecular Parameters of Interest

Electronic effects
  Ionization constants
    Sigma substituent constant
  Distribution constant
  Resonance effect
  Field effect
  Molecular orbital indices[a]
    Atomic/electron net charge
    Nucleophilic superdelocalizability
    Electrophilic superdelocalizability
    Free radical superdelocalizability
    Energy of the lowest empty molecular orbital
    Energy of the highest occupied molecular orbital
    Frontier self-atom polarizability
    Frontier atom–atom polarizability
    Intermolecular coulombic interaction energy
    Electric field created at point (A) by a set of charges on a
      molecule
Hyprophobic parameters
  Partition coefficients
  P_i substituent constants
  R_M value in liquid–liquid chromatography
  Elution time in high-pressure liquid chromatography (HPLC)
  Solubility
  Solvent partition coefficients
  pK_a
Steric effects
  Intramolecular steric effects
    Steric substituent constant
    Hyperconjugation correction
  Molar volume
  Molar refractivity, R substituent constants
  Molecular weight
  Van der Waals radii
  Interatomic distances
Substructural effects
  Three-dimensional geometry
  Fragment and molecular properties
  Chain lengths

[a]Calculated or theoretical parameters.

Hansch and co-workers. This group's major contribution was to propose that earlier observations of the importance of the relative lipophilicity to biologic activity be incorporated into the formal LFER approach to provide a general QSAR model for biological effects. As a suitable measure of lipophilicity, the partition coefficient (log P) between 1-octanol and water was proposed, and it was demonstrated that this was an approximately additive and constitutive property and that it was therefore calculable, in principle, from a molecular structure. A probabilistic model for the Hansch equation can be expressed as

$$\log (1/C) = k\,\pi^2 = k'\,\pi + p\,\sigma + k''$$

$$\text{or } \log (1/C) = -k\,(\log P)^2 + k'\,(\log P) + p\,\sigma + k''$$

$$Ba_i = + a_j x_{ij} + 1$$
$$j$$

where C is the dose that elicits a constant biological response (e.g., ED$_{50}$ and LD$_{50}$), π is the substituent lipophilicity, log P is the partition coefficient, σ is the substituent electronic effect of Hammet, and k, k', p, and k'' are the regression coefficients derived from the statistical curve fitting. The reciprocal of the concentration reflects the fact that higher potency is associated with lower dose, and the negative sign for the π^2 or $(\log P)^2$ term reflects the expectation of an optimum lipophilicity, designated π_0 or log P$_0$.

The statistical method used to determine the coefficients is multiple linear regression. A number of statistics are derived in conjunction with such a calculation, which allow the statistical significance of the resulting correlation to be assessed. The most important of these are s, the standard deviation, r^2, the coefficient of determination or percentage of data variance accounted for by the model (r, the correlation coefficient, is also commonly cited), and F, a statistic for assessing the overall significance of the derived equation, values, and confidence intervals (usually 95%) for the individual regression coefficients in the equation. Also very important in multiparameter equations are the cross-correlation coefficients between the independent variables in the equation. These must be low to ensure true "independence" or orthogonality of the variables, a necessary condition for meaningful results.

In a like manner, there are a number of approaches for using structural and substructural data and correlating these to biological activities. Such approaches are generally classified as regression analysis methods, pattern recognition methods, and miscellaneous other methods (e.g., factor analysis, principal components, and probabilistic analysis).

The regression analysis methods which use structural data have been the most productive and useful. "Keys"—or fragments of structure—are assigned weights as predictors of an activity, usually in some form of the Free–Wilson model, which was developed at virtually the same time as the Hansch. According to this method, the molecules of a chemical series are structurally decomposed into a common moiety (or core) that may be substituted in multiple positions. A series of linear equations of the form

are constructed where BA is the biological activity, Xj is the jth substituent with a value of 1 if present and 0 if not, aj is the contribution of the jth substituent to BA, and 1 is the overall average activity. All activity contributions at each position of substitution must sum to zero. The series of linear equations thus generated is solved by the method of least squares for the aj and 1. There must be several more equations than unknowns and each substituent should appear more than once at a position in different combinations with substituents at other positions. The favorable aspects of this model are the following:

1. Any set of quantitative biological data may be employed as the dependent variable.

2. No independently determined substituent constants are required.

3. The molecules comprising a sample of interest may be structurally dismembered in any desired or convenient manner.

4. Multiple sites of variable substitution are readily handled by the model.

There are also several limitations: A substantial number of compounds with varying substituent combinations are required for a meaningful analysis, the derived substituent contributions give no reasonable basis for extrapolating predictions from the substituent matrix analyzed, and the model will break down if nonlinear dependence on substituent properties is important or if there are interactions between the substituents.

Pattern recognition methods comprise yet another approach to examining structural features and/or chemical properties for underlying patterns that are associated with differing biological effects. Accurate classification of untested molecules is again the primary goal. This is carried out in two stages. First, a set of compounds, designated the training set, is chosen for which the correct classification is known. A set of molecular or property descriptors (features) is generated for each compound. A suitable classification algorithm is then applied to find some combination and weight of the descriptors that allow perfect classification. Many different statistical and geometric techniques for this purpose have been used. The derived classification function is then applied in the second step to compounds not included in the training set to test predictability. In published work these have generally also been other compounds of known classification. Performance is judged by the percentage of correct predictions. Stability of the classification function is usually tested by repeating the procedure several times with slightly altered, but randomly varied, sets or samples.

The main difficulty with these methods is in "decoding" the QSAR in order to identify particular structural fragments responsible for the expression of a particular activity. In addition, even if identified as "responsible" for activity, far harder questions for the model to answer are whether the structural fragments so identified are "sufficient" for activity, whether it is always "necessary" for activity, and to what extent its expression is modified by its molecular environment. Most pattern recognition methods use as weighting factors either the presence or absence of a particular fragment or feature (coded 1 or 0) or the frequency of occurrence of a feature. They may be made more sophisticated by coding the spatial relationship between features.

Applications in Toxicology

SAR methods have been developed to predict a number of toxicological endpoints [genotoxicity, carcinogenesis, dermal irritation and sensitization, lethality)LD$_{50}$ values), biological oxygen demands, and teratogenicity] with varying degrees of accuracy, and models for the prediction of other endpoints are under development. Additionally, both U.S. EPA and U.S. FDA have models for mutagenicity/carcinogenicity that they utilize to "flag" possible problem compounds.

It should be expected that qualitative models are more "accurate" than quantitative ones, and that the more possible mechanisms associated with an endpoint, the less accurate (or more difficult) a prediction.

Further Reading

Gad, S. C. (1993). *In Vitro Toxicology*. Raven Press, New York.
Gad, S. C. (1998). *Statistics: An Experimental Design for Toxicologists*, 3rd ed. CRC Press, Boca Raton, FL.

—*Shayne C. Gad*

Related Topics

Dose–Response Relationship
Hazard Identification
LD$_{50}$/LC$_{50}$
Median Lethal Dose
Molecular Toxicology
Risk Assessment, Human Health

Strychnine

- CAS: 57-24-9
- SYNONYMS: Kwik-Kil; Strychnos; Mouse-Rid; Mouse-Tox
- CHEMICAL CLASS: Naturally occurring alkaloid
- CHEMICAL STRUCTURE:

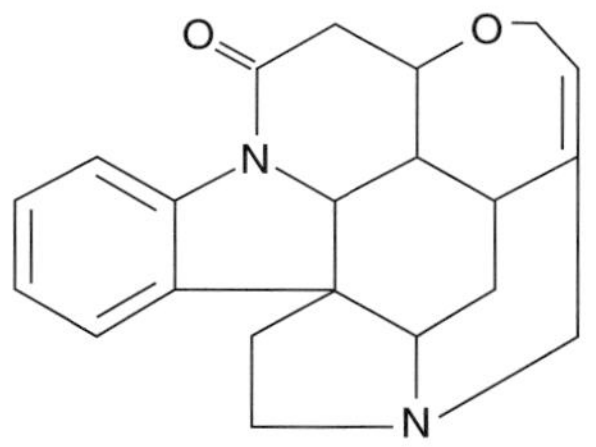

Use
Strychnine is used chiefly in poison baits for rodents and sometimes birds.

Exposure Pathways
The primary pathways for accidental exposure to strychnine are inhalation and ingestion. Ocular and dermal exposures may also occur.

Toxicokinetics
Strychnine is rapidly absorbed from the gastrointestinal tract, nasal mucosa, and parenteral sites. It is readily metabolized in the liver by microsomal enzymes. The highest concentrations of strychnine are found in the liver, kidneys, and blood. Urinary excretion of strych-

nine is inversely proportional to the dose. About 15% will appear unchanged in the urine within 24 hr. Strychnine has a half-life of about 10 hr.

Mechanism of Toxicity

The exact mechanism of strychine's action in the nervous system is unclear but it is thought that the inhibitory action of the neurotransmitter glycine at Renshaw cell-motor axon synapses is blocked by strychnine. This essentially decreases excitatory thresholds and produces tetanic convulsions in response to sensory stimuli. While the main locus for strychnine's neurotoxicity is the spinal cord, the medulla also appears affected. Effects on other organ systems appear to be secondary to these actions in the central nervous system (CNS). Strychnine competitively blocks the binding of glycine to membranes isolated from spinal cord.

Human Toxicity: Acute

Within 15–30 min after ingestion of strychnine, the patient may experience restlessness, apprehension, heightened acuity of perception, hyperreflexia, and muscle stiffness of the face and legs. Violent convulsions can follow these symptoms or may occur in the absence of these previous symptoms. As poisoning progresses, the convulsions may become more violent and the intervals shorter.

Human Toxicity: Chronic

Significant cumulative toxicity is not recognized because both detoxication and excretion are comparatively rapid.

The ACGIH TLV-TWA is 0.15 mg/m^3. The LD_{Lo} (oral) human is 30 mg/kg. Toxicity has been reported at 0.1 mg/100 ml in blood concentrations.

Clinical Management

Treatment is basically symptomatic and supportive with emphasis on controlling neuromuscular hyperactivity. The patient should be moved to fresh air, and eyes and skin should be decontaminated immediately with water. Emesis is not recommended because the probability of seizures is high. Gastric lavage procedures should be used as deemed necessary depending on the patient's level of consciousness and the amount of strychnine ingested. Activated charcoal should be used immediately to minimize absorption.

Animal Toxicity

Strychnine is a compound of high acute toxicity. The oral LD_{50} value in rats is about 15 mg/kg. Parenteral routes of exposure are more toxic; LD_{50} values in laboratory rodents range from about 1 to 4 mg/kg.

—Janice Reeves and Carey Pope

Related Topics

Neurotoxicology: Central and Peripheral
Pesticides

Styrene

- CAS: 100-42-5
- SYNONYMS: Ethenylbenzene; phenylethylene; vinylbenzene
- CHEMICAL CLASS: Hydrocarbon
- CHEMICAL STRUCTURE:

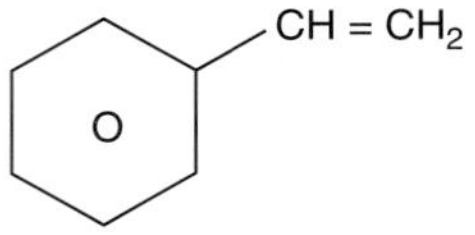

Uses

Styrene is used in the manufacture of plastics, protective coatings, copolymer resins, styrenated polyesters, and as a chemical intermediate. Synthetic rubber (styrene butadiene) accounts for approximately 10% of total styrene use.

Exposure Pathways

The closed system techniques currently used in styrene monomer and copolymer resin production limit worker exposures to a TWA exposure of generally less than 10 ppm, which is below the ACGIH TLV of 50 ppm. However, in open systems used to manufacture some reinforced plastics (e.g., shower stalls and boats), worker exposures to liquid styrene and resins via inhalation and dermal contact may pose a health hazard. It is unlikely that the residual styrene monomer contained in consumer products poses a significant health risk to the general public.

Toxicokinetics

Styrene is rapidly absorbed by all routes of exposure (i.e., gastrointestinal tract, respiratory tract, and skin). In humans, almost all of the absorbed styrene is metabolized to mandelic acid (approximately 70%) and phenylglyoxylic acid (approximately 30%). Other metabolites include 4-vinylphenol, hippuric acid, and the glucuronide of styrene glycol. Metabolism occurs primarily in the liver and to a lesser extent in extrahepatic tissues (e.g., kidney, intestine, and lung). Once absorbed, styrene is rapidly distributed throughout the body. Tissue distribution studies in rats and mice indicate that styrene and/or its metabolites accumulate in the liver, kidneys, heart, subcutaneous fat, lung, brain, and spleen. In both species, fat contained the highest concentration of styrene and/or its metabolites; this observation suggests that fat may act as a reservoir for these compounds. Approximately 90–97% of the styrene absorbed by humans is eliminated as urinary metabolites; only a small fraction of the absorbed dose is eliminated in expired air or urine as the parent compound.

Mechanism of Toxicity

The mechanism of styrene intoxication is unknown.

Human Toxicity

Acute inhalation exposures may result in irritation of the nasal mucosa, eyes, and skin as well as depression of the central nervous system (CNS). Symptoms of CNS depression include nausea, drowsiness, and ataxia. The disagreeable odor of styrene, which is detectable at approximately 0.1–0.5 ppm, serves as a good warning aid. However, olfactory fatigue may occur at high concentrations.

Clinical Management

Acute exposures are likely to be associated with CNS depression and, at very high doses, pulmonary irritation. Removal from exposure and ventilatory support are the initial priorities. Alert individuals ingesting more than 2 or 3 mg/kg should be given syrup of ipecac. Because hydrocarbon pneumonitis is a significant risk with styrene ingestion, intubation should precede lavage in those individuals at risk of aspiration because of obtundation.

Animal Toxicity

Sublethal exposures of animals to styrene result in irritation of the eyes and upper respiratory tract, ototoxicity, and CNS depression. The oral LD_{50} for styrene in male and female rats is approximately 5000 mg/kg. The LC_{50} in rats exposed to styrene for 4 hr is 2770 ppm; the LC_{50} in mice exposed to styrene for 2 hr is 4940 ppm.

—*Ralph Parod*

Related Topics

Combustion Toxicology
Respiratory Tract

Sudan Grass

- ◆ SYNONYMS: *Sorghum sudanense, Sorghum vulgare* var *sudanense, Holcus sudanensis,* Gramineae family; grass sorghum; shattercane; Johnson grass

- ◆ DESCRIPTION: Sudan grass is an annual grass with stems up to 9 ft tall that branch from the base. Leaf blades are up to 0.5 in. wide and 12 in. long. The plant flowers in a 12-in.-long, erect, loose panicle that is approximately half as wide as it is tall. It also forms a grass or grain-like glume.

    Sudan grass is native to Sudan. It is cultivated and naturalized widely in the United States but is most common in the southern states. The grass has spread as a weed in waste places, railroad yards, highway medians, and cultivated fields. As seedlings, virtually all species of Sorghum resemble one another and are frequently confused with young corn plants.

Exposure Pathway

Ingestion of any part of the plant would be the common route of exposure. The green, aerial portions, especially the leaves, stems, and canes are toxic.

Toxicokinetics

Limited information on absorption is available. After ingestion, the onset of symptoms of cyanide toxicity is expected in 30 min to 2 hr; however, symptoms could be delayed (see Cyanide). Small doses of cyanide are converted to thiocyanate by an enzymatic reaction catalyzed by rhodanese, an enzyme widely distributed in tissues, with the highest concentration in the liver. The rhodanese system can detoxify large amounts of cyanide but cannot respond quickly enough to prevent fatalities.

Mechanism of Toxicity

The sorghum group contains the cyanogenic glycoside dhurrin. The hydrolysis of this compound yields hydrocyanic acid. The glycosides themselves are harmless. Hydrolysis is complete in 10 min at 20°C. The reaction takes place slowly in an acid pH, but an alkaline medium hastens the process. A delay in symptoms after ingestion would be explained by a slow hydrolysis when transportation occurs from the acidic stomach medium to the alkaline medium in the duodenum. Released and absorbed hydrocyanic acid forms a stable complex with ferric iron and cytochrome oxidase, inhibiting the activity of the enzyme and aerobic metabolism. Cells with cytochrome oxidase are unable to utilize the available oxygen and suffer from hypoxia. Therefore, the nerve cells can no longer obtain oxygen and the respiratory center ceases to function.

Forage sorghum can accumulate levels of nitrates that can also produce poisoning. Photosensitization has been reported in sheep from a photodynamic pigment in some species.

Human Toxicity

Sudan grass poisonings in humans are not reported. Symptoms are expected to be similar to those exhibited in animals. Patients would be treated as for other cyanogenic plant ingestions.

Clinical Management

In symptomatic patients, decontamination should be deferred until other basic and advanced life-support measures have been instituted. Induction of emesis is not recommended. Subtoxic amounts do not require emesis; moreover, the potentially rapid progression of clinical course contraindicates it. Lavage can be used with significant ingestions. Activated charcoal may be effective despite the low molecular weight of cyanide compounds. The cyanide antidote kit should only be administered in those persons with significant symptoms (impaired consciousness, seizures, acidosis, and unstable vital signs). Arterial blood gases, electrolytes, serum lactate, whole blood cyanide levels, and methemoglobin levels should be monitored and treated as necessary. Hyperbaric oxygen can be used for those with severe symptoms not responding to normal supportive and antidotal treatment.

Animal Toxicity

Livestock cyanide poisoning may result from plant consumption. Mucous membranes of the eyes and mouth may appear congested. Gastric contents, if examined immediately, have a characteristic benzaldehyde odor, resulting from benzaldehyde production from aglycone breakdown of certain cyanogenic glycosides. Potential symptoms expected are hyperpnea to dyspnea, excitation, gasping, staggering, paralysis, prostration, convulsions, coma, and death. Cherry red blood may be noted.

Nitrogen poisoning can also result from the nitrates in these plants. In ruminant digestion, nitrates are converted to nitrites, which are about 10 times more toxic. They are the more immediate cause of poisoning. Symptoms of nitrite toxicity include cyanosis, severe dyspnea, trembling, and weakness with a chocolate brown discoloration of the blood.

In prolonged cases of Sudan grass ingestion, chronic cystitis, ataxia, and urinary bladder fibrosis may be seen. Some environmental factors that increase the cyanogenic potential of the plant are high nitrogen and low phosphorus in soil and age of plants (young growth has the highest potential for toxicity). Many years of selective breeding have resulted in hybrids having lower potential for developing hydrogen cyanide.

—*Lanita B. Myers*

Sulfites

Inorganic sulfites and bisulfites (such as sodium sulfite, Na_2O_3S) are used in photography, the bleaching of wool, and the preserving of foods (e.g., meats and egg yolks), beverage, and medications. They act as effective antioxidant compounds. Sulfite-induced bronchospasm (sometimes leading to asthma) was first noticed as an acute sensitivity to metabisulfites, which were sprayed on restaurant salads (and salad bars) and used in wine. Approximately 450,000 (5%) of the 9 million asthmatics in the United States may be sulfite sensitive. Sulfites normally are detoxicated rapidly to inorganic sulfate by the enzyme sulfite oxidase. In sensitive individuals, there is apparently a deficiency in this enzyme, making these individuals supersensitive to sulfites. (The U.S. FDA has taken the position that the addition of sulfite to food is safe only when properly disclosed on the food label.)

—*Shayne C. Gad*

Related Topics

Food Additives
Food and Drug Administration

Sulfur Dioxide

- CAS: 7446-09-5
- SYNONYMS: Sulfurous anhydride; sulfurous oxide
- CHEMICAL CLASS: Oxidant gas
- CHEMICAL STRUCTURE:

$$O = S = O$$

Uses

Sulfur dioxide is a primary component of air pollution. It is used commercially to preserve fruits and vegetables and as a disinfectant in food production. It is also used in bleaching straw and textiles.

Exposure Pathways

Contact with mucous membranes (eyes and nose) and inhalation are possible routes of exposure.

Toxicokinetics

Absorption is dependent on the level of exposure. Sulfur dioxide is highly soluble. It is readily absorbed and distributed throughout the body. It is metabolized and excreted through urine. Elimination from the respiratory tract is slow and may persist a week after exposure.

Mechanism of Toxicity

Most inhaled sulfur dioxide is detoxified through molybdenum metabolism in the liver. On moist skin or mucous membranes, it is converted to sulfurous acid, a direct irritant.

Human Toxicity

Sulfur dioxide is irritating to the eyes, mucous membranes, and respiratory tract. High levels of exposure produce cardiac arrest. Moderate exposure produces pulmonary edema. Low exposure results in systemic acidosis. It is mutagenic at 5700 ppb. The LC_{Lo} is 1000 ppm/10 min. The PEL is 5 ppm/8 hr.

Clinical Management

If skin or eye exposure occurs, the affected areas should be flushed with water for about 15 min. If ingested, the stomach contents should be diluted with water or milk. Gastric lavage or emesis should not be attempted. Pain should be treated without numbing the central nervous system. Open airways and steady blood pressure should be maintained. Prednisolone (2 mg/kg/day) should be given for 10 days.

Animal Toxicity

In rats, sulfur dioxide accelerates aging and produces heart, lung, and kidney damage. The inhalation LD_{50} is 2520 ppm/1 hr in rats. The LD_{50} is 3000 ppm/ 30 min in mice.

—*Jayne E. Ash and Shayne C. Gad*

Related Topics

Absorption
Pollution, Air
Respiratory Tract

Sulfur Mustard

♦ CAS: 550-60-2; 39472-40-7; 68157-62-0

♦ SYNONYMS: Distilled mustard; HD; H sulfide; bis(2-chloroethyl; 1-chloro-2(β-chloroethylthio)ethane β, β′-dichloroethyl sulfide; bis(2-chloroethyl) sulfide; β,β′-dichlorodiethyl sulfide; mustard gas; blister agent; 2^12^1-dichloroethyl sulfide; dichloroethyl sulfide; Iprit; Kampstoff; Lost; Schwefellost; yellow cross; yperite; Senfgas; S-Lost; S-yperite

♦ CHEMICAL CLASS: Vesicant; alkylating agent

♦ MOLECULAR FORMULA: $C_4H_8Cl_2S$

♦ CHEMICAL STRUCTURE:

$$Cl-\underset{\underset{H}{|}}{\overset{\overset{H}{|}}{C}}-\underset{\underset{H}{|}}{\overset{\overset{H}{|}}{C}}-S-\underset{\underset{H}{|}}{\overset{\overset{H}{|}}{C}}-\underset{\underset{H}{|}}{\overset{\overset{H}{|}}{C}}-Cl$$

Uses

Sulfur mustard is used as a chemical warfare agent, blister agent, and alkylating agent.

Exposure Pathways

Ocular, percutaneous, inhalation, ingestion, and injection routes are possible. Effects may be local, systemic, or both and appear hours after exposure. Sulfur mustard is both a vapor and a liquid threat to all exposed skin and mucous membranes.

Toxicokinetics

Because of its high lipid solubility, sulfur mustard quickly penetrates the lipid barrier membrane of the cell. It produces skin vesication, eye injury, and damage to the respiratory tract. Mustard is a lipophilic liquid that penetrates mucosal surfaces easily. Penetration is dependent on volatility and is proportional to temperature. On human skin, approximately 80% of applied mustard evaporates and 20% penetrates the skin. Of that 20%, about 10% binds to skin sites and the remaining 10% is absorbed systemically. Although mustard penetrates the skin rapidly, there is a latent period of a few hours before clinical changes are observed. The 10% mustard retained in the skin binds rapidly to proteins. It has also been demonstrated that 80% of the inhaled vapor is absorbed in the nose.

Sulfur mustard is also an alkylating agent which reacts with a variety of compounds found in the tissues producing a cytotoxic effect. The hematopoietic tissue, such as bone marrow, lymph nodes, and spleen, are especially sensitive. Its rate of detoxification is very low and repeated exposures are cumulative due to sensitization. Sulfur mustard slowly hydrolyzes in water with a half-life of 5 min at 37°C forming hydrochloric acid and thiodiglycol. Both sulfur and nitrogen mustards in a polar solvent such as water form a cyclic onium cation (sulfonium and immonium) by intramolecular cyclization and a free chloride anion. The cyclized form is responsible for the varied effects of mustard. It has been suggested that mustard is oxidized via sulfoxides to sulfones and the compounds formed from alkylation are excreted in the urine. The mustards, both sulfur and nitrogen, combine predominantly with the thiol group and are excreted conjugated as cysteinyl derivatives. Conversion to the onium ion may block the formation of the reactive alkylating agent episulfonium ion from sulfur mustard analog 2-choroethylethyl sulfide. Mustard reacts with tissue within minutes. Blood, tissue, and blister fluid do not contain mustard. Exposure does not occur by contact with body fluids or tissues.

Mechanism of Toxicity

Mustard was demonstrated to alkylate all of the nitrogens and oxygens in DNA. The DNA cross-linking and depurination followed by chain scission appears to be related to cytotoxicity. Alkylation of guanine in the 0–6 position has been suggested as a mutagenic lesion. Mustard can similarly cause toxicity in many systems owing to depletion of glutathione, which leads to an increase in oxyradicals and electrophiles. Glutathione and antioxidant enzymes, such as superoxide dismutase, catalase, and glutathione peroxidase, are scavengers of free radicals and oxyradicals, thus protecting cells from injury. Diminished glutathione levels in the tissue may trigger free radical chain reactions leading to membrane lipid peroxidation resulting in loss of membrane fluidity, increased membrane permeability, inactivation of several membrane enzymes, and DNA strand breaks which contribute to cell death. Mustard also possesses cholinergic activity, which may be responsible for the gastrointestinal effects and miosis.

Human Toxicity

The topical effects of mustard occur in the eyes, airways, and skin. Ingestion may cause direct injury to the gastrointestinal tract and systemically absorbed mustard may affect the bone marrow, gastrointestinal tract, and the central nervous system (CNS). The vapor and aerosol are less toxic to the skin or eyes than the liquid form. The eyes are more sensitive to the action of mustard than any other parts of the body. Depending on the severity of exposure, the onset of ocular effects is from 1 to 3 hr up to 4–12 hr, and recovery may take from 1 to 2 weeks up to several months. The symptoms vary from lacrimation, "sand sensation" in the eyes, swollen and edematous conjunctiva and lids to blepharospasm, blurring of vision, marked hyperemia and edema of the conjunctiva, edema of the corneal epithelium, miosis, and a mucoserous discharge. Following severe exposure, headache and edema of the conjunctiva with miosis occur. Damage may progress to dense corneal opacification with deep ulceration and vascularization. The ocular median incapacitating dose (ICt_{50}) has been estimated to be 200 mg-min/m^3 and the maximum allowable Ct for eyes has been estimated as 2 mg-min/m^3. Marginal eye effects have occurred at 12–70 mg-min/m^3 and mild reddening at 70 mg-min/m^3.

Blisters do not form in the eyes. Swelling and loosening of corneal epithelial cells may lead to corneal edema and clouding with leukocytes. Scarring may follow between the iris and lens, restricting pupillary movements, and may predispose victims to glaucoma. Panophthalmitis, perforation of the cornea, and loss of the eye can also occur.

The median inhalation dose in man for death (LCt_{50}) has been estimated as 1500 mg-min/m^3. When inhaled, the upper respiratory tract (nose, throat, and trachea) is inflamed following a latency period of a few hours. There may be irritation or burning of the nose, epistaxis, sinus pain or irritation, and irritation or soreness of the pharynx. Symptoms include hoarseness, which may progress to aphonia. Damage to the trachea and upper bronchi leads to productive cough. Involvement of the lower respiratory tract is accompanied by fever, dyspnea, moist rales, bronchopneumonia, and necrosis of the smaller airways with hemorrhagic edema. Pulmonary edema is not a feature of mustard exposure although hemorrhagic pulmonary edema may occur in severe cases. The primary airway lesion is necrosis of the respiratory mucosa, followed by damage to the airway musculature.

Systemic effects include injury to the bone marrow, lymph nodes, and spleen, as indicated by a decreased white blood cell count, resulting in increased susceptibility to local and systemic infections. Loss of appetite, diarrhea, and apathy may also occur. Ingestion of liquid mustard in food or water produces nausea and vomiting, pain, diarrhea, and prostration.

Respiratory failure is the common cause of death following the inhalation of mustard. This may result from mechanical obstruction by pseudomembranes which developed from necrosis of the airway mucosa and resultant inflammation or from secondary pneumonia. Mustard-induced bone marrow suppression may be a contributing factor in the later, septic deaths from pneumonia.

The percutaneous median incapacitating dose (ICt_{50}) in man has been estimated as 2000 mg-min/m^3 at 70–80°F. Wet skin enhances mustard absorption, thus the ICt_{50} is reduced in hot, humid weather. At 90°F, the ICt_{50} is estimated as 1000 mg-min/m^3. The maximum allowable Ct for skin has been estimated as 5 mg-min/m^3. The minimum effective dose in man for mild to moderate erythema at 90°F is 50 mg-min/m^3.

The warm moist skin of the perineum, external genitalia, arm pits, antecubital fossae, and neck are particularly susceptible to the effects of local mustard vapor exposure. There is a latent period of from 6 to 12 hr before any local manifestations such as erythema develop. The erythema gradually becomes brighter, resembling sunburn, and is accompanied by itching and mild burning lasting for several days. Except for mild vapor burns, the erythema is followed by blistering. Small vesicles develop within the erythematous areas which enlarge and coalesce to form bullae. These bullae are large, domed, thin-walled, translucent yellowish blisters surrounded by erythema. The blister fluid is clear, thin, straw colored at first and later becomes yellow, and tends to coagulate. The fluid does not contain mustard, is nonirritating, and is not a vesicant. If the blister does not rupture, it resorbs in about a week.

Following liquid exposures or extremely high doses, a central zone of coagulation necrosis may develop with blister formation at the periphery. Healing takes longer, and secondary infections are more prone to occur. These disabling skin lesions may also be accompanied by systemic effects including anorexia, nausea, vomiting, depression, and fever.

The effects of mustard on the CNS are not well defined. However, individuals exposed to even small amounts of mustard have been reported to appear sluggish, apathetic, and lethargic. The minor psychological problems have been known to linger for a year and even longer.

Chronic daily exposure to mustard can cause sensitization; chronic lung impairment (cough, shortness of breath, and chest pain); cancer of the mouth, throat, and respiratory tract.

Mustard burns of the skin are usually followed by a persistent brown pigmentation except at the site of actual vesication where there may be depigmentation. Repeated burns may lead to hypersensitivity. That is, reexposure will cause erythema with or without edema and pronounced itching and burning within 1 hr and at lower concentrations than previously required. Vesication heals more rapidly. Frequent manifestations of reexposure in sensitized individuals include a morbilliform rash and eczematoid dermatitis surrounding old lesions.

Clinical Management

The victim must be removed from the source of contamination quickly by adequately protected attendants and then decontaminated (solution of sodium hypochlorite or liquid household bleach or Fullers' earth). Administer oxygen and/or artificial respiration if dyspnea is present or breathing has stopped.

The skin should be treated with calamine or other soothing lotion or cream to reduce burning and itching. Large blisters should be unroofed and covered with a sterile dry dressing if the casualty is ambulatory or left uncovered if the casualty is not ambulatory. Denuded unroofed areas should be irrigated with saline or soapy water and covered with a topical antibiotic (e.g., silver sulfadiazine or mefanide acetate). Multiple or large areas of vesication require hospitalization and whirlpool irrigation. Systemic analgesics are indicated especially

prior to manipulation of the patient or irrigation of the burn areas. A systemic antipruritic (trimeprazine) may also be used.

Treatment of ocular injury includes thorough irrigation, application of homatropine (anticholinergic) ophthalmic ointment, and topical antibiotics several times daily. Vaseline or similar products should be applied regularly to the edges of the eyelids to prevent them from sticking together. Topical analgesics may be useful in severe blepharospasm for examination of the eye but should be used sparingly. Systemic analgesics should be used for eye pain. Sunglasses may reduce discomfort from photophobia, and the victim must be reassured that complete healing and restoration of vision will result.

Steam inhalation and cough suppressants may relieve upper airway symptoms (sore throat, nonproductive cough, and hoarseness). Appropriate antibiotic therapy should only be instituted following confirmation of infection by positive sputum tests (gram stain and culture).

Intubation should be accomplished prior to the development of laryngeal spasm or edema so that adequate ventilation is established and suction of necrotic and inflammatory debris can be facilitated. Oxygen may be required as well and early use of CPAP or PEEP may be beneficial. Bronchoscopy may be required if pseudomembrane has developed to permit suction of the necrotic debris by direct vision.

Bronchodilators or steroids may also be used to relieve bronchospasm.

Death may occur between the 5th and 10th day postexposure because of pulmonary insufficiency complicated by a compromised immune response from mustard-induced bone marrow damage.

Atropine (0.4–0.6 mg intramuscularly or intravenously) or other anticholinergic or antiemetic drugs may be used to control nausea and vomiting.

If the bone marrow is depressed, sterilization of the gastrointestinal tract by nonabsorbable antibiotics may reduce the possibility of infection from enteric organisms. Bone marrow transplants or blood transfusion may be indicated. The recent introduction of granulocyte colony-stimulating factor may offer hope in the management of bone marrow depression.

A casualty of mustard exposure also requires the general supportive care given to a severely ill patent as well as the specific care given to a burn patient. This

TABLE S-27
Acute Toxicities of Sulfur Mustard in Various Species
by Various Routes of Exposure

Route of exposure/species	
Inhalation vapor (10-min exposure)	LCt$_{50}$ (mg-min/m^3)
Mouse	1200
Rat	1000
Guinea pig	2000
Rabbit	2800
Cat	700
Dog	700
Monkey	800
	LD$_{50}$ (mg/kg)
Percutaneous liquid	
Guinea pig	20
Mouse	92
Rat	5
Rabbit	40
Dog	20
Intravenous	
Mouse	8.6
Rabbit	1.1
Dog	0.2
Rat	0.7
Subcutaneous	
Guinea pig	20
Mouse	20
Rat	1.5
Rabbit	20

includes the liberal use of systemic analgesics and anti-pruritics and maintenance of fluid and electrolyte balance. Parenteral food supplements and vitamins may also be beneficial.

Animal Toxicity

The acute toxicity of sulfur mustard has been determined in many animal species and by many routes of administration. The acute data are presented in Table S-27. Topically in animals, irritation, erythema, and edema occur.

—Harry Salem and Frederick R. Sidell

(The views of the authors do not purport to reflect the position of the U.S. Department of Defense. The use of trade names does not constitute official endorsement or approval of the use of such commercial products.)

Related Topics

Nerve agents
Neurotoxicology: Central and Peripheral

Nitrogen Mustards
Organophosphates

Surfactants, Anionic and Nonionic

◆ SYNONYMS:

Hand dishwashing detergents—Dove; Sunlight; Palmolive; Dermassage; Joy; Ivory; Ajax

Nonindustrial automatic dishwashing detergents (granules and liquid)—Cascade Dishwasher Detergent; Electrosol Dishwasher Detergent; Palmolive Dishwasher Detergent

◆ CHEMICAL CLASS: Surfactants

Uses

These detergents are used to clean dishes in the home. Although not commonly recommended, hand dishwashing detergents have been used for their emetic effect to treat accidental poisonings when syrup of ipecac is not available.

Exposure Pathways

Ingestion of detergents by children is especially common, followed by dermal and ocular exposures.

Mechanism of Toxicity

Detergents are not absorbed, thus the main effect of toxicity comes from contact of surfactants on the tissues. Hand dishwashing detergents are noncorrosive and generally have only a mild irritating effect on mucosal tissues. Automatic dishwashing detergents generally have a stronger irritating effect on mucosal tissues because, in addition to surfactants, they often contain alkaline builders.

Household products rarely produce gastrointestinal burns.

Human Toxicity

Ingestion of hand dishwashing detergents typically produces mild nausea, vomiting, and/or diarrhea. Ocular exposure may produce mild burning and transient irritation.

Ingestion of household automatic dishwasher detergents may also produce mild oral and stomach irritation; however, these detergents have the potential to produce stronger mucosal irritation due to their more alkaline composition. Granular detergents have the ability to reside for longer periods of time in the oral cavity and esophagus; thus, the irritant effects may be more pronounced. Ocular effects are often more injurious, with a high potential for corneal burns and abrasions.

Clinical Management

Any detergent ingestion should be handled with dilution. This decreases the potential for mucosal irritation by diluting the surfactants. The effects of detergent ingestion are transient; therefore, fluid and electrolyte status is rarely affected. Ocular exposures should be treated with copious irrigation, and an ocular exam should follow if irritation persists.

Animal Toxicity

Gastrointestinal irritation and ocular irritation is expected.

—Anne E. Bryan

Related Topics

Detergents
Sensory Organs
Skin

2,4,5-T

- CAS: 93-76-5
- SYNONYM: 2,4,5-Trichlorophenol
- CHEMICAL CLASS: Chlorinated phenoxyacetic acids. A closely related compound is 2,4-D (2,4-dichlorophenoxyacetic acid; see 2,4-D).
- CHEMICAL STRUCTURE:

Use
2,4,5-T is manufactured for use as a broad-spectrum herbicide.

Exposure Pathways
Exposure to 2,4,5-T is by the oral and dermal routes.

Toxicokinetics
2,4,5-T is absorbed dermally in humans. Radioactivity was found in all tissues examined as well as in milk and fetuses after single oral administration of 0.17–41 mg/kg [^{14}C]2,4,5-T to pregnant rats. 2,4,5-T is eliminated largely unchanged. No metabolites are known, although it is possible that a small amount is eliminated as a glucuronide conjugate. The volume of distribution after a single oral dose of 5 mg/kg varies as follows: in humans, 0.079 liters/kg; in rats, 0.14 liters/kg; and in dogs, 0.22 liters/kg. 2,4,5-T is bound extensively to the plasma protein which could limit renal clearance of the herbicide; 2,4,5-T is also bound to renal cortex microsomal and cytosol fractions. 2,4,5-T given orally to volunteers (100–150 mg) was readily absorbed and gradually eliminated from blood plasma, showing a first-order elimination rate; more than 80% of a dose was excreted in urine in intact form within 72 hr. Clearance of 2,4,5-T from plasma and body of dogs, mice, and humans is slower than that in rats.

Mechanism of Toxicity
The effect of 2,4,5-T was investigated in *in vitro* studies with sublethal concentrations of 2,4,5-T, which showed an inhibitory effect on calcium-dependent ATP-ase. *In vivo* exposure to various sublethal concentrations of 2,4,5-T during 96 hr caused a significant inhibition of microsomal calcium-dependent ATPase. 2,4,5-T inhibits renal anion transport. Exposure of cells to 2,4,5-T resulted in a dose-dependent inhibition of DNA synthesis.

Human Toxicity
2,4,5-T in pure form is considered to be of relatively low toxicity. Limited data are available on exact toxic doses. Intravenous injection of up to 28 mg/kg of 2,4-D has been well tolerated, while a dose of 50 mg/kg

produced toxicity. Death has resulted following ingestion of 80 mg/kg.

Common findings after acute ingestion included miosis, coma, fever, hypotension, emesis, tachycardia, and muscle rigidity. Complications may include respiratory failure, pulmonary edema, and rhabdomyolysis. Ingestions cause burning of the mouth, esophagus, and stomach. Irritation of skin, eyes, nose, and throat may also occur. Tachycardia is common. Cardiac arrhythmias occurred in one suicide case. Pulmonary edema has been reported. Respiratory paralysis and bradypnea are common in large ingestions. Vertigo, headache, malaise, and paresthesias have been reported occasionally in occupational handlers. Higher doses may produce muscle twitching and spasms, followed by profound muscle weakness and unconsciousness. Individual idiosyncrasies may be involved in reported neuropathies. Rhabdomyolysis may occur. Myotonia (stiffness of legs) has been observed in severely poisoned persons. Vomiting and diarrhea have been reported. Elevated LDH, SGOT (AST), and SGPT (ALT) have been reported. Albuminuria, hemoglobinuria, and azotemia may occur. Acute exposure may cause irritation of the skin. Chloracne from chlorodioxin contaminants in 2,4,5-T has been reported in heavily exposed workers. 2,4,5-T itself is not believed to be carcinogenic or teratogenic in humans; these effects, produced by technical grades of the chemical are believed due to the dioxin that is present as an impurity.

Classification of Carcinogenicity

Evidence in humans is limited; overall summary evaluation of carcinogenic risk to humans is group 2B: The agent is possibly carcinogenic to humans.

Clinical Management

These herbicides can be measured in plasma and urine by high-performance liquid chromatography. Chlorophenoxy compounds do not affect blood cholinesterase activities.

Emesis may be indicated in recent substantial ingestion unless the patient is or could rapidly become obtunded, comatose, or convulsing. It is most effective if initiated within 30 min. Activated charcoal/cathartic: a charcoal slurry, aqueous or mixed with saline cathartic or sorbitol should be administered. A baseline CBC, electrolytes, and renal and hepatic function tests should be obtained. Urine should be tested for protein, RBCs, and myoglobin. Urine output should be monitored.

LDH, SGOT (AST), and alkaline phosphatase should be followed to detect liver injury, and CPK should be followed to detect muscle damage. Urine pH, arterial pH, and bicarbonate should be measured to detect acidosis. Respiratory depression, hypotension, and metabolic acidosis should be treated. Adequate urine flow should be maintained with intravenous fluids if victim is dehydrated. The patient should be monitored closely for cardiac arrhythmias, hyperthermia, and seizures.

If exposed via inhalation, the victim should be moved to fresh air and monitored for respiratory distress. If cough or difficulty in breathing develop, evaluation for respiratory tract irritation, bronchitis, or pneumonitis should be performed. Humidified supplemental oxygen (100%) should be administered with assisted ventilation as required.

Exposed eyes should be irrigated with copious amounts of tepid water for at least 15 min. If irritation, pain, swelling, lacrimation, or photophobia persist, the patient should be seen in a health care facility.

If clothing is contaminated, it should be removed and discarded. Affected skin should be washed vigorously, including hair and nails; soap washings should be repeated.

Animal Toxicity

2,4,5-T in pure form is considered to be of relatively low toxicity. 2,4,5-T itself is not believed to be carcinogenic or teratogenic in animals; these effects, produced by technical grades of the chemical, are believed to be due to the dioxin that is present as an impurity. In 2-year feeding trials no effect was observed in rats receiving 30 mg/kg diet nor in 90-day trials in beagle dogs at 60 mg/kg diet. Single oral doses of 100 mg/kg body weight of 2,4,5-T fed to pigs caused anorexia, vomiting, diarrhea, and ataxia; at autopsy, hemorrhagic enteritis and congestion of liver and kidney were found. 2,4,5-T (containing no detectable 2,3,7,8-tetrachlorodibenzo-*p*-dioxin) affected chromosomes of bone marrow cells of mongolian gerbil (*Meriones unguiculatus*) that received five consecutive daily intraperitoneal injections by causing significant increases in chromatid gaps, chromatid breaks, and fragments after total doses of 250 mg/kg or more but not after 150 mg/kg or less.

The no-effect levels for embryotoxicity for commercial 2,4,5-T were as follows: rat, 25 mg/kg/day; mouse, 20 mg/kg/day; hamster, 40 mg/kg/day; and monkey, 40 mg/kg/day.

Oral LD$_{50}$s were as follows: mouse, 389 mg/kg; rat, 500 mg/kg; guinea pig, 381 mg/kg; dog, >100 mg/kg. The percutaneous LD$_{50}$ in rats was >5000 mg/kg.

—Robin Guy

Related Topics

Chlorophenoxy Herbicides
Pesticides
Pollution, Water

Tabun

- CAS: 77-81-6
- SYNONYMS: GA; ethyl *N,N*-dimethylphosphoramidocyanidate ethyl dimethylphosphoramidocyanidate; dimethylaminoethoxy-cyanophosphine oxide; dimethylamidoethoxyphosphoryl cyanide; ethyldimethylaminocyanophosphonate; ethyl ester of dimethylphosphoroamidocyanidic acid; ethyl phosphorodimethylamidocyanidate; G agent; nerve gas; nerve agent
- CHEMICAL CLASS: Tabun is a nonpersistent anticholinesterase compound or organophosphate (OP) nerve agent.
- MOLECULAR FORMULA: $C_5H_{11}N_2O_2P$
- CHEMICAL STRUCTURE:

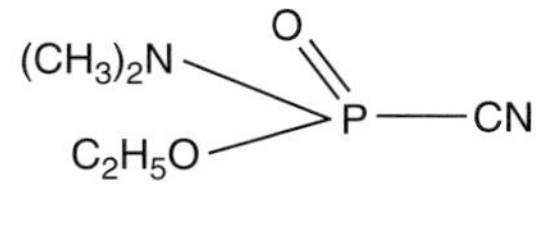

$$(CH_3)_2N \diagdown \quad \overset{O}{\underset{\|}{P}} \diagup CN$$
$$C_2H_5O \diagup$$

Use

Tabun is a nerve agent used in chemical warfare.

Exposure Pathways

Casualties are caused primarily by inhalation but can occur following percutaneous and ocular exposure as well as by ingestion and injection.

Toxicokinetics

Tabun is absorbed both through the skin and via respiration. Nerve agents inhaled as vapors or aerosols enter the systemic circulation resulting in toxic manifestations from seconds to 5 min following inhalation.

The enzyme OP hydrolase hydrolyzes tabun, sarin, soman, and diisopropyl fluorophosphate at approximately the same rate.

Mechanism of Toxicity

Tabun and other nerve agents are organophosphorus cholinesterase inhibitors. They inhibit the enzymes butyrylcholinesterase in the plasma, acetylcholinesterase on the red blood cell, and acetylcholinesterase at cholinergic receptor sites in tissues. These three enzymes are not identical. Even the two acetylcholinesterases have slightly different properties, although they have a high affinity for acetylcholine. The blood enzymes reflect tissue enzyme activity. Following acute nerve agent exposure, the red blood cell enzyme activity most closely reflects tissue enzyme activity. However, during recovery, the plasma enzyme activity more closely parallels tissue enzyme activity.

Following nerve agent exposure, inhibition of the tissue enzyme blocks its ability to hydrolyze the neurotransmitter acetylcholine at the cholinergic receptor sites. Thus, acetylcholine accumulates and continues to stimulate the affected organ. The clinical effects of nerve agent exposure are caused by excess acetylcholine.

The binding of nerve agent to the enzymes is considered irreversible unless removed by therapy. The accumulation of acetylcholine in the peripheral and central nervous systems leads to depression of the respiratory center in the brain, followed by peripheral neuromuscular blockade causing respiratory depression and death.

The pharmacologic and toxicologic effects of the nerve agents are dependent on their stability, rates of absorption by the various routes of exposure, distribution, ability to cross the blood–brain barrier, rate of reaction and selectivity with the enzyme at specific foci, and their behavior at the active site on the enzyme.

Red blood cell enzyme activity returns at the rate of red blood cell turnover, which is about 1% per day. Tissue and plasma activities return with synthesis of new enzymes. The rates of return of these enzymes are not identical. However, the nerve reactivation can be accomplished therapeutically by the use of oximes prior to aging. Aging is the biochemical process by which the agent–enzyme complex becomes refractory to oxime reactivation. The toxicity of nerve agents may include direct action on nicotinic acetylcholine receptors (skeletal muscle and ganglia) as well as on muscarinic acetylcholine receptors and the central nervous system.

Recently, investigations have focused on organophosphate nerve agent poisoning secondary to acetylcholine effects. These include the effects of nerve agents on γ-aminobutyric acid neurons and cyclic nucleotides. In addition, changes in brain neurotransmitters such as dopamine, serotonin, noradrenaline, and acetylcholine following inhibition of brain cholinesterase activity have been reported. These changes may be due in part to a compensatory mechanism in response to overstimulation of the cholinergic system or could result from direct action of nerve agent on the enzymes responsible for noncholinergic neurotransmission.

Human Toxicity

Following inhalation exposure, the median lethal dosage (LCt_{50}) in man has been estimated to be 135 mg-min/m^3 at a respiratory minute volume (RMV) of 15 liters/min for a duration of 0.5–2 min and 200 mg-min/m^3 at a resting RMV of 10 liters/min. For percutaneous vapor the LCt_{50} is estimated to be between 20,000 and 40,000 mg-min/m^3, while for liquid tabun, the percutaneous human LD_{50} is estimated to be 1–1.5 g per man. The permissible airborne exposure concentration of tabun for an 8-hr workday or a 40-hr workweek is an 8-hr TWA of 0.0001 mg/m^3. The number and severity of signs and symptoms following tabun exposure are dependent on the quantity, rate, and route of entry. Very small doses to the skin may cause local sweating and tremors with few other effects. Individuals intoxicated with tabun display approximately the same sequence of signs and symptoms regardless of the route of exposure. Signs and symptoms following vapor exposure include runny nose, tightness of chest, dimness of vision and miosis (pin-point pupils), difficulty in breathing (dyspnea), drooling and excessive sweating, nausea, vomiting, cramps, involuntary defecation and urination, twitching, jerking, staggering, headache, confusion, drowsiness, coma, and convulsions. Death follows cessation of respiration. Death following inhalation and liquid in the eye occurs from 1 to 10 min following exposure. If skin absorption is sufficient to be lethal, death may occur within 1 or 2 min or be delayed for 1 or 2 hr.

Clinical Management

Management of nerve agent intoxication consists of decontamination, ventilation, administration of antidotes, and supportive therapy.

The three therapeutic drugs for treatment of nerve agent intoxication are atropine, pralidoxime chloride, and diazepam.

Atropine, a cholinergic blocking or anticholinergic drug, is effective in blocking the effects of excess acetylcholine at peripheral muscarinic sites. The usual dose is 2 mg, which may be repeated at 3- to 5-min intervals. Pralidoxime chloride (protopam chloride; 2-PAM CL) is an oxime used to break the agent–enzyme bond and restore the normal activity of the enzyme. Abnormal activity decreases and normal strength returns to skeletal muscles, but no decrease in secretions is seen following oxime treatment. The usual dose is 1000 mg (iv or im), which may be repeated two or three times at hourly intervals, intravenously or intramuscularly. Diazepam, an anticonvulsant drug, is used to decrease convulsive activity and reduce brain damage that may occur from prolonged seizure activity. It is suggested that all three of these drugs be administered at the onset of severe effects from nerve agent exposure, whether or not seizures occur. The usual dose of diazepam is 10 mg (im).

Miosis, pain, dim vision, and nausea can be relieved by topical atropine in the eye. Pretreatment with carbamates may protect the cholinesterase enzymes before nerve agent exposure. Pyridostigmine bromide is available as a pretreatment for nerve agent exposure. It is available in 30-mg tablets; tablets should be administered every 8 hr. When used prior to exposure, it should be followed by atropine and pralidoxime chloride after exposure. LD_{50}s are increased several fold and survival is increased if the nerve agent is soman (GD). It may be useful before tabun exposure.

Supportive therapy may include ventilation via an endotracheal airway if possible and suctioning of excess secretions from the airways.

Animal Toxicity

Tabun is similar in action to sarin (GB); however, it is about half as toxic as sarin by inhalation and is more irritating to the eyes at low concentrations.

Small doses of nerve agents in animals can produce tolerance. They have also been demonstrated to produce neuropathies, myopathies, and delayed neurotoxicity in addition to their classical cholinergic effects. In rats, acute administration of nerve agents in subconvulsive doses produced tumors and hindlimb adduction. In animals nerve agents can also cause behavioral as well as cardiac effects.

The cause of death is attributed to anoxia resulting from a combination of central respiratory paralysis, severe bronchoconstriction, and weakness or paralysis of the accessory muscles for respiration.

Signs of nerve agent toxicity vary in rapidity of onset, severity, and duration of exposure. These are dependent on the specific agent, route of exposure, and dose. At the higher doses, convulsions and seizures indicate central nervous system toxicity. Following nerve agent exposure, animals exhibit hypothermia resulting from the cholinergic activation of the hypothalamic thermoregulatory center. In addition, plasma concentrations of pituitary, gonadal, thyroid, and adrenal hormones are increased during organophosphate intoxication.

Table T-1 lists the LCt_{50}s (mg-min/m³) reported following the inhalation of tabun as well as acute toxicities by other routes of exposure in various animal species.

—Harry Salem and Frederick R. Sidell

(The views of the authors do not purport to reflect the position of the U.S. Department of Defense. The use of trade names does not constitute official endorsement or approval of the use of such commercial products.)

Related Topics

Behavioral Toxicology
Cholinesterase Inhibition
Nerve Agents
Neurotoxicology: Central and Peripheral
Organophosphates
Psychological Indices of Toxicity

TABLE T-1
Acute Toxicities of Tabun in Various Species
by Various Routes of Exposure

Route of exposure/species	LCt_{50} (mg-min/m³)
Inhalation (10-min exposure)	
Guinea pig	3,930
Cat	2,500
Rat	3,040
Rabbit	8,400
Dog	4,000
Monkey	2,500
Mouse	450
	LD_{50} (mg/kg)
Percutaneous	
Rat	18
Rabbit	2.5
Dog	30
Monkey	9.3
Mouse	1.0
Guinea pig	35
	LD_{50} (µg/kg)
Intravenous	
Cat	47
Rat	66
Rabbit	63
Dog	85
Mouse	150
Intraperitoneal	
Rat	490
Mouse	604
Subcutaneous	
Dog	284
Rat	162
Rabbit	375
Mouse	250
Monkey	70
Guinea pig	120
Hamster	245
Intramuscular	
Chicken	118
Monkey	34
Mouse	440
Rat	800
Oral	
Rat	3,700
Dog	200
Rabbit	16,300

TCDD

- CAS: 1746-01-6
- SYNONYM: 2,3,7,8-Tetrachlorodibenzo-*p*-dioxin
- CHEMICAL CLASS: Chlorinated dioxins (see Dioxins)
- CHEMICAL STRUCTURE:

Use

TCDD has been found as a contaminant in some herbicides and defoliants, including Agent Orange (a defoliant used in Vietnam). Agent Orange is a 50:50 mixture of the *n*-butyl esters 2,4-D and 2,4,5-T.

Exposure Pathways

Dermal exposure, inhalation, or ingestion are all possible routes of exposure. TCDD is concentrated in the food chain.

Toxicokinetics

TCDD accumulates in fatty tissues. It has a very long biological half-life and is resistant to metabolism.

Mechanism of Toxicity

TCDD binds noncovalently to the aryl hydrocarbon receptor.

Human Toxicity

Dermal exposure has produced chloracne, porphyrinuria, and porphyria cutanea tarda. TCDD is a possible carcinogen, teratogen, and mutagen.

Clinical Management

Affected skin should be washed with soap and water. If ingested, syrup of ipecac should be given, then gastric lavage performed, followed by a saline cathartic. In case of muscular or cardiac disturbances, intravenous lidocaine (50–100 mg) should be administered.

Animal Toxicity

Toxic effects in animals include severe weight loss/anorexia, liver diseases, chloracne, and gastric ulcers. TCDD is a suspected teratogen, carcinogen, and mutagen in animals. Oral LD_{50}s are 1 mg/kg in the guinea pig and 20 mg/kg in the rat.

—*Jayne E. Ash and Shayne C. Gad*

Related Topics

Immune System
Toxic Torts

Tellurium (Te)

- ◆ CAS: 13494-80-9
- ◆ SELECTED COMPOUNDS: Tellurium dichloride, $TeCl_2$ (CAS: 10025-71-5); tellurium oxide, TeO_2 (CAS: 7446-07-3)
- ◆ CHEMICAL CLASS: Metals

Uses

Tellurium is used in semiconductors and in "daylight" vapor lamps. It is also used in the manufacturing of rubber and certain metal alloys and as a coloring agent in glass and ceramics.

Exposure Pathways

Tellurium is ingested with foods such as nuts, fish, and certain dairy products. Many fatty foods contain tellurium, and some plants, like garlic, accumulate tellurium from the soil. Neither drinking water nor ambient air contain significant amounts of tellurium. Skin contact is not a significant exposure pathway.

In industrial settings, inhalation may be a significant exposure pathway. Airborne concentrations of tellurium are higher in the vicinity of metallurgical industries. Like selenium, tellurium is obtained as a byproduct of copper, lead, and zinc refining. It is produced mainly from the tailings of bismuth.

Toxicokinetics

Tellurium is poorly absorbed from the gastrointestinal tract; in contrast, tellurous acid is absorbed dermally. Tellurium concentrates in a variety of organs, primarily in the bones and kidneys, followed by the liver and the adipose tissue. Tellurium is metabolized in the body by the reduction to tellurides and then biotransformed to dimethyltelluride (a reaction similar to the biotransformation of selenium), which is volatile and can be exhaled. Most tellurium is excreted in urine and bile.

Human Toxicity

The toxicity of tellurium is dependent on the oxidation state. The tellurites, $(TeO_3)^{-2}$, are the most toxic compared to tellurates, $(TeO_4)^{-2}$, or elemental tellurium.

Acute toxicity from inhalation results in the relatively nonspecific symptoms of nausea, sweating, and loss of sleep in some and drowsiness in others. As with selenium, a garlic odor is evident. Kidney damage and fatty degeneration of the liver have been noted in severe cases. In two fatal cases, cyanosis and garlic breath were prominent before coma and death. Fatty degeneration and edema were noted in both cases. A glucose suspension of tellurium was once administered intramuscularly to treat syphilis. Only a metallic taste and garlic breath were noted.

The ACGIH TLV-TWA for tellurium and its compounds is 0.1 mg/m^3.

Clinical Management

Vitamin C (ascorbic acid) reduces the characteristic garlic breath; however, it may also adversely affect the kidneys when an excess amount of tellurium is present. BAL (British Anti-Lewisite; 2,3-dimercaptopropanol) is contraindicated since it enhances the toxicity of tellurium.

Animal Toxicity

Tellurium has not undergone a bioassay for carcinogenesis. Administration of tellurium to pregnant rats produced offspring with hydrocephalus. Experimental animals failed to grow and lost hair.

—Arthur Furst and Shirley B. Radding

Related Topic

Metals

Terbutaline

- CAS: 23031-25-6
- SYNONYM: Brethine
- PHARMACEUTICAL CLASS: A selective β_2-adrenergic agonist

- CHEMICAL STRUCTURE:

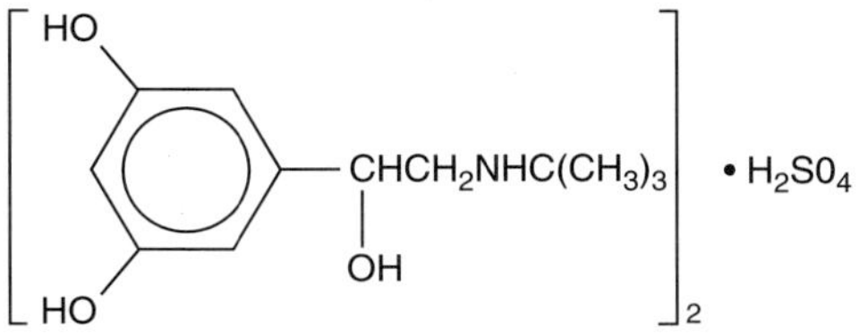

Uses

Terbutaline is used as a bronchodilator and for the prevention of premature labor. Unlabeled use includes treatment of hyperkalemia.

Exposure Pathways

Ingestion is the most common route of accidental and intentional exposure to terbutaline. Inappropriate overuse of the inhalation aerosol may also occur. Terbutaline is available as an inhalation aerosol, as tablets (2.5 and 5 mg), and as a solution for subcutaneous injection (1 mg/ml).

Toxicokinetics

Taken orally, terbutaline is poorly and incompletely absorbed from the gastrointestinal tract, with approximately 47% of the unchanged drug found in the feces. Administered subcutaneously, it is well absorbed with peak serum levels in 20 min. The bioavailability and biotransformation of terbutaline depend greatly on route of administration. With oral dosing there is significant first-pass biotransformation by sulfate and glucuronide conjugation in the liver and gut wall, with only 15% of absorbed terbutaline available as unchanged drug. With inhalation and parenteral dosing, the majority of the drug is available as unchanged terbutaline. The volume of distribution is 1.47 liters/kg. The percentage of protein binding is 15%. With oral dosing terbutaline is eliminated primarily as sulfate (70%) and glucuronide (30%) conjugates. Approximately 10–15% is cleared in the urine as unchanged drug. With parenteral and inhalation exposure, the majority of the drug is cleared in the urine as unchanged terbutaline (68 and 60%, respectively). The elimination half-life is 12–20 hr.

Mechanism of Toxicity

The primary mechanism of terbutaline is the stimulation of adenyl cyclase, which catalyzes cyclic adenosine monophosphate (AMP) from adenosine triphosphate (ATP). In the liver, buildup of cyclic AMP stimulates glycogenolysis and an increase in serum glucose. In skeletal muscle, this process results in increased lactate production. Direct stimulus of sodium/potassium ATPase in skeletal muscle produces a shift of potassium from the extracellular space to the intracellular space. Relaxation of smooth muscle produces a dilation of the vasculature supplying skeletal muscle, which results in a drop in diastolic and mean arterial pressure (MAP). Tachycardia occurs as a reflex to the drop in MAP or as a result of β_1 stimulus. β-Adrenergic receptors in the locus ceruleus also regulate norepinephrine-induced inhibitory effects, resulting in agitation, restlessness, and tremor.

Human Toxicity

The toxic events of terbutaline overdose follow its β-adrenergic agonist activity. The effects of terbutaline overdose are usually mild and benign; however, they can be prolonged. Cardiovascular effects are usually limited to a sinus tachycardia and widened pulse pressure. Although there may be a drop in diastolic pressure, the systolic pressure is maintained by increased cardiac output from the tachycardia. Evidence of myocardial ischemia after terbutaline overdose has been infrequently reported. Transient hypokalemia may occur, caused by a shift of extracellular potassium to the intracellular space. A transient metabolic acidosis can be seen due to increased lactate production. Restlessness, agitation, and tremors are common in terbutaline overdose.

Clinical Management

Basic and advanced life-support measures should be utilized as necessary. Terbutaline overdoses rarely require treatment beyond gastrointestinal decontamination. Activated charcoal effectively binds terbutaline. The hypokalemia produced reflects a transient shift in potassium location rather than a true deficit of potassium. Therefore, only rarely is there a need for external replacement therapy. A conservative approach to the tachycardia is recommended. In the rare event of complications, intravenous propranolol rapidly and effectively reverses the symptoms of terbutaline poisoning.

—*Henry A. Spiller*

Terfenadine

- CAS: 506-79-08
- SYNONYM: Seldane
- PHARMACEUTICAL CLASS: A butyrophenone-derivative H-1 receptor antagonist
- CHEMICAL STRUCTURE:

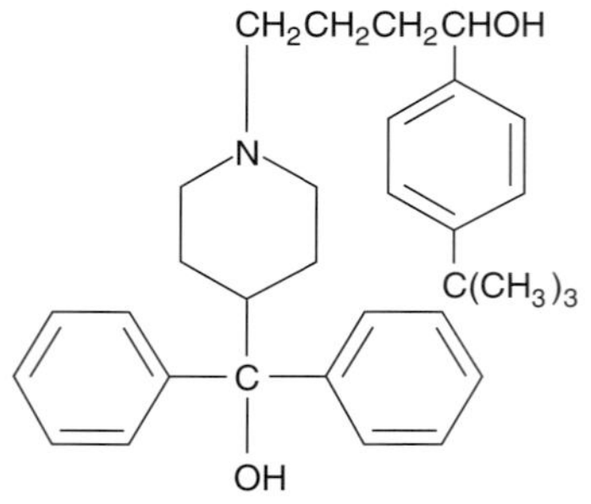

Use

Terfenadine is indicated for the symptomatic relief of seasonal allergic rhinitis.

Exposure Pathways

Ingestion is the route of both accidental and intentional exposures to terfenadine.

Toxicokinetics

Approximately 70% of an oral dose of terfenadine is absorbed rapidly and achieves peak plasma concentration in 1 or 2 hr. Terfenadine is metabolized extensively in the liver by cytochrome P450 microsomal enzyme system. The carboxylic acid metabolite has approximately one-third the antihistaminic activity of terfenadine. The distribution of terfenadine and its metabolites into human body tissue and fluids has not been determined. The drug is 97% protein bound. Terfenadine and its metabolites are excreted in the feces (via biliary elimination) and in the urine. Approximately 60 and 40% of an oral dose of the drug is excreted in the feces and urine, respectively, within 24–48 hr, principally as metabolites. The pharmacokinetic half-life is approximately 8.5 hr; however, the pharmacologic effects may persist much longer.

Mechanism of Toxicity

Signs and symptoms following acute terfenadine overdosage have generally been mild. However, cardiotoxic effects following an overdose have been reported. Although the mechanism is not well understood, evidence in animal models suggests the cardiotoxic effects may result at least in part from blockade of the potassium channel involved in repolarization of cardiac cells. It is also postulated that H-3 receptors (mediating a regulatory feedback mechanism) may be involved in the development of cardiovascular toxicity. Anticholinergic effects appear to be unlikely causes of the cardiac effects.

Human Toxicity

Unlike other antihistamines, terfenadine is considered "nonsedating" and lacks the anticholinergic properties. Serious cardiac effects, including prolongation of the QT interval, arrhythmias (i.e., ventricular tachycardia, torsades de pointes, ventricular fibrillation, and heart block), arrest, hypotension, palpitations, syncope, seizures, and death, have been reported in patients receiving terfenadine. These cardiotoxic effects have been reported in patients receiving terfenadine doses of 360 mg and are more likely at doses exceeding 600 mg. Patients with impaired liver function and geriatric patients may be at risk for the development of cardiovascular toxicity. Concomitant use of azole derivatives (i.e., itraconazole and ketoconazole) and macrolide antiinfectives (i.e., erythromycin) with astemizole may also increase the risk of toxicity. These antiinfectives interfere with microsomal enzyme system resulting in an increased serum concentration.

Clinical Management

Basic and advanced life-support measures should be utilized as necessary. Appropriate gastrointestinal decontamination procedures should be administered based on the history of the ingestion and the patient's level of consciousness. Close electrocardiographic monitoring should be instituted for a minimum of 24 hr.

—*Carla M. Goetz*

Tetrachloroethane

- ◆ CAS: 79-34-5
- ◆ PREFERRED NAME: Acetylene tetrachloride
- ◆ CHEMICAL CLASS: Chlorinated hydrocarbons; haloalkanes
- ◆ CHEMICAL STRUCTURE:

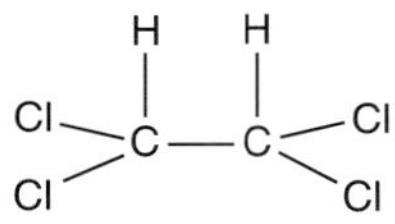

Uses

Tetrachloroethane is used as a solvent for fats, oils, waxes, resins, rubber, copal, phosphorus, and sulfur. It is used in herbicides, pesticides, and soil sterilizers and is also used in biochemical testing of the gastrointestinal tract, liver, and kidneys.

Exposure Pathways

Ingestion, inhalation, and dermal contact are possible routes of exposure.

Toxicokinetics

Tetrachloroethane is eliminated slowly. One hour after exposure, 97% is found in lungs. After 3 days, 50% is eliminated through expiry and 28% through urine.

Mechanism of Toxicity

Fatty tissue degeneration and dehydration are mechanisms of toxicity. It is metabolized by P450. Fission of carbon chloride and oxidation occurs. Tetrachloroethane increases lipids and triglycerides in the liver and reduces ATP levels.

Human Toxicity

Persons with existing liver trouble or obesity have a greater susceptibility. Acute symptoms include eye/nasal irritation, headache, and nausea. This may progress to cyanosis, central nervous system (CNS) depression, and unconsciousness or coma within hours. Chronic symptoms include CNS depression and gastric/hepatic damage. The PEL is 5 ppm/8 hr. The ACGIH TLV is 2 ppm.

Clinical Management

Exposed skin should be washed with soap and water. If inhaled, respiratory therapy should be provided. If ingested, emesis should be induced or gastric lavage performed. Levels of fluid excretion and blood pressure should be maintained.

Animal Toxicity

Tetrachloroethane is corrosive to the eyes and skin. It inhibits production of antibodies in rabbits. It is a teratogen and causes liver damage. The oral LD_{50} is 0.20 ml/kg in the rat.

—Jayne E. Ash and Shayne C. Gad

Related Topics

Alkyl Halides
Pollution, Water

---

*Tetrachloroethylene**

♦ CAS: 127-18-4
♦ SYNONYMS: Tetrachloroethene; 1,1,2,2-tetrachlroethylene; perchloroethene; perchloroethylene; PCE; PERC
♦ CHEMICAL CLASS: Chlorinated olefinic hydrocarbon

Uses

PERC is manufactured by direct chlorination of ethylene or a petroleum hydrocarbon stream. It has been

* Information Source: Hazardous Substance Data Bank, National Library of Medicine, Bethesda, MD, July 1996.

used extensively in the dry cleaning industry for metal cleaning and degreasing, for processing and finishing textiles, as an extraction solvent, in chemical processing, as heat exchange fluid, as a grain fumigant, for fluorocarbon manufacturing processes, and in typewriter correction fluid. PERC is a clear liquid of high volatility. Its vapor is almost six times as dense as air and has a chloroform-like odor detectable from 5 to 50 ppm in air.

Exposure Pathways

Major human exposure has been in the dry cleaning industry and in industries employing degreasing procedures. In addition, inhalation of contaminated urban air (especially near point sources such as dry cleaners), drinking contaminated water from contaminated aquifers, and drinking water distributed in pipelines with vinyl liners may offer additional exposure opportunities.

Toxicokinetics

Comparison of the urinary trichloro compound levels with PERC in the environment revealed that while the metabolite levels increased essentially linearly, up to PERC concentrations of 100 ppm, leveling off was apparent in the metabolite excretion when the exposure to PERC was more intense (e.g., more than 100 ppm), indicating that the capacity of humans to metabolize this chlorinated hydrocarbon is rather limited. A tentative calculation indicated that at the end of an 8-hr shift with exposure to tetrachloroethylene at 50 ppm (TWA), 38% of the PERC absorbed through the lung would be exhaled unchanged, <2% would be metabolized to be excreted in the urine, while the rest would remain mostly in the fat stores of the body to be eliminated later.

Absorption typically takes place through inhalation of the volatile solvent but may also take place through dermal exposure or ingestion of contaminated drinking water.

Metabolism is saturable and relatively slow with only a small percentage of the administered dose excreted as metabolites, the major one being trichloroacetic acid. Following exposure to PERC, trichloroacetic acid and trichloroethanol have been found in the urine of humans and animals. Additionally, oxalic acid, dichloracetic acid, and ethylene glycol have been reported in the urine of exposed animals. Other reported biotransformation products include inorganic chloride and *trans*-1,2-dichloroethylene in expired air.

Once in the bloodstream, PERC tends to distribute to body fat. In human tissue at autopsy, ratios of fat to liver concentrations are greater than 6:1. An autopsy after a fatal PERC exposure revealed an eightfold greater concentration in the brain compared with blood. PERC reached near steady-state levels in the blood of human volunteers within 2 hr of continuous exposure.

The respiratory half-life for elimination of PERC has been estimated at 65–70 hr and is a result of the very slow elimination of PERC from fat stores. The half-life of elimination of trichloro metabolites of PERC is estimated as being 144 hr. This long half-life of elimination has serious implications with regard to the accumulation of PERC during chronic or multiple exposure situations.

Mechanism of Toxicity

PERC is metabolized to trichloroacetic acid and other trichloro metabolites in the liver. Trichloroacetic acid has been shown to produce peroxisome proliferation in mice. This may have implications for the apparent increase in liver tumors in mice. PERC also has been shown to distribute rapidly to the central nervous system (CNS) and is known to have an affinity for the lipophilic cellular membranes in the brain.

Human Toxicity

Contact with liquid PERC results in eye and skin irritation, and persistent exposure will result in defatting of the skin and subsequent dermatitis. Acute inhalation exposure has been reported to cause CNS depression, alcohol intolerance, liver necrosis, kidney injury, malaise, headache, dizziness, fatigue, tinnitus, visual field reduction, sensory disturbances, lightheadedness, sweating, a staggering gait, inebriation, mental dullness, and death by anesthesia. Cardiac arrhythmias, peripheral neuropathies, proteinuria, hematuria, and oliguric renal failure have also been associated with PERC exposure.

Studies relating to chronic exposure of those working in dry cleaning plants have reported some CNS effects, some liver function abnormalities, renal dysfunction, and central and peripheral neurotoxicity. Other effects from chronic exposure to PERC include cardiac arrhythmias, reduced color perception, impaired memory, impaired vision, confusion, disorientation, fatigue, personality changes, and agitation.

There are some suggestions of increased liver and urinary tract tumors in exposed humans, but the data are inadequate to make any definite conclusions. Other studies of occupationally exposed workers suggest that there are increased cancer risks for lung, cervix, skin, liver, esophagus, urinary tract, and for leukemia. Chromosomal abnormalities have been reported in the circulating lymphocytes of exposed workers.

IARC has classified PERC as a probable human carcinogen based on positive findings in animals and suggestive, although inconclusive, findings in humans.

Animal Toxicity

Animal studies have confirmed the effects on liver and kidneys and have identified neurotoxic effects and electroencephalographic abnormalities in rats. Chronic long-term animal studies have demonstrated the tumorigenic potential of PERC as well as its potential for producing renal damage and changes in brain biochemistry among other effects. Lifetime studies in mice given PERC orally or by inhalation showed highly significant increases in hepatocellular carcinomas in a dose-dependent manner. Mononuclear cell leukemia has also been reported in both sexes of Fisher 344 rats as a result of inhalation of PERC. Mutagenicity data offer little in terms of confirmatory evidence since these studies have shown both positive and negative findings in the same assays. PERC is considered to be an animal carcinogen.

—*R. A. Parent, T. R. Kline,*
and D. E. Sharp

Related Topics

Peroxisome Proliferators
Pollution, Water

Tetrachlorophenoxyacetic Acid

- CAS: 5416-64-8
- PREFERRED NAME: 2,3,5,6-Tetrachlorophenoxyacetic acid
- CHEMICAL CLASS: Chlorophenoxy compound
- CHEMICAL STRUCTURE:

Uses

Tetrachlorophenoxyacetic acid is used as a plant growth regulator, herbicide, and tumor inhibitor.

Exposure Pathways

Dermal contact, inhalation, and ingestion are possible routes of exposure.

Toxicokinetics

Tetrachlorophenoxyacetic acid is rapidly absorbed, distributed, and eliminated after metabolism. It does not have an accumulated effect.

Mechanism of Toxicity

There are few effects at low, repeated doses. Tetrachlorophenoxyacetic acid increases blood urea nitrogen. It is a defatting agent in the liver and decreases mucous in the lungs and intestines.

Human Toxicity

Effects include chloracne. Intoxication is characterized by dizziness, nausea, vomiting, diarrhea, respiratory complications, aching, myotoxia, fatigue, and weakness. Some kidney damage is possible.

Clinical Management

If ingested, emesis or gastric lavage plus activated charcoal should be used. Generous fluid replacement and alkalinization of urine should be ensured. The patient should be observed for metabolic acidosis, hyperthermia, hyperkalemia, and myoglobinuria.

Animal Toxicity

Tetrachlorophenoxyacetic acid produces cleft palate and liver and kidney damage in rats. Gastrointestinal irritation also results. At toxic doses, symptoms include tenseness, stiff extremities, muscular weakness, ataxia, and paralysis. Tetrachlorophenoxyacetic acid inhibits the growth of adenocarcinoma in rats.

—Jayne E. Ash and Shayne C. Gad

Tetrahydrofuran

- CAS: 109-99-9
- SYNONYMS: Cyclotetramethylene oxide; diethylene oxide; THF; tetramethylene oxide
- CHEMICAL CLASS: Substituted epoxide
- CHEMICAL STRUCTURE:

Uses

Tetrahydrofuran is a solvent used in natural and synthetic polymers and resins such as polyvinyl chloride and vinylidene chloride copolymers. It is also used in the manufacture of lacquers, glues, paints, and inks.

Exposure Pathways

Industrial exposures to tetrahydrofuran are most likely to occur by inhalation with possible skin and eye contact. Accidental ingestion is also possible.

Toxicokinetics

Little is known concerning the complete toxicokinetics of tetrahydrofuran. Following inhalation, when healthy volunteers were exposed to 100 or 400 ppm tetrahydrofuran in air, the percentage of expired tetrahydrofuran was 25–35%. The elimination half-life of tetrahydrofuran was 30 min in individuals exposed to 200 ppm for 3 hr. Some tetrahydrofuran is absorbed in the nasal cavity due to its solubility and inspiratory flow rate. Perhaps some tetrahydrofuran uptake in the nasal tissue is dependent on its reaction with tissue substrates. Some tetrahydrofuran can be metabolized in the nasal cavity. Tetrahydrofuran blood concentrations were higher at 1 hr postexposure than immediately after cessation of exposure. *In vitro* studies indicated that tetrahydrofuran was first hydroxylated by microsomal enzymes. High concentrations (10^{-2} m) of tetrahydrofuran inhibited the *in vitro* activity of rat hepatic cytochrome P450 by 80%. Tetrahydrofuran has been noted to enhance the toxic action of a number of compounds and stimulate the rapid absorption of reactive metabolites. Some of the tetrahydrofuran is excreted in the exhaled breath, while the various metabolites of tetrahydrofuran are excreted in the urine.

Mechanism of Toxicity

Irritation of the upper respiratory tract is attributed to the solubility of tetrahydrofuran in the mucous membranes causing irritation of the sensory nerve endings. The direct action of tetrahydrofuran on the skin and eyes is the result of irritation to these tissues.

Human Toxicity

Exposure to tetrahydrofuran has been reported to cause irritation of the skin, eyes, and respiratory tract. Individuals exposed to high concentrations of tetrahydrofuran have complained of nausea, dizziness, tennitus, headache, and central nervous system (CNS) depression. Narcosis has been observed in humans exposed to tetrahydrofuran around 25,000 ppm. The probable oral lethal dose in humans is estimated to be between 50 and 500 mg/kg.

In order to minimize the potential for irritation to the upper respiratory tract and to provide a margin of safety for possible tetrahydrofuran-induced systemic effects, an ACGIH TLV-TWA of 200 ppm and a STEL of 250 ppm have been recommended. The margin of safety afforded by the TLV for tetrahydrofuran is cur-

rently under review based on animal studies indicating respiratory epithelium changes at 200 ppm. NIOSH established an IDLH value of 20,000 ppm, which is the LEL for tetrahydrofuran.

Clinical Management

Exposures by inhalation should be monitored for respiratory tract irritation, bronchitis, or pneumonitis. Humidified supplemental 100% oxygen should be administered. Following ingestion, milk or water should be used to dilute the tetrahydrofuran in the stomach. A charcoal slurry with saline cathartic should be administered. Gastric lavage may be indicated. Treatment of CNS depression is symptomatic. Renal and hepatic function should be monitored. Exposed eyes should be irrigated with copious amounts of water for at least 15 min. If irritation, pain, swelling, lacrimation, or photophobia persist, the patient should be seen in a health care facility. From dermal exposure, the affected skin should be washed thoroughly with soap and water. If irritation persists, a health care facility should be contacted.

Animal Toxicity

Although tetrahydrofuran has long been known as an organic solvent, only a few animal studies have been conducted. The oral LD_{50} in rats for tetrahydrofuran is 2.3 ml/kg. A single 4-hr inhalation study of tetrahydrofuran in rabbits at 100–12,000 ppm produced a transient dose-related decrease of tracheal ciliary activity. Concentrations of tetrahydrofuran >25,000 ppm produced anesthesia. Tetrahydrofuran was irritating to rabbit skin when applied topically in solutions exceeding a 20% concentration. Male rats exposed to 5000 ppm tetrahydrofuran for 12 weeks at 4 hr/day showed signs of systemic intoxication, skin and respiratory tract irritation, liver function disturbance, and abnormalities in glucose metabolism. Some rats exhibited slight respiratory tract irritation at 200 ppm. In another study, male rats exposed to 200, 1000, or 2000 ppm for 18 weeks at 6 hr/day demonstrated an increase in muscle acetylcholinesterase activity in a concentration-dependent manner. Rats exposed to 200 ppm tetrahydrofuran, 4 hr/day, 5 days/week, exhibited damage to the nasal and tracheal epithelian. At 1000 ppm tetrahydrofuran, severe damage to the same structures was observed.

Rats and mice were exposed for 6 hr/day, 7 days/ week on Gestation Days 6–19 for rats and 6–17 for mice at 600, 1800, or 5000 ppm tetrahydrofuran. Pregnant mice that inhaled 5000 ppm tetrahydrofuran died, while those exposed to 1800 ppm were sedated. Some treatment-related effects were reduced fetal body weight and reduced ossification of the sternebrae. The maternal no-observed-adverse-effect level (NOAEL) for both species was 1800 ppm; the NOAEL for developmental toxicity was 1800 ppm in rats and 600 ppm in mice.

Tetrahydrofuran was not mutagenic in *Salmonella typhimurium* TA100 at 50 μl per plate and it failed to induce sex-linked recessive lethals in *Drosophila melanogaster* by ingestion or injection. In cultured Chinese hamster ovary cells, there was no indication of induction of chromosomal aberrations or sister chromatid exchanges.

The National Toxicology Program completed prechronic oral gavage and inhalation studies of tetrahydrofuran in male and female Fischer 344 rats and B6C3F1 mice. No chronic gavage study was scheduled, but tetrahydrofuran has been scheduled for chronic inhalation bioassay.

—*Edward Kerfoot*

Tetranitromethane

- ◆ CAS: 509-14-8
- ◆ Synonym: TNM
- ◆ Chemical Class: Aliphatic nitro compounds
- ◆ Molecular Formula: CN_4O_8
- ◆ Chemical Structure:

$$
\begin{array}{c}
NO_2 \\
| \\
O_2N - C - NO_2 \\
| \\
NO_2
\end{array}
$$

Uses
Tetranitromethane is used as an oxidizer in rocket propellants, as a diesel fuel additive, and as an explosive.

It is also used as a biochemical agent to nitrate tyrosine proteins.

Exposure Pathways
Tetranitromethane is an oily liquid with a vapor pressure less than that of water. It occurs as an impurity in 2,4,6-trinitrotoluene (TNT). The primary routes of occupational exposure are inhalation and dermal contact.

Toxicokinetics
No data exist regarding the absorption of tetranitromethane; however, based on toxicity reported in humans and animals, it is clear that it is readily absorbed by the oral route and through inhalation. Rats administered single oral doses exhibited dose-related methemoglobinemia at 90 min, suggesting that the metabolism of tetranitromethane results in the formation of nitrites. Methemoglobinemia was not observed following intravenous or inhalation exposures, suggesting that the blood effects seen in oral studies resulted from nitrase reduction in the gut. No data are available regarding the distribution of absorbed tetranitromethane. No elimination data are available for this compound.

Mechanism of Toxicity
The mechanism of toxicity for tetranitromethane is not known. Methemoglobinemia formation reported following oral administration may be a result of reduction of tetranitromethane in the gut. Nasal lesions observed in lifetime inhalation studies of rats and mice were attributed to the significant irritating properties of the material.

Human Toxicity
Tetranitromethane is a strong irritant of the eyes and mucous membranes. Workers exposed to the heated tetranitromethane have complained of irritation of the eyes and respiratory tract. Complaints following long-term exposures have included headache, drowsiness, and respiratory distress. Salivation, coughing, pulmonary edema, and methemoglobinemia have also been reported. Deaths due to methemoglobinemia and respiratory failure have been reported following exposure to crude TNT.

Clinical Management
If contact with the liquid occurs, affected areas should be flushed thoroughly with water for at least 15 min.

The victim should be observed for burns or resulting irritation. In case of inhalation, the victim should be moved to fresh air, an airway established, and respiration maintained as necessary. The patient should be monitored for irritation and pulmonary edema. If ingestion occurs, emesis should be induced if the victim is conscious. Gastric lavage may be indicated if the victim is unconscious or convulsing. Treatment for methemoglobinemia and/or monitoring for possible liver and kidney injury may be required.

Animal Toxicity

Tetranitromethane is highly toxic to mice and rats by the oral and inhalation routes. In rats, an inhalation LC_{50} of 17.5 ppm and an oral LD_{50} of 130 mg/kg have been reported. The corresponding values in mice are 54.4 ppm and 375 mg/kg. Effects of overexposure in animals include eye and respiratory irritation, pulmonary edema, lung injury, bronchopneumonia, liver and kidney injury, and, in cats, methemoglobinemia. In lifetime inhalation studies, tetranitromethane caused nasal lesions indicative of chronic irritation of the nasal cavity. In addition, tetranitromethane caused increased incidences of alveolar and bronchiolar neoplasms in rats and mice and lung carcinoma in rats.

—Daniel Steinmetz

Related Topic

Respiratory Tract

Tetrodotoxin

- ♦ CAS: 4368-28-9
- ♦ SYNONYMS: Balloon fish toxin; blowfish toxin; globefish toxin; puffer fish toxin; swellfish toxin; toad fish toxin; TTX; tarichatoxin; triturus embryonic toxin; MTX (maculotoxin). Other sources of the toxin include fugu, freshwater puffer (*Tetraodon leiurus*) of Thailand, goby fish (gobius criniger), California newt (tarichatorosa), eastern salamander, *T. pivularis*, *T. granulosa*, *Notophthalmus viridescens*, *Cynops pyorhogaster*, *C. ensicaudus*, *Triturus vulgaris*, *T. cristatus*, *T. alpestris*, *T. marmoratus*, *Paramesotriton hongkongensis*, blue-ringed octopus (*Halalochlaena maculosa*) of Australia, Atelopus species of frogs in Central America, trumpet shell (*Charonia sauliae*), Japanese wory shell, frog shell (*Tutufa lissostoma*), and the xanthid crab.
- ♦ PHARMACEUTICAL CLASS: Nonprotein, heat-stable neurotoxin
- ♦ MOLECULAR FORMULA: $C_{11}H_{17}N_3O_8$

Use

The fish muscle is nontoxic during the nonreproductive season and considered a delicacy in Japan.

Exposure Pathway

Exposure occurs through ingestion of flesh, viscera, or skin containing tetrodotoxin. The viscera contains the highest concentration.

Toxicokinetics

Tetrodotoxin is readily absorbed from the gastrointestinal tract within 5–10 min. Effects can occur in 5 min to approximately 3 hr. The toxin can also be absorbed through the skin. Nontoxic fish become toxic when kept in close contact with toxic fish. The peak concentration is reached in 20 min. Tetrodotoxin is widely distributed in the body with concentrations highest in the kidneys and heart and lowest in the brain and blood. An appreciable amount is excreted in the urine in an unchanged form. The half-life of the parent compound varies from 30 min to 3 hr.

Mechanism of Toxicity

Tetrodotoxin is a nonprotein, heat-stable neurotoxin that is 160,000 times more potent than cocaine on isolated nerves. Neuromusculature transmission is interrupted by tetrodotoxin in the motor neurons and on the muscle membrane. It has a direct blocking action on skeletal muscle fibers. This toxin possesses local anesthetic-like properties that block the fast sodium channel during depolarization.

Human Toxicity

An oral dose of 10 μg/kg may be fatal. The fatality rate for puffer fish in Japan is 61.5% of approximately 50 cases annually. Ingestion of one California newt has resulted in death. The *C. newt* contains approximately 250 μg of toxin. It has been suggested that human toxicity from tetrodotoxin is the result of a reversible effect on myelinated nerve fibers along the entire length of the axon by lowering the conductance of sodium currents at nodes of Ranvier. Paresthesia begins shortly after ingestion (10–45 min) and vomiting is common. Lightheadedness and weakness also occur early. The weakness develops first in the hands and arms and then in the legs. Seizures are rare. Ascending paralysis develops with respiratory depression (6–24 hr postingestion). Hypotension due to vasodilation occurs. Cardiac dysrhythmias (bradycardia, asystole, and depressed AV node conduction) have been reported. Salivation, muscle twitching, hypothermia and/or diaphoresis, pleuritic chest pain, dysphagia, petechial hemorrhages, hematemesis, difficulty in speaking, pallor, blistering, and desquamation can be seen. Death is usually secondary to muscle weakening leading to respiratory paralysis. The prognosis is good if the patient survives the first 24 hr.

Clinical Management

No antidote is available. Symptomatic and good supportive care is mandatory. Syrup of ipecac should be avoided due to the rapid development of symptoms. Gastric lavage with activated charcoal is recommended. Endoscopic removal of gastric fish debris has been performed. All patients should be admitted. Blood pressure, heart rate, and respirations should be monitored and supported. Hypotension can be treated with norepinephrine, epinephrine, or dopamine. Hypertension can be treated with nitroprusside, labetalol, or hydralazine. Anticholinesterase drugs such as edrophonium or neostigmine have been tried early in serious cases. Experimental reversal of the blockade results when an increase in release of acetylcholine at the neuromuscular junction is induced by anticholinesterase drugs. Laboratory determination of tetrodotoxin is not useful.

Animal Toxicity

Intraperitoneal LD_{50}s of 8 and 10 μg/kg have been reported in mice. An oral LD_{50} of 200 μg/kg has been reported in cats.

—*C. Lynn Humbertson*

Related Topic

Neurotoxicology: Central and Peripheral

Thalidomide

◆ CAS: 50-35-1

◆ SYNONYMS: K-17; 2-phthalimidoglutarimide; *N*-phthalylglutamic acid imide; *N*-(2,6-dioxo-3-piperdyl)-phthalimide; Talimol; Sedalis; Kevadon; Distavil; NSC-66847

◆ PHARMACEUTICAL CLASS: Piperidinedione derivative

◆ CHEMICAL STRUCTURE:

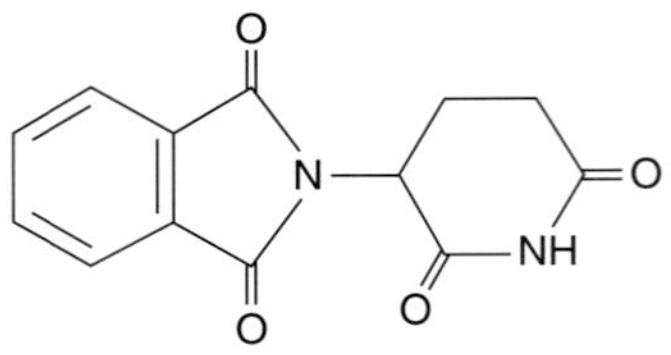

Uses

Thalidomide was formerly used as a sedative/hypnotic. It is currently used as an immunosuppressant in the treatment of erythema nodosum leprosum, graft-versus-host disease, macular degeneration, oral ulcers in AIDS patients, and other disorders.

Exposure Pathway

Exposure is exclusively by ingestion.

Toxicokinetics

All available data are derived from therapeutic dosing. Peak plasma levels occur 2–6 hr following oral doses. Bioavailability in animals varies from 67 to 93%. There is conflicting data on whether thalidomide undergoes hepatic metabolism. It is a nonpolar compound that is extensively bound to plasma proteins and has a volume of distribution of about 120 liters/kg in healthy adults. Less than 1% of a dose is excreted unchanged in the urine, suggesting that elimination is largely nonrenal.

The serum elimination half-life is approximately 8 or 9 hr after a single oral dose.

Mechanism of Toxicity

Thalidomide has significant teratogenic effects in humans, and it also affects the central and peripheral nervous systems through unknown mechanisms. Evidence of a toxic arene oxide metabolite is unsubstantiated.

Human Toxicity

Thalidomide was first marketed as a sedative/hypnotic agent in Germany in 1956 and in the United Kingdom 2 years later. It was subsequently withdrawn from the market in 1961 after more than 12,000 reports of fetal abnormalities, particularly phocomelias. Other teratogenic effects include eye and ear abnormalities, esophageal and duodenal atresias, and defects in internal organs such as the heart and kidneys. Congenital defects of the kidneys and nervous system may persist throughout life.

Adverse reactions include dose-related peripheral neuropathy (primarily sensory), nausea, vomiting, constipation, dry mouth, headache, and erythematous rashes. Dose-related central nervous system depression is relatively common. Thalidomide is contraindicated in pregnancy and in women of childbearing age.

Clinical Management

Patients with thalidomide overdose should receive supportive care with attention to airway maintenance. There are no antidotes and no data to support measures to enhance excretion of thalidomide.

—S. Rutherfoord Rose

Related Topics

Developmental Toxicology
Toxicity Testing, Developmental

Thallium (Tl)

- ◆ CAS: 7440-28-0
- ◆ SELECTED COMPOUNDS: Thallium chloride, TlCl (CAS: 7791-12-0); thallium nitrate, TlNO$_3$ (CAS:

10102-45-1); thallium sulfate, Tl$_2$SO$_4$ (CAS: 7446-18-6)

- ◆ CHEMICAL CLASS: Metals

Uses

Thallium is a by-product of iron, cadmium, and zinc refining. It is used in metal alloys, imitation jewelry, optical lenses, artists' pigments, semiconductors, ceramics, and X-ray detection devices. It has limited use as a catalyst in organic chemistry.

In the past, thallium (chiefly thallium sulfate) was used as a rodenticide and insecticide. Its use as a rodenticide was outlawed in 1965 due to its severe toxicity (a source of accidental and suicidal human exposures). Medicinally, it has been used as a depilatory and in the treatment of venereal disease, skin fungal infections, and tuberculosis.

Exposure Pathways

Ingestion and dermal contact have been the primary exposure pathways. For the general population, exposures have resulted from thallium contamination in food or use of thallium in depilatory creams.

Toxicokinetics

Thallium is easily absorbed from the gastrointestinal tract, the lungs, and the skin. After absorption, thallium is carried by the RC and is distributed (like potassium) into the intracellular spaces. Thallium mimics potassium ions in a variety of biochemical and biological functions. Unlike potassium, however, once thallium enters the cells, it is released slowly. It can concentrate in the liver and kidneys. Since it is soluble at physiological pH, it does not form complexes with bone. Most thallium is excreted in the urine, but it is excreted slowly and can be detected months after exposure.

Mechanism of Toxicity

Thallium's mechanism of toxicity is related to its ability to interfere with potassium ion functions. Various potassium-dependent enzymes and proteins have a higher affinity for thallium than the potassium ion; thus, the normal functions of the biological factors are inhibited. At very low concentrations, thallium can stimulate NA$^+$/K$^+$-activated ATPase. At higher doses it can uncouple mitochondrial oxidative phosphorylation. It inhibits succinic acid dehydrogenase and alkaline phosphatase. Toxic doses interfere with protein synthesis. Thallium also has a high reactivity with sulf-hydryl groups and, thus, inhibits the normal reactions of groups with -SH bonds.

Human Toxicity

Thallium is one of the most toxic of all metals. It is a cumulative poison with an estimated lethal dose of 8–12 mg/kg in humans. It is difficult to predict the outcome of thallium poisoning. With high exposure, death results very soon.

Regardless of the entry route, the major symptoms of thallium poisoning are gastrointestinal stress, neurological problems, and hair loss. Hair loss throughout the body is common and begins a little over a week after exposure. Gastrointestinal symptoms include abdominal pain and bleeding and ulceration of the colon. Neurological signs appear within a few days of exposure. Pain develops, fingers become numb, motor weakness is noted, and lower limbs may become paralyzed. The eyes become inflamed and retrobulbar neuritis with some loss of central vision follows. Intraocular hemorrhage, formation of cataracts, and optic nerve atrophy can occur.

Myocardia damage with EKG changes can result, and first hypotension and then hypertension can occur. Although thallium can concentrate in the kidneys, renal damage occurs in some cases (it is not generally extensive).

Thallium crosses the placental barrier and can be active in the last trimester of pregnancy. Loss of hair and nail deformation are noted in exposed newborns.

The ACGIH TLV-TWA for thallium (elemental and soluble compounds) is 0.1 mg/m^3 with a skin exposure warning.

Clinical Management

For acute exposure, ipecac should be administered and lavage performed. Charcoal binds the thallium ion. Potassium diuresis somewhat helps excretion but may have side effects. Hemodialysis is also recommended with potassium administration. Chelating agents are not effective. Since calcium metabolism is disturbed, supplementary calcium is indicated.

Animal Toxicity

Among animal species, the toxicity of thallium acetates, nitrates, and sulfates varies. In rats, the LD_{50} ranges from 15 to 30 mg/kg. The oxide is slightly less toxic

(LD_{50}, 70 mg/kg). Thallium has not been shown to be carcinogenic, although rats that were chemically exposed developed papillomas and exhibited inflammatory proliferation in the forestomach. Thallium causes malformation in chicks; however, teratological studies in animals produced ambiguous results.

—*Arthur Furst and Shirley B. Radding*

Related Topics

Metals
Sensory Organs

Theophylline

- ◆ CAS: 58-55-9
- ◆ SYNONYMS: 1,3-Dimethylxanthine; anhydrous theophylline; Elixophyllin SR; Somophyllin; Theophyl; Theolair; Slo-Bid; Slo-phyllin; Theodur. Aminophylline is the ethylenediamine salt of theophylline.
- ◆ PHARMACEUTICAL CLASS: A naturally occurring methylxanthine derivative structurally related to caffeine
- ◆ CHEMICAL STRUCTURE:

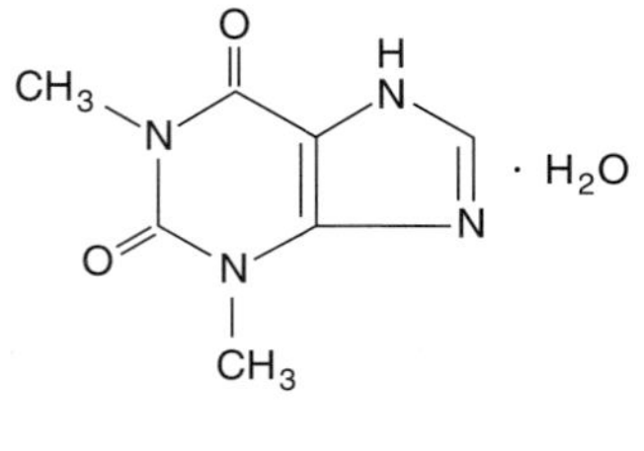

Uses

Theophylline is used as a bronchodilator in the treatment of asthma and reversible bronchospasm associated with chronic bronchitis and emphysema. Unlabeled use includes treatment of sleep apnea in neonates.

Exposure Pathways

Ingestion of sustained-release products is the most common route of both accidental and intentional exposure to theophylline. Theophylline is available in oral and intravenous dosage forms. Aminophylline is available in oral, rectal, and intravenous dosage forms.

Toxicokinetics

In therapeutic oral dosing, theophylline is well absorbed, producing peak serum levels in 2 hr. However, overdose with the commonly available sustained-release formulations produces a delayed absorption pattern, with peak levels as late as 16 hr. The matrices of these sustained-release formulations may agglutinate, with the potential to form pharmacobezors, further altering and delaying the absorption phase. In adults and children, theophylline is metabolized in the liver by oxidation and N-demethylation, producing 3-methylxanthine, 1,3-dimethyluric acid, and 1-methyluric acid. In premature neonates, minimal biotransformation occurs, with the main metabolite being caffeine. The average volume of distribution is 0.45 liters/kg. Protein binding is 40%. The elimination half-life varies by age. The average half-lives by age are as follows: adults, 6 or 7 hr; children 6 months to 13 years, 3.5–4 hr; children less than 6 months, 7 hr; and neonates, 20 hr.

Mechanism of Toxicity

The mechanism of action is unknown. Suggested theories of action include increased cellular cyclic adenosine monophosphate levels via inhibition of phosphodiesterase, increased turnover of monoamines in the central nervous system, inhibition of prostaglandins, and antagonism of adenosine receptors. Theophylline causes a release of endogenous catecholamines. There is a positive inotropic and dose-dependent chronotropic response. Hypokalemia, hypercalcemia, and hyperglycemia are caused by a mechanism regulated by the β-adrenergic system. Methylxanthines are weak diuretics by inhibition of renal tubular sodium resorption.

Human Toxicity

Theophylline has a narrow therapeutic index, with 12–25% of overdose patients developing serious or life-threatening symptoms. Age >60 years and chronic use are risk factors for increased morbidity and mortality.

In acute overdose, peak serum levels >100 μg/ml may be predictive of arrhythmias and seizures. In chronic overdose, peak serum levels are poorly correlated with incidence of life-threatening events but levels >40 μg/ml are suggestive of increased risk. The use of sustained-release formulations and the presence of pharmacobezors in the gut may make it difficult to determine peak serum levels. Sinus tachycardia is the most common cardiac sign of theophylline toxicity. Ventricular and supraventricular tachycardia, ectopic beats, hypotension, and cardiac arrest may occur. Metabolic acidosis, hypokalemia, hypercalcemia, and hyperglycemia may be seen. Tremulousness and agitation frequently occur. Intractable seizures may occur in severe intoxications, probably secondary to adenosine receptor antagonism in the brain. Onset of seizures is a poor prognostic indicator. Persistent vomiting is commonly seen and may interfere with attempts at therapy.

Clinical Management

Basic and advanced life-support measures should be utilized as necessary. Activated charcoal effectively adsorbs theophylline and should be employed in both acute and chronic overdoses. Multiple-dose activated charcoal (MDAC) has been shown to significantly increase drug clearance and reduce serum theophylline levels. In most cases, it is the mainstay of therapy. If persistent vomiting interferes with the administration of MDAC, antiemetics and ranitidine may be effective. Ventricular dysrhythmias may respond to lidocaine. Tachyarrhythmias as well as ventricular dysrhythmias unresponsive to lidocaine may respond to beta blockers or verapamil. Beta blockers should be used with caution in persons with a history of asthma. Seizures should be treated with benzodiazepines or phenobarbital. Intractable seizures may require pentobarbital. Phenytoin is ineffective in theophylline-induced seizures. Extracorporeal removal (ECR) may improve outcome if instituted before the onset of life-threatening symptoms. Hemoperfusion with a charcoal cartridge is more effective than hemodialysis but is less available. ECR should be considered in acute overdose patients with levels >100 μg/ml, patients older than 60 years with levels >50 μg/ml, and chronic overdose patients with levels >40–60 μg/ml, especially in the elderly. Due to the routine use of sustained-release preparations, early serum level measurements may not be representative of the peak level.

Animal Toxicity

Methylxanthines are not commonly used in animals. Limited information on toxicity exists. Tachyarrhythmias, hypotension, and seizures may be seen.

—*Henry A. Spiller*

Thiamine

♦ CAS: 67-03-8

♦ SYNONYMS: Vitamin B₁; aneurine hydrochloride; thiamine hydrochloride; thiadoxine; thiamol; Vitamin B hydrochloride

♦ PHARMACEUTICAL CLASS: Water-soluble vitamin

♦ MOLECULAR FORMULA: $C_{12}H_{17}N_4OS \cdot ClH \cdot Cl$

♦ CHEMICAL STRUCTURE:

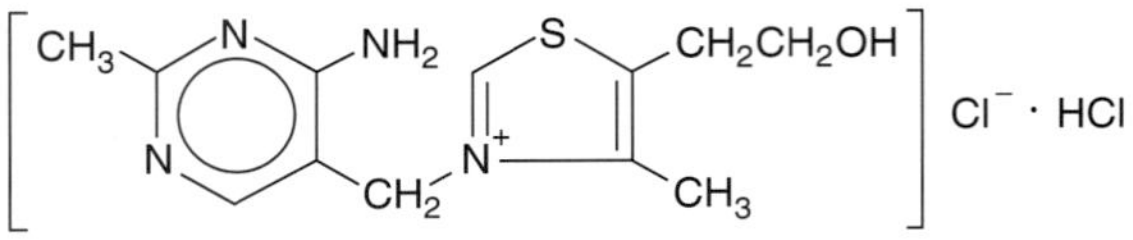

Uses

Thiamine is a nutritional supplement used during periods of deficiency (beriberi). Thiamine needs increase during diseases of the small intestine, malabsorption, congenital metabolic dysfunction, liver disease, alcoholism, and during pregnancy and lactation.

Exposure Pathways

Routes of exposure are oral, intravenous, and intramuscular. Dietary sources include cereal grains, the hull of rice, yeast, peas, beans, pork, and beef.

Toxicokinetics

Thiamine is readily absorbed from the gastrointestinal tract mainly in the duodenum. It is hepatically metabolized and widely distributed to almost all body tissue. Thiamine is renally excreted almost entirely as metabo-

lites. Excess thiamine (beyond daily body need) is excreted unchanged and as metabolites in the urine.

Mechanism of Toxicity

Unknown

Human Toxicity

Acute toxic effects are not expected even with doses of 50–100 times the recommended daily allowance; however, hypersensitivity reactions have been reported. Chronic large doses may cause headache, irritability, insomnia, weakness, tremors, and tachycardia.

Clinical Management

Acute ingestions seldom require treatment. Chronic excessive use should be discontinued and any toxic effects treated symptomatically.

Animal Toxicity

Acute toxicity not expected. It would be unlikely for animals to be given chronic thiamine overdoses. Very large parenteral doses of thiamine have produced neuromuscular and ganglionic blockage in animal studies.

—*Denise L. Kurta*

Thiazide Diuretics

♦ REPRESENTATIVE COMPOUNDS: Available thiazide diuretics (United States) include bendroflumethiazide, benzthiazide, chlorothiazide, chlorthalidone, hydrochlorothiazide, hydroflumethiazide, indapamide, methychlorthiazide, metolazone, polythiazide, quinethazone, and trichlormethiazide.

♦ SYNONYMS: Bendroflumathiazide—Naturetin; benzthiazide—Exna; chlorothiazide—Diuril, Sodium Diuril; chlorthalidone—Hygroton; hydrochlorothiazide—HCTZ, Esidrix, HydroDIURIL; hydroflumethiazide—Diucardin,

Saluron; indapamide—Lozol; methychlorthiazide—Enduron, Aquatensen; metolazone—Zoroxolyn, Mykrox; polythiazide—Renese; quinethazone—Hydromox; trichlormethiazide—Metahydrin, Naqua

◆ PHARMACEUTICAL CLASS: Diuretic; antihypertensive

◆ CHEMICAL STRUCTURE:

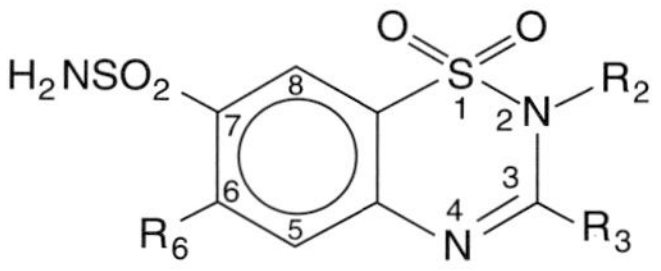

Uses

Thiazide diuretics are used in the management of edema, management of hypertension, treatment of diabetes insipidus, prophylaxis of renal calculus formation, and treatment of electrolyte disturbances associated with renal tubular acidosis.

Exposure Pathway

Ingestion is the most common route of both accidental and intentional exposure to the thiazide diuretics. They are available in an oral dosage form. Chlorothiazide is also available in a parenteral dosage form.

Toxicokinetics

Thiazides are absorbed in varying degrees from the gastrointestinal tract. The onset of action occurs within 2 hr with peak effects occurring in 3–6 hr. Following intravenous administration of chlorothiazide, the onset of action occurs within 15 min with a peak effect occurring in 30 min and a duration of 2 hr. The duration of effect following a single oral dose of a thiazide is dependent on the rate of excretion, which is lowest for chlorothiazide and hydrochlorothiazide (6–12 hr) and highest for chlorthalidone (24–72 hr). Thiazides are not metabolized to a significant extent. The volume of distribution varies with the individual agent—from as low as 0.83 liters/kg for hydrochlorothiazide to as high as 4.17 liters/kg for chlorthalidone. Thiazides cross the placenta and are distributed into breast milk. They are primarily excreted in the urine unchanged by glomerular filtration and active secretion. Excretion may be delayed in patients with uncompensated congestive heart failure or impaired renal function. The elimination half-life is variable: 1.5 and 2.5 hr for chlorothiazide and hydrochlorothiazide, respectively, and up to 44 hr for chlorthalidone.

Mechanism of Toxicity

Thiazides act on the cortical diluting segment of the nephron or early distal tubule where they inhibit the reabsorption of sodium, chloride, and water. The resulting diuresis causes a decrease in extracellular fluid volume and plasma volume. As the reabsorption of potassium is also inhibited, hypokalemia may develop. Electrolyte abnormalities may precipitate cardiac arrhythmias. Hypotensive activity following chronic use may occur partly by direct arteriolar dilation. Thiazide actions at pancreatic sites can induce hyperglycemia. The mechanism of central nervous system depression is unknown.

Human Toxicity

Overdose may result in diuresis, gastrointestinal upset, lethargy, and coma. Hypokalemia and hypomagnesemia may occur and may precipitate cardiac arrhythmias; hyponatremia may occur in older age groups. Diuretic blood levels are not clinically useful in monitoring for toxicity; a minimum toxic or lethal dose has not been defined. Ingestion of amounts significantly exceeding maximum daily doses have been tolerated in children.

With chronic exposure, hypochloremic alkalosis may occur. Hyperuricemia, hyperglycemia, hyperlipidemia, hypokalemia, and hyponatremia may also occur.

Clinical Management

Gastrointestinal decontamination procedures should be used as deemed appropriate to the patient's level of consciousness and history of ingestion. Cathartics should be avoided because of their ability to promote fluid and electrolyte losses. Basic and advanced life-support measures should be utilized as necessary. There are no antidotes. No systematic studies have been conducted to evaluate the use of hemodialysis or hemoperfusion as means of enhancing elimination. Fluid and electrolyte imbalances should be evaluated and corrected. Potassium losses should be replaced while continually monitoring the EKG. Sodium and chloride losses should be replaced slowly to prevent neurological damage.

Animal Toxicity

The LD$_{50}$ for chlorothiazide in the dog is >1 g/kg.

—*Elizabeth J. Scharman*

Related Topic

Gastrointestinal System

Thioacetamide

- CAS: 62-55-5
- SYNONYMS: Acetothioamide; ethanethioamide; thiacetamide; TAA
- MOLECULAR FORMULA: C$_2$H$_5$NS

Exposure Pathways

Inhalation, dermal, oral, and ocular are possible routes of exposure.

Toxicokinetics

Thioacetamide is readily absorbed through the skin. The initial oxidative conversion is mediated by rat liver microsomes and a flavin-containing monooxygenase system to form thioacetamide-S-oxide, a proximate metabolite. The mechanism for a second oxidative state to the reactive dioxygenated sulphene intermediate has not been resolved. The highest levels of radioactivity were observed in the liver following oral administration of [^{3}H]thioacetamide (in diet) to male rats. Approximately 80% of [^{35}S]thioacetamide was excreted in the urine of rats within 24 hr of intravenous administration.

Mechanism of Toxicity

Thioacetamide acts as an indirect hepatotoxin and causes parenchymal cell necrosis. The potential role of thioacetamide in the initiation phase of carcinogenesis may be associated with an increase in nucleoside triphosphate activity in the nuclear envelope with a cor-

responding increase in RNA transport activity. Alterations in the transport phenomenon of nuclear RNA sequences are considered an early response to carcinogens.

Human Toxicity: Acute

Inhalation may cause irritation of the respiratory tract characterized by rhinitis, tracheitis, and pulmonary edema. High concentrations may result in central nervous system depression and death from respiratory paralysis. Skin contact may cause irritation and ocular contact may be associated with palebral edema, keratitis, and corneal defects. Ingestion may cause nausea, vomiting, headache, convulsions, and unconsciousness.

Human Toxicity: Chronic

Prolonged exposure by inhalation may result in headache, irritability, nausea, and vomiting. Repeated contact with skin may cause dermatitis and prolonged ocular contact may cause conjunctivitis. NTP lists thioacetamide as an anticipated carcinogen.

Clinical Management

Basic and advanced life-support measures should be utilized as necessary. Gastric decontamination may be accomplished by lavage or emesis. Sodium bicarbonate solution should be used to reduce acidity. The use of amyl nitrite by inhalation for 15–30 sec of every minute may be indicated in severe poisonings.

Animal Toxicity

There is sufficient evidence of carcinogenicity in animals. Repeated dietary administration has produced liver cell tumors in mice and bile duct and liver tumors in rats. Cirrhosis has also been observed in both species.

—*Shayne C. Gad*

Thioxanthenes

- REPRESENTATIVE COMPOUNDS: Chlorprothixene (CAS: 113-59-7); thiothixene; flupenthixol (CAS: 2413-38-9); zuclopenthixol (CAS: 53772-83-1)

SYNONYMS: Chlorprothixene—α-2-chloro-9-(3-dimethylaminopropylidene) thiaxanthen, Taractan; thiothixene—*N,N*-dimethyl-9-(3-(4-methylpiperazin-1-yl) propylidene) thioxanthene-2-sulfonamide, Navane; flupenthixol—2-(4-(3-(2-trifluoromethylthioxanthen-9-ylidene) propyl) piperazin-1-yl)ethanol dihydrochloride;

Zuclopenthixol—*Z*-2-(4-(3-(2-chloro-10*H*-dibenzo-(*b,e*) thiin-10-ylidene)propyl) piperazin-1-yl)ethanol

◆ PHARMACEUTICAL CLASS: Neuroleptic agent, antipsychotic, major tranquilizer

◆ CHEMICAL STRUCTURE:

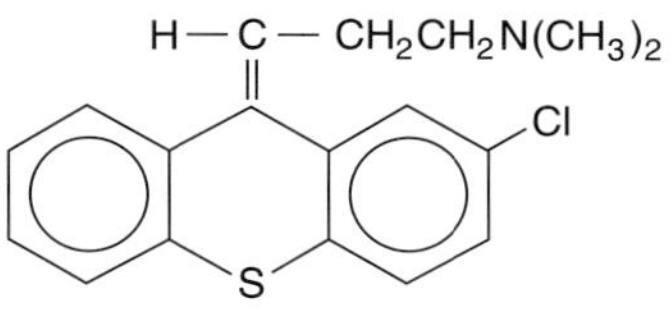

Chlorprothixene

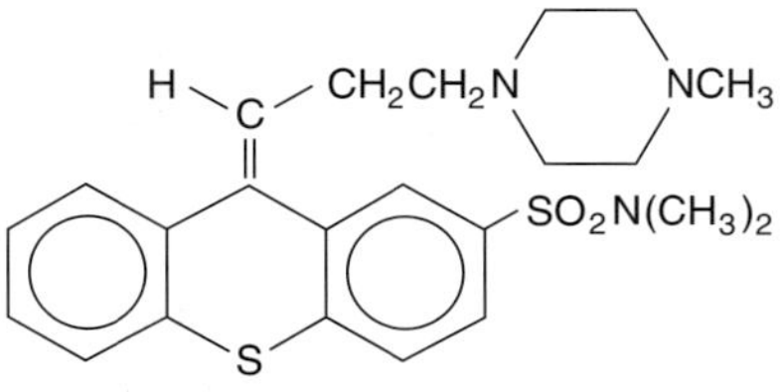

Thiothixene

Uses

Thioxanthenes are used in the treatment of psychosis, including schizophrenia, senile psychosis, pathological jealousy, and borderline personality disorder. Other uses include the treatment of pain, postoperative neuralgia, sedation, anxiety neurosis, childhood behavior problems, and depression.

Exposure Pathways

Thioxanthenes are available in injectable and oral dosage forms. The principal exposure pathway is intentional ingestion by adults or accidental ingestion by small children.

Toxicokinetics

Thioxanthenes are readily but incompletely absorbed due to first-pass metabolism in the gut wall. Peak absorption occurs in 1 to 2 hr. Thioxanthenes are extensively metabolized in the liver through glucuronic acid conjugation, N-dealkylation, and sulfoxidation. Thioxanthenes are widely distributed throughout the body, including the central nervous system. They are highly protein bound (>99%), with a volume of distribution ranging from 11 to 23 liters/kg. The main metabolites are excreted in both the urine and feces. There is some enterohepatic circulation. The elimination half-life ranges from 8 to 12 hr up to 34 hr.

Mechanism of Toxicity

Thioxanthenes work primarily by blocking postsynaptic dopamine-mediated neurotransmission by binding to dopamine (DA-1 and DA-2) receptors. In addition to significant antidopaminergic action, the thioxanthenes also possess weak anticholinergic and serotonergic blockade, moderate α-adrenergic blockade, and quinidine-like effects.

Human Toxicity

Clinical signs of toxicity most frequently reported include extrapyramidal effects, sedation, coma, and rarely hypotension and cardiac arrhythmias. Neuroleptic malignant syndrome has been reported after therapeutic use and acute intoxication. The most commonly reported dystonic reactions include akathisias, stiff neck, stiff or protruding tongue, and tremor. Anticholinergic effects, including dry mouth, blurred vision, and tachycardia, may occur. Cardiac effects include prolonged Q-T interval and mild hypotension. Hypokalemia has also been noted. Seizures are rarely seen. Adverse reactions following therapeutic use include tardive dyskinesia, sedation, dysphoria, photosensitivity, anorexia, nausea, vomiting, constipation, diarrhea, and dyspepsia.

Clinical Management

All basic and advanced life-support measures should be implemented. Gastric decontamination should be performed. Thioxanthenes are readily absorbed by activated charcoal. Aggressive supportive care should be instituted. Dystonic reactions respond to intravenous benztropine or diphenhydramine. Oral therapy with diphenhydramine or benztropine should be continued for 2 days to prevent recurrence of the dystonic reaction. For patients suffering from neuroleptic malignant syndrome, dantrolene sodium and bromocriptine have been used in conjunction with cooling and other supportive measures. Arrhythmias should be treated with lidocaine or phenytoin. Diazepam is the drug of choice for seizures. Hemodialysis and hemoperfusion have not been shown to be effective.

Animal Toxicity

Reported signs of toxicity in animals have included sedation, dullness, photosensitivity, weakness, anorexia, fever, icterus, colic, anemia, and hemoglobinuria. Treatment consists of gastric decontamination and aggressive supportive care.

—Douglas J. Borys

Thiram

- ◆ CAS: 137-26-8
- ◆ SYNONYM: Tetramethylthiuram disulfide
- ◆ CHEMICAL CLASS: Dithiocarbamate fungicide
- ◆ CHEMICAL STRUCTURE:

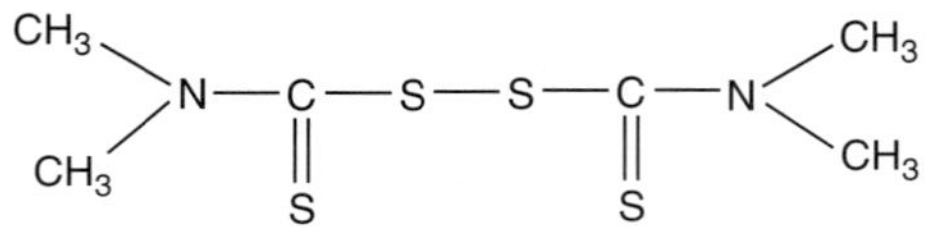

Use

Thiram is used for foliage and seed treatment to prevent fungal growth.

Exposure Pathways

Inhalation, ingestion, and dermal exposures are all possible.

Toxicokinetics

Absorption of thiram across the skin and gastrointestinal lining is slower than absorption of the organochlorine and ester insecticides. Carbon disulfide is produced by biotransformation of thiram. Thiram is well distributed throughout the body. Thiram and metabolites are primarily excreted in the urine.

Mechanism of Toxicity

There are several biochemical mechanisms reported for central nervous system (CNS) toxicity of disulfiram. Some sulfhydryl enzymes (e.g., amino acid oxidases) are inhibited by dithiocarbamates. Small amounts of thiram are converted to carbon disulfide, a known neurotoxicant. Disulfiram is an inhibitor of dopamine-β-hydroxylase and thereby interferes with norepinephrine biosynthesis: This can lead to tissue norepinephrine depletion and dopamine accumulation. As with other dithiocarbamates, thiram inhibits acetaldehyde oxidase and interferes with the metabolism of ethanol. Thiram is not an acetylcholinesterase inhibitor.

Human Toxicity: Acute

Contact dermatitis and sensitization have been reported with thiram exposure. Exposure to thiram may cause weakness, ataxia, and hypothermia. Nausea, vomiting, diarrhea, and irritation to the eyes, nose, and skin may also be noted.

Human Toxicity: Chronic

Sensitization and cross-sensitization to other dithiocarbamates (e.g., disulfuram) may occur. The ethyl analog of thiram (i.e., thiuram) has reportedly caused peripheral neuropathy.

The ACGIH TLV-TWA is 5.0 mg/m³. The LC_{Lo} (inhalation) humans is 30 mg/m³.

Clinical Management

There are no specific antidotes for poisoning by thiram. Basic life-support measures should be instituted. Oxygen therapy is effective in relieving the distress of antabuse-like reactions. Patients should be moved to fresh air and monitored for respiratory distress. Emesis is most effective if initiated within 30 min of ingestion. Activated charcoal may be used to block absorption. Intravenous fluids are useful in restoring extracellular fluid volume.

Animal Toxicity

Thiram is a compound of moderate to low acute toxicity. The oral LD_{50} values for thiram exhibit a broad range (about 600–4000 mg/kg). Dermal LD_{50} values in laboratory species are $\geqq 2$ g/kg. Blood disorders (e.g., leukopenia and reduced hematopoiesis) were reported in rabbits following repeated exposures to thiram. Thiram is mutagenic in some assays but was not carcinogenic in rodent carcinogen bioassays even at maximum tolerated doses. Thiram can cause peripheral neuropathy in rodents, presumably because of carbon disulfide production *in vivo*. High doses of thiram produced embryotoxicity and teratogenic effects in mice.

—*Janice Reeves and Carey Pope*

Threshold Limit Value

A threshold Limit Value (TLV) is the level of an airborne substance, physical agent, or other environmental conditions to which nearly all workers may be exposed routinely without adverse health effects. TVLs are based on industrial experience with these agents and on experimental exposures of humans and animals.

TLVs are not precise boundaries between safe and hazardous conditions. Individual response to an exposure varies greatly and may be affected by factors such as age, gender, smoking, illness, genetic makeup, personal hygiene, and the use of alcohol, medications, and other drugs. Therefore, no single exposure level will protect all workers. Also, TLVs derived for the workplace are not appropriate limits for protecting the general population, which is believed to have a higher proportion of susceptible persons than those actively employed.

Historically, the term "threshold limit value" was used synonymously with "maximum acceptable concentration," but current usage in the United States refers to the exposure guidelines published by the American Conference of Governmental Industrial Hygienists (ACGIH, Cincinnati, Ohio). The acronym TLV generally refers to the guideline levels set by the ACGIH and is a registered trademark of the ACGIH. The ACGIH publishes annually a listing of TLVs for chemical sub-

stances, including dusts, gases, and vapors, and for physical agents, such as heat, cold, hand–arm vibration, ionizing radiation, noise, radiofrequency and microwave radiation, magnetic fields, and light. The ACGIH Biological Exposure Indices (BEIs), reference values for chemical markers of overall exposure measured in biological specimens from a worker (e.g., blood, urine, and exhaled air), are listed with the TLVs. TLVs and BEIs are derived by independent committees of government scientists in consultation with experts from the private sector.

The TLVs for airborne chemical substances may be given for several exposure durations. The TLV-time-weighted average (TLV-TWA) is the TLV for the time-weighted average concentration of a contaminant over an 8-hr period. The TLV-TWA assumes an 8-hr workday and a 40-hr workweek. The TLV-short-term exposure limit (TLV-STEL) is a 15-min time-weighted average concentration which should not be exceeded at any time during the workday. The TLV-ceiling (TLV-C) is an exposure level which should never be exceeded, even for a very short time period.

The proper application of the TLVs requires an understanding of the information from which each was derived. This important background is available in the *Documentation of the Threshold Limit Values and the Biological Exposure Indices*.

Further Reading

American Conference of Governmental Industrial Hygienists (ACGIH). *Threshold Limit Values for Chemical Substances and Physical Agents and Biological Exposure Indices*. ACGIH, Cincinnati, OH.
American Conference of Governmental Industrial Hygienists (ACGIH). *Documentation of the Threshold Limit Values and the Biological Exposure Indices*. ACGIH, Cincinnati, OH.
Cook, W. A. (1956). Symposium on threshold limits, present trends in MACs. *Am. Ind. Hygiene Assoc. Q.* **17**, 273–274.
Shrenk, H. H. (1947). Interpretation of permissible limits. *Am. Ind. Hygiene Assoc. Q.* **8**, 55–60.
Sterner, J. H. (1955). Threshold limits—A panel discussion. *Am. Ind. Hygiene Assoc. Q.* **16**, 27–39.

—*Charles Feigley*

Related Topics

American Conference of Governmental Industrial Hygienists

Thyroid Extract

- CAS: 8028-36-2
- SYNONYMS: Dry thyroid; dessicated thyroid; Thyreoidin; Thyroid Strong
- CHEMICAL CLASS: Natural hormone that provides levothyroxine (T_4) and liothyronine (T_3)

Uses

Thyroid extract is used in the treatment of hypothyroidism, myxedema, and cretinism. It is also used as a diagnostic agent and for suppression of pituitary thyroid-stimulating hormone. Synthetic derivatives are preferred because of uniform potency.

Exposure Pathway

Ingestion is the route of exposure in both accidental and intentional exposures.

Toxicokinetics

Thyroid extract is partially absorbed from the gastrointestinal tract. Up to 79% of a therapeutic dose is absorbed. Approximately 99% is protein bound. Thyroid extract contains both levothyroxine (T_4) and liothyronine (T_3). T_4 is deiodinated in the liver, kidney, and tissues to form active T_3 and inactive T_3. The half-life of T_4 is 5.3–9.4 days. T_3 has a half-life of 2.5 days.

Mechanism of Toxicity

Thyroid hormones are necessary for metabolism, growth, and development. The main effect of thyroid hormones is increased metabolic rate, increased oxygen consumption, and increased metabolism of carbohydrates. Forty percent of T_4 is converted to T_3. T_3 is three to five times more toxic than T_4. T_3 increases aerobic mitochondrial function, causing an increased rate of utilization of high-energy phosphates. This causes a stimulation of myosin adenosine triphosphatase (ATPase) and reduces tissue acetic acid. The toxic effects are extensions of the pharmacologic properties of T_4 and T_3.

Human Toxicity: Acute

Ingestion of small amounts of thyroid extract produces few symptoms, if any. Symptoms of thyroid toxicity include increased heart rate, nausea, vomiting, diarrhea, restlessness, and fever. Fever is generally an early symptom of toxicity, while restlessness and tachycardia develop days after exposure.

Human Toxicity: Chronic

The concern with chronic exposure is the development of thyrotoxicosis. The development of thyrotoxicosis with acute exposure is rare. Thyrotoxicosis should be suspected in patients exhibiting tachycardia, cardiac arrhythmias, hypertension, tremors, and seizures. Coma and circulatory collapse can be seen in severe thyrotoxicosis.

Clinical Management

Basic and advanced life-support measures should be utilized as needed. Appropriate gastric decontamination should be performed such as emesis, lavage, and/or activated charcoal. EKG and blood pressure monitoring should be utilized in severe cases. Measurements of T_3 and T_4 levels should be obtained frequently in large ingestions until levels have normalized. Propranolol can be used to treat hypertension, tachycardia, and cardiac arrhythmias. Extracorporeal means of elimination are ineffective.

Animal Toxicity

Animals that ingest thyroid extract are at risk for thyroid toxicity. Signs of toxicity include vomiting, diarrhea, tachycardia, tachypnea, a decreased level of consciousness, and restlessness.

—*Bridget Flaherty*

Related Topic

Endocrine System

Tin (Sn)

- ◆ CAS: 7440-31-5
- ◆ SELECTED COMPOUNDS: Stannous oxide, SnO (CAS: 21651-19-4); triethyltin, $Sn(C_2H_5)3$ (CAS: 29680-38-4); triphenyltin hydroxide, $Sn(C_6H_6)OH$ (CAS: 76-87-9)
- ◆ CHEMICAL CLASS: Metals

Uses

Tin has many uses, including protective coatings, tin plate, and cans. Alloys such as bronze, brass, and solders also contain tin. Minor uses include dyes, ceramics, flame retardants, and pigments. Stannous fluoride is often used in toothpaste to prevent cavities. Organotin compounds have been used as antifouling agents, as pesticides, and as stabilizers in plastics.

Exposure Pathways

Inhalation, dermal contact, and ingestion are all potential exposure pathways. In industrial areas, tin is inhaled from polluted air. Organotin compounds can be absorbed dermally, stannous fluoride can be swallowed from toothpaste, and elemental tin may be ingested with food. Large amounts of tin must be ingested before levels of absorbed tin are detectable.

Toxicokinetics

Inorganic tin compounds are not easily absorbed from the gastrointestinal tract. Inhaled tin first resides in the lungs and then is transferred to the liver and kidneys. Absorbed compounds are carried by the red blood cells. Inorganic tin is mainly excreted in urine. Organic tin compounds are more easily absorbed from the gastrointestinal tract and skin, concentrated in the blood and urine, and excreted in the bile.

There is little information on the effect of tin on enzymes. Organic tin compounds can inhibit the hydrolysis of adenosine triphosphate, resulting in uncoupling of oxidative phosphorylation.

Mechanism of Toxicity

Organic tin compounds inhibit mitochondrial oxidative phosphorylation and brain glucose oxidation. Very little data are available on inorganic tin.

Human Toxicity

Inhaled tin leads to a mild pneumoconiosis known as stannosis. Orally absorbed inorganic tin produces nonspecific symptoms, including nausea, diarrhea, muscle twitching, and even paralysis. Organic tin compounds are much more toxic than inorganic tin compounds. Tetraethyltin is converted to triethyltin, which is a potent skin irritant and neurotoxin. It produces depression with loss of memory and aggressive behavior. It also produces cerebral edema and encephalopathy. Triphenyltin is an immunodepressant. Some organic tin compounds are unusually toxic to the central nervous system.

The ACGIH TLV-TWA for tin metal, tin oxide, and inorganic compounds (except SnH_4) is 2 mg/m³. The TLV for tin oxide is 1 mg/m³. The TLV for organic tin compounds is 0.1 mg/m³ with a skin exposure warning.

Clinical Management

Supportive measures must be taken; there is not a specific antidote or chelating agent for tin.

Animal Toxicity

Currently, there is not any conclusive evidence that tin compounds are mutagenic. These compounds are neither carcinogenic nor teratogenic. Ascribed tin concentrates in the kidneys, liver, and bone of experimental animals.

—Arthur Furst and Shirley B. Radding

Related Topics

Metals
Organotins
Pollution, Water

Tissue Repair

Introduction

The exposure to a variety of chemicals, including drugs and environmental or occupational pollutants, can cause tissue damage. When injury and associated acute inflammation response result in cell necrosis and extracellular matrix damage, tissue repair takes place, replacing dead cells with normal tissue.

Tissue repair occurs in well-characterized steps which mainly involve regeneration of specialized cells by proliferation of the surviving ones, formation of granulation tissue, and tissue remodeling. The relative contribution of each of these processes to tissue repair depends on the organ injured and the nature as well as the severity of the injury. In cutaneous repair, if damage is moderate, there is complete restoration of the epithelium, which is referred to as restitutive repair with no sequelae. However, in skeletal muscle damage, for example, specialized cells cannot regenerate and the repair process is characterized by the deposition of connective tissue, resulting in loss of organ function. This pathological repair, in contrast to restitutive repair, occurs when there is chronic injury and the amount of fibrosis is important. This is well exemplified by the response of the liver to chronic exposure to some chemical agents (e.g., carbon tetrachloride). Here, there is persistent injury and a chronic inflammatory response that leads to fibrosis and eventually to cirrhosis. In this section, the general aspects and mechanisms of tissue repair as studied mainly in wound healing are reviewed, and repair in specialized tissues, such as the liver, kidney, and lung, are described since they are frequently sites of chemically induced injury.

Basic Mechanisms of Tissue Repair

Generalities

Tissue injury, which can create tissue loss, gives rise to the tissue repair process that takes place in well-characterized steps. After clot formation, inflammatory cells invade the injured tissue; then, fibroblasts migrate, proliferate, and synthesize extracellular matrix components, participating in the formation of granulation tissue. Finally, specialized cells regenerate by proliferation of surviving cells, and tissue remodeling implicates extracellular matrix degradation, decrease of cellularity, and development of the scar. Repair by scar formation develops when resolution fails to occur in an acute inflammatory process, when there is ongoing tissue necrosis in chronic inflammation, and when parenchymal cell necrosis cannot be repaired by regeneration. Indeed, the mechanisms of healing depend on the regenerative ability of damaged parenchymal cells.

The cells of the body can be divided into three groups—labile, stable, and permanent—on the basis of their mitotic capacity. Labile cells normally divide actively throughout life to replace cells that are being continually lost from the body. Examples of labile cells include basal epithelial cells of all epithelial linings (e.g., epidermis, hair follicle cells, and epithelium of ducts) and hematopoietic stem cells in the bone marrow. Injury to a tissue containing labile parenchymal cells is followed by rapid regeneration. Stable cells typically have a long life span and are therefore characterized by a low rate of division. The parenchymal cells of most solid glandular organs—e.g., liver (hepatocytes), kidney (renal tubular epithelium), and lung (alveoli)—and mesenchymal cells (e.g., fibroblasts, endothelial cells, and adipocytes) are examples of stable cells. Regeneration in tissues composed of stable cells requires that enough viable tissue remains to provide a source of parenchymal cells for regeneration and that there be an intact connective tissue framework in the area of necrosis. Although stable cells have a long resting phase, they can divide rapidly upon demand. The liver, a tissue with stable cells (less than one mitosis for every 15,000 cells), regenerates after a loss of 75% of its mass. After acute liver injury in experimental animals and in humans, a significant zonal necrosis of hepatocytes can occur associated with partial collapse of the matrix and development of scar tissue. However, the regenerative response of hepatocytes enables restoration of the liver architecture. Finally, permanent cells have no capacity for mitotic division in postnatal life. Examples of such cells are neurons in the central and peripheral nervous systems and skeletal and cardiac muscle cells. Injury to permanent cells is always followed by scar formation; thus, no regeneration is possible (Fig. T-1).

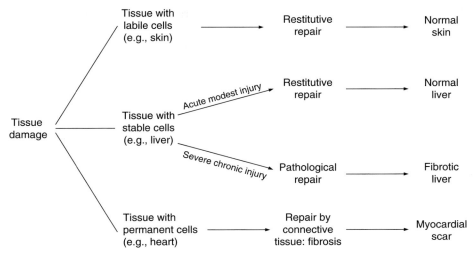

FIGURE T-1. *Damage and repair in tissues with labile cells, stable cells, and permanent cells.*

During cutaneous wound repair, different situations may be observed: The fetus possesses the unique ability to heal skin wounds without scar formation; after birth, the repair process ends with scar tissue; or, in some situations (e.g., second- and third-degree burn injury), healing may be complicated by the development of hypertrophic scars. In organs, when the tissue injury is not chronic, and when tissue loss does not destroy too many specialized cells of the organ, a scar that does not impair function is formed. However, the stimuli inducing cell necrosis and granulation tissue formation can remain, and then organ fibrosis develops. In summary, it is evident that the outcome is affected by the nature, severity, and duration of the injury. A common response after injury, at many sites, includes platelet aggregation and fibrin deposition with formation of a blood clot. This is followed by an influx of inflammatory cells, essentially mononuclear cells and granulocytes. The inflammatory exudate, including fibrin, blood, and necrotic tissue, is liquefied by lysosomal enzymes derived from neutrophils and removed by lymphatics and/or macrophage phagocytosis. Platelets and inflammatory cells release many cytokines and growth factors, which induce the migration and the proliferation of fibroblasts; furthermore, granulation tissue formation requires an in-growth of new blood vessels.

Cell Migration and Proliferation

In response to injury, resident fibroblasts migrate with capillaries into the injured tissue. The specific factors inducing migration are not well-known and, *in vivo*, migration and proliferation are often linked, so it is difficult to clearly define the specific role of either process. It is well accepted, however, that platelet-derived growth factor (PDGF) and basic fibroblast growth factor (FGF) are potent chemotactic agents for fibroblastic cells. FGF is produced mainly by macrophages and endothelial cells. Connective tissue growth factor (CTGF), secreted by endothelial cells, is both mitogenic and chemotactic for smooth muscle cells and fibroblasts. Transforming growth factor-β_1 (TGF-β_1), which induces the production of CTGF, is able to induce migration of monocytes and fibroblasts by itself. Insulin-like growth factor-I (IGF-I) promotes the migration of endothelial cells and allows the neovascularization of the granulation tissue.

The platelet degranulation that takes place after injury results in the release of significant amounts of PDGF, TGF-β, epidermal growth factor (EGF), and IGF-I, as well as tumor necrosis factor-α (TNF-α), CTGF, and FGF—all potent mitogens for fibroblastic cells. Contradictory results have been published concerning the role of TGF-β but the discrepancies may be explained by differences in doses and in target cells used. TGF-β stimulates fibroblast division at low concentrations but stimulates differentiation (i.e., increased collagen and matrix production) at high concentrations. Basic FGF is also a potent endothelial cell mitogen. New blood vessel formation involves the initial destruction of capillary basement membranes, the migration and division of endothelial cells, and then reformation of capillary structures. Basic FGF stimulates all these activities in cultured endothelial cells, suggesting that it may play a central role in neovascularization and thus in the formation of the granulation tissue. Vascular endothelial growth factor acts synergistically with basic FGF to stimulate endothelial cell function. In skin, keratinocyte growth factor (KGF) secreted by fibroblasts is a specific, potent chemoattractant and mitogen for keratinocytes, with minimal effects on cells of mesodermal origin, unlike other members of the FGF family. KGF may constitute the dermal/epidermal signal stimulating reepithelialization.

In summary, a plethora of cytokines are released by platelets and other cells during the initial events after injury. Each cytokine may play different roles but cytokines are known to generally act in a coordinated sequence. Furthermore, it is necessary to keep in mind that complex interactions between cytokines and extracellular matrix components modify their reciprocal activities.

Myofibroblastic Differentiation

During the repair process, fibroblasts acquire the morphological and biochemical characteristics of smooth muscle cells. These modified fibroblasts, named myofibroblasts, express microfilaments bundles of actin, or stress fibers, which are thought to be the force-generating element involved in contraction, as in smooth muscle cells. Myofibroblasts are interconnected by gap junctions and are connected to the extracellular matrix by the fibronexus, a transmembrane complex involving intracellular microfilaments in apparent continuity with extracellular fibronectin fibers.

Moreover, myofibroblasts express cytoskeletal proteins, which are known to play a key role during the process of cell contraction. By using cytoskeletal protein expression, different phenotypes have been characterized which coexpress in addition to cytoplasmic actin isoforms: (a) vimentin; (b) vimentin and desmin; (c) vimentin and α-smooth muscle actin; (d) vimentin, desmin, and α-smooth muscle actin; (e) vimentin, α-smooth muscle actin, and smooth muscle myosin heavy chains; and (f) vimentin, α-smooth muscle actin, desmin, and smooth muscle myosin heavy chains. α-smooth muscle actin (α-SMA) is the actin isoform typical of contractile vascular smooth muscle cells and is expressed by virtually all myofibroblastic cell populations *in vivo*. Little is known about the mechanisms

leading to the development of fibroblastic features similar to those of smooth muscle cells and to their persistence in some pathological conditions. As with smooth muscle cells, extracellular matrix components and cytokines are good candidates for modulating fibroblast phenotype and cytoskeletal protein expression.

The extracellular matrix not only represents a structural support but also appears to play a central role as a source of signals influencing growth and differentiation of different cell types, including fibroblasts. For example, it is now well accepted that matrix components such as proteoglycans are involved in cell phenotype modifications. Interestingly, proteoglycans, mainly heparan sulfates, are produced by mast cells, and the presence of numerous mast cells has been associated with the repair process. In all fibroblastic populations tested, an increase in α-SMA expression has been observed after heparin treatment. Thus, proliferation and α-SMA synthesis by fibroblasts appear to be distinct phenomena during the formation and progression of granulation tissue. The action of heparin and its nonanticoagulant derivatives on the expression of α-SMA in fibroblasts both *in vitro* and *in vivo* requires the active replication of these cells and may be related to the phenotype changes which occur during the evolution of fibrocontractive diseases. We can assume that heparin facilitates the presentation of differentiation or maturation factors present in serum-to-cell receptors.

The role of macrophages in the first steps of granulation tissue formation is well-known, and several studies indicate that granulocyte/macrophage-colony stimulating factor (GM-CSF) is important in this process. *In vivo*, GM-CSF increases α-SMA expression in granulation tissue surrounding osmotic minipumps implanted subcutaneously and filled with GM-CSF. However, in cultured fibroblasts, GM-CSF cannot induce α-SMA expression, which indicates that it does not act directly *in vivo*. After GM-CSF local treatment, the appearance of α-SMA-rich myofibroblasts is preceded by a characteristic cluster-like accumulation of macrophages. This suggests that factors secreted by macrophages induce typical myofibroblastic differentiation with the induction of α-SMA expression.

Among factors inducing α-SMA, TGF-β is probably the most efficient, both *in vivo* and *in vitro*. Other cytokines and growth factors, such as PDGF, basic FGF, and TNF-α, despite their profibrotic activity, do not induce α-SMA in myofibroblasts. Contrary to TGF-β_1, interferon-γ (IFN-γ) decreases the α-SMA protein and mRNA expression *in vitro* and could be a good candidate for decreasing myofibroblastic activities during excessive scarring.

Extracellular Matrix Deposition

Among extracellular matrix components, different types of collagens, glycoproteins, and proteoglycans are involved in tissue repair mechanisms. In the early phases of healing, fibronectin (a glycoprotein) plays a key role in the formation of the granulation tissue. Fibronectin is chemotactic for fibroblasts and promotes the organization of endothelial cells into capillary vessels. The fibroblast is the main cell type implicated in the production of extracellular matrix components during the repair process.

Recently, it has been shown that α-SMA-expressing myofibroblasts are the main cells producing collagen during pulmonary fibrosis. Fibrillar collagens are synthesized by fibroblasts in the form of a precursor, procollagen. After posttranslational modifications (hydroxylation and glycosylation) and triple-helix assembly, precursor procollagen molecules are secreted from the cell. During extracellular processing of procollagen to collagen, the propeptides are removed from the major collagen triple-helical domain by specific endoproteinases.

Supramolecular aggregates of fibrillar collagen usually contain more than one type of collagen, and such heterotypic fibrils are arranged in different patterns in different tissues. Heterotypic fibrils containing type I, III, and/or V collagens are expressed in a number of tissues of mesenchymal origin. Fibrillar collagens polymerize to form fibrils that serve as stabilizing scaffolds in extracellular matrices. Young fibroblasts in granulation tissue form type III collagen that is later replaced by stronger cross-linked type I collagen. Fibrillar collagen fibers are flexible but inelastic and are responsible for much of the tensile strength of scar tissue. Basement membranes are delicate structures, found at the interface between cells and stroma, that contain collagen type IV, laminin, nidogen, and heparin sulfate proteoglycan.

Basement membranes are stable structures that, in normal circumstances, have a slow turnover. In certain situations, however, such as tissue repair, neovascularization requires rapid basement membrane degradation and reassembly. Whereas tensile strength is provided by members of the collagen family, the ability to recoil after transient stretching is provided by elastic fibers. Contrary to what is classically believed, it seems that elastin, in some situations, plays an important role during the early stages of tissue repair.

Extensive *in vitro* and *in vivo* studies have shown that a variety of cytokines are capable of modulating the metabolism of the extracellular matrix. Although some discrepancies are reported in the literature, it is accepted that interleukins (IL-1 and IL-4), EGF, basic FGF, PDGF, and IGF-I upregulate fibroblast collagen production.

The most extensively studied growth factor capable of increasing extracellular matrix deposition is TGF-β. This factor stimulates extracellular matrix deposition by mechanisms which involve both enhanced matrix synthesis and reduced matrix degradation. At least two members of the TGF-β family, TGF-β_1 and TGF-β_2, have been shown to enhance type I collagen gene expression. The elevation of type I collagen mRNA levels is accompanied by increased transcription of the genes by activation of the corresponding promoters. TGF-β enhances the steady-state levels of other collagen mRNAs, including types III and IV. In addition, TGF-β stimulates the production of fibronectin, elastin, and proteoglycans and enhances their incorporation into the extracellular matrix by mesenchymal cells in culture. TGF-β is also known to increase expression of integrins, which are the cell surface receptors through which various cells interact with extracellular matrix components, such as fibronectin and collagens. Furthermore, TGF-β blocks matrix degradation by decreasing the synthesis of proteases and increasing the synthesis of protease inhibitors. TGF-β also induces its own production by cells, thus amplifying its biological effects. It is clear that TGF-β is a key cytokine that initiates and terminates tissue repair and whose sustained production underlies the development of tissue fibrosis.

Remodeling and Scar Formation

Important cellular changes occur in the tissue repair process from granulation tissue to scar formation. As shown in cutaneous tissue, as the wound closes and evolves into a scar there is an important decrease in cellularity and, in particular, myofibroblasts disappear. The process(es) responsible for this cellular loss is not well defined but it has been shown that, in late phases of wound healing, many myofibroblasts show changes compatible with apoptosis. This type of cell death could be responsible for the disappearance of myofibroblasts. This hypothesis has now been tested by means of morphometry at the electron microscopic level and by *in situ* labeling of fragmented DNA. Results indicate that the number of myofibroblastic and vascular cells undergoing apoptosis increases as the wound closes. Apoptotic changes are also seen in both pericytes and endothelial cells. Macrophages containing phagolysosomes are seen near fibroblastic cells, suggesting that this is the major route of removal of apoptotic bodies. Results support the assumption that apoptosis is the mechanism of granulation tissue evolution into a scar.

The nature of the stimuli causing the death of granulation tissue and fibroblastic and vascular cells is not defined. It has been shown that the supernatant of TGF-β-treated fibroblasts induces apoptosis in transformed fibroblasts. TGF-β-induced elimination of transformed fibroblasts by their untransformed counterparts could represent a potential mechanism present during granulation tissue fibroblastic cell disappearance.

During normal wound healing, apoptosis affects target cells consecutively rather than producing a single wave of cell disappearance. These observations are in line with the gradual resorption of granulation tissue after wound closure. In a recent work, it was observed that covering granulation tissue with a skin flap results in a massive apoptotic process. Total skin flap induces, in an accelerated way, the same phenomena that develop gradually during the normal evolution of a wound into a scar. Moreover, signs of extracellular matrix remodeling were observed. Particularly, a wide range in the diameter of collagen fibrils was noted, as usually seen during collagen breakdown.

In the scar, continuous removal of collagen by collagenases is balanced by synthesis of new collagen by fibroblasts. Furthermore, initially, fibroblasts produce a mix of type I and III collagens; when remodeling of the scar occurs, type I progressively replaces type III. Contraction constitutes an important phase of repair by scar formation. Contraction decreases the size of the scar and enables the surviving cells of the organ to function with maximal effectiveness. Myofibroblasts participate actively in this phenomenon (Table T-2).

Repair in Specific Tissues

The basic mechanisms of wound healing apply to all tissues, but each organ contains specialized cells and unique extracellular matrices that impart some specificity to the repair process after injury.

TABLE T-2
Sequence of Main Events and Factors Involved
in Tissue Repair

Phase	Factors involved
Initial phase	
Fibrin clot formation	Fibronectin, TGF-β, PDGF,
Platelet aggregation	FGFb, CTGF, IGF-I,
Inflammatory cell recruitment	TNF-α, KGF
Cell migration	
Intermediate phase	
Cell regeneration	TGF-β, TGF-α, EGF, PDGF,
Synthesis of extracellular matrix	FGFb, IL-1, IL-4, TNF-α,
	GM-CSF
Final phase	
Extracellular matrix remodeling	Proteases, apoptosis

Liver

The outcome after liver injury depends on the nature, extent, and chronicity of the injury. Thus, after liver damage there may be complete parenchymal regeneration, scar formation, or a combination of both. Many kinds of chemically induced acute liver injury can result in death of centrilobular or periportal hepatocytes, and repair is effected by regeneration of surviving local cells or by possible recruitment of cells from a stem compartment in periportal areas. Here, fibrogenesis is usually considered to be transient.

In general, rapid restoration of normal architecture takes place if the extracellular matrix is not markedly affected. When there is a significant loss of parenchymal cells and of the extracellular matrix, and the stimulus is chronic, the repair process is by regeneration and excessive fibrosis occurs.

The factors involved in the regeneration and fibrosis as well as their cellular origin have been the subject of intense investigation. The many growth factors and cytokines described for the wound healing process are also involved in liver repair. In addition to systemic growth factors such as hepatic growth factor, PDGF and other factors secreted locally by infiltrating inflammatory cells, and later by the sinusoidal stellate cells, modulate the repair process. Among the many cytokines described, TGF-β has attracted the most attention. In very early stages after injury, active TGF-β is localized at injured sites and is thought to originate from a latent form found in blood and platelet granules in sites of damage. The activated TGF-β has been shown to promote recruitment of inflammatory cells and of fibroblasts, which in turn produce TGF-β and other factors (e.g., PDGF, EGF, TNF, and IL-1) intensifying the cytokines tissue distribution. Studies show that high TGF-β tissue content can stimulate release of mitogenic factors for fibroblasts and capillary endothelial cells. After this early phase of repair, TGF-β promotes secretion of extracellular matrix components by parenchymal and nonparenchymal cells.

Recent work clearly indicates that the hepatic stellate cell (also known as Ito cell or fat storing cell) is the main source of extracellular matrix protein in the liver, but cooperation with hepatocytes and other sinusoidal cells (Kupffer and endothelial) is necessary for stimulation of secretion of matrix proteins. The proliferation and phenotypic modulation of the hepatic stellate cell to a myofibroblast-like cell are key events in fibrogenesis and repair. However, other potentially fibrogenic cells in the liver include portal fibroblasts and cells proximal to central veins (second-layer fibroblasts), which are activated in different types of fibrosis.

While TGF-β is the main factor in promotion and accumulation of extracellular matrix components in normal wound healing in the liver, it also participates in remodeling of the new matrix through modulation of protease activity.

In summary, in many types of liver injury, the process of fibrogenesis is generally transient, and damaged tissue recovers virtually entirely. By contrast, in chronic injury, fibrogenesis leads to excessive accumulation of extracellular matrix components and can progress to fibrosis and cirrhosis.

Recent observations have indicated that stimulation of liver tissue repair after chemically induced injury may be the underlying mechanism for an autoprotection process. Autoprotection is defined as the protection when a lethal dose of a chemical is given after a nonlethal dose of the same agent is administered. Apparently, dividing liver parenchymal cells can be less susceptible to the toxic effect of a number of chemical agents.

Kidney

The kidney, like other organs (e.g., liver and lung), is an active site of uptake, activation, reabsorption, and excretion of chemical agents, making the renal tubular cells highly susceptible to toxic injury. Toxic damage together with ischemia toxins can result in significant alterations of renal structure and function that can cause acute renal failure. It is well established that after toxic acute renal failure, regenerative mechanisms become rapidly operative to reestablish tubular epithelium integrity. When damage to the renal tubular epithelium is reversible, nephronal segments undergo a cellular synthetic repair of the epithelial cells, mainly characterized by repletion of cellular ATP content. Also, cells reassemble the cytoskeletal filaments, restoring membrane polarity.

When injury to the renal tubular epithelium is irreversible, cellular proliferation takes place, which results in re-epithelialization of the denuded extracellular basement membrane and differentiation of regenerative immature into mature tubular cells. If the tubular basement membrane is disrupted in the course of the toxic injury, additional fibroblast proliferation occurs with increased deposition of extracellular matrix components, and some collapse of the tubular lumen can occur. The regenerative capacity of the renal tubule is maximal in cortical tubules and minimal in the medullary tubule. In contrast to renal tubules, glomeruli do not exhibit regenerative capacity, so the damage that induces necrosis of glomerular endothelial or epithelial cells undergoes repair by fibrosis. Many studies have indicated that repair after tubular injury depends mainly on the release of local cytokine growth factors that interact with extracellular matrix components, initiating the regenerative phase and the differentiation of the tubular cells. In the early phase of repair, TGF-β and PGDF expression in endothelial and mesangial cells has been observed. These cytokines induce monocyte–macrophage chemotaxis but are not mitogenic for renal tubular epithelial cells. A significant role of EGF, IGF, and HGF was also established in the regenerative phase, and macrophages have been the main cellular source of these factors. Autocrine expression of FGF by regenerating tubular cells seems to also play an important role in repair regulation. The potential participation of atrial natriuretic factor in the protection of tubular epithelium integrity also has been invoked.

Lung

The lung may be damaged by noxious airborne or blood-borne agents and, like other tissues, the outcome after injury varies according to the duration and extent of damage. The lung, unlike other organs, normally has abundant empty spaces to fulfill its function. Many kinds of injury result in loss and distortion of the airspace, with subsequent impairment of function. In pulmonary tissue, the epithelial lining of trachea and bronchi has high regenerative capacity and, if injury is not extensive, repair is effected by the regeneration of adjacent epithelia. There is an initial inflammatory response, and the alveolar exudate is rapidly cleared by neutrophils and macrophages. Following injury, the alveolar type II pneumocytes, which are reserve cells, migrate and proliferate to later undergo differentiation to type I pneumocytes. If acute tissue injury is extensive, and includes ruptured alveolar basement membranes, re-epithelialization is delayed and a fibroproliferative response occurs within the alveolar airspace, intersitium, and microvessels. Mesenchymal cells from alveolar septa proliferate and undergo phenotypic modulation to myofibroblasts, secreting significant amounts of extracellular matrix components.

The pattern of epithelial repair has been linked to the development of chronic pulmonary lesions. With agents that result in early epithelial repair, there follows little fibrogenic response, while with agents that induced delayed epithelial cell proliferation, severe fibrosis may ensue. Then, delayed or impaired epithelial repair is an important factor in the development of pulmonary fibrosis.

Many studies have been conducted on the factors that modulate lung repair and, as with repair following injury to other tissues, TGF-β, PDGP(S), FGF(S), and GM-CSF have been shown to play important roles in granulation tissue formation. *In vitro* studies indicate that KGF stimulates type II pneumocyte proliferation—a finding that needs to be corroborated in *in vivo* lung repair. Extensive production of TGF-β and of TNF-α may induce pulmonary fibrosis.

Besides their involvement in fibrogenesis, these cytokines also seem to modulate the expression of α-SMA by alveolar myofibroblasts. While much has been learned about factors promoting the fibroproliferative response after injury, few data are available on the fate of granulation tissue in alveolar airspaces. Recent results suggest that cell death participates in remodeling associated with tissue repair. Peptides present in the lung repair phase following injury can induce death of both fibroblasts and endothelial cells. It is noteworthy that peptides obtained during the injury phase did not cause cell death. Apparently, the mode of fibroblast death differed from typical apoptosis.

Further Reading

Bennett, N. T., and Schultz, G. S. (1993). Growth factors and wound healing: Biochemical properties of growth factors and their receptors. *Am. J. Surg.* **165**, 728–737.

Desmoulière, A., and Gabbiani, G. (1996). The role of myofibroblast in wound healing and fibrocontractive diseases. In *The Molecular and Cellular Biology of Wound Repair* (R. A. F. Clark, Ed.), 2nd ed., pp. 391–423. Plenum, New York.

Humes, H. D., Lake, E. W., and Liu, S. (1995). Renal tubule cell repair following acute renal injury. *Miner. Electrolyte Metab.* **21**, 353–365.

Inuzuka, S., Veno, T., and Tanikawa, K. (1994). Fibrogenesis in acute liver injuries. *Pathol. Res. Pract.* **190**, 903–909.

Kovacs, E. J. (1991). Fibrogenic cytokines: The role of immune mediators in the development of scar tissue. *Immunol. Today* **12**, 17–23.

Mauviel, A., and Uitto, J. (1993). The extracellular matrix in wound healing: Role of the cytokine network. *Wounds* **7**, 137–152.

Mehendale, H. M., Roth, R. A., Gandolfi, A. J., Klaunig, J. E., Lemasters, J. J., and Curtis, L. R. (1994). Novel mechanisms in chemically induced hepatotoxicity. *FASEB J.* **8**, 1285–1295.

Snyder, L. S., Hertz, M. I., Peterson, M. S., Harmon, K. R., Marinelli, W. A., Henke, C. A., and Herman, P. B. (1991). Acute lung injury. Pathogenesis of introalveolar fibrosis. *J. Clin. Invest.* **88**, 663–673.

Witschi, H. (1991). Role of the epithelium in lung repair. *Chest* **99**(3, Suppl.), 22S–25S.

—*Beatriz Tuchweber, Alexis Desmoulière, and Giulio Gabbiani*

Related Topics

Cell Proliferation
Kidney
Liver
Mechanisms of Toxicity
Respiratory Tract
Skeletal System
Skin

Titanium (Ti)

- CAS: 7440-32-6
- SELECTED COMPOUNDS: Titanium dioxide, TiO_2 (CAS: 13463-67-7); titanium tetrachloride, $TiCl_4$ (CAS: 7550-45-0)
- CHEMICAL CLASS: Metals

Uses

Titanium metal is lightweight and has high strength; thus, it is used in aircraft and other structures where cost is not a major factor. It also resists corrosion, making it especially useful in surgical implants and prostheses. Titanium fibers are used as an asbestos substitute. Titanium's most widely used compound, titanium dioxide, is used as a white pigment in paints and plastics and as a food additive to whiten flour, dairy products, and candies. It is also used in cosmetics.

Exposure Pathways

Ingestion is the primary exposure pathway. Corn oil, butter, and white wheat products are perhaps the main sources of titanium. In industrial settings, inhalation is an important pathway. Titanium is not absorbed dermally.

Toxicokinetics

Approximately 3% of ingested titanium is absorbed. The lungs are the main depot for inhaled titanium; it tends to remain in the lungs for long periods. Titanium is also found in the kidneys, the liver, and some fat tissue. Titanium crosses the blood–brain and placental barriers. Absorbed titanium is excreted in the urine.

Human Toxicity

Titanium dioxide appears to be relatively nontoxic. It has been used extensively in food products without apparent adverse effects. Generally, titanium dioxide is considered physiologically inert by all routes; however, if relatively high concentrations of titanium dioxide dusts are inhaled, toxicological actions are noted. A weak fibrosis of the lung tissue occurs but is not fatal. Occupationally, it is classified as a nuisance particulate with an ACGIH TLV-TWA of 10 mg/kg.

The liquid titanium tetrachloride is corrosive to skin and membranes of the eye. This may be due to liberation of hydrochloric acid on hydrolysis.

Clinical Management

Because of the low toxicity of titanium dioxide, there have not been any reports of therapy. Generally, titanium dioxide is biologically nonreactive when administered orally or intravenously.

Animal Toxicity

Titanium is neither mutagenic nor carcinogenic. Titanocene (an organic compound), however, induced fibrosarcoma when injected intramuscularly in rats. This same compound was carcinogenic against the Ehrlich ascites tumor in mice. There have been reports of tumors induced with the pure metal. Titanium dioxide did not induce tumors when administered orally; however, a few lung tumors were detected after titanium dioxide dust was inhaled by rats.

—*Arthur Furst and Shirley B. Radding*

Related Topic

Metals

Tobacco

- ◆ CAS: 54-11-5 (nicotine)
- ◆ SYNONYMS: *Nicotiana tabacum* (cultivated tobacco); *Nicotiana rustica*; methylpyridylpyrrolidine
- ◆ DESCRIPTION: Tobacco products contain dried tobacco leaves, which are used to take advantage of the psychoactive effects of the alkaloid nicotine (see Nicotine). Snuff has a pH of 7.8–8.2. Cigarettes are acidic. Chewing tobacco has alkali added and is basic.
- ◆ MOLECULAR FORMULA: $C_{10}H_{14}N_2$ (nicotine)
- ◆ CHEMICAL STRUCTURE: (nicotine)

Use

Tobacco products do not have a therapeutic use and can produce physiologic addiction. Tobacco enemas have been used to treat intestinal parasites.

Exposure Pathways

Tobacco is smoked, nasally insufflated, or chewed to make the nicotine bioavailable for absorption.

Toxicokinetics

The absorption of the nicotine in tobacco is incomplete after ingestion. Rectal administration via an enema may bypass the first-pass metabolism and result in higher serum levels and toxicity. Snuff is well absorbed nasally. Cigarette tobacco contains 15–20 mg nicotine/g of tobacco and cigars contain 15–40 mg nicotine. Cigarette butts contain 25% of the total cigarette nicotine content. By the 1980s, cigarettes contained 15 mg tar and 1.3 mg nicotine. Snuff is made from powdered tobacco leaf and contains from 4.6 to 32 mg/g nicotine in the moist material; dry snuff contains 12.4–15.6 mg/g.

Peak plasma levels occur 15–30 min after ingestion and 2–10 min after smoking cigarettes. Nicotine undergoes a large first-pass effect during which the liver metabolizes 80–90%. Smaller amounts are metabolized in the lungs and kidneys. The metabolites include isomethylnicotinium ion, nornicotine, cotinine, and nicotine-1-N-oxide. Protein binding ranges from 4.9 to 20%. The presence of significant amounts of nicotine in the gastrointestinal tract after intravenous dosing suggests that passive diffusion or enterohepatic circulation occurs. The apparent volume of distribution in animals is approximately 1 liter/kg. In one clinical study, it was 2 liters/kg in smokers and 3 liters/kg in nonsmokers. Nicotine passes into breast milk in small quantities. Nicotine and its metabolites are excreted in the urine. At a pH of 5.5 or less, 23% is excreted unchanged. At a pH of 8, only 2% is excreted in the urine. Nicotine can be found in the urine of nonsmokers.

Mechanism of Toxicity

Nicotine is stimulating to the autonomic nervous system ganglia and neuromuscular junction. The most prominent effects relate to stimulation of the adrenal medulla, central nervous system (CNS), cardiovascular system (release of catecholamines), gastrointestinal tract (parasympathetic stimulation), salivary and bronchial glands, and the medullary vomiting center. There is subsequent blockade of autonomic ganglia and the neuromuscular junction transmission, inhibition of catecholamine release from the adrenal medulla, and CNS depression.

Human Toxicity: Acute

Nicotine is highly toxic. Ingestion of more than one cigarette or three cigarette butts, one cigar, or a pinch of snuff is toxic. Symptoms begin within 30–90 min of ingestion and persist for 1 or 2 hr after mild exposure and 18–24 hr after severe intoxication. Vomiting usually occurs within minutes of absorption, which helps to decrease the severity of intoxication. Abdominal pain and delayed diarrhea are possible. CNS symptoms include headache, dizziness, agitation, incoordination, convulsions, and/or coma. Cardiovascular effects seen include initial hypertension followed by hypotension, tachycardia then bradycardia, and cardiac arrhythmias. Respiratory symptoms include initial tachypnea followed by dyspnea, increased bronchial secretions, respiratory depression, cyanosis, and/or apnea. Infants are especially sensitive to the effects of tobacco.

Human Toxicity: Chronic

Chronic use of snuff has caused oropharyngeal cancer. Tobacco–alcohol ambylopia is seen in chronic smokers who are malnourished and alcoholic. Green tobacco sickness occurs in young workers who do not smoke but work with wet, uncured tobacco. Withdrawal symptoms can occur when use of a tobacco product is stopped. The occurrence of various cancers and decreased cardiovascular function are increased with tobacco use. These effects may also occur via passive inhalation of cigarette or cigar smoke (see Tobacco Smoke).

Clinical Management

Syrup of ipecac-induced emesis should be avoided since seizures or lethargy can occur rapidly. Activated charcoal should be administered. Seizures should be treated with diazepam or phenytion. Atropine can be used to control signs of excess parasympathetic stimulation. If hypotension does not respond to intravenous fluids, dopamine or norepinephrine may be indicated. Antacids should be avoided since nicotine has greater absorption in an alkaline media. Vital signs and level of consciousness should be monitored closely. Further care is symptomatic and supportive. Nicotine laboratory determination is only of diagnostic value and does not direct therapy.

Animal Toxicity

A dose of 10 mg/kg (buccally) is fatal in dogs. Symptoms include initial hyperexcitability, hyperpnea, salivation, vomiting, diarrhea, then depression, incoordination, and paralysis.

—*C. Lynn Humbertson*

Tobacco Smoke

Tobacco smoke (cigarette smoke) may present perhaps the greatest current public health threat. It is estimated that inhalation of this complex mixture of organic compounds is related to 400,000–450,000 deaths annually in the United States.

There is conclusive evidence that the tars occurring in cigarette smoke can lead to lung cancer; the chief factors are age of individual at initiation of smoking, extent of inhalation, and amount smoked per day. Polonium, a radioactive element, is known to occur in cigarette smoke. In addition, more than 100 compounds have been identified, including nicotine, cresol, carbon monoxide, pyridine, and benzopyrene (a carcinogen).

Cigarette smoke has been implicated in acute respiratory illness and chronic obstructive lung disease, but the effect of exposure to mainstream cigarette smoke has yielded ambiguous results in humans and in animal models. In humans, the number of alveolar monocyte/macrophage cells (MOs) is increased three- to fivefold in smokers compared to nonsmokers. This may be a result of increased production of IL-1 by the resident alveolar MOs, resulting in enhanced influx of other inflammatory cells (polymorphonuclear cells and peripheral blood mononuclear cells) into the lung. In addition to the increased numbers of MOs, the MOs present appear to be in an activated state, as evidenced by an increase in cytoplasmic inclusions, increased enzyme levels, altered surface morphology, and enhanced production of oxygen radicals. However, despite their apparent activated state, these MOs seem to have decreased phagocytic and bactericidal activity. Although the primary site of exposure of the immune system to cigarette smoke is the lung, selected immune parameters have been shown to be altered in smokers. Decreased serum immunoglobulin levels and decreased natural killer cell activity have been reported. Concentration-dependent leukocytosis (increased numbers of T and B cells) is well defined in smokers when compared to nonsmokers. However, the question of whether there is a relationship between smoking and lymphocyte function is debatable.

Numerous immunological studies have been conducted in animals exposed to cigarette smoke that demonstrate suppression of antibody responses, biphasic lymphoproliferative capacity (enhanced, then suppressed with continued exposure), and enhanced susceptibility to murine sarcoma virus and influenza virus. Animal studies cannot precisely replicate human exposure conditions because of the route of exposure and the rapid chemical changes that occur to the components of tobacco smoke upon its generation.

Widely available in tobacco products and in certain pesticides, nicotine has diverse pharmacologic actions and may be the source of considerable toxicity. These toxic effects range from acute poisoning to more chronic effects. Nicotine exerts its effects by binding to a subset of cholinergic receptors, the nicotinic receptors. These receptors are located in ganglia, at the neuromuscular junction, and also within the central nervous system (CNS), where the psychoactive and addictive properties most likely reside. Smoking and "pharmacologic" doses of nicotine accelerate heart rate, elevate blood pressure, and constrict blood vessels within the skin. Because the majority of these effects may be prevented by the administration of α- and β-adrenergic blockade, these consequences may be viewed as the result of stimulation of the ganglionic sympathetic nervous system. At the same time, nicotine leads to a sensation of "relaxation" and is associated with alterations of electroencephalographic (EEG) recordings in humans. These effects are probably related to the binding of nicotine with nicotinic receptors within the CNS, and the EEF changes may be blocked with an antagonist, mecamylamine.

Acute overdose of nicotine has occurred in children who accidentally ingest tobacco products, in tobacco workers exposed to wet tobacco leaves, or in workers exposed to nicotine-containing pesticides. In each of these settings, the rapid rise in circulating levels of nicotine leads to excessive stimulation of nicotinic receptors, a process that is followed rapidly by ganglionic paralysis. Initial nausea, rapid heart rate, and perspiration are followed shortly by marked slowing of heart rate with a fall in blood pressure. Somnolence and confusion may occur, followed by coma; if death results, it is often the result of paralysis of the muscles of respiration.

Such acute poisoning with nicotine fortunately is uncommon. Exposure to lower levels for longer duration, in contrast, is very common and the health effects of this exposure are of considerable epidemiologic concern. In humans, however, it has been impossible so far to separate the effects of nicotine from those of other components of cigarette smoke. The complications of smoking include cardiovascular disease, cancers (especially malignancies of the lung), chronic pulmonary disease, and attention deficit disorders in children of women who smoke during pregnancy. Nicotine may be a factor in some of these problems. For example, an increased propensity for platelets to aggregate is seen in smokers and this platelet abnormality correlates with the level of nicotine. Nicotine also places an increased burden on the heart through its acceleration of heart rate and blood pressure, suggesting that nicotine may play a role in the onset of myocardial ischemia. In addition, nicotine also inhibits apoptosis and may play a direct role in tumor promotion and tobacco-related cancers.

It seems more clear that chronic exposure to nicotine has effects on the developing fetus. Along with decreased birth weights, attention deficit disorders are more common in children whose mothers smoke cigarettes during pregnancy, and nicotine has been shown to lead to analogous neurobehavioral abnormalities in animals exposed prenatally to nicotine. Nicotinic receptors are expressed early in the development of the nervous system, beginning in the developing brain stem and later expressed in the diencephalon. The role of these nicotinic receptors during development is unclear; however, it appears that prenatal exposure to nicotine alters the development of nicotinic receptors in the CNS. These changes may be related to subsequent attention and cognitive disorders in animals and children.

—*Shayne C. Gad*

Related Topics

Carcinogenesis
Cardiovascular System
Developmental Toxicology
Immune System
International Agency for Research on Cancer
Neurotoxicology: Central and Peripheral
Respiratory Tract

Toluene

- ♦ CAS: 108-88-3
- ♦ SYNONYMS: Methylbenzene; phenylmethane; toluol (DOT); Antisal 1a; methacide; methylbenzol; NCI C07272; tolueen (Dutch); toluen (Czech); tolueno (Spanish); toluolo (Italian); tolu-sol
- ♦ CHEMICAL CLASS: Aromatic hydrocarbon

Uses

Toluene is a clear, flammable liquid with a sweet odor that is widely used in both the chemical and the pharmaceutical industries. In terms of production, it is the 24th highest volume chemical in the United States. It is derived mainly from petroleum refining and only a small percentage of that produced is used directly. Most toluene is added to automobile or aviation gasoline mixtures (benzene > xylene > toluene) to increase octane ratings. Toluene is an excellent organic solvent and is used extensively in the manufacture of benzene derivatives, caprolactam, saccharin, medicines, dyes, perfumes, TNT, toluenediisocyanates (polyurethane resins), toluene sulfonates (detergents); as a solvent for scintillation counting; in paints and coatings, gums, resins, rubber; and as a diluent and thinner in nitrocellulose lacquers, plastic toys, and model airplanes. Toluene is also used extensively in the production of glues and is responsible for the narcosis and permanent brain damage seen in "glue sniffers."

Exposure Pathways

Automobile emissions contribute the majority of toluene that is found in the atmosphere. Toluene is the most prevalent aromatic hydrocarbon in the air, with detected levels ranging from 0.14 to 59 ppb. Toluene has also been detected in surface water and treated wastewater effluents at levels generally below 10 μg/liter. Toluene is readily biodegradable and will not bioconcentrate to a great degree. In a study of edible aquatic organisms, 95% of the tissues sampled had levels <1 ppm.

Because toluene is fairly volatile, exposure for humans would occur principally by inhalation. Dermal exposure may also be significant, especially in an industrial setting, where skin may be exposed for long periods of time. Oral exposure is the least probable route and would occur primarily as a result of accidental poisoning or suicide.

Toxicokinetics

Toluene is readily absorbed from the lung and gastrointestinal tract, although studies in animals suggest absorption occurs more slowly in the gastrointestinal tract. Slow absorption also occurs through skin. Studies of humans and animals indicate that inhaled toluene distributes to tissues that are high in fat content (e.g., body fat, bone marrow, and brain) or well supplied with blood (e.g., liver). It seems reasonable that similar distribution would occur for other routes of exposure.

In both humans and animals, toluene is rapidly excreted as both the unchanged compound in expired air and a metabolite in the urine. Toluene is converted in the liver to water-soluble hippuric acid and conjugated cresols, which are then excreted in the urine. This conversion has been demonstrated in man and animals exposed via inhalation, although it is expected to occur for other exposure routes as well. Another excretion route for toluene is exhalation of the unchanged chemical. This excretion route might be expected to operate for all exposure routes but be more effective for exposures via inhalation.

Mechanism of Toxicity

Although the exact biochemical mechanism of toxicity has not been identified for toluene, it is known that the primary toxic effect of toluene is dysfunction of the brain and central nervous system (CNS; narcosis). The main function of neurons is to conduct electrochemical signals to one, several, or thousands of other cells. The normal physiology of these neurons is, in turn, largely dependent on the integrity of the cell membrane, which polarizes and depolarizes during the transmission of these signals. Thus, the most probable mechanism of toxicity is the unique sensitivity of the cell membranes of neurons to the solvent property of toluene, which disrupts the normal transmission of nerve impulses.

Human Toxicity

Much of the information on toxicity of toluene to humans comes from studies of solvent abuse (such as glue sniffing) and during exposure in the workplace (e.g., painters and printers). Interpretation of the data can be difficult due to the fact that these individuals are simultaneously exposed to mixtures of other chemicals. Both acute experimental and occupational exposures to toluene in the range of 100–1500 ppm (approximately 325–5600 mg/m^3) have elicited dose-related CNS alterations, such as fatigue, confusion, and incoordination, as well as impairments in reaction time and perceptual speed. At 200–500 ppm, headache, nausea, eye irritation, loss of appetite, a bad taste, lassitude, and incoordination are reported but not accompanied by significant laboratory or physical findings. For high acute exposures (approximately 30,000 ppm), initial lightheadedness and exhilaration is followed by progressive development of narcosis and central nervous system depression. With long-term exposure, blood abnormalities, psychomotor disorders, changes in the lens of the eye, immune system changes, kidney effects, menstrual disorders, and birth defects have been observed in some, but not all, studies of workers or abusers, and the possible confounding effect of mixed chemical exposure is mentioned in most. Liver effects, which figure prominently in animal studies, have not been observed in occupationally exposed individuals. Based on epidemiological studies, there is no evidence that toluene can cause cancer in humans.

Clinical Management

Persons who have been overcome by toluene fumes or gases should be removed from the area of exposure to fresh air. Should breathing become labored or shallow, medical intervention (e.g., artificial respiration) may be necessary. Following accidental or intentional ingestion, vomiting should not be induced and prompt medical attention should be obtained. Liquid toluene spills on exposed skin should be immediately dried with an absorbent towel and then washed with soap and water.

Animal Toxicity

Toxicity to the embryo or fetus and teratogenic effects have been rarely observed in animal studies. These effects were seen in only one experiment in which the dose was high enough to be toxic to the mother as well. More frequently, when maternal toxicity was not present, fetal toxicity or teratogenicity was not found. Growth inhibition of rat pups born during inhalation exposure to toluene through two generations has been observed.

CIIT conducted a 2-year inhalation toxicology study in Fischer 344 rats exposed to atmospheric toluene. The concentrations used were 30, 100, or 300 ppm (113, 377, or 1130 mg/m^3) for 6 hr per day, 5 days per week. The only finding was a dose-related reduction in hematocrit values (number of red blood cells) in female rats exposed to 100 and 300 ppm toluene. This may not be considered a significant toxic effect. Therefore, a NOAEL was set at the highest exposure level—300 ppm (equivalent to 29 mg/kg/day).

NTP also conducted a 2-year inhalation study in mice and rats (doses of 600 or 1200 ppm in rats and 120, 600, and 1200 ppm in mice, 6.5 hr per day, 5 days per week). Lesions of the nasal cavity (in rats) and abnormal growth (hyperplasia) of the bronchial lining (in mice) were seen, but no deaths nor significant body weight change were observed during the course of the study. There was no evidence of cancer induction in this study.

In a recent study by NTP, rats and mice were given oral doses of toluene, ranging from 312 to 5000 mg/kg, 5 days per week for 13 weeks. General toxic effects were seen in both species at 2500 mg/kg, which included decreased movement or prostration, tearing and salivation, and body tremors. A few animals died at this dose. There were changes in organ weights and microscopic pathologic changes of several organs at 1250 mg/kg in rats. Organ weight changes, but not pathologic changes, were seen at 2500 mg/kg in mice but not at lower doses. The only adverse effect seen at the lowest dose was increased liver and kidney weights at 625 mg/kg in rats.

Although there was no evidence of cancer in the CIIT or NTP studies, and most mutagenicity tests have been negative, U.S. EPA considers the data inadequate to classify toluene relative to its carcinogenicity; it is rated D (not classified, inadequate evidence in animals) in the current weight-of-evidence system.

Under U.S. EPA current guidelines for risk assessment, the acceptable exposure dose for humans (or reference dose) is 0.2 mg/kg/day. For an average human weighing 70 kg, this dose is equivalent to approximately 1/2000th of an ounce.

Under the Safe Drinking Water Act, the maximum contaminant level (MCL) is the standard criteria for drinking water and the maximum contaminant level goal (MCLG) is the goal. The MCL and MCLG for toluene in drinking water is 1000 μg/liter, based on health protective limits developed from the CIIT study. OSHA recommends workplace air concentrations do not exceed 100 ppm; ACGIH recommends 50 ppm (based on potential skin exposure).

—*Stephen Clough*

Related Topics

Neurotoxicology: Central and Peripheral
Pollution, Air
Pollution, Water
Sensory Organs
Skin

Toluene Diisocyanate

- ◆ CAS: 584-84-9
- ◆ PREFERRED NAME: Toluene 2,4-diisocyanate
- ◆ SYNONYMS: 2,4-Diisocyanatotoluene; TDI; nacconate 100
- ◆ MOLECULAR FORMULA: $CH_3C_6H_3(NCO)_2$
- ◆ CHEMICAL STRUCTURE:

Uses

Toluene diisocyanate is used in the production of polyurethane foams, elastomers, and dyes. It is a crosslinking agent for nylon 6.

Exposure Pathways

Dermal contact, inhalation, and ingestion are possible routes of exposure.

Toxicokinetics

Toluene diisocyanate has rapid linear absorption via inhalation with persistence in bodily tissues at low levels for up to 2 weeks. It is metabolized into 2,4-diaminotoluene in man.

Mechanism of Toxicity

Toluene diisocyanate is a cross-linking agent and a sensitizer.

Human Toxicity

Toluene diisocyanate is a strong irritant to the eyes, skin, and respiratory system. It is a lacrimating agent and a strong dermal and pulmonary sensitizer. It is also a suspected carcinogen. The ACGIH TLV for toluene diisocyanate 0.005 ppm.

Clinical Management

Affected eyes should be irrigated with running water. Contaminated areas of skin should be washed with soap and water. Patients asymptomatic of respiratory effects should receive oxygen and ventilatory support.

Animal Toxicity

The RD_{50} (50% respiratory depressive concentration) in mice is 0.4 ppm. The LC_{50} (inhalation) is 14 ppm/4 hr in rats and 10 ppm/4 hr in mice.

—Shayne C. Gad and Jayne E. Ash

Related Topic

Respiratory Tract

Toluidine

- ◆ SYNONYMS: Aminotoluene; *m*-toluidine (CAS 108-44-1); *o*-toluidine (CAS 95-53-4); *p*-toluidine (CAS 106-49-0)

- ◆ CHEMICAL CLASS: Substituted aromatic
- ◆ CHEMICAL STRUCTURE:

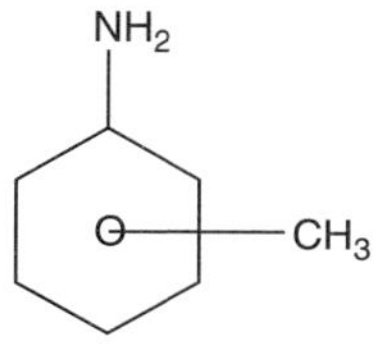

Uses

Toluidine is used to produce dyes for textiles and other substances and as an accelerator in vulcanization. It is also used in organic synthesis.

Exposure Pathways

Inhalation and dermal contact are possible routes of exposure.

Toxicokinetics

In urine, 26% of *o*-toluidine is excreted after 24 hr. Excretion for *m*-toluidine and *p*-toluidine over that same period is 10%.

Mechanism of Toxicity

Toluidine interferes with enzymes associated with the detoxification process and monoxygenase system. It defats membranes.

Human Toxicity

o-Toluidine and *p*-toluidine are carcinogens. The ACGIH TLV is 2 ppm for all forms.

Clinical Management

All contaminated areas should be washed, including inside ear canals and under nails. The exposed person should be monitored for methemglobinemia. If contamination is 30% or less, bed rest is recommended. If contamination is over 30%, the patient should be observed and given oxygen therapy. If contamination is over 50%, the exposed person should be given intravenous glucose solution. If contamination is ε60%, methylene blue should be administered.

Animal Toxicity

Toluidine is an irritant primarily due to defatting. It is a mild skin irritant and moderate eye irritant in rabbits. Oral LD_{50}s in rats are 450 mg/kg for *m*-toluidine, 670 kg/mg for *o*-toluidine, and 3360 mg/kg for *p*-toluidine. Oral LD_{50}s in mice are 740 mg/kg for *m*-toluidine, 520 mg/kg for *o*-toluidine, and 330 mg/kg for *p*-toluidine.

—Shayne C. Gad and Jayne E. Ash

Toxicity, Acute

By definition, acute toxicity studies are conducted to determine the total adverse biological effects caused during a finite period of time following the administration of single, frequently large doses of a chemical, physical agent (dust and fibers), or some form of energy (ionizing or nonionizing radiation). The objectives of such studies are to discover any adverse health effect that could be attributed to the agent under investigation, including any immediate biochemical, physiological, and/or morphological changes; and any delayed changes suggesting some secondary injury to body organs/tissues, as well as the death of the animal. In general, the effects observed in the experimental subjects, usually animals, are directly related to the amount of the substance administered. It is a common misconception that acute toxicity studies are designed only to express the potency of an agent in terms of the median lethal dose (LD_{50}), a value representing the estimated dose causing death of 50% of the population of test subjects exposed under the defined conditions of the test. Nothing could be further from the truth; acute toxicity studies encompass a number of experiments. The usual battery of acute toxicity tests is shown in Table T-3.

Since information concerning the toxicity of the agent must be obtained before humans are exposed to it, this necessitates the use of animal models as surrogates or substitutes for the human. Several strains of rodents (mice and rats) are used routinely to determine the acute toxicity of new agents. Experiments are conducted using both males and females of the species because of known sex differences in response(s) to various agents. Indeed, an extensive body of data has been acquired over several decades using rodent species, thereby permitting chemical-to-chemical, inter- and intraspecies comparisons. It is desirable to have acute toxicity data from nonrodent (rabbit, guinea pig, dog, and monkey) species, particularly if the results from the mouse and rat differ greatly, suggesting distinct species differences in response(s). In contrast, should similar toxicity be seen in a number of experimental animal species, the same toxicity might be produced in the human at some as yet unknown dosage to the agent.

The agent should be administered via the route by which the species might be expected to obtain the toxicant. In general, the usual routes of exposure for the human include ingestion, inhalation, or contact with the skin. Accidental exposure to industrial or home products might also include having them splashed into the eyes or onto the skin. However, if an agent being tested is a therapeutic agent, exposure might require inhalation, intravenous, intramuscular, or subcutaneous routes of administration.

Why are acute toxicity studies necessary? With any new agent, there will be workers exposed to relatively high concentrations during its manufacture, handling, packaging, and use. Accidents may occur not only in the workplace but also during the transportation (ship, rail, and truck) of the agent, with exposure of bystanders in the immediate vicinity of the accident and risk to personnel involved with the accident or cleanup of the spillage (e.g., police, firefighters, emergency response teams, and sanitation crews). There is also the potential of accidental and/or intentional exposure of the product user, members of his/her family, neighbors, children, and pets. It is essential to know just how toxic or nontoxic the agent may be in the most bizarre circumstances of exposure (e.g., ingesting or inhaling the agent or getting it on the skin). How much is safe? How little is too much? The information obtained from acute toxicity studies can be found printed in a material safety data sheet required by law to be prepared for each and every product manufactured and to be available to the public, to industries using the products, and to health and safety professionals.

Determination of the Lethal Dose

The LD_{50} is a statistical estimate of the acute lethality of an agent administered to a specified sex, age, and strain of a species of animal. The value provides a

TABLE T-3
A Battery of Tests for the Evaluation of Acute Toxicity

Test	Description	Study period
Lethality	LD$_{50}$ (LC$_{50}$) or an estimated value	24 hr
	Surviving animals—close observation permits determination of duration of toxicity; recovery; development of secondary toxicity; changes in hematology, blood chemistry, urinalysis; changes in organ/tissue function	14 days
Primary irritation	Skin	
	Exposure	24 hr
	Evaluation	24, 48, and 72 hr
	Eye	
	Exposure	1.0 sec
	Evaluation	1.0, 24, 48, and 72 hr
Sensitization	Repeated (5 days/week) dermal application	14 days
	Rest period of 10–14 days	
	Challenge dose at Days 28–30	
	Evaluation	24, 48, and 72 hr
Photoallergic and phototoxic reactions	Repeated treatment (oral, iv, dermal) for 10–14 days	
	Rest period for 14–21 days	
	Retreatment with ultraviolet light on shaved skin patch	
	Evaluation	24, 48, and 72 hr

measure of the relative toxicity of an unknown agent compared to other agents administered by the same route to the same species, strain, age, and sex of the animal. As listed in Table T-3, the LD$_{50}$ value, an indicator of lethal potency, is frequently the first biological parameter determined for a new chemical, the agent being administered via the route by which the human might acquire a high concentration, and animal mortality being assessed in the 24-hr period after treatment. Given that people might acquire the chemical by different routes, it might be necessary to carry out two experiments, choosing two of the three possible routes (ingestion, inhalation, or dermal) of administration in anticipation of quite different values. Although accurate determinations of the lethal potency are no longer required (designing experiments in which 60–100 animals might be used in the classical determination of the LD$_{50}$), some insight into the potency, even a rough estimate of the range of acute toxicity, is essential. Regulatory agencies are still concerned about massive spills and the impact of these on the health of local populations. The Bhopal methylisocyanate incident revealed just how little information was available on acute toxicity associated with inhalation, dermal, and ocular exposure to this agent. While criticism has been

leveled that the LD$_{50}$ values are just numbers on which to hang one's hat, they are valid predictors of acute toxicity, albeit only of mortality and not of morbidity or long-term adverse health effects.

Animal rights activists have repeatedly challenged the need for the LD$_{50}$ determination in terms of the inflicting of injury to the test animals and the needless waste of large numbers of animals to obtain a number that is only a rough estimate. A properly designed study will yield much more information than just "a number." While, by definition, 50% of the animals will die, close observation of these animals during the first 12-hr period after treatment may reveal several biological clues to possible mechanisms by which the toxicant may be causing an effect—clues that are valuable to the clinical toxicologist in attempts to alleviate human suffering. However, 50% of the animals will survive the treatment, and these survivors are a repository of biological effects elicited by the test agent. These effects can be studied over the next 14-day period to assess the short or long duration of toxicity; the rapid or slow recovery; the appearance of any additional, delayed, or secondary toxic effects; changes in hematology, serum biochemistry, and urinalysis; and changes in organ/tissue function (liver, kidney, and nervous system) measured by relatively noninvasive techniques, all without having to destroy the animals. When the animals are euthanized at 14 days after treatment, organs/tissues will be obtained for detailed microscopic examination to correlate observed biological effects and/or injury with possible morphological changes. Thus, the animals surviving the toxic insult are a veritable treasure trove of information concerning the mechanism(s) of the chemical-induced toxicity.

Given the spectrum of observed and measured biological effects, one important aim of the acute toxicity study is to develop a quantitative relationship between the intensity of a measured response or adverse health effect and the concentration(s) of agent administered. Assuming that the dosage of agent can be "delivered" to the test animals accurately with minimal variability, this leaves interanimal variability in response, one major reason that the classical LD_{50} determination uses 8–10 animals per treatment group. The biological responses and variability are usually presented in graphic form, with the X-axis representing the range of dosage and the Y-axis reflecting the biological response in some quantitative manner (Fig. T-2). From such graphs, a

dosage-related appearance of target organ toxicity may be determined, with some organs/tissues responding to low levels of the agent (Fig. T-2A) and others responding only at elevated concentrations (Fig. T-2B). In addition, the slope of the dose–effect relationship for each organ/tissue can be determined, indicating whether or not small changes in dosage produce marked biological changes (a steep slope, potent agent; Fig. T-2A), or the reverse, where large increases in dosage are accompanied only by weak to modest changes in responses (a shallow slope, weakly toxic agent; Fig. T-2C).

Range-Finding for the Lethal Dose

Given the arguments about the needless waste of vast numbers of animals to determine "accurate" LD_{50} values for chemicals where such precision is not essential, methods are used to arrive at the appropriate lethal dose with only six animals required. An initial dosage (mg/kg body wt) is arbitrarily chosen and administered to a pair of suitable animals (rodents) with subsequent close observation for effects. If little or no toxicity is

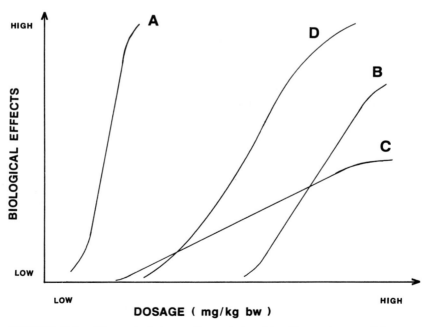

FIGURE T-2. *Theoretical dose–effect relationships illustrating possible different target organ (A–D) responses over a wide range of agent dosages, indicating the importance of the slope of the relationship to predict whether or not a small or large change in dosage is required to induce marked, moderate, or weak biological effects.*

observed in a predetermined time period (24 hr), the second pair of animals receives a dosage 50% (3/2) higher than the initial dosage. If, once again, no toxicity is observed, the dosage administered to a third pair of animals may be double the original dosage selected. If, however, this 50% higher dosage is lethal to one or both of the animals, some concentration between these two values will be selected and administered to the third pair of animals. If the originally chosen concentration causes severe toxicity and mortality, the dosage administered to a second pair will be downscaled to order of 66% (2/3) of the initial dosage.

Such experiments permit an investigator to determine the smallest lethal dose. How close may this be to the actual LD_{50} of the agent? Extensive studies have shown that reasonably precise lethal doses can be determined by the range-finding approach, with the approximate lethal dose for 86% of those chemicals tested in this manner being within 30% of the known LD_{50} values determined by the classical approach. However, one must appreciate the fact that 14% of the chemicals were outside of this range; thus, the technique is not infallible. For most chemicals, this approach would be adequate and most regulatory agencies no longer demand accurate determinations. There are, of course, exceptions.

Primary Irritation Studies

Products freely available for purchase by the public, including cosmetics, pesticides, household products (e.g., cleaners, detergents, waxes, and polishes), and health care products (e.g., soaps, bath gels, shampoos, creams, and mouthwashes), must undergo testing for dermal and ocular irritancy or corrosion based on industrial, medical, and regulatory experience that accidents can and will happen with these products, e.g., the agent(s) being splashed at some time on the skin and/or into the eyes. Regulatory agencies require this information to meet regulations for classification, transportation, packaging, and labeling.

The Skin Test

Beginning with the knowledge that no animal species has skin comparable to that of the human, the white laboratory rat is generally used in initial studies of dermal irritation, although the guinea pig, rabbit, and miniature swine have relatively sensitive skin and are used extensively along with the dog and the rhesus monkey. As with the human, the absorptive properties of skin vary considerably from region to region on the body surface. The choice of test species may be dictated by (1) the physical properties of the agent, (2) the solubility of the agent in water- or organic-based solutions or emulsions, (3) the inherent toxicity of the agent (low toxicity requiring the application of large volumes), and (4) known sensitivies of animal skin compared to that of the human based on previous experience with similar agents. The choice of test species may also be governed by the experimental design of the test: (1) applying a number of different concentrations of agent and the need for comparable control sites where the skin is similar, and (2) applying the test agent to intact or abraded skin, with the latter having slight damage done to the stratum corneum (the dead cell layer on the surface) by mechanical means to mimic the loss of this protective barrier from cuts, scratches, and abrasions encountered on human skin, particularly in industrial settings.

The test substance (liquid, solid, paste, or emulsion) is usually applied to a shaved test area (6.0 cm^2 or 1.0 in.^{2}) of skin on the back in a uniform layer, with the site being covered by a gauze patch taped in place to prevent the animal from licking the material off the site during normal grooming. The trunk may be wrapped with an impervious plastic sheet; this practice is particularly useful when multiple sites are treated. At the end of the test period, usually 24 hr, the coverings are removed, the residual test agent is wiped off with gauze wetted with warm soapy water, and the test areas are evaluated for (1) edema (puffiness and swelling) and (2) erythema (redness and inflammation). This evaluation is repeated at 48 and 72 hr after treatment. The OECD has introduced a 4-hr rabbit covered patch test in place of the 24-hr study.

While this is a highly subjective test, a number of scoring systems have been developed to measure the degree of puffiness and redness visually. The scoring system assigns a number (0–4) for edema and for erythema, ranging from none through very slight, slight, moderate, to severe effects. Average numerical scores for the primary irritation index (PDI) are obtained from the results on five or six animals per treatment level over the time period (24, 48, or 72 hr) to provide information on the potency of the test agent to cause irritation, the speed or slowness of recovery from effects, and the estimated irritation values providing guidance to regulatory agencies for warnings that should be printed on product labels to protect the consumer.

In Vitro Irritation: Skin

Calculation of the PDI from the numerical scores has demonstrated the variability in the Draize skin test, showing both interspecies and site differences. Suitable *in vitro* testing procedures have proven useful in standardizing and quantifying the responses for certain predefined endpoints of toxicity such as changes in cell permeability and cytototoxicity. A number of *in vitro* test systems have been developed based on the use of cell-free, macromolecular complex mixtures (Skintex), cultured microorganisms, mammalian cell lines (human fibroblasts–keratinocytes), human tissue equivalents such as Testskin, and isolated skin tissue from swine or human cadavers (Table T-4). The chemistry involved with most of these techniques usually examines cell viability, cellular metabolism, cell membrane integrity,

keratinization and/or stratification (growth, spreading, and layering), with the quantitative endpoints being dye penetration, leakage of cell proteins, cell staining, cell death, and cell number and colony growth.

The development of human tissue equivalent systems, in which a skin-like layer of cells forms a membrane between an upper and lower chamber, has proven extremely useful, the agent being added to the upper medium chamber and penetration into or through the cell layers are monitored by removal of medium samples from the lower chamber for analysis or by specific stains applied to the cell layer to detect cellular damage by microscopic study. Isolated dermis from human cadavers or from swine can be used in double-chamber systems in a similar manner to that described for tissue equivalents. Advanced techniques using isolated porcine skin flaps *in situ* with an intact subdermal blood supply are seeing greater use: Samples of the blood can be collected at different time intervals for the analysis of the penetration of the test agent. Human volunteers, of course, are regularly used to assess chemicals previously screened by the previously described methods.

TABLE T-4
In Vitro Methods for Testing Skin Toxicity

Skintex System: cell-free complex mixture of various macromolecules, with physicochemical changes occurring on interaction with irritants, resulting in changes in light transmission

Cell cultures: microorganisms and mammalian cell lines, examining
 Neutral red dye penetration
 Protein leakage, colorimetric assay
 Coomassie blue or Kenacid blue for cell proliferation and protein synthesis

Testskin: human tissue equivalents (human keratinocytes on collagen or collagen–glycosaminoglycan matrix with human fibroblasts, forming an epidermis for examining
 Dye penetration (Neutral red)
 Dye exclusion
 Cellular damage by agents
 Chemical penetration
 Cell growth and development
 Cellular metabolism

Isolated tissues: prepared for a double chamber; agent applied to the top and sampled from the bottom to test penetration through the skin; can use cadaver skin or isolated animal (swine) skin as test tissue

Human volunteers: dermal penetration of the test agent and measurement of quantity residing on the surface after a predetermined time interval

---

The Eye Test

Damage to the eye is an all too common consequence of an accidental splashing of industrial chemicals, home and health care products, pesticides, solvents, etc., resulting in painful and frequently permanent injury. The Draize eye test, first described in 1944, has become a target of animal welfare groups, antivivisectionists, and concerned scientists who claim that it is not required, is inhumane, causes unnecessary pain to the test animal, generates a subjective result even with the available detailed scoring systems used, and is prone to interlaboratory variability in results such that the test is meaningless. The thought of knowingly placing some highly irritating agent in an animal's eye and causing pain is abhorrent. In fact, if the dermal irritancy test is positive, there is little scientific basis for carrying out the eye test since the agent will almost certainly be positive in the eye. Hence, the dermal irritancy test will screen out the highly toxic agents. However, between the highly damaging, strong acids or bases and completely innocuous agents lie a wide variety of seemingly neutral, slightly acidic or basic soaps, detergents, shampoos, cosmetic creams, and lotions, all of which may show minimal effects on the skin but still be irritating if accidentally introduced into the eye. Literally thousands of products must be tested annually. For the sake of occupational, bystander, and consumer safety, these products must still be subjected to an eye irritation test.

The basic ocular irritation test in the rabbit is described in detail under Eye Irritancy Testing, but it is important to point out that the number of test animals can be reduced from the usual six at each exposure level to two or, at most, three animals per dose without sacrificing much accuracy. Many test series have shown 88–91% accuracy with two animals per treatment group. The agent, instilled in the pouch formed by the lower eyelid, is held in place for 1 sec and then released. The treated eye is not washed, allowing the animal's own tear secretions to flush out the material. The untreated eye serves as a control. Both eyes are examined at 1, 24, 48, and 72 hr after treatment; the irritation (or damage) to the cornea, the conjunctiva, and the iris is scored numerically in a subjective manner. The test is open in that the experiment can be terminated at 72 hr if there is no evidence of irritation, but observed effects can be assessed for a longer time period.

In Vitro Irritation: Eye

Considerable effort has been made by industry and both national and international regulatory bodies to replace the eye irritancy test with suitable *in vitro* assays for such toxicity endpoints as cytotoxicity, corneal opacity, and inflammation, replacing the subjective nature of the assessment with objective and quantitative measurements (Table T-5). What has evolved from the myriad of test tube and cell culture assay systems and *in vitro*, isolated eye or corneal test systems currently undergoing development and validation is one or more battery of tests, none of them giving a complete answer to the question, but each contributing some quantifiable information for a selected endpoint. These test batteries will be used as screening devices to identify the strong to moderate irritants and the nonirritants, leaving those agents showing suspicious or equivocal results to be tested in animals. These *in vitro* test systems will aid significantly in reducing the number of animals subjected to the eye test, but they will never totally replace it.

In defense of the Draize eye test, it must be pointed out that, to date, none of the *in vitro* alternative tests have proven applicable or acceptable to regulatory

TABLE T-5
In Vitro Methods for Testing Eye Toxicity

Cytotoxicity
  Immortalized mammalian cell lines such as HeLa, V79, human keratinocytes and mouse fibroblasts, and canine kidney cells to study
    Dye uptake/exclusion—viable cells
    Dye penetration—cell integrity
    Dye penetration—membrane damage
Opacity
  Eyetex assay: formation of high-molecular-weight protein aggregates causing reduced light transmission
  Isolated bovine cornea: prepared in a two-compartment chamber; chemical-induced damage causing changes in light transmission through the cornea
    An increase signifying loss of corneal cells
    A decrease indicating opacity
    Fluorescein dye uptake assessing cell damage
Inflammation
  Bovine corneal cup method: inflammatory response releasing chemotactic factors and then reacted with neutrophils
  Bovine corneal cup assay: inflammatory response releasing specific mediators (histamine, serotonin, prostaglandins, leukotrienes, and thromboxanes) that can be collected in the bath medium and quantitated by chemical assay
  Rat vaginal tissue assay: similar to bovine corneal cup assay with release of specific mediators
  Fertile chicken egg choriollantoic membrane (CAM) assay: scoring for vascular changes in the membrane blood vessels with fluorescein dye as well as necrotic damage

agencies. However, the European Economic Community has taken the lead in this field by proposing to ban the European sale of all cosmetics in 1998 that have been tested in animals, subject, of course, to member country acceptance of and adherence to the regulation passed. This decree will encourage further *in vitro* test development, validation of existing test systems, and government acceptance of these in lieu of *in vitro* eye irritancy tests.

Skin Sensitization Studies

A dermal reaction is seen whereby exposure to a certain chemical causes little effect following initial contact with it but, with repeated (daily, weekly, or even once a month) exposure of the skin, an effect, usually an erythema or red spot, is seen that occurs earlier in time, is more severe, and persists for a longer duration. Subsequent exposures, even though weeks or years apart, result in what appears to be an allergy-like, delayed reaction at the site of exposure or even on parts of the body where no exposure has occurred. The pattern of development of this skin condition, frequently found in the workplace, is suggestive of an allergy.

Invariably, the test animal used for skin sensitization studies is the guinea pig. A number of different test procedures have been developed, but all are similar in that repeated, daily low doses of the test agent are injected intradermally or applied on closely shaved skin over a 14-day period. This treatment period encourages the development of an immunological response in the animal. Following a suitable 10- to 14-day resting period, a challenge dose, usually a lower concentration than was used as a sensitizing dose, is applied to a fresh, untreated site and the reactions are scored on the basis of the incidence of animals responding by showing edema and erythema and on the severity of these responses, using the subjective scoring system described previously, over a period of 24, 48, and 72 hr after the challenge dose. A greater irritation (edema and erythema) after the challenge dose, when compared to that after the sensitizing dose administered to a control animal, is indicative of chemical sensitization.

It is known that certain chemicals, upon penetration into or through the skin, act as haptens and react strongly with certain cellular proteins to form antigenic complexes, which the circulatory macrophages recognize as "foreign" to the body and to which they will develop antibodies. With time, the entire immune system becomes "alerted" to this antigen, and subsequent exposure to the same or a closely related chemical results in an allergic response.

Photoallergic and Phototoxic Reactions

These skin conditions, found in the workplace and in some cases of therapeutic treatment, involve the interaction of certain wavelengths (275–325 nm) of ultraviolet (UV) light that can penetrate skin to the depth of the subdermal blood capillaries and a host of drugs (e.g., sulfonamides, tetracyclines, thiazides, phenothiazines, chlordiazepoxide, cyclamates, hexachlorophene, and griseofulvin) and chemicals (coal tar derivatives and dyes), resulting in the formation of highly reactive intermediates. In the phototoxic situation, current theories suggest that, in the presence of light of suitable wavelength (UV < 320 nm), the chemical molecules are converted into reactive intermediates that can cause direct local cellular toxicity displayed as delayed erythema and hyperpigmentation, followed by a desquamation (shedding or scaling) of the skin. With photoallergic skin reactions, the light-induced activation of the chemical results in the strong binding of some reactive intermediate to cellular and blood plasma proteins to produce antigens that stimulate antibody formation and recognition by the complete immune system as being foreign to the host body. Photoallergic reactions are seen as an immediate urticaria or red rash thought to be antibody mediated and as a delayed eczematous condition considered to be associated with a cell-mediated immune response.

The guinea pig or rabbit are species of choice for studying ultraviolet light-induced chemical toxicity in the skin. In most cases, small amounts of the test agent will be administered orally or by intravenous injection for 10–14 days. Following a resting period of 14–21 days, the animals will receive the same dosage of test agent via the same route of administration, with exposure to light of an appropriate wavelength on an area of closely shaved skin. Scoring the edema and erythema is done by the subjective numerical system described previously. Such animal studies are complicated by the necessity of finding the correct wavelength of UV light to activate the particular chemical being tested; The narrower the band on either side of the specific wavelength, the more intense the biological effect that will be seen. More frequently, one sees screening tests for photoallergy and phototoxicity being included in toxicity data submissions since, as the test systems and diagnostic techniques improve, more of these toxic effects are being detected in the workplace, in the home, and in therapeutic settings with patients receiving certain medications.

Further Reading

Bruner, L. H. (1992). Ocular irritation. In *In Vitro Toxicity Testing* (J. M. Frazier, Ed.), pp. 149–190. Dekker, New York.

Chan, P. K., and Hayes, A. W. (1994). Acute toxicity and eye irritancy. In *Principles and Methods of Toxicology* (A. W. Hayes, Ed.), pp. 579–647. Raven Press, New York.:

DeLeo, V. A. (1992). Cutaneous irritancy. In *In Vitro Toxicity Testing* (J. M. Frazier, Ed.), pp. 191–203. Dekker, New York.:

Ecobichon, D. J. (1997). *The Basis of Toxicity Testing*, 2nd ed., pp. 43–86. CRC Press, Boca Raton, FL.

Gordon, V. C., Harvell, J., Bason, M., and Maibach, H. I. (1994). In vitro methods to predict dermal toxicity. In *In Vitro Toxicology* (S. C. Gad, Ed.), pp. 47–55. Raven Press, New York.

Patrick, E., and Maibach, H. I. (1994). Dermatotoxicology. In *Principles and Methods of Toxicology* (A. W. Hayes, Ed.), pp. 767–803. Raven Press, New York.

Sina, J. F., and Gautheron, P. D. (1994). Ocular toxicity assessment in vitro. In *In Vitro Toxicology* (S. C. Gad, Ed.), pp. 21–46. Raven Press, New York.

—*Donald J. Ecobichon*

Related Topics

Analytical Toxicology
Dose–Response Relationship
Eye Irritancy Testing
Federal Food, Drug, and Cosmetic Act
Hazard Identification
LD$_{50}$/LC$_{50}$
Levels of Effect in Toxicological Assessment
Photoallergens
Skin
Toxicity, Chronic
Toxicity, Subchronic
Toxicity Testing

Toxicity, Chronic

Chronic toxicity studies may be defined as those involving the characterization of adverse health effects following the long-term, repeated administration of a test substance over a significant portion of the life span of the test animal species. In general, the term usually denotes a study conducted for longer than 3 months. However, depending on the test species being used, a 24- to 72-hr exposure period could represent a chronic study for aquatic insects between hatching and flying, a span of some months for birds, or a number of years if dogs or monkeys were to be used. Originally, in regulatory terms, a chronic mammalian study

signified a duration of 2 years, approximately 70–80% of the life span of a laboratory rodent. However, the length of chronic studies has been shortened in the past decade from 2 years to 1 year and currently stands at 6-months duration. This reduction was based partly on results from 286 repeated-dose toxicity studies in which new findings were noted after 6 months for only seven compounds. All significant findings were detected within 6 months for 91% of studies in rats, 98% of those in dogs, and 87% of investigations using monkeys. To accommodate animals with longer life spans or the shortening of the study duration, the dosage regimen is usually adjusted upward so that the level obtained in a lifetime will be acquired in the shorter time interval. There are, of course, major problems inherent in overloading the animals' physiological capacity to distribute, biotransform, and excrete the excessive amounts administered. These difficulties lead to interference in function and to secondary toxic effects seen in other organs.

In chronic studies, it is essential to distinguish between a study defining the shape and nature of the dose–effect relationship for some or any toxicological endpoint and one in which the primary objective is to evaluate the presence or absence of a particular toxicological effect, e.g., neurotoxicity or carcinogenesis.

Experimental Design

With the longer duration of chronic studies and the labor-intensive nature of the investigations involving the employment of a number of individuals to care for the animals, to obtain samples of biological fluids (blood and urine), to carry out various analyses (hematology, blood biochemistry, urinalysis, quantitation of tissue residues of test agent, etc.), and to prepare and examine histological slides of various body tissues, careful attention must be paid to the design of the study. Such investigations can become very expensive, particularly if they have to be repeated because of an oversight, a mistake, or the appearance of unexpected toxicity. It is important to develop an experimental protocol based on the following questions:

1. How many animals will be acquired?

2. How many dosage levels should be used?

3. When does the "lesion" or toxicity begin?

4. How rapidly does the toxicity progress toward signs and symptoms?

5. Does the toxicity disappear (slowly or rapidly) when exposure is stopped?

6. How long should the study be conducted?

7. How can the main "theme" of the study be retained when other, unexpected toxicity is observed, including excessive mortality within a single treatment group, etc.?

A piece of paper and a pencil are the most valuable tools, at this stage of the study, to develop responses to the question "what if" this or that might happen during the investigation and attempting to anticipate what might happen with repeated administration of the agent. Study designs should be as open-ended as possible, retaining flexibility to react to the unforeseen, unpredicted events, as well as to those that are anticipated. An example of an experimental protocol is shown in Fig. T-3.

All too frequently, chronic studies are carried out according to the guidelines of national or international reg-

ulatory agencies rather than according to good scientific principles. The regulatory guidelines are only to guarantee a minimum of requirements, information, results, etc., standardized or harmonized within and between national and international governmental bodies.

There may be scientific justification in carrying out a study in a particular manner not commonly ascribed to by such agencies. Few agencies would react unfavorably to a design defended by valid scientific principles.

The most difficult task in developing the protocol for a chronic toxicity study is the selection of an appropriate range of dosages to be used, based on limited knowledge about the effects of long-term exposure of an animal model to the test agent. Some prior knowledge of the shape of the dose–effect relationship is required and is most frequently obtained from subchronic studies conducted beforehand (see Toxicity, Subchronic). The objective is to use dosages of the test agent, administered by inhalation, in the diet, in drinking water, or by oral gavage that will cause minimal adverse health effects within the study period but will not cause excessive mortality. The maximum tolerated dose (MTD) is frequently used as the highest dose. This exposure level is defined as the highest dose that causes no more than a 10% decrease in body weight and does not produce mortality, clinical signs of toxicity, or pathologic lesions that would be predicted to shorten the animal's natural life span. Usually, two

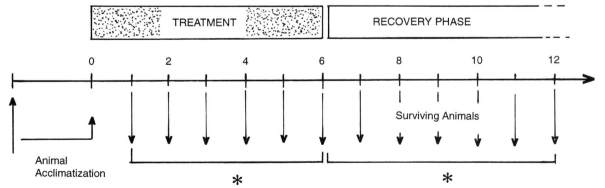

FIGURE T-3. *The design of a chronic (6-month, 180-day) study; the planning chart enables the investigator to determine the total number of animals required based on the number of dosage levels and the number of treated animals required for euthanasia at each selected time interval (30 days). Periodic selection of representative subgroups of each population (controls, low, intermediate, and high levels of test agent) would permit both a dosage- and time-related study of the development of toxicant-related lesions as well as changes in physiological and/or biochemical tests of organ and tissue function/injury. Included in the design is a posttreatment recovery phase open to a further 6 months to assess the permanence or reversibility of the toxicant effects. *Animals (representative subgroups) euthanized at predetermined time intervals for physiological, biochemical, and morphological study.*

lower levels, an intermediate (MTD/4) and a low (MTD/8) level will be selected in anticipation of observing a gradation in both appearance and severity of effect(s). Suitable numbers of control, untreated animals must be carried throughout the study, resulting in a four-dose design.

More appropriate for carcinogenicity studies in which cancer is the only toxic endpoint of interest, the MTD has caused considerable controversy when used in chronic toxicity studies. Many scientists believe that the highest dose should be above the MTD to be certain of eliciting some quantifiable deleterious effect(s), e.g., to demonstrate that the model "works." Other scientists feel just as strongly that the highest dose should be lower than the MTD. In these scenarios, either a dose-related increase in adverse health effects or little or no toxicity may be detected independent of minimal body weight changes. The latter scenario poses a number of problems in interpreting the results for regulatory purposes. In other study designs, the lower dose levels may be fractions of the selected highest dose either equally spaced, as is shown in Fig. T-4 (50 and 25%, respectively, of the highest dose chosen), or unequally spaced (20 and 1.0% of the highest dose chosen).

A basic principle of toxicology is that there is a correlation between biological effects and the level(s) of exposure—a dose–effect relationship. With a three-dose design (O, X/2, X), straight-line relationships can always be determined. However, this may not reflect the true situation, whereas with a four-dose design, the usual curvilinear (concave or convex) relationship will be seen. Prior to embarking on the chronic study, an initial trial period of 2 or 3 weeks duration should be carried out at the selected dosage range, even to the point of conducting dose-dependent kinetic and tissue distribution studies on a few animals at each treatment level in order to assess whether or not the test animals will tolerate the dosages selected. Adjustments to the preselected dosages can be made at this time without compromising the remainder of the study. Frequently, some downward adjustment of the highest dose may be necessary when excessive toxicity is observed.

The duration of the study may be dictated by guidelines from certain national or international regulatory bodies. As was indicated earlier, chronic studies were previously of 2 years duration. The exorbitant costs involved plus the concept that the same toxicity could be detected and quantified in a shorter period of time by giving higher dosages resulted in a reduction in the duration to a 6-month time period. This is still a contentious issue among toxicologists, with many maintaining that 6 months is only a fraction (20%) of the life span of a rodent and that toxicity may not appear until the animal is older than 12 months, when geriatric dysfunction begins to occur.

In earlier chronic studies, the animals were allowed to proceed until obvious toxicity was seen or the animals became moribund—these animals were euthanized in a humane manner for study. However, by that time, the toxicity was well advanced and, of course, the question of when the toxicity began or subtle changes in organ function occurred, the usual signals of impending toxicity, could not be answered. Such questions can only be answered by the periodic (every 30 days) selection of representative subgroups from each treatment group and from control animals for euthanasia and an in-depth study of biochemical, physiological, and morphological indices of toxicity.

Such an approach does not preclude the periodic sampling of blood from animals or the collection of urine and feces for analysis using techniques that are not life-threatening. Such a protocol will permit the investigator to identify pre-toxic changes in organ function and morphology as well as to determine when, in a dose-dependent manner, toxicity appears initially,

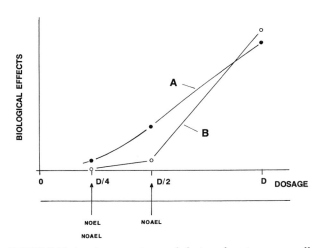

FIGURE T-4. *An experimental design showing unequally spaced dosages and theoretical results used for extrapolation to estimate the no-observed-adverse-effect level (NOAEL) or the no-observed-effect level (NOEL). In curve A, some slight degree of toxicity was observed at the lowest dose administered, permitting only the estimation of an NOAEL. In curve B, no toxicity was observed at the lowest dose and slight toxicity at the intermediate dose, permitting the estimation of a NOEL and a NOAEL.*

both as obvious signs and symptoms and as morphological changes.

How does one know whether or not the toxicity persists following termination of the exposure or that the signs and symptoms disappear slowly or quickly? Are there any long-lasting effects? Is the tissue damage reversible or irreversible? To answer these questions, additional animals should be incorporated into the study protocol so that, at the end of the treatment period, there is a reasonable population of animals remaining—sufficient to participate in a recovery phase study. Once again, small representative subgroups from each treatment group and controls will be euthanized and subjected to detailed study at predetermined time intervals(e.g., at 30, 60, and 90 days).

How many animals will be needed to provide biochemical, physiological, and morphological data for each of the previously mentioned planned intervals of subgroup selection (during and after treatment) as well as for the unexpected toxicity and mortality that almost certainly will be encountered if the study is being conducted properly? What constitutes a representative subgroup? If one accepts the premise that 5 animals of each sex, selected at each time interval for euthanasia, are representative of the population being studied, the number of animals required can be calculated quickly. A larger number of controls, e.g., 10 animals, would be required at each interval so that variability in the population as a whole can be monitored. Ten to 15 additional animals should be included in each treatment group in anticipation that some mortality may occur during treatment. Thus, for a 6-month chronic toxicity study having three dosage levels plus control animals, with subgroups being euthanized at 30-day intervals during treatment and at 30-day intervals over a 3-month recovery phase, an investigator might consider a minimum of 150 male and 150 female animals undergoing treatment with 90 control animals of each sex (a total of 480 animals). The number of animals could be reduced by spacing out the time intervals of subgroup selection but perhaps at the risk of missing some subtle change in one or more parameters being assessed, thereby not recognizing the appearance of the toxic effect(s).

A wide range of biochemical and physiological parameters should be monitored throughout the entire study, both during and after treatment (Fig. T-5). The moribund animals or those euthanized at preselected intervals will undergo an extensive morphological ex-

amination, both gross and microscopic, in order to identify possible organs or tissues where the test agent may exert an effect. Changes in body weight, food, and water consumption can provide information concerning the tolerance/aversion of the test animal to the agent in the food or water. A reduction in body weight, particularly if it is dose related, over the study period is a simple but effective indicator of the animals' well-being; any sharp deviation will alert the investigator to a possible chemical-related event. If the animal does not feel well, it will not eat sufficient food to maintain normal growth and development, this being particularly critical in small rodents that have an elevated basic metabolic rate. Taste aversion to the test agent can be identified quickly by measuring the amount of food or water ingested in a 24-hr period. A spectrum of biological markers—general tests of hematology, blood serum biochemistry, and urinalysis—should be planned before the experiment is started, picking parameters that, if they are seen to change, will point in a meaningful way to some organ/tissue that may be affected by the test agent. All of these tests should be broad enough in scope to detect the unexpected as well as the anticipated toxicity. During this prolonged treatment period, various noninvasive tests of neurological competence (behavioral, sensory perception, motor function, and learning skills) can be conducted in addition to organ (liver, cardiac, and pulmonary) function tests that will pose minimal risk to the animals' health. At euthanasia, the animals will be dissected and a wide range of organs will be removed, fixed, sliced, and stained appropriately for light and electron microscopic examination. An attempt will be made to correlated morphological changes or damage with the biochemical and physiological changes observed and quantified.

Carcinogenicity Studies

Chronic studies include one additional endpoint of toxicity: any carcinogenicity related to exposure to the test agent. Traditionally, and mainly because tumor formation is seen in older animals, carcinogenicity studies in rodents are conducted for a 2-year period, separate from the shorter term, 6-month chronic studies. The dosage range used for carcinogenic assessment is lower than that used for chronic toxicity because the

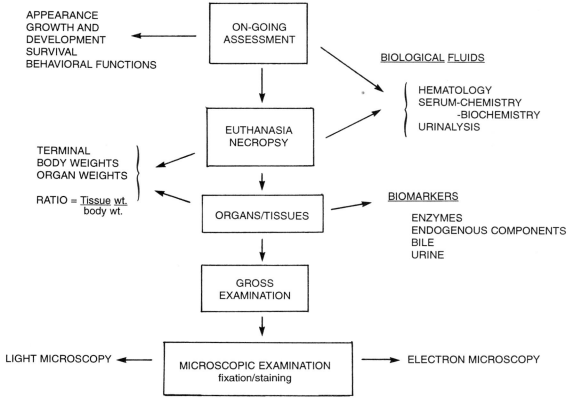

FIGURE T-5. *A flowchart depicting the various parameters or endpoints of toxicity to be monitored during the chronic study, both during and after treatment, as well as those studied following euthanasia and necropsy. The routine, periodic assessment of chosen parameters may detect the onset of impending toxicity, proving invaluable for the detection of developing lesions and as predictors of target–organ toxicity.*

endpoint is tumor formation, independent of any other sort of toxicity.

Considerable controversy has arisen among scientists over the selection of the highest dosage to be used in carcinogenic studies. One faction suggests a value known as the MTD—the highest dose of the test agent during a chronic study that can be predicted not to alter the animals' normal longevity from effects other than carcinogenicity. This dosage level is also known as the "minimally toxic dose." With this dosage, selected from a chronic study as a reference point, two lower dosage levels, equally or unequally spaced, can be calculated so that, it is hoped, the lowest level of exposure that will cause tumor formation can be determined. Another faction claims that doses near the MTD may cause significant cell mortality and/or compensatory changes (mitogenesis) that would make the damaged organ/tissue more susceptible to tumor formation, and

that these levels are far higher than those that the human would encounter. Other investigators recommend that the top dose be some appropriate multiple of the expected human exposure. While this controversy has not been settled, retrospective review of a large number of chemical-induced carcinogenicity studies has revealed that two-thirds of the carcinogens would have been detected even if the estimated MTD had not been included but that, in many studies, some site-specific carcinogenic effects would not have been observed. Among the remaining one-third of the studies, approximately 80% had elevated rates of site-specific tumors at lower doses as well. Most carcinogenic effects observed at the highest dose were also present at reduced incidences at lower doses (MTD/4 and MTD/2), although the results may or may not be statistically significant. The choice of dosages for long-term carcinogenicity studies will remain a contentious issue.

Interpretation of Results

The main objective of any chronic study is to supply a database that will provide assurance to public concerns about the safety of chemicals found in the human environment, the sources usually being air, food, and water, and the results of such studies being used in safety evaluation, risk assessment, and risk manage-ment decisions. The objective of long-term toxicity testing, usually in rodents, is to assess the potential chronic toxicity of a chemical, including carcinogenicity, effects that would not be evident in subchronic studies. Pertinent to the studies is the development of a dose–effect relationship that ranges from no observed effect through minor changes to overt toxicity and the determination of dosages at which these observations occur. These values will be used by regulatory agencies to determine safe levels of exposure stated as a maximum allowable concentration, recommended maximum levels, reference dose, virtually safe dose, tolerance or acceptable daily intake, etc.

Endpoints of toxicity and the severity of observed adverse health effects obtained from a four-dosage range study may be represented by arbitrarily determined dosage values such as the lowest-observed-adverse-effect level, the no-observed-adverse-effect level (NOAEL), or the no-observed-effect level (NOEL) (Fig. T-4). These values, of course, must be derived at the end of the chronic toxicity study based on the observations. It is unlikely that all three values would be obtained from a study. Usually, one of these indices might be determined with a degree of reliability in the estimated value. Either the NOAEL or NOEL can be used by regulatory bodies to establish reasonable, estimated values for the indices mentioned in the previous paragraph.

Further Reading

Arnold, D. L., Blumenthal, H., Emmerson, J. L., and Krewski, D. (1984) The selection of doses in chronic toxicity/carcinogenicity studies. In *Current Issues in Toxicology* (H. C. Grice, Ed.), pp. 9–49. Springer-Verlag, New York.
Ecobichon, D. J. (1997). *The Basis of Toxicity Testing*, 2nd ed., pp. 87–116. CRC Press, Boca Raton, FL.
Haseman, J. K., and Lockhart, A. (1994). The relationship between use of the maximum tolerated dose and study sensitivity for detecting rodent carcinogenicity. *Fundam. Appl. Toxicol.* 22, 382–391.
Stevens, K. R., and Mylecraine, L. (1994). Issues in chronic toxicology. In *Principles and Methods of Toxicology*, pp. 673–695. Raven Press, New York.
Stringer, C. (1992). Safety workshops—ICH. *Regul. Affairs J.* 3, 350–356.

—*Donald J. Ecobichon*

Related Topics

Analytical Toxicology
Carcinogenesis
Dose–Response Relationship
LD_{50}/LC_{50}
Levels of Effect in Toxicological Assessment
Maximum Allowable Concentration
Maximum Tolerated Dose
Toxicity, Acute
Toxicity, Subchronic
Toxicity Testing

Toxicity, Subchronic

While acute exposure to high concentrations of chemicals can occur in any environment (the outdoors, the home, and the workplace), individuals are more frequently exposed over much longer periods of time to agents at levels lower than those that might prove fatal. Acute exposure studies will not identify those adverse health effects, both immediate and/or delayed, that might arise as a consequence of longer term, lower level exposure. The simulation of such exposure requires the development of more carefully designed experiments in which larger numbers of animals are used, lower levels of the potential toxicant are administered by a suitable route of exposure over a longer time period, and a number of preselected biochemical, physiological, and morphological endpoints of toxicity are monitored and quantified throughout the study period. In general, a subchronic study is one conducted over a 21- to 90-day period, using surrogate animal species to mimic conditions anticipated to be found in human exposure.

These longer term studies are designed to examine the nature of the toxic effects from lower dosages at the organ, tissue, and cellular level in order to determine possible mechanisms of toxicant action. The repeated administration/exposure to the agent will permit examination of possible cumulative effects as body burdens of the agent and/or biotransformation products (metabolites) are acquired with time. Subchronic studies will allow the investigator to ascertain the variation in response(s) by making close observations of a continuum of biological changes and/or unique events occurring over a wide range of dosage levels in both sexes and ages of different animal species. This will permit the identification of the appropriate dosage at which biochemical, physiological, and morphological changes, both macroscopic and microscopic, occur in relation to the level and/or duration of exposure. The last objective of such studies is to be able to predict the long-range adverse health effects in the test animal species, using the results to extrapolate whether or not toxicity might be expressed in the human at some as yet unknown level of exposure.

The old adage holds true that "the more species of animals in which the same biological response(s) to an agent can be produced, the greater is the chance that, at some dosage, the same effect might occur in the human." Invariably, subchronic studies are conducted in at least two species—one rodent species, with a choice of the mouse, rat, or possibly the hamster, and a nonrodent species, frequently the dog (purebred beagle), the rabbit, or, occasionally, a strain of monkey (rhesus or macaque). With such diverse species being studied, distinct variations in response(s) related to physiological (distribution, storage, and excretion) or biochemical (e.g., biotransformation rate and type of metabolites formed) differences should be anticipated, thus, it is hoped, permitting a better appreciation of how the human might respond.

Experimental Design

The design of subchronic studies is extremely important not only because of the longer time period involved but also because such studies are labor-intensive, involving a number of people in caring for the animals, obtaining blood samples, carrying out analyses on samples (hematology, blood chemistry, and urinalysis) or examining morphological specimens (e.g., preparing and staining slides of tissue sections and light and electron microscopic evaluation). Since these studies become very expensive, it is important to set up an experimental design before the studies are begun that asks the following questions:

1. How many animals are required?

2. How many dosage levels should be used?

3. When does the "lesion" or toxicity begin?

4. How rapidly does the toxicity progress toward signs and symptoms?

5. Does the toxicity disappear (slowly or rapidly) when exposure is stopped?

6. How can the main "theme" of the study be retained (or required) when other, unexpected toxicity is observed, including excessive mortality within a single treatment group, etc.?

7. How long should the study be conducted?

Any study design should be open-ended, allowing for unforeseen and unpredicted events that frequently appear in longer term studies as well as those events predicted. A piece of paper and a pencil are the most valuable tools at this stage of the design, when one must ask the question "what if" this or that might happen during the study. A simple experimental design is shown in Fig. T-6.

Having selected the agent for study, the first question is what dosage range should be used and how many dosage levels should be necessary. Generally chosen from the dosages used in acute toxicity studies, one should always use three dosage levels: a high level guaranteed to elicit toxicity in the animal model and two lower (intermediate and low) dosages in the hope that a gradation in the appearance and severity of toxicity will be observed. Comparable control (untreated) animals must be carried through the study as well. A major objective of the study is to establish a relationship between biological effects and a level of exposure, e.g. a dose–effect relationship that may be linear or, more likely, curvilinear.

The duration of the study may be dictated by guidelines from specific regulatory agencies. In general, sub-

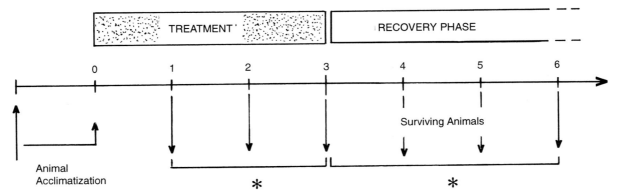

FIGURE T-6.   *The design of a subchronic (3-month, 90-day) study; the planning chart enables the investigator to determine the total number of animals required based on the number of dosage levels and the number of treated animals required for euthanasia at each selected time interval (30 days). Periodic selection of representative subgroups of each population (controls, low, intermediate, and high levels of test agent) would permit both a dosage- and time-related study of the development of toxicant-related lesions as well as changes in physiological and/or biochemical tests of the organ and tissue function/injury. Included in the design is a posttreatment recovery phase open to a further 90 days to assess the permanence or reversibility of the observed effects. *Animals (representative subgroups) euthanized at predetermined time intervals for physiological, biochemical, and morphological study.*

chronic feeding studies, particularly if the agent may be incorporated into the diet, drinking water, or given by oral gavage, are of 90-day duration. Inhalation or dermal exposure studies may be from 21 to 90 days in duration.

It is unlikely that the investigator can predict when toxicity will appear based on the acute toxicity results. One can allow the subchronic study to proceed until sick animals are observed before euthanizing them in a humane manner but, by that time, the toxicity will be in an advanced state. The question of when does toxicity begin to appear can only be answered by the periodic (every 7 or 30 days in 21- or 90-day studies, respectively) selection of representative subgroups from each treatment group for euthanasia and in-depth study of biochemical, physiological, and morphological indices of toxicity. Such a design allows the investigator to identify pretoxic changes in organ function and morphology as well as to determine when, in a dose-dependent manner, toxicity appears.

If exposure is stopped, does the toxicity persist or do the signs and symptoms slowly or quickly disappear? Is the tissue damage reversible or irreversible? By incorporating additional animals into the treatment groups at the beginning, there is a good chance that, at the end of the treatment period, there will be sufficient animals surviving to allow a recovery phase to be studied, again selecting small, representative subgroups for euthanasia and detailed study at predetermined time intervals (30, 60, or 90 days).

How many animals are needed to cover all the eventualities mentioned previously—three dosage levels, periodic euthanasia of subgroups during treatment and after termination of exposure, possible expected and unexpected toxicity, some mortality among the animals, related or unrelated to treatment, etc.? What constitutes a representative group or subgroup? How many untreated control animals should be included? If one accepts the premise that 5 animals of each sex, selected at each time interval for euthanasia, are representative of the population under study, then the number of animals required can be calculated quickly with a few (10–15) additional animals being included for "safety." By the time the study is under way, the animal cost is the least expensive item in the investigation. Thus, one should not be reticent at including more animals. For a 90-day feeding study, an investigator would conservatively consider 150 male and 150 female rodents to adequately protect the study from the vagaries of Murphy's law (if anything can happen, it will), including adequate numbers of control and "spare" animals undergoing treatment.

A wide range of parameters can and should be measured during the entire study. Some, as simple as body weight, growth/development, and food and water consumption, are noninvasive and pose no risk to the animals. A change in body weight, particularly in small rodents having normally high basal metabolic rates, is a simple but effective indicator of general well-being: A sharp decrease alerts the investigator to perhaps a chemical-related appetite depression, although the effect may be as simple as an aversion to the taste of the test agent in the diet rather than toxicity. Frequently, one can see a gradation in growth curves between the control animals and those in the three treatment groups. A spectrum of biological markers—general tests of hematology, blood serum chemistry, and urinalysis—should be planned before the start of the study along with other specific physiological and biochemical markers. These parameters should be based on anticipated target organ toxicity but, of course, remain broad enough in scope to detect the unexpected toxicity as well. Such a scheme is shown in Fig. T-7. During the treatment period, various noninvasive tests of neurological competence (behavior, sensory perception, motor function, and learning skills) can be conducted along with liver, kidney, cardiac, and pulmonary function tests that pose minimal risk to the survival of the test animal.

Information Management

As can be appreciated, with more than 300 animals in a subchronic study being monitored periodically for

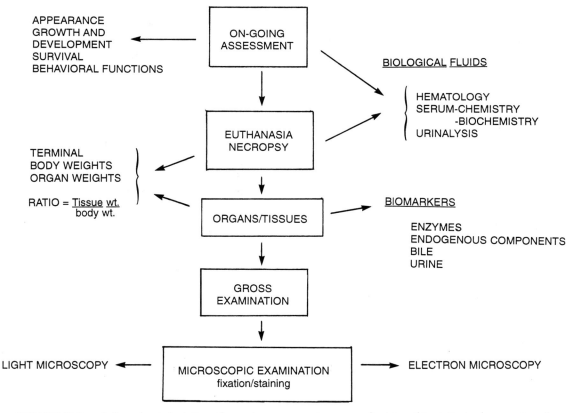

FIGURE T-7. *A flowchart depicting the various parameters or endpoints of toxicity to be monitored during the subchronic study, both during and after treatment, as well as those studied following euthanasia and necropsy. The routine, periodic assessment of chosen parameters may detect the onset of impending toxicity, proving invaluable for the detection of developing lesions and as predictors of target-organ toxicity.*

any adverse health effect, one can accumulate literally thousands of continuous and terminal data points on biochemical, physiological, and morphological parameters which become a significant burden to the investigator and the staff. Management of these data is crucial. Most of the data flow control has become automated on computer.

Good Laboratory Practices regulations insist on the appropriate management of animal data so that quality assurance/quality control (QA/QC) personnel can, at any time, select an animal number and track it and its data throughout the study until it either dies or is euthanized. The laboratory carrying out the study will have a QA/QC staff, while the company for whom the study is being done will have its QA/QC personnel check the study in progress and at the completion of the report before it is submitted to another QA/QC evaluation carried out by the regulatory agency.

Interpretation of Results

Depending on the eventual use of the chemical, the results of a well-designed and conducted subchronic study may provide all of the information required for the agent (e.g., for a drug that will be used for only a limited time period of treatment). However, if exposure is anticipated to be longer or if the individual may be exposed to the agent (pesticides, food additives, industrial chemicals, etc.) for a reasonable portion of a lifetime, the results of the subchronic study may only justify making a decision on the need for additional or more extended and perhaps specific studies to determine more clearly the toxicological profile of the agent. In the time period of the study, some preliminary evidence may have been obtained to show that if treatment had persisted for a longer time at the same or at lower dosages, some specific target organ toxicity might have become manifest. In such a situation, the subchronic study results have at least established a dosage range for administration over a significant proportion of the animal species life span (e.g., the characteristic chronic toxicity study).

Changes observed in body weight gain, organ weight, hematological and biochemical data, organ function, etc. should be subjected to trend analysis with those parameters measured in control animals. These findings should be correlated with the pathological and histopathological data. Since a basic tenet of toxicology is that there should be some correlation between observed biological effect(s) and the dosage of test chemical, much effort is expended in establishing a trend in a dose–effect relationship for each parameter being measured.

Subchronic studies are of limited value for predicting the toxic effects of life span exposure. The nature and degree of toxicity vary; the sensitivity performance and metabolic capability of organs/tissues change with aging and the spontaneous occurrence of other diseases. No predictions can be made from subchronic exposure concerning the mutagenic, teratogenic, or carcinogenic potential of the test agent, and any effects on reproduction can only be related to primary effects on the testes and ovaries.

Given the known uncertainties that arise from qualitative and quantitative differences as well as similarities in toxicological effects observed in animals and humans, interpretation of the dose-related effects must be done cautiously. If a no-observed-adverse effect level of dosage can be determined from the experiment, this may be used to establish values for acceptable daily intake or a reference dose for setting tolerances of additives in food, for residue levels of unintentional contaminants, or for acceptable levels of exposure (threshold limit values or maximum acceptable concentrations) to chemicals in the workplace. However, beyond these indices, extrapolation to what might occur during a lifetime of exposure is extremely risky.

Further Reading

Ecobichon, D. J. (1997). *The Basis of Toxicity Testing*, 2nd ed., pp. 87–116. CRC Press, Boca Raton, FL.

Wilson, N. H., and Hayes, J. R. (1994). Short-term repeated dosing and subchronic toxicity studies. In *Principles and Methods of Toxicology* (A. W. Hayes, Ed.), pp. 649–672. Raven Press, New York.

—*Donald J. Ecobichon*

Related Topics

Analytical Toxicology
Dose–Response Relationship
Good Laboratory Practices
Levels of Effect in Toxicological Assessment
Toxicity, Acute

Toxicity, Chronic
Toxicity Testing

Toxicity Testing, Alternatives

Since the early 1980s, public perception of the value of (and benefits from) animal testing has been strongly influenced by what is now called the animal rights movement. This concern has done a great deal of good because it has caused careful consideration of why and how testing is performed, with significant alterations being made in practices across the range of safety assessment. Its impact has been uneven, however, in the degree of sensitivity of the issue of animal testing in different industries and organizations. Some organizations no longer perform any animal testing, having it conducted externally if it is required. Others simply are quiet about it.

The guiding principles subscribed to by many in the field are the four R's. The historical beginnings of this concept date to 1959 when Russell and Burch first proposed what have come to be called the three R's of humane animal use in research: replacement, reduction, and refinement. These three principles have served as the conceptual basis for reconsideration of animal use in research. To these has been added a fourth principle: responsibility.

Replacement means utilizing methods which do not use intact animals in place of those that do. For example, veterinary students may use a canine cardiopulmonary resuscitation simulator, Resusci-Dog, instead of living dogs. Cell cultures may replace mice and rats that were fed new products to discover substances poisonous to humans. In addition, using the preceding definition of animal, an invertebrate (e.g., a horseshoe crab) could replace a vertebrate (e.g., a rabbit) in a testing protocol.

Reduction refers to the use of fewer animals. For instance, changing practices allow toxicologists to estimate the lethal dose of a chemical with as few as one-tenth the number of animals used in traditional tests. In biomedical research, long-lived animals such as primates may be used in multiple sequential protocols assuming that the protocols are not deemed inhumane or scientifically conflicting. Designing experimental protocols with appropriate attention to statistical inference can lead to either decreases or increases in the numbers of animals used. Through coordination of efforts among investigators, several tissues may be simultaneously taken from a single animal. Reduction can also refer to the minimization of any unintentionally duplicative experiments, perhaps through improvements in information resources.

Refinement entails the modification of existing procedures so that animals are subjected to less pain and distress. Refinements may include administration of anesthetics to animals undergoing otherwise painful procedures; administration of tranquilizers for distress; humane destruction prior to recovery from surgical anesthesia; and careful scrutiny of behavioral indices of pain or distress followed by cessation of the procedure or the use of appropriate analgesics. Refinements also include the enhanced use of noninvasive imaging technologies that allow earlier detection of tumors, organ deterioration, or metabolic changes and the subsequent early euthanasia of test animals.

Responsibility is the fourth R, which was not in Russell and Burch's initial proposal. To toxicologists this is the cardinal R. They may be personally committed to minimizing animal use and suffering and to doing the best possible science of which they are capable, but at the end of it all, toxicologists must stand by their responsibility to be conservative in ensuring the safety of the people using or exposed to the drugs and chemicals produced and used in our society. This is particularly true for medical devices and includes in it the element of ensuring adherence to regulatory requirements and standards.

Toxicology, and particularly the portion of it associated with the assessment of commercial products for safety, is philosophically a conservative scientific practice. If a choice is to be made as to whether to accurately identify human hazard or to overpredict any hazard, with both predictions having a degree of uncertainty, the latter course will be chosen. Likewise, if an evaluation process which is dated but very familiar is challenged by a new technology which is scientifically, economically, and ethically superior but with which there is no precedent or prior history of use, the former will be selected. Both of these two choices are made with reference to what (1) is regulatorily accepted (i.e., codified in law) and (2) what has to date been the standard in litigation defense. The key assumptions currently underlying safety assessment are (1) that other organisms can serve as accurate predictive models of toxicity in humans, (2) that selection of an appropriate model to use is essential to the accurate prediction of adverse health effects in humans, and (3) that an understanding of the strengths and weaknesses of any particular model is required before the relevance of specific findings to humans can be established. When we refer to models, we usually mean test organism, though in fact the manner in which parameters are measured (and which parameters are measured) to characterize an endpoint of interest is also a critical part of the model (or, indeed, may actually constitute the model). To an increasing degree, both *in vivo* (intact, higher organism) and *in vitro* models are used, though the degree of utilization of *in vitro* models has lagged behind its potential use.

Mechanisms of chemical toxicity are largely identical in humans and animals. Our increased understanding of mechanisms on the molecular and cellular level has caused some of the same people who question the general principle of predictive value of animal tests to suggest that the state of knowledge is such that mathematical models or simple biochemical or cell culture systems could always be used in place of intact animals to accurately predict or warn of toxicities in humans. This last suggestion also misses the point that the final expressions of toxicity in humans or animals are frequently the summation of extensive and complex interactions occurring at cellular and biochemical levels. For example, although it was once widely believed (and still is believed by many animal rights activists) that *in vitro* mutagenicity tests would replace animal bioassays for carcinogenicity, this is clearly not the case on either scientific or regulatory grounds. Although there are differences in the responses of various species (including humans) to carcinogens, the overall predictive value of such results (when tempered by judgment) is clear.

Increasingly, alternative models that use other than intact higher organisms are being used in toxicology for a number of reasons. These reasons include desires for specificity of response, use of small quantities of test materials, and expedited development, all of which are particularly important in the biotechnology industries. Well-reasoned use of *in vitro* or other alternative test model systems is essential to the development of a product safety assessment program that is both effective and efficient.

The "ideal" test to answer a safety assessment question should have an endpoint measurement that provides data such that dose–response relationships can be obtained. Furthermore, any criterion of effect must be sufficiently accurate in the sense that it can be used to reliably resolve the relative toxicity of two test chemicals that produce distinct yet similar responses (in terms of hazard to humans). In general, it may not be sufficient to classify test chemicals into generic toxicity categories. For instance, if a test chemical falls into an intermediate toxicity category but is borderline to the next, more severe toxicity category, it should be treated with greater concern than another test chemical that falls at the less toxic extreme of the same immediate category. Therefore, it is essential for a test system to be able to place test chemicals in an established toxicity category as well as to rank materials relative to others in that category.

The endpoint measurement of the ideal test system must be objective. This is important so that a given test chemical will yield similar results when tested using the standard test protocol in different laboratories. If it is not possible to obtain reproducible results in a given laboratory over time or between various laboratories, then the historical database against which new test chemicals are evaluated will be time or laboratory dependent. If this is the case, then there will be significant limitations on the application of the test system because it could potentially produce conflicting results. From a regulatory point of view, this possibility would be highly undesirable. Along these lines, it is important for the test protocol to incorporate internal standards to serve as quality controls. Thus, test data could be represented utilizing a reference scale based on the test system response to the internal controls. Such normalization, if properly documented, could reduce interest variability.

From a practical point of view, there are several additional criteria that the ideal test should meet. Alternatives to current *in vivo* test systems basically should be designed to evaluate the observed toxic response in a manner as closely predictive of the outcome of interest in humans as possible. In addition, the test should be fast enough so that the turnaround time for a given test chemical is reasonable for the intended purpose (very rapid for a screen and timely for a definitive test). The speed of the test and the ability to conduct tests on several chemicals simultaneously will determine the overall productivity. The test should be inexpensive so that it is economically competitive with current testing practices. Finally, the technology should be easily transferred from one laboratory to another without excessive capital investment (relative to the value of the test performed) or the need for special skills for test implementation.

The point is that these characteristics of the ideal test system provide a general framework for evaluating alternative test systems in general. No test system is likely to be ideal. Therefore, it is necessary to weigh the strengths and weaknesses of each proposed test system in order to reach a conclusion as to the effectiveness of a particular test.

In recent years, tremendous progress has been made in our understanding of mechanisms of biological action down to the molecular level. This has translated to multiple modifications and improvements to *in vivo* testing procedures which now give us tests which (1) are more reliable, reproducible, and predictive of potential hazards in humans; (2) use fewer animals; and (3) are considerably more humane than earlier test forms. Since 1971 *in vitro*/alternative test systems have been proposed, developed, and validated to at least some extent. Yet the perception persists that little has changed in how safety assessment is performed by or for industry.

In both theory and practice, *in vivo* and *in vitro* tests each have potential advantages, as summarized in Tables T-6 and T-7. It should be noted that the relative weight assigned these advantages will differ depending on the information required and how it is to be used.

Can the proper tests be selected, especially when a decision must be made between using with an existing test system or adopting a new one? What are the available options?

The division between test system models is more complex than *in vivo* and *in vitro*, of course. There are a range of options under *in vitro*, each with its own advantages and disadvantages, as shown in Table T-8. Each of these levels will need to be considered.

TABLE T-6
Rational for Using *in Vivo* Test Systems

- Evaluate actions/effects on intact animals and assess organ/tissue interactions
- Allow either neat chemicals or complete formulated products (complex mixtures) to be evaluated
- Yield data on the recovery and healing processes
- Are currently predominantly the required statutory tests for agencies worldwide
- Afford quantitative and qualitative evaluations using a scoring system that is generally capable of ranking materials according to their relative hazards
- Are amenable to modifications to meet the requirements of special situations (such as multiple dosing or exposure schedules)
- Allow the use of an extensive available database and have cross-reference capabilities for evaluation of relevance to human situation
- Are associated with ease of performance and relatively low capital costs in many cases
- Are generally both conservative and broad in scope, providing for maximum protection by erring on the side of overprediction of hazard to humans

TABLE T-7
Limitations of *in Vivo* Testing Systems That Serve as a Basis for Seeking *in Vitro* Alternatives for Safety Assessment Tests

- May involve complications and/or confounding or masking of findings
- May assess only the short-term site of application or immediate structural alterations produced by agents; however, specific *in vivo* tests may only be intended to evaluate acute local effects, so this may be a purposeful test system limitation
- Require stringent technician training and monitoring (particularly because of the subjective nature of evaluation)
- May not perfectly predict results in humans if the objective is to exclude or identify severe acting agents
- Structural and biochemical differences between test animals and humans that make extrapolation from one to the other difficult
- Lack of standardization
- Variable correlation with human results
- Large biological variability between experimental units (i.e., individual animals)
- Large, diverse, and fragmented databases that are not readily comparable
- Require a comparatively longer time to express/evaluate endpoints
- Require comparatively larger quantities of test material
- May be conducted using either a single endpoint (e.g., lethality and corrosion) or a so-called "shotgun" or multiple-endpoint approach (e.g., a 13-week oral toxicity study)
- Are the accepted norm for evidence in courts of law for litigation cases

TABLE T-8
Levels of Models for Toxicity and Research

Level/model	Advantages	Disadvantages
In vivo (intact higher organism)	Full range of organismic responses similar to target species	Cost Ethical/animal welfare concerns Species-to-species variablity
Lower organisms (earthworms, fish)	Range of integrated organismic responses	Frequently lack responses typical of higher organisms Animal welfare concerns
Isolated organs	Intact but isolated tissue and vascular system Controlled environmental and exposure conditions	Donor organism still required Time-consuming and expensive No intact organismic responses Limited duration of viability
Cultured cells	No intact animals directly involved Ability to carefully manipulate system Low cost Ability to study a wide range of variables	Instability of system Limited enzymatic capabilities and viability of system No (or limited) integrated multicell and/or organismic responses
Chemical/biochemical systems	No donor organism problems Low cost Long-term stability of preparation Ability to study a wide range of variables Specificity of response	No de facto correlation to *in vivo* system Limited to investigation of a single defined mechanism
Computer simulations	No animal welfare concerns Speed and low per-evaluation cost	May not have predictive value beyond a narrow range of structures Expensive to establish

It should be noted that, in addition to potential advantages, *in vitro* systems per se also have a number of limitations that can contribute to their not being acceptable models. Findings from an *in vitro* system which either limit their use in predicting *in vivo* events or make them totally unsuitable for the task include there being wide differences in the doses needed to produce effects or differences in the effects elicited. Some reasons for such findings are detailed in Table T-9.

At the same time, there are substantial potential advantages in using the *in vitro* system. The scientific advantages of using cell or tissue culture in toxicological testing are isolation of test cells or organ fragments from homeostatic and hormonal control, accurate dosing, and quantitation of results. It is important to devise a suitable model system which is related to the mode of toxicity of the compound. Tissue and cell culture has the immediate potential to be used in two very different ways by industry. First, it has been used to examine a particular aspect of the toxicity of a compound in relation to its toxicity *in vivo* (i.e., mechanistic or explanatory studies). Second, it has been used as a form of rapid screening to compare the toxicity of a group of compounds for a particular response. Indeed, the pharmaceutical industry has used *in vitro* test systems in these two ways for years in the search for new potential drug entities. The extension of these approaches to safety assessment is a much more recent occurrence.

The theory and use of screens in toxicology have previously been reviewed by the author. Mechanistic and explanatory studies are generally called for when a traditional test system gives a result that is either unclear or for which the relevance to the real-life human exposure situation is unclear. *In vitro* systems are particularly attractive for such cases because they can focus on very defined single aspects of a problem or pathogenic response, free of the confounding influence of the multiple responses of an intact higher level organism.

TABLE T-9
Possible Interpretations When *in Vitro* Data Do Not Predict Results of *in Vivo* Studies

Chemical is not absorbed at all or is poorly absorbed in *in vivo* studies.

Chemical is well absorbed but is subject to "first-pass effect" in the liver.

Chemical is distributed so that less (or more) reaches the target tissue than would be predicted on the basis of its absorption.

Chemical is rapidly metabolized to an active or inactive metabolite that has a different profile of activity and/or different duration of action than the parent drug.

Chemical is rapidly eliminated (e.g., through secretory mechanisms).

Species of the two test systems used are different.

Experimental conditions of the *in vitro* and *in vivo* experiments differed and may have led to different effects than expected. These conditions include factors such as temperature or age, sex, and strain of animal.

Effects elicited *in vitro* and *in vivo* by the particular test substance in question differ in their characteristics.

Tests used to measure responses may differ greatly for *in vitro* and *in vivo* studies, and the types of data obtained may not be comparable.

The *in vitro* study did not use adequate controls (e.g., pH, vehicle used, volume of test agent given, and samples taken from sham-operated animals), resulting in "artifacts" of methods rather than results.

In vitro data cannot predict the volume of distribution in central or in peripheral compartments.

In vitro data cannot predict the rate constants for chemical movement between compartments.

In vitro data cannot predict the rate constants of chemical elimination.

In vitro data cannot predict whether linear or nonlinear kinetics will occur with specific dose of a chemical *in vivo*.

Pharmacokinetic parameters (e.g., bioavailability, peak plasma concentration, and half-life) cannot be predicted based solely on *in vitro* studies.

In vivo effects of chemical are due to an alteration in the higher order integration of an intact animal system which cannot be reflected in a less complex system.

Further Reading

Gad, S. C. (1993). *In Vitro Toxicology*. Raven Press, New York.

Gad, S. C., and Kapis, M. B. (1993). *Non-Animal Techniques in Biomedical and Behavioral Research and Testing*. Lewis, Ann Arbor, MI.

Salem, H. (1995). *Animal Test Alternatives*. Dekker, New York.

—*Shayne C. Gad*

Related Topics

Ames Test
Analytical Toxicology
Animal Models
Dominant Lethal Tests
Dose–Response Relationship
Host-Mediated Assays
In Vitro Test
In Vivo Test
Mouse Lymphoma Assay
Toxicity, Acute
Toxicity, Chronic
Toxicity, Subchronic

Toxicity Testing, Aquatic

Introduction

Freshwater and marine environments contain complex ecosystems such as ponds, rivers, lakes, and estuaries. Each of these ecosystems contains unique biota that may be represented by several thousand species. These biota, both flora and fauna, are often exposed to a variety of toxicants, including those that result from anthropogenic activities. In some cases, toxicity and environmental damage can occur. The study of these adverse effects on freshwater and marine biota and on the ecosystems that contain them is aquatic toxicology.

Aquatic toxicology differs from mammalian toxicology in several aspects. The primary goal of aquatic toxicology is to assess the effect of toxicants on the many diverse populations and communities of plants and animals inhabiting marine and freshwater environments. The biota are cold-blooded and the physical and chemical characteristics of the aquatic environment have a significant effect on their sensitivity to toxicants. The aquatic test species of interest, unlike in mammalian studies, can be used directly. The objective of mammalian toxicology is to assess effects on humans whose sensitivity to toxicants is less affected by their environment than aquatic organisms. The dose of the toxicant used in mammalian toxicology can be measured more accurately, the mechanisms of toxic action are better understood, and the test methods are more established.

Various species of aquatic life, particularly fish, have been used in toxicity experiments for more than 130 years. One of the earliest reported studies was conducted with fish in 1863 and the first proposed standard test species was the goldfish in 1917. Toxicity tests have been conducted with increasing frequency since the 1960s due to the numerous environmental regulations that have been enacted and the increasing availability of standardized test methods, the first of which were published in 1960 for animal test species and in 1970 for algae.

Many test methods are available for aquatic toxicity testing (Table T-10). They differ in cost, precision, complexity, and the skill needed to conduct them. Nevertheless, their objectives are similar: They are conducted to determine relative potency among chemicals and relative susceptibility among different species and life stages, and to identify other variables that influence the overall outcome of exposure. Toxicity tests are conducted usually to meet regulatory guidelines for the

TABLE T-10
Available Laboratory Testing Methods for
Aquatic Toxicology

Single species tests
    Trout
    Fathead minnows
    Daphnia
    Bluegills
    Algae
Multispecies tests
    Experimental streams
    Ponds
    Microcosms
Bioconcentration tests
Effluent toxicity tests
Sediment toxicity tests
Phytotoxicity tests
    Algae
    Vascular plants
    Duckweed

---

use and discharge of commercial chemicals such as pesticides or nonpesticides. In addition, toxicity results are used to derive national water-quality standards to protect aquatic life and to determine the environmental effects of municipal and industrial effluents.

Aquatic toxicologists do not use all the available toxicity tests for any single toxicant. Instead, a tiered approach is used to provide a systematic and comprehensive process of deriving the toxicity data needed to assess the environmental hazard of a chemical. This approach consists of conducting short-term screening tests prior to using predictive studies that are more complex and time-consuming. This sequential evaluation provides an efficient use of resources and tends to eliminate unnecessary testing. The decision points and testing phase depend on the quality and quantity of data needed for the test substance of interest. The types of toxicity tests used in the tiered approach are discussed briefly.

Single-Species Toxicity Tests: Methodologies

There are two basic types of aquatic single-species toxicity tests: acute and chronic. Acute toxicity tests have been the "workhorse" of aquatic toxicologists for many years. These tests are relatively simple, take little time, and are cost-effective, and a large historical database exists for many chemicals and effluents. Acute toxicity tests are most often used to screen toxicity quickly or to determine the relative sensitivities of different test species. Mortality is the effect monitored during the test duration of 48 hr (invertebrates) or 96 hr (fish). In a typical acute toxicity test, 5–10 organisms are exposed under static conditions in glass test beakers to five test concentrations. A control is included. The test concentrations and control are conducted in triplicate. Daily observations are made on survival and dead organisms are removed.

At test termination, the concentration that kills 50% of the test organisms (LC_{50} value) is determined using probit analysis or graphical interpolation. Unlike in chronic toxicity tests, there is no test solution renewal, the organisms are unfed, and there is no analytical verification of the test concentrations. Furthermore, cumulative, chronic, and sublethal effects of a chemical usually are not evaluated in acute toxicity tests, although frequently behavioral changes and lesions caused by a chemical can be determined.

Chronic toxicity tests are more complex and time-consuming than acute studies and for these reasons are conducted less frequently. The methodologies for these tests differ considerably, unlike for acute tests, because they are designed for the specific life histories of the various test species. Chronic toxicity tests may be for a full life cycle (egg–egg), partial life cycle (embryo–larval), and partial life history (egg–death). Full life cycle tests are uncommon with fish due to the long durations that are necessary(1 or 2 years). Partial life cycle tests with fish can be as short as 7 days or as long as 60 days. The early life stage of fish (embryo/larva) is usually the most sensitive period in a fish's life cycle and, consequently, partial life cycle tests are used as surrogates for the full life cycle studies. Chronic tests may be conducted for more than one complete life cycle if algal and invertebrate species are used since their life cycles are shorter than those of fishes. Lethal and sublethal effects are monitored in chronic toxicity studies, and these effects include changes in growth, reproduction, behavior, physiology, and histology.

Fathead minnows and the freshwater invertebrate, *Daphnia magna,* are common species used in chronic toxicity tests. A brief discussion of the study design used in the daphnid study follows. The *D. magna* were exposed to the cationic surfactant in a 21-day static-renewal toxicity test. Ten 250-ml glass beakers containing 200 ml of the test solution were used for each of the five test concentrations and controls. A solvent was used in the study, which necessitated the use of a solvent control. Seven of the beakers at each concentration contained one daphnid (less than 24 hr old) and the remaining three contained five organisms. Growth, survival, and reproduction were monitored in those beakers containing one daphnid, whereas only survival was monitored in the remaining test chambers. The test species were fed daily a combination of algae, trout chow, and alfalfa. The test solutions were renewed daily. Prior to renewal, survival and number of young were determined. Surviving adults then were transferred to test chambers containing "fresh" solution at the same levels of toxicants to which the organisms had been exposed previously. At test termination, the effects of the surfactant on growth, reproduction, and survival were compared for daphnids exposed to the various toxicant concentrations and those in the solvent control. It can be seen that the first significant effect

level of the surfactant was 0.76 mg/liter based on changes in mean length, total young, and mean brood size.

Toxicity tests may be static, continuous flow, or, as in the previous example, static renewal based on the toxicant dosing technique. Static and flowthrough procedures are more widely used in toxicity tests conducted with pure chemicals and animal test species. Chronic toxicity tests conducted with effluents are usually static renewal, and those with algae are static. There is no change or renewal of the test substance and dilution water in a static test. This design is the simplest and least expensive; however, the toxicant concentrations may decrease due to adsorption and biodegradation. The test solutions and dilution water are renewed periodically, usually daily in a static-renewal test. In a continuous-flow test, the dilution water and test substance are continuously or intermittently renewed. The exposure concentrations remain fairly constant and dose–response relationships can be well defined.

A variety of aquatic toxicity test methods have been published for single species (Table T-11) and several have been standardized. Test method development is an ongoing process, however, which continues to increase the efficiency of these methods and often results in alternative study designs.

TABLE T-11
Availability of Standardized or "Consensus" Toxicity Test Methods

Variable	Organization				
	ASTM	APHA et al.	U.S. EPA	OECD	EPA/COE
Test substance					
Chemicals	•	•	•	•	
Effluents	•		•		
Sediment	•				•
Test types: acute					
Invertebrates	•	•	•	•	•
Fish	•	•	•	•	•
Test types: chronic					
Algae	•	•	•	•	
Duckweed	•		•		
Invertebrates	•	•	•	•	
Fish	•	•	•		
Microcosm	•	•	•		
Bioconcentration	•				

Experimental Conditions

In general terms, toxicity tests are conducted in a laboratory or a room controlled for light and temperature. The test solutions containing the test species are monitored for pH, temperature, dissolved oxygen, and hardness. The test organisms are exposed for a predetermined duration, which varies depending on the type of test and test species. Daily observations on lethal and sublethal effects are made, and several calculations, such as the LC_{50} value, highest no-observed-effect concentration (NOEC), and lowest-observed-effect concentration (LOEC), are determined based on the most sensitive effect parameter of interest. Although toxicity tests have similarities, as discussed later, variations among test animals, instrumentation, and methods influence the outcomes and utility of the assessments.

Test Chambers

The types of test chambers used in toxicity tests depend on the test species. Various sizes of beakers, aquaria, jars, bowls, and petri dishes have been used. The test chambers usually are constructed of material such as glass, Teflon, and certain plastics that minimize leaching of toxicants and adsorption of the test substance.

Test Concentrations

Xenobiotic concentrations used in an acute toxicity test are based routinely on results obtained from a pretest or range-finding test. The test concentration range for a chronic test is based on the results of an acute test conducted prior to the chronic test. There are no standard guidelines for conducting these preliminary tests. Generally, 5–10 organisms are exposed to several test concentrations, which are usually an order of magnitude apart. The dilution water and exposure conditions—that is, water temperature, hardness, and pH—in range-finding tests usually are similar to those in the definitive test.

The test substances used in toxicity tests have been in most cases pure chemical compounds and municipal and industrial effluents. However, toxicity tests are being conducted more frequently with dredged spoil materials (prior to ocean disposal), hazardous waste leachates, and contaminated sediments due to increasing regulatory concern for their potential environmental impacts. The test organisms are exposed in the defini-

tive test to give concentrations chosen in a geometric progression. The test concentrations and control are replicated at least threefold. The test compound is added to the dilution water, which may be well water, reconstituted water, dechlorinated tap water, uncontaminated river water, and natural or artificial seawater. The dilution water is well aerated and undesirable organisms are removed before use.

An organic solvent is used to dissolve substances with minimal water solubility. Several have been used and include triethylene glycol, dimethyl sulfoxide, acetone, and dimethyl formamide. The LC_{50} values for these solvents are between 9000 and 92,500 mg/liter. The concentration of the solvent in the test water should not exceed 0.5 ml/liter or should not be more than 1/1000 of the LC_{50} value of the solvent. When an organic solvent is used, a solvent control is included in the study.

Toxicant delivery systems are used to deliver on a once-through basis the various test concentrations to the test chambers in continuous-flow toxicity tests. The serial proportional diluter is the most common design used to mix the dilution water with the test substance to produce the desired test concentrations. The construction materials in toxicant delivery systems, like those for the test chambers, should not be rubber, certain plastics, or metallic.

The test concentrations are confirmed analytically during chronic toxicity tests. Analyses are performed at least weekly for each test concentration and control for tests of 7 days duration or longer. In tests of shorter duration, analyses usually are conducted on alternate days. Analytical verification of the test concentrations in range-finding and acute toxicity tests seldom is performed, and the results from these tests generally are based on nominal concentrations.

Test Species

Historically, animal test species have been used more frequently than plant species and freshwater species more frequently than marine species. These trends can be seen in several data summaries (Tables T-12 and T-13).

Most toxicity tests are conducted with single cultured test species such as those listed in Table T-14. The more commonly used freshwater species, particularly in tests used for regulatory compliance, are fathead minnows (*Pimephales promelas*), several daphnid spe-

TABLE T-12
Most Commonly Reported Tested Species in the
ACQUIRE Database

Common name	Species	Percentage of reported data
Rainbow trout	*Salmo gairdneri*	8.6
Bluegill	*Lepomis macrochirus*	5.6
Fathead minnow	*Pimephales promelas*	4.9
Water flea	*Dephnia magna*	3.9
Carp	*Cyprinus carpio*	2.4
Coho salmon	*Oncorphynchus kisutch*	2.1
Goldfish	*Carassius auratus*	1.9
Channel catfish	*Ictalurus punctatus*	1.7
Mosquito fish	*Gambusia affinis*	1.5
Brook trout	*Salvelinus fontinalis*	1.0
Green algae	*Scenedesmus quadricauda*	0.9
Water flea	*Daphnia pulex*	0.8

cies (*D. magna* and *Ceriodaphnia dubia*), and green algae (*Selenastrum capricornutum*). Common marine species are sheepshead minnows (*Cryprinodon variegatus*), mysid shrimp (*Mysidopsis bahia*), and a diatom, *Skeletonema costatum*.

The species in Table T-14 were selected based on several criteria, primarily ease of culture, commercial availability, and size. The test species are acclimated

TABLE T-13
Types of Tests and Test Species Used in Deriving Toxicity Data for Submissions under TSCA Section 4 as of 1988

Test type	Species	Number of tests
Acute toxicity	Fathead minnow (F)[a]	37
	Rainbow trout (F)	27
	Sheepshead minnow (M)	12
	Daphnids (F)	23
	Midge (F)	12
Partial life cycle	Fathead minnow (F)	1
	Rainbow trout (F)	7
Full life cycle	Daphnics (F)	20
	Mysid shrimp (M)	5
Phytotoxicity	Alga (F)	23
	Alga (M)	3
Bioconcentration	Bluegill (F)	1
	Fathead minnow (F)	1
	Rainbow trout (F)	1
	Mussel (M)	1
	Oyster (M)	2

[a] F, freshwater; M, marine.

for a specific time prior to testing to eliminate diseased organisms. Generally, a minimum of 10 animals are exposed in static and static-renewal tests and 20 in a flowthrough test to each test concentration and control. The recommended loading density for the test species is between 0.5 and 0.8 g/liter in static tests and between 1 and 10 g/liter in continuous flowthrough tests.

Reference toxicants often are used to determine the "health" of the test species. There is no widely used reference toxicant; several that have been used include dodecyl sodium sulfate (anionic surfactant), sodium chloride, sodium pentachlorophenol, and cadmium chloride.

Sensitivity is a criterion that is used in the choice of a test species. The sensitivity of the species in Table T-14 relative to one another as well as to indigenous flora and fauna in the ecosystem is a matter of contention. There is no single test species and no group of test species consistently most sensitive to toxicants or most reliable for extrapolation to all other organisms. Most toxic effects reported for a variety of test substances have been species specific. Therefore, acute toxicity tests are conducted first with a variety of freshwater and marine test species to determine the most sensitive plant and animal. These sensitive species then are used in all subsequent chronic testing.

Calculations

The results of acute toxicity tests are reported as the LC_{50} and EC_{50} values and their 95% confidence intervals. Probit analysis is the most commonly used to determine LC_{50} values. Graphical interpolation can be used to estimate the LC_{50} value where the proportion of deaths versus the test concentration is plotted for each observation time.

The NOEC and the LOEC are the usual calculations reported from chronic toxicity tests. The NOEC is the highest concentration in which the measured effect is not statistically different from that of the control. The LOEC is the lowest concentration at which a statistically significant effect occurred. These concentrations are based on the most sensitive effect parameters—that is, hatchability, growth, and reproduction. The statistical procedure for these calculations combines the use of analysis of variance techniques and multiple comparison tests. In some cases, the maximum acceptable toxic concentration (MATC) is reported from chronic toxicity results. The MATC is a concentration (x) that is within the range of the NOEC and LOEC: NOEC $\geq$

TABLE T-14
Freshwater and Marine Species Used in Toxicity Tests

Freshwater	Saltwater
Fish	Fish
Salmo gairdneri (rainbow trout)	*Cyprinodon variegatus* (sheepshead minnow)
Salvelinus fontinalis (brook trout)	*Fundulus heteroclitus* (mummichog)
Ictalurus punctatus (channel catfish)	*Menidia berylina* (silverside)
Pimephales promelas (fathead minnow)	*Gasterosteus aculeatus* (threespine stickleback)
Lepomis macrochirus (bluegill)	*Leiostomus santhurus* (spot)
Carassius auratus (goldfish)	Invertebrates
Invertebrates	*Acartis tonsa* (copepod)
Daphnia magna (daphnid)	*Neanthes* sp. (polychaeta)
Daphnia pulex (daphnid)	*Callinectes* sp. (crab)
Ceriodaphnia dubia (daphnid)	*Penaeus* spp. (pink shrimp)
Gammarus lacustris (amphipod)	*Palaemonetes* spp. (grass shrimp)
Chironomus sp. (midge)	*Crassostrea virginica* (oyster)
Physa integra (snail)	*Arbacia punctulata* (sea urchin)
Cambarus sp. (crayfish)	Plants
Plants	Algae
Algae	*Skeletonema costatum* (diatom)
Selenastrum capricornutum (green)	*Thalassiosire pseudonana* (diatom)
Chlorelia vulgaris (green)	*Champia parvula* (red)
Microcystis aeruginosa (blue green)	
Navicula spp. (diatom)	
Vascular	
Lemna minor (duckweed)	
Lemna gibba (duckweed)	
Myriophyllum spicatum (water milfoil)	
Ceratophyllum demersum (coontail)	

$x <$ LOEC. The first-effect concentration can be expressed as the geometric mean of the two terms.

Variability/Precision

Toxicity tests conducted with freshwater and marine species are considered relatively precise and reliable based on current information concerning interlaboratory and intralaboratory comparisons of toxicity results. Generally, the LC_{50} values from acute toxicity tests conducted under similar experimental conditions vary less than threefold. This has been observed for metals, effluents, reference toxicants, and different organic compounds. Coefficients of variation (CV) for acute and chronic toxicity tests conducted with daphnic species and chemicals and effluents are between 27 and 39%. The CV values for several reference toxicants and acute daphnic studies ranged between 10 and 72% and from 47 to 83% for chronic toxicity tests with algae.

Multispecies Toxicity Tests

The results of the "traditional" acute single-species toxicity tests conducted in the laboratory cannot be used alone to predict effects on natural populations, communities, and ecosystems. The cultural species in laboratory tests are different from those in most ecosystems. Conditions such as the size of the test species, its life stage, and nutritional state can have an effect on toxicity. Furthermore, the experimental conditions in laboratory tests cannot duplicate the complex interacting physical and chemical conditions of ecosystems, such as seasonal changes in water temperature, dissolved oxygen, and suspended solids. In addition to these environmental modifying factors, aquatic life is usually exposed simultaneously to numerous potential toxicants (mixtures). Although the toxicities of binary and ternary mixtures have been evaluated for some chemicals in laboratory toxicity tests, the resultant information has predictive limitations.

Because of the deficiencies of single-species toxicity tests, alternative approaches are evolving to address ecosystem structural and functional processes. Multi-species tests include the use of laboratory microcosms, outdoor ponds, experimental streams, and enclosures. There are no standardized procedures for these tests. They are conducted with plant and animal species obtained from laboratory cultures and biota collected from natural sources. They can be conducted indoors or outdoors. The toxic effects, in addition to those used for single-species tests, are determined for structural parameters, such as community similarity, diversity, and density, and for functional parameters, such as community respiration and photosynthesis. Effects on these parameters are reported as the NOEC and LOEC.

Sediment Toxicity Tests

In the past, toxicity tests have been conducted primarily with water column dwelling or planktonic organisms, with the objective of controlling water pollution. However, it has been realized that sediments act as "reservoirs" for chemicals that can adversely affect benthic aquatic life and, at times, also affect planktonic life. This concern has led to the development of sediment quality criteria to protect aquatic life. Test methods have been developed to support the derivation of these criteria and to support other related regulatory activities (e.g., Superfund site evaluations and ocean disposal of dredged materials).

Most sediment toxicity tests have been conducted in the laboratory with single species of freshwater and marine benthnic organisms such as amphipods and midges, but in some cases planktonic species also have been used. Most tests conducted to date have been acute and have been of 10 days duration or less. Sediment toxicity tests are conducted with the solid phase or the pore water (interstital water). Methods have been published describing the collection and preparation techniques.

Test guidelines for marine sediment and freshwater sediment also have been reported. Standardized methods are available for freshwater invertebrates and freshwater and marine amphipods.

The availability of reliable test methods for contaminated sediments is relatively recent and the test method development process continues. A variety of issues remain to be solved before these types of studies will be considered as effective as those with planktonic species. Among the more important of these issues are validation of the single-species test results and determination of variations in species sensitivity.

Effluent Toxicity Tests

Toxicity tests are used in the National Pollutant Discharge Elimination System, permitting one to determine the toxicity effects of municipal and industrial effluents and storm water overflows on aquatic life. A summary of the experimental conditions in several of the available test methodologies appears in Table T-15. The methodologies differ slightly from those used for pure chemicals. For example, the choices of the dilution water and the effluent collection technique are important considerations. In most cases water collected from the receiving water above the outfall is used for dilution, and composite samples of effluent are used. The test species—an algae, invertebrate, and fish—are usually exposed to five effluent dilutions for 4–7 days. The tests are static renewal except those for algae, which are static. The calculations reported are the LC_{50} value, the NOEC, and the LOEC, which are expressed as percentage effluent. The cause(s) of toxicity in the effluent—that is, specific effluent constituents—can be identified using comparative toxicity testing and chemical fractionation techniques.

The freshwater invertebrate, *C. dubia,* is a test species commonly used in effluent toxicity evaluations. The *C. dubia* used in a study are obtained from a laboratory culture. Effluent collected within 72 hr from the source is used after temperature acclimation. The static-renewal test usually is conducted in a laboratory located off-site from the effluent source but the tests may be conducted on-site using a mobile bioassay facility. The test is conducted at 25°C, 10–20 mE/m²/sec, and under a photoperiod of 16-hr light/8-hr dark. Five test concentrations are used that include undiluted effluent (100%) and four dilutions such as 50, 25, 12, and 6%. The effluent is diluted with either a high-quality laboratory water or water collected from the receiving water above the effluent outfall. The control is composed of 100% dilution water. For each test concentration and the control, 10 30-ml plastic test chambers containing 15 ml of the test solution are used. Each test chamber contains one daphnid and daily observations on mortality and young production are made during the 7-day test. The organisms are fed daily a combination of yeast, trout chow, and algae.

TABLE T-15
Comparison of Several Experimental Variables in Chronic Toxicity Tests Conducted with Effluents

Test type	Duration (days)	Number of test concentrations	Test species	Age of test organisms	Total test species exposed	Number of replicates	Temperature (°C)	Light intensity (mE/m²/sec)
Static renewal	7	5	Fathead minnow (freshwater fish)	<24 hr	30–60	3–4	25 ± 1	10–20
Static renewal	7	5	Sheepshead minnow (marine fish)	<24 hr	30–60	3–4	25 ± 2	10–20
Static renewal	7	5	*Ceriodaphnia dubia* (freshwater invertebrate)	<24 hr	10	10	25 ± 1	10–20
Static renewal	7	5	*Mysidopsis bahia* (marine invertebrate)	<24 hr	40	8	25–27	10–20
Static	4	5	*Selenastrum capricornutum* (freshwater; green alga)	4–7 days	1 × 104 (initial)	3	25 ± 1	86 ± 8.6
Static	4	5	*Skeletonema costatum* (marine; diatom)	4–7 days	2 × 104 (initial)	3	20 ± 2	60 ± 6

Surviving organisms are transferred daily to renewed test solutions. The NOEC and LOEC values are determined based on the adverse effects on survival and reproduction occurring during the 7-day test.

Phytotoxicity

The majority of aquatic toxicity tests have been conducted with animal test species since they once were thought to be more sensitive than plants. This generalization is not supported technically, based on a review of the data for most toxicants. Nevertheless, only recently have phytotoxicity tests been conducted routinely with a limited number of species of algae and vascular plants.

A variety of test methods are available to determine the phytotoxic effects of chemicals and effluents (Table T-16). The freshwater algal species most frequently used has been the microalga, *S. capricornutum*, for which a relatively large database exists. Marine species used include the diatom, *S. costatus,* and the red macroalga, *Champia parvula.*

Acute toxicity tests seldom are conducted with algae. The chronic toxicity tests conducted with microalgae are for 3 or 4 days duration although exposures can be less than 1 day if effects on photosynthesis are measured. These static exposures occur in a liquid nutrient-enriched medium under conditions of controlled pH, temperature, and light. Inhibitory and stimulatory effects on population growth are monitored during the exponential growth phase. Five test concentrations and a control are included in each study. The most common calculation reported is the 96-hr EC_{50} value (concentration that reduces growth 50%) but algistatic (completely stops growth) and algicidal (lethal) concentrations also have been reported. In addition, the SC_{20} concentration (stimulatory concentration) is reported if growth stimulation is observed. The SC_{20} value represents the concentration that increases algal growth 20% above that of the algal population in the control.

Floating and rooted macrophytes are used less frequently in toxicity tests than algae. The duckweeds, freshwater floating species, are more commonly used than most due to their small size and rapid growth. Several published methods are available describing their use, particularly *Lemna minor* and *L. gibba.* Tests with these species are usually of 4–14 days duration, during which effects on frond number and chlorophyll content are monitored. The results are expressed as an EC_{50} value and the NOEC. The tests are conducted, like algae, in a nutrient-enriched medium. The test chambers can be fruit jars, plastic cups, test tubes, and Erlynmeyer flasks. The key research issue that remains to be investigated before the duckweeds will be more widely accepted as suitable test species is their sensitivity relative to other aquatic plant and animal test species.

TABLE T-16
Experimental Conditions in Several Phytotoxicity Tests Conducted with Algae and Duckweed

TABLE T-16
Experimental Conditions in Several Phytotoxicity Tests Conducted with Algae and Duckweed

Test type	Duration (days)	Number of test concentrations	Tests species	Number of replicates	Temperature (°C)	Light intensity ($\mu E/m^2/sec$)
Algae						
Static	4	5	*Selenastrum capricornutum* (F)[a]	3	20–24	300 ± 25
			Scenedesmus quadricauda (F)			
			Clorella vulgaris (F)			
			Skeletonema costatum (M)			
Static	3	5	*Selenastrum capricornutum* (F)	3	21–25	120 ± 20%
			Scenedesmus subspicatus (F)			
Static	4	5	*Selenastrum capriconutum* (F)	3	20–24	30–90
			Microsystis aeruginosa (F)			
			Anabaena flos-aquae (F)			
			Navicula pelliculosa (F)			
			Skeletonema costatum (M)			
			Dunaliella tertiolecta (M)			
Duckweed						
Flowthrough	7	5	*Lemna minor* (F)	4	22.8 ± .6	2700
Static	14	5	*Lemna gibba* (F)	3	25 ± 2	100
Static	4	5	*Lemna minor* (F)	3–6	27 ± 2	6456

[a] F, freshwater; M, marine.

The use of rooted macrophytes such as pondweeds (*Potamogeton* spp.), waterweeds (*Elodea* and *Hydrilla*), the water hyacinth (*Eichhornia crassipes*), coontail (*Ceratophyllum demersum*), and water milfoil (*Myriophyllum* spp.) in toxicity tests is less common than that for algae and duckweeds due to their large size and slow growth. There are no standard or commonly used test methods for these species. Consequently, there is a need for their development and validation. The experimental techniques that have been used vary considerably. Recently, seeds from aquatic macrophytic vegetation have been used to assess the toxicities of chemicals and effluents. These studies are usually of 4–7 days duration, and the effect parameters are seed germination, root elongation, and early seedling growth. The use of whole plant rooted macrophytes and their seeds in toxicity tests will increase in the future as sediment quality criteria to protect aquatic life and wetlands increase in regulatory importance. However, for this to occur test method development and validation and determination of species sensitivity will be needed.

Bioconcentration

A bioconcentration study is conducted to derive information on the ability of an aquatic species to concentrate a toxicant in its tissues. This uptake and accumulation can be hazardous to the organism as well as to other aquatic life utilizing the test species as a food source. Bioconcentration tests are usually conducted with single chemicals and single species of algae, fish, and bivalve mollusks. A variety of fish have been used, including the fathead minnow, bluegill, rainbow trout, sheepshead minnow, and several species of oysters, scallops, and mussels.

There are several test designs that can be used to estimate the bioconcentration potential of a compound. Typically, one group of the test species is exposed to the toxicant for an uptake and depuration phase. A control is included in which the test species is not exposed to the toxicant. In assessing the concentration of the test chemical in the organism, the literature contains examples of measuring total residues and measuring only the parent compound, dependent primarily on the methodology used. The uptake phase is usually for 28 days or until a steady state is attained. The depuration period lasts until the concentration in the test species is 10% of the steady-state concentration in the tissue. During both phases, the test water and test species are analyzed daily for the test chemical. All results from a bioconcentration study are based on measured concentrations. The uptake rate, depuration rate, and bioconcentration factor (BCF) typically are reported. The relevance of the BCF value to the survival of the organism and to ecosystem dynamics is an issue that has received and will continue to receive significant scientific attention.

Further Reading

Hammons, A. S. (1981). *Methods for Ecological Testing.* Ann Arbor Science, Ann Arbor, MI.

Richardson, M. (1995). *Environmental Toxicology Assessment.* Taylor & Francis, Bristol, PA.

—*Shayne C. Gad*

Related Topics

Analytical Toxicology
Bioaccumulation
Bioconcentration
Biomarkers, Environmental
Ecological Toxicology, Experimental Methods
Effluent Biomonitoring
Environmental Toxicology
Microtox
Photochemical Oxidants
Pollution, Water
Risk Assessment, Ecological

Toxicity Testing, Behavioral

Introduction

Behavioral toxicology, a part of the larger domain of neurotoxicology, uses a wide range of methods to evaluate changes in behaviors of model organisms (in modern toxicology, largely rats and mice) as a means of identifying and studying adverse effects of chemicals on the nervous system. As such, behavioral toxicology tests should be considered functional testing of a complex organ system.

Behavior may be defined as anything an organism does—any move an organism makes. The behavior of an organism at any moment is the result of the external environment, the past history of the organism, and the internal environment (e.g., biochemical or electrical processes and hormonal levels) within the organism. The behaviorist studies the functional relationship between the behavior of an organism and these variables. The behavioral toxicologist includes exposure to a chemical as one of these variables. An aspect of the environment that controls behavior in a functional manner is termed a stimulus. A unit of behavior, defined by the experimenter, is termed a response. There are two types of responses: respondent and operant. In addition, either type of response may be unconditioned (unlearned) or conditioned (learned).

Respondent behaviors include such actions a smooth muscle contraction, autonomic responses, glandular secretions, and elicited motor responses such as reflexes. Unlearned respondents are used frequently in observational batteries and include such measures as orienting to stimuli or reflex startle to intense stimuli. In the Soviet Union, learned respondents are used extensively in toxicology. The famous experiment of Pavlov, in which dogs learned to salivate at the sound of a bell after numerous pairings of the bell being followed by presentation of food, is an example of a conditioned respondent. Respondents are paired with an eliciting stimulus in a one-for-one relationship (i.e., light vocalization).

Operant responses, on the other hand, have no single eliciting stimulus but occur within the context of many environmental stimuli. The consequences of a certain behavior affect the probability that this behavior will be produced again. Locomotor activity is often given as an example of an unlearned operant because there is no attempt on the part of the experimenter to condition a particular type of response. An extremely powerful tool at the disposal of the behavioral toxicologist is that of operant conditioning. Operant conditioning takes advantage of the control that the immediate outcome of behavior has in determining the subsequent frequency of similar behavior. If the outcome of a particular behavior increases its frequency, it is termed a positive reinforcer (i.e., food). If it decreases the frequency, it is called negative reinforcer (i.e., shock). The great strength of this technique is that it may be used to teach a large variety of tasks with wide complexity. Questions can be asked about attention, learning, memory, sensory function, and general well-being of the subject. Many techniques discussed in this chapter rely on the principles of operant conditioning. Behavioral assessment would be impossible without a thorough understanding of these principles on the part of the investigator.

The behavior of organisms can be divided into motor function, sensory function, learning and memory, and performance on intermittent schedules of reinforcement. These classes are somewhat arbitrary, and virtually all behavioral tests measure more than one of these functions. For example, motor function affects almost all testing, intact sensory function is necessary for learning, performance on intermittent schedules most certainly has a learning component, and so forth. Often, different functions are not separable by one test or type of test; therefore, it is imperative to study several types of behavior to determine the function(s) that is affected. It should be pointed out that none of the procedures described in the following sections should be examined in isolation because all are part of a comprehensive investigation of the potential behavioral effects of a toxicant.

The examples of tests presented in each section are certainly not an exhaustive list but were chosen because they are often used or because they promise to contribute substantially to the understanding of behavioral toxicity.

The front end of this tier approach is a neurobehavioral observational screen, the tool of choice for initial identification of potentially neurotoxic chemicals. The use of such screens, other behavioral tests methods, or what are generally called clinical observations does, however, warrant one major caution or consideration. That is, short-term (within 24 hr of dosing or exposure) observations are insufficient on their own to differentiate between pharmacological (reversible in the short term) and toxicological (irreversible) effects. To so differentiate, it is necessary to either use additional means of evaluation or have the period during which observations are made extended through at least 3 or 4 days.

Screening Batteries

As pointed out previously, one of the agendas that has emerged for behavioral toxicity involves the screening of new chemicals for potential neurotoxicity. The behavioral tests utilized for such purposes are often referred to as apical tests because they require the integrated function of several organ systems, including the nervous system. Such batteries typically have included two behavioral components; a functional observational battery (FOB) and an evaluation of motor activity. Table T-17 depicts the component tests of the FOB developed by Gad, which include an array of measures of both unconditioned operant and respondent behaviors. Such batteries have been shown to exhibit utility for screening potential neurotoxicity, i.e., hazard identification and elaboration. Those components of the FOB directed to cholinergic functions exhibited sensitivity to the effects of the anticholinesterase carbaryl, whereas few such signs of cholinergic disturbances were evident in the presence of the nonanticholinesterase pesticide chlordimeform.

Motor activity is frequently included in screening batteries both as a measure of motor function and as an apical test. Motor activity, generally considered an unconditioned behavior, exists at some baseline level and is a complex behavior that includes numerous components such as ambulation, rearing, grooming, and sniffing. As discussed previously, toxicants may alter motor activity by affecting any or all of its component behaviors.

From the standpoint of screening and hazard identification, a notable point relating to the interpretation of data from FOB and motor activity studies is whether the effects observed in response to toxicant exposure

TABLE T-17

Example of Behavioral Procedures Included in a Functional Observation Battery

Home cage and open field	*Manipulative*	*Physiological*
Posture	Ease of removal	Body temperature
Convulsions	Ease of handling	
Palpebral closure	Palpebral closure	Body weight
Lacrimation	Approach response	
Piloerection		
Salivation	Touch response	
Vocalizations	Finger-snap response	
Time to first step		
Rearing	Tail-pinch response	
Urination		
Defecation	Righting reflex	
Gait	Catalepsy	
Arousal	Hindlimb foot splay	
	Forelimb grip strength	
	Hindlimb grip strength	

represent a direct effect of the toxicant on the nervous system or are secondary to changes in other systems since such apical tests rely on the functional integrity of multiple systems. In some circumstances, the fact that the toxic effect is ultimately expressed in behavior may minimize the importance of the direct vs indirect source of the effect. It also should be noted in the interpretation of toxicant-induced changes in FOBs and motor activity measures that the concurrent presence of body weight loss or decline in food or water intake does not by any means necessarily indicate that the behavioral changes are the result of malaise or sickness as these measures may change independently of each other.

Motor Function

Deficits in motor function are frequently produced in humans as a result of their toxic exposure to a chemical. Heavy metals such as mercury, lead, and manganese; insecticides such as chlordecone (Kepone) or organophosphorus compounds; and air pollutants such as carbon disulfide all produce changes in motor function. Gross assessment of motor function should be performed as part of an initial toxicity screen. Batteries that include observational assessment of muscle tone, body posture, equilibrium, and gross coordination have been suggested. The next level of testing includes such techniques as ability to stay on a rotating rod or quantification of hindlimb splay. The former requires an automated apparatus as well as animal training and practice in order to reduce test variability to acceptable levels. The latter technique is simpler, involving the placement of ink on the paws of rodents after which they are dropped from a specific height. Quantification of hindlimb splay does not require training of the animal and is fast and easy. Another screening procedure for assessment of neuromotor function requires that a rodent grasp a bar attached to a strain gauge after which the animal is pulled on manually until it lets go. Assessment of swimming ability is also suitable for incorporation into screening tests. The rodent is placed in a pool of water and such measures as swimming movement, position in the water, and ability to keep the head above water are assessed. These analyses may reveal motor deficits that are not apparent during locomotion on land.

Little work has been focused on more sophisticated tests for assessing neuromuscular function. One promising procedure employs operant techniques to train an animal to depress a lever within a specific force band and time period, thus allowing assessment of the effects of toxic agents on fine motor control or strength.

The test that probably is used most extensively in screening for nervous system toxicity is locomotor activity, in large part because no training is required and activity can be measured rapidly. Locomotor activity represents the functional output of many systems of the body, including but certainly not exclusively motor systems. In addition, although such measurements may appear straightforward, there are many variables that must be considered. Motor activity is not a single activity but consists of many acts, such as horizontal and vertical movement, sniffing, rearing, grooming, and scratching. With some types of measuring devices, even tremor may be monitored. There are, therefore, many methods of monitoring, including scoring the classes of movement by observation, measuring horizontal movement only, measuring vertical displacement with devices that gauge force generated against the floor, and combinations of these measurements. Even within a class of automated devices, there is large variability in the configuration of each apparatus and in the method of measurement. With different kinds of apparatuses, different behaviors can be measured.

When a toxicant is introduced, activity may increase, decrease, or remain unchanged depending on choice of apparatus, age of the animal, relative novelty and complexity of the environment, and many other variables. Although a change in an animal's activity as a result of its exposure to a toxicant indicates a change in the function of its nervous system, interpretation is not straightforward. The change can be due to the toxicant's primary effect on nervous system function or to its effect on some other system that results in a secondary effect on nervous system function. Certainly, extrapolation from activity measurements in rodents to such phenomena as "hyperactivity" in children is unwarranted, both because of lack of consistency in the experimental work and because such syndromes in humans do not consist exclusively, or necessarily, of increases in motor activity.

Sensory Function

Sensory disturbances often result from human exposure to toxic agents, both as vague symptoms reported by the patient and as clearly demonstrable deficits in sensory function. Deficits in visual, auditory, and tactile functions have been reported for a variety of toxicants, including metals (methylmercury and lead), acrylamide, solvents, and pesticides. A variety of techniques, from very simple to extremely sophisticated, have been utilized to assess sensory function in animals exposed to toxicants. Probably the grossest of these is the orienting response, which consists of observing whether the animal turns toward a crude stimulus (e.g., click, light, or touch). Such a procedure is subjective, nonspecific, and insensitive and indicates only the possibility of gross sensory impairment. The auditory startle reflex and discrimination learning tests are often viewed as tests of sensory function. However, there are many other systems involved in these tests; therefore, sensory effects may not be discriminable from motor effects, learning and memory, and attention abilities.

An extremely promising technique for sensory system evaluation is modulation of reflex startle by presentation of a low-intensity stimulus immediately prior to a high-intensity stimulus that elicits the startle response. Such a technique may be used to estimate sensory threshold, and sensory deficits may be differentiated from nonsensory, such as motor, deficits. This technique is, therefore, specific and reasonably sensitive. It has the advantage of being inexpensive and rapid and requires no training of the animal.

Operant training of an animal allows a very detailed evaluation of sensory function. Such techniques are time-consuming and sometimes expensive, but they are useful for careful characterization of toxicant effects for which there is good evidence of sensory impairment. The species chosen for testing must have sensory function as similar to humans as possible. For visual system testing, for example, the rodent is usually not an appropriate model because its visual system differs in fundamental ways from that of humans.

Animals can be trained to report reliably and in great detail about their sensory perception. This is accomplished through "psychophysical" techniques; that is, sensory function is determined by behavioral means. Such methodology is appropriate for determination of no-effect levels and for detailed characterization of toxic effects. Conditioned suppression is a useful technique for estimating sensory thresholds. A steady baseline rate of responding (such as a lever pressing or licking) is established by use of an intermittent schedule of reinforcement. A test stimulus is presented to an animal several times during its ongoing behavior and signals a specific latency (usually 2 or 3 min) to an unavoidable electric shock. The animal decreases its rate of response (suppresses) during the stimulus in anticipation of the shock, which indicates that the animal detects the stimulus. This technique can be employed to estimate threshold and to detect changes in threshold produced by a toxicant.

Stebbins characterized the thresholds for detection of sound over the range of frequencies normally detectable in the monkey and the effect of an ototoxic agent on these thresholds. This was done by training the monkey to keep its hand in contact with a sensor until it detected the onset of a tone and to break its contact upon detection of the tone. Intensity of the tone was then varied for each frequency tested to determine the intensity at which the monkey was unable to detect the tone. Stebbins was thus able to follow the development of hearing loss produced by an ototoxic antibiotic, from initial high-frequency loss to later low-frequency loss. These changes in hearing in the monkey were correlated with the pattern of receptor loss in the inner ear.

A psychophysical procedure was also used to determine the spatial visual function of monkeys exposed chronically to methylmercury but showed no overt signs of poisoning. In this experiment, the monkey faced two oscilloscopes, one blank and one displaying vertical bars. The monkey had access to two levers, one corresponding to each oscilloscope. The task was to respond on the lever corresponding to the scope on which the bars appeared. The oscilloscope displaying the bars varied randomly from trial to trial. The frequency and darkness of the bars was varied in a systematic manner, allowing a determination of the spatial visual function of each monkey. Monkeys exposed to methylmercury were found to have deficits of high- but not low-frequency spatial vision. Similar behavioral techniques have been used to characterize visual and somatosensory impairment produced by acrylamide. Such studies demonstrate the power of operant techniques in detection of very subtle sensory deficits, which may be the only discernible effects of a toxicant at low level exposure.

Learning and Memory

Loss of memory and inability to concentrate are symptoms frequently reported as a result of human exposure to toxicants such as polychlorinated biphenyls, solvents, methylmercury, and pesticides. Furthermore, developmental exposure may produce mental retardation or learning impairment. It is therefore of great value to test such abilities in animals as markers of toxic effect. There are many techniques available for assessment of learning and memory. Aside from gross screening procedures, this area has probably received the most attention from behavioral toxicologists. Techniques range in complexity from those appropriate for screening to characterization of specific deficits. A screening procedure that is often considered a test of learning is habituation, which is a progressive decrease in reactivity to repeated presentations of a stimulus. Reactivity can be measured in terms of response of the whole organism, as in startle or orienting, or in terms of habituation of a discrete reflexive response, such as blinking. Obviously, habituation must be differentiated from motor effects, fatigue, and sensory adaptation. It is a measure of gross integration of the nervous system and may not involve the higher centers.

A learned behavior that is obviously of adaptive advantage to an animal is its ability to avoid a substance that it ingested shortly before the onset of an illness or adverse effect. This conditional taste aversion can be used to measure toxicity, for example, by pairing a novel taste (a sugar treat, for example) with administration of a toxicant. If the animal feels ill soon afterward, it will avoid the novel substance in the future. This technique has proved to be sensitive to the effects of neurotoxic agents.

At the next level of sophistication, avoidance procedures (utilizing negative reinforcement) are frequently used. Passive avoidance procedures require the animal (rodent) to refrain from leaving a specific area in order to avoid a shock to the feet. Active avoidance requires the animal to move from a specific area at the onset of a cue in order not to be shocked. These procedures are greatly affected by the baseline level of arousal and ongoing motor activity of the animal. It may often be the case that a toxicant produces an effect on one and not on the other of these avoidance tests or affects the behavior in opposing ways, depending on whether the animal is more or less active than the control animal. These tests, therefore, are considered rather nonspecific.

Discrimination tasks have proved useful in detecting effects of toxicants on learning and memory. The procedure most often employed is termed a "forced choice" because the animal is presented with two or more stimuli simultaneously and must indicate its choice by some operant response. These tasks are typically one of two

types; spatial and nonspatial. With spatial discrimination, the animal must respond to a certain position (i.e., left) in order to be reinforced. A nonspatial task requires responding to a specific stimulus (pattern, color, or direction of a tone) regardless of position. Different operants may be utilized in discrimination testing. For rodents, mazes of various sorts are often employed, whereas for other species (as well as for rodents) operants besides locomotion are utilized.

Primates are often tested in a Wisconsin General Testing Apparatus. The monkey faces a panel on which stimuli are placed. A reinforcement, such as a raisin, is placed in a recessed well under the correct stimulus. The monkey's response consists of displacing one of these stimuli; if the choice is correct the reinforcement is collected. Automated apparatuses are used with all laboratory species. Typically the response consists of pressing one of several available levers or push buttons in order to signal the choice. Levine developed a technique for rodents in which a photocell beam is interrupted with the nose as an operant. The technique requires no training by the investigator and may be used with young animals.

Discrimination tasks have proved to be sensitive to impairment resulting from exposure to lead. The difficulty of the task may be an important variable on the effects of a toxicant on performance.

Once the task is learned, a discrimination reversal paradigm provides additional information on the animal's learning ability. The previously correct stimulus becomes the incorrect one so that the animal is required to learn a response opposite from the one previously learned. The discrimination reversal paradigm may often be more sensitive to neurotoxicity than simply acquisition of discrimination tasks, as has been found in monkeys exposed to lead early in life.

There are several other means to test spatial orientation or memory that require little or no training of the animal. An apparatus appropriate for use with rodents is the radial arm maze. Typically, this maze consists of a central arena from which radiate a number of arms like spokes of a wheel. The end of each arm is baited with a reinforcement, and the animal simply has to find all the reinforcements within a certain period of time. The most economical strategy is to enter each arm only once. There obviously need not be a memory component to this task, depending on the strategy adopted by the animal (i.e., "always turn left"). Similarly, motor impairment confounds this task because the number of reinforcements collected in a specified time is the typical dependent variable. The neurotoxicant trimethyltin has been found to disrupt a rodent's ability to perform this test. A somewhat analogous task used for primates is the Hamilton Search Task. A row of boxes, each containing a reinforcement, is presented to the monkey. The monkey can collect the reinforcement from each box by lifting the lid; again, the most economical approach is to lift each lid only once. This test differs from the radial arm maze in that a delay is instituted between responses during which the boxes are withdrawn from the monkey's reach, thus making memory more likely a component of the performance. (It is possible to institute a delay in the radial arm maze as well, but this is most often not done.) Monkeys exposed to lead postnatally required more trials to learn to perform this task than did their controls.

There are several operant tasks that offer the opportunity to separate an animal's learning from its performance of a known task. Repeated acquisition is such a task and requires the animal to learn a new sequence of lever presses each session. The learning baseline may be more sensitive to disruption by a toxicant than the performance of an already acquired sequence.

A task that tests attention and short-term memory is matching to sample. Monkeys are most typically used for these tasks, although other species are also capable of learning them. In a nonspatial matching-to-sample task, for example, the animal is presented with a stimulus (color, pattern, or object) that is then withdrawn. Following this, a set of stimuli are presented, and the animal indicates which of these is identical in some dimension to the sample stimulus. Delays of various durations may be instituted between the presentation of the sample and test stimuli to test short-term memory. Such tasks have been found to be sensitive to effects produced by lead in monkeys who were exposed to it in early life.

Intermittent Schedules of Reinforcement

Performance generated by intermittent schedules of reinforcement has played an important role in behavioral pharmacology and is proving a useful tool in behavioral toxicology. On an intermittent schedule, an animal is not reinforced for every response but for a number of responses according to certain "rules." Most intermittent schedules are based on reinforcing the organism as a function of the number of responses emitted, some temporal requirement for emission of responses, or a combination of these. For example, a fixed ration (FR) schedule requires the animal to emit a fixed number of responses in order to be reinforced. A fixed interval (FI) schedule, on the other hand, requires that a certain fixed length of time elapse before a response is reinforced. Although only one response need be emitted at the end of the interval for reinforcement, the organism typically emits many responses during the interval. Interval schedules generally generate a lower rate of responding than do ratio schedules. The FI schedule generates a characteristic pattern of responding for which a variety of parameters may be analyzed. These parameters are potentially sensitive to disruption by psychoactive agents. Another schedule of some utility in behavioral toxicology is the differential reinforcement of low rate (DRL) schedule in which the animal is required to wait a specified time between responses in order to be reinforced.

Intermittent schedules may also be maintained by negative reinforcement, usually by a brief mild electric shock. The most popular of these is continuous or "Sidman" avoidance in which each response postpones a shock by a fixed amount of time. By spacing its successive responses within this time interval, the animal may postpone shock indefinitely. This schedule is particularly useful as a comparison to behavior generated by positive reinforcement if a toxicant is suspected of producing anorexia. Simple intermittent schedules such as these have been used fairly widely in behavioral toxicology and have proved to be sensitive to the effects of a number of industrial and environmental toxicants.

Intermittent schedules of reinforcement can be combined to form more complicated schedules such as multiple schedules of reinforcement. For example, if fixed ratio and fixed interval schedules are presented to an animal in succession during a single test session, the resulting multiple schedule is termed a multiple FR-FI schedule. Each component of the multiple schedule is independent and occurs in the presence of a different external discrimination stimulus that signals the schedule component in effect. Schedule components are typically presented in an alternating fashion, first one schedule and then the other; this allows the investigator to collect data on both types of behavior almost simultaneously. This schedule in particular has proved to be useful in detecting behavioral toxicity.

Multiple schedules offer the investigator an opportunity to study behavior controlled by different variables, which may be differentially sensitive to the effects of a toxicant. For example, toluene produced a decrease in test animals' response rate in the FR component and an increase in their response rate in the DRL component of a multiple schedule. Furthermore, the relative sensitivity of the two components was different. Similarly, the animals' response in the FI component of a multiple FR-FI was sensitive to disruption by methyl *n*-amyl ketone, whereas their response in the FR component was not. The FI component of the multiple FI-FR schedule was more sensitive to disruption in both monkeys and rodents who sustained developmental lead exposure.

Schedules of reinforcement may be used to monitor toxic effects other than or in addition to direct effects on the central nervous system (CNS). These may include peripheral nervous system toxicity or damage to some other organ systems resulting in general malaise or the animal's feeling "sick." For example, acrylamide, an organic solvent that produces a "dying back" axonopathy, produced decreases in animals' FR response rate. The FR schedule typically produces high response rates and thus may be sensitive to impaired motor function. Rats exposed to ozone decreased their responding on an FI schedule, which was interpreted as a decrease in their motivation as a result of the general discomfort produced by ozone.

Social Behaviors

Animals, particularly mammals, engage in a wide variety of social, sexual, and maternal (or paternal) behaviors that are multidimensional and extremely complex. Despite the obvious importance of social behavior in humans, very little research has been focused on the effects of toxicants on social interactions, and the utility of such interactions in behavioral toxicology is unknown. The reason for this may be the enormous number of variables, which necessitates focusing on only a few parameters to the exclusion of all others. Moreover, many of these behaviors are specific to certain species (e.g., grooming, pup retrieval, and submissive gestures), raising the question of the validity of extrapolation to human behavior.

Each of these components of behavior can be combined into an increasingly complex set of testing para-

digms and, as such, then can be used to evaluate a distinct potential toxic event. An example of this is behavioral teratology.

Behavioral teratology is defined as a separate component of behavioral toxicology primarily in its focus on behavioral modifications resulting from toxic exposures during early development. In general, such studies track the outcome of such exposures over the postnatal and possibly into the juvenile and early adult stages of the life cycle. Outcome measures almost invariably include the development of physical landmarks and reflexes, and also generally include assessment of one or more behavioral functions. Often attempts are made to evaluate multiple behavioral functions, such as motor function and activity, and sensory capabilities and learning in the same experiment. In addition, testing for species-specific behaviors, such as aggression, play, and vocalization, may be included.

Testing during infancy, postnatal, and juvenile periods of development sometimes requires modifications of procedures that are utilized with adults or even the development of new paradigms. In other cases, behavioral paradigms identical to those used in more mature subjects may be used, albeit with parametric modifications. One example of the former is a procedure that has been widely used in behavioral teratology studies as an assessment of olfactory and motor capabilities and is referred to as "homing behavior," a behavior used by rat pups to locate the nest should it be displaced. In such a test, a rat pup, the typical experimental subject for most experimental behavioral teratology studies, is placed in the center of a rectangular apparatus in which one side contains clean bedding material and the other side contains bedding from the pup's home cage. The time taken for the pup to orient to or to reach the home cage bedding constitutes the dependent variable of interest. Since this performance depends on both olfactory capabilities and the development of appropriate motor skills, it represents a type of apical evaluation. It has demonstrated that olfactory discriminations can be learned by rat pups. Pairing aversive electric shock with a distinctive odor leads pups to avoid the odor.

—Shayne C. Gad

Related Topics

Analytical Toxicology
Behavioral Toxicology

Toxicity Testing, Carcinogenesis

Introduction

Carcinogenesis and the broader realm of risk assessment (as it applies to toxicology) have in common that, based on experimental results on a nonhuman species at some relatively high dose or exposure level, an attempt is made to predict the level of impact in humans at much lower levels. This entry examines the assumptions involved in these undertakings, reviews the aspects of design and interpretation of animal carcinogenicity studies, takes a critical look at low-dose extrapolation models and methods, and presents the framework on which risk assessment is based.

Carcinogenicity studies are the longest and most expensive of the extensive battery of toxicology studies required for the registration of pharmaceutical products in the United States and in other major countries. In addition, they are often the most controversial with respect to interpretation of their results. These studies are important because, as noted by IARC, "in the absence of adequate data on humans, it is biologically plausible and prudent to regard agents for which there is sufficient evidence of carcinogenicity in experimental animals as if they presented a carcinogenic risk to humans."

Bioassay Design

Carcinogenicity bioassays have two possible objectives, though (as will be shown) the second is now more important and (as our understanding of carcinogenesis has increased) is increasingly crowding out the first.

The first objective is to detect possible carcinogens. Compounds are evaluated to determine if they can or cannot induce a statistically detectable increase of tumor rates over background levels, and only by happenstance is information generated which is useful in risk assessment. Most older studies have such detection as their objective. Current thought is that at least two species must be used for detection.

The second objective for a bioassay is to provide a range of dose–response information (with tumor incidence being the response) so that a risk assessment may be performed. Unlike detection, which requires only one treatment group with adequate survival times (to allow expression of tumors), dose response requires at least three treatment groups with adequate survival. The selection of dose levels for this case will be discussed later. However, given that the species is known to be responsive, only one species of animal need be used for this objective.

To address either or both of these objectives, three major types of study designs have evolved. First is the classical skin painting study, usually performed in mice. A single, easily detected endpoint (the formation of skin tumors) is evaluated during the course of the study. Though dose response can be evaluated in such a study (dose usually being varied by using different concentrations of test material in volatile solvent), most often detection is the objective of such a study. Though others have used different frequencies of application of test material to vary dose, there are data to suggest that this only serves to introduce an additional variable. Traditionally, both test and control groups in such a test consist of 50–100 mice of one sex (males being preferred because of their very low spontaneous tumor rate). This design is also used in tumor initiation/promotion studies.

The second common type of design is the original NCI bioassay. The announced objective of these studies was detection of moderate to strong carcinogens, although the results have also been used in attempts at risk assessment. Both mice and rats were used in parallel studies. Each study used 50 males and 50 females at each of two dose levels (high and low) plus an equal-sized control group. The NTP has recently moved away from this design because of a recognition of its inherent limitations.

Finally, there is the standard industrial toxicology design, which uses at least two species (usually rats and mice) in groups of no fewer than 100 males and females each. Each study has three dose groups and at least one control. Frequently, additional numbers of animals are included to allow for interim terminations and histopathological evaluations. In both this and the NCI design, a many organs and tissues are collected, processed, and examined microscopically. This design seeks to address both the detection and dose–response objectives with a moderate degree of success.

Selecting the number of animals to use for dose groups in a study requires consideration of both biological (e.g., expected survival rates and background tumor rates) and statistical factors. The prime statistical consideration is reflected in Table T-18. It can be seen in this table that if, for example, we were studying a compound which caused liver tumors and were using mice (with a background or control incidence of 30%), we would have to use 389 animals per sex per group to be able to demonstrate that an incidence rate of 40% in treatment animals was significant compared to the controls at the $p = 0.05$ level.

Perhaps the most difficult aspect of designing a good carcinogenicity study is the selection of the dose levels to be used. At the start, it is necessary to consider the first underlying assumption in the design and use of animal cancer bioassays—the need to test at the highest possible dose for the longest practical period.

The rationale behind this assumption is that although humans may be exposed at very low levels, detecting the resulting small increase (over background) in the incidence of tumors would require the use of an impracticably large number of test animals per group. This point is illustrated in Table T-19, which shows, for instance, that although only 46 animals (per group) are needed to show a 10% increase over a zero background (i.e., a rarely occurring tumor type), 770,000 animals (per group) would be needed to detect a 0.1% increase above a 5% background. As we increase dose, however, the incidence of tumors (the response) will also increase until it reaches the point where a modest increase (e.g.,10%) over a reasonably small background level (e.g., 1%) could be detected using an acceptably small-sized group of test animals. There are, however, at least two real limitations to the highest dose level. First, the test rodent population must have a survival rate after receiving a lifetime (or 2 years) of regular doses to allow for meaningful statistical analysis. Second, we really want the metabolism and mechanism of action of the chemical at the highest level tested to be the same as at the low levels where human exposure would occur. Unfortunately, we usually must select the high dose level based only on the information provided by a subchronic or range finding study, but selection of too low a dose will make the study invalid for detection of carcinogenicity and may seriously impair the use of the results for risk assessment.

There are several solutions to this problem. One of these has been the rather simplistic approach of the NTP Bioassay Program, which is to conduct a 3-month

TABLE T-18
Sample Size Required to Obtain a Specified Sensitivity at $p < 0.05$ Treatment Group Incidence

Background tumor incidence (%)	P^a	0.95	0.90	0.80	0.70	0.60	0.50	0.40	0.30	0.20	0.10
0.30	0.90	10	12	18	31	46	102	389			
	0.50	6	6	9	12	22	32	123			
0.20	0.90	8	10	12	18	30	42	88	320		
	0.50	5	5	6	9	12	19	28	101		
0.10	0.90	6	8	10	12	17	25	33	65		
	0.50	3	3	5	6	9	11	17	31	68	
0.05	0.90	5	6	8	10	13	18	25	35	76	464
	0.50	3	3	5	6	7	9	12	19	24	147
0.01	0.90	5	5	7	8	10	13	19	27	46	114
	0.50	3	3	5	5	6	8	10	13	25	56

Incidence rate (%) — Required sample size

[a]Power for each comparison of treatment group with background tumor incidence.

TABLE T-19

Average Number of Animals Needed to Detect a Significant Increase in the Incidence of an Event (e.g., Tumors and Anomalies) over the Background Incidence (Control) at Several Expected Incidence Levels Using the Fisher Exact Probability Test ($p = 0.05$)

Background incidence (%)	Expected increase in incidence (%)					
	0.01	0.1	1	3	5	10
0	46,000,000[a]	460,000	4,600	511	164	46
0.01	46,000,000	460,000	4,600	511	164	46
0.1	47,000,000	470,000	4,700	520	168	47
1	51,000,000	510,000	5,100	570	204	51
5	77,000,000	770,000	7,700	856	304	77
10	100,000,000	1,000,000	10,000	1,100	400	100
20	148,000,000	1,480,000	14,800	1,644	592	148
25	160,000,000	1,600,000	16,000	1,840	664	166

[a]Number of animals needed in each group—controls as well as treated.

range-finding study with sufficient dose levels to establish a level that significantly (10%) decreases the rate of body weight gain. This dose is defined as the maximum tolerated dose (MTD) and is selected as the highest dose. Two other levels, generally one-half MTD and one-quarter MTD, are selected for testing as the intermediate and low dose levels. In many early NCI studies, only one other level was used.

The dose range-finding study is necessary in most cases, but the suppression of body weight gain is a scientifically questionable benchmark when dealing with establishment of safety factors. Physiologic, pharmacologic, or metabolic markers generally serve as better indicators of systemic response than body weight. A series of well-defined acute and subchronic studies designed to determine the "chronicity factor" and to study onset of pathology can be more predictive for dose setting than body weight suppression.

Also, the NTP's MTD may well be at a level where the metabolic mechanisms for handling a compound at real-life exposure levels have been saturated or overwhelmed, bringing into play entirely artifactual metabolic and physiologic mechanisms.

Selection of levels for the intermediate and lower doses for a study is easy only in comparison to the selection of the high dose. If an objective of the study is to generate dose–response data, then the optimal placement of the doses below the high is such that they cover as much of the range of a response curve as possible and yet still have the lowest dose at a high enough level that one can detect and quantify a response. If the objective is detection, then having too great a distance between the highest and next highest dose creates a risk to the validity of the study. If the survival in the high dose is too low, yet the next highest dose does not show nonneoplastic results (i.e., cause other than neoplastic adverse biological effects) such as to support it being a high enough dose to have detected a strong or moderate carcinogen, the entire study may have to be rejected as inadequate to address its objective. Statistical guidelines have been proposed (for setting dose levels below the high) based on response surfaces. In so doing they suggest that the lowest dose be no less than 10% of the highest.

Although it is universally agreed that the appropriate animal model for testing a chemical for carcinogenicity would be an animal whose metabolism, pharmacokinetics, and biological responses were most similar to humans, economic considerations have largely constrained practical choices to rats and mice. The use of both sexes of both species is preferred on the grounds that it provides for (in the face of a lack of understanding of which species would actually be most like humans for a particular agent) a greater likelihood of utilizing the more sensitive species. Use of the mouse is both advocated and defended on these grounds and because of the economic advantages and the species' historical utilization. There are those who believe that the use of the mouse is redundant and represents a diversion of resources while yielding little additional information, citing a "unique contribution" for mouse data in 273 bioassays of only 13.6% of the cases (i.e., 37 cases). Others question the use of the mouse based on the belief that it gives artifactual liver carcinogenesis results. One suggestion for the interpretation of mouse bioassays is that in those cases in which there is only an increase in liver tumors in mice (or lung tumors in strain A mice) and no supporting mutagenicity findings (a situation characteristic of some classes of chemicals), the test compound should not be considered an overt carcinogen. This last aspect, however, is even more strongly focused on the strain of mouse that is used than on the use of the species itself.

The NCI NTP currently recommends an F1 hybrid cross between two inbred strains, the C57B1/6 female and the C3H male, the results being commonly designated as the B6C3F1. This mouse was found to be very successful in a large-scale pesticide testing program in the mid-1960s. It is a hardy animal with good survival, easy to breed, disease resistant, and has been reported to have a relatively low spontaneous tumor incidence. Usually at least 80% of the control mice are still alive at a 24-month termination.

Unfortunately, while it was originally believed that the spontaneous liver tumor incidence in male B6C3F1 mice was 13.7%, it actually appears to be closer to 32.1%. The issue of spontaneous tumor rates and their impact on the design and interpretation of studies will be discussed more fully later. Thus, use of a cross of two inbred mouse strains is also a point of controversy. Haseman and Hoel (1979) have presented data to support the idea that inbred strains have lower degrees of variability of biological functions and tumor rates, making them more sensitive detectors and quantitators. These authors also suggest that the use of a cross from two such inbred strains allows one to more readily detect tumor incidence increases. On the other hand, it has been argued that such genetically homogeneous strains do not properly reflect the diversity of metabolic functions present in the human population (particularly functions that would serve to detoxify or act as defense mechanisms).

Study length and the frequency of treatment are design aspects that must also be considered. These are aspects in which the objective of detection and dose–response definition conflict.

For the greatest confidence in a "negative" detection result, an agent should be administered continuously for the majority of an animal's life span. The NTP considers 2 years to be a practical treatment period in rats and mice, although the animals currently used in such studies may survive an additional 6–12 months. Study lengths of 15–18 months are considered adequate for shorter lived species such as hamsters. An acceptable exposure/observation period for dogs is considered to be 7–10 years, an age equivalent to about 45–60 years in humans. For dietary treatments, continuous exposure is considered desirable and practical. With other routes, practical considerations may dictate interrupted treatments. For example, inhalation treatment for 6–8 hr per day, 5 days per week is the usual practice. Regimens requiring special handling of animals, such as parenteral injections, are usually on a 5-days per week basis. With some compounds intermittent exposures may be required because of toxicity. Various types of recovery can occur during exposure-

free periods, which may either enhance or decrease chances of carcinogenicity. In view of the objective of assessing carcinogenicity as the initial step, intermittent exposure on a 3- to 5-days per week basis is considered both practical and desirable for most compounds.

Following cessation of dosing or exposure, continued observation during a nontreatment period may be required before termination of the experiment. Such a period is considered desirable because (1) induced lesions may progress to more readily observable lesions, and (2) morphologically similar but noncarcinogenic proliferative lesions that are stress-related may regress. Neoplastic or "neoplastic-like" legions that persist long after removal of the stimulus are considered serious consequences from the hazard viewpoint. Many expert anatomical pathologists, however, believe they are able to diagnose and determine the biological nature of tumorous lesions existing at the time of treatment without the added benefit of a treatment-free period.

In determining the length of an observation period, several factors must be considered: period of exposure, survival pattern of both treated and control animals, nature of lesions founds in animals that have already died, tissue storage and retention of the chemical, and results of other studies that would suggest induction of late-occurring tumors. The usual length of a treatment-free observation period is 3 months in mice and hamsters and 6 months in rats. An alternative would be to terminate the experiment or an individual treatment group on the basis of survival (e.g., at the point at which 50% of the group with the lowest survival has died).

The arguments against such prolonged treatment and maintenance on study revolve around the relationship between age and tumor incidence. As test animals (or humans) become older, the background ("naturally occurring") incidence of tumors increases and it becomes increasingly difficult to identify a treatment effect from the background effect. Salsburg has published an analysis of patterns of senile lesions in mice and rats, citing what he calls the principle of biological confounding: "If a particular lesion (e.g., pituitary tumor) is part of a larger syndrome induced by the treatment, it is impossible to determine whether the treatment has 'caused' that lesion."

This could lead to a situation in which any real carcinogen would be nonidentifiable. If the usual pattern of old age lesions for a given species or strain of animals includes tumors, then almost every biologically active treatment can be expected to influence the incidence of tumors in a cluster of lesions at a sufficiently high dose.

Reconsidering our basic principles of experimental design, it is clear that we should try to design bioassays so that any carcinogenesis is a clear-cut, single event, unconfounded by the occurrence of significant numbers of lesions due to other causes (such as age). One answer to this problem is the use of interim termination groups. When an evaluation of tumor incidences in an interim sacrifice sample of animals indicates that background incidence is becoming a source of confounding data, termination plans for the study can be altered to minimize the loss of power.

A number of other possible confounding factors can enter into a bioassay unless design precludes them. These include (1) cage and litter effects, which can be avoided by proper prestudy randomization of animals and rotation of cage locations; (2) vehicle (e.g., corn oil has been found to be a promoter for liver carcinogens); and (3) the use of the potential hazard route for man (e.g., dietary inclusion instead of gastric intubation).

Bioassay Interpretation

The interpretation of the results of even the best designed carcinogenesis bioassay is a complex statistical and biological problem. In addressing the statistical aspects, we shall have to review some biological points which have statistical implications as we proceed.

All such bioassays are evaluated by comparison of the observed results in treatment groups with those in one or more control groups. These control groups always include at least one group that is concurrent, but because of concern about variability in background tumor rates, a historical control group is also considered in at least some manner.

The underlying problem in the use of concurrent controls alone is the belief that the selected population of animals are subject both to an inordinate degree of variability in their spontaneous tumor incidence rates and that the strains maintained at separate breeding facilities are each subject to a slow but significant degree of genetic drift. The first problem raises concern that, by chance, the animals selected to be controls for any particular study with be either "too high" or "too low" in their tumor incidences, leading to either a false-positive or false-negative statistical test result when test

animals are compared to these controls. The second problem leads to concern that, over the years, different laboratories will be using different standards (control groups) against which to compare the outcome of their tests, making any kind of relative comparison between compounds or laboratories impossible.

Further Reading

Depass, L., and Gad, S. C. (Eds.) (1994). *Safety Assessment for Pharmaceuticals*, pp. 167–184. Van Nostrand Reinhold, New York.:
Waalkes, M. P., and Ward, J. M. (1994). *Carcinogenesis*. Raven Press, New York.

—*Shayne C. Gad*

Related Topics

Analytical Toxicology
Animal Models
Carcinogen Classification Schemes
Carcinogen–DNA Adduct Formation and DNA Repair
Carcinogenesis
Dose–Response Relationship
International Agency for Research on Cancer
Mouse Lymphoma Assay
Mutagenesis
National Toxicology Program
Risk Assessment, Human Health
Toxicity Testing, *In Vivo*

Toxicity Testing, Dermal

Evaluation of materials for their potential to cause dermal irritation and corrosion due to acute contact has been common for industrial chemicals, cosmetics, agricultural chemicals, and consumer products since at least the 1930s (generally, pharmaceuticals are only evaluated for dermal effects if they are to be administered topically–and then by repeat exposure tests, which will not be addressed here). As with acute eye irritation tests, one of the earliest formal publications of a test method (although others were used) was that of Draize *et al.* in 1944. The methods currently used are still basically those proposed by Draize *et al.* and, to date, have changed little since 1944. Though (unlike their near relatives, the eye irritation tests) these methods have not particularly caught the interest or spotlight of concern of the animal welfare movement, there are efforts under way to develop alternatives that either do not use animals or are performed in a more humane and relevant (to human exposure) manner.

Among the most fundamental assessments of the safety of a product or, indeed, of any material that has the potential to be in contact with a significant number of people in our society are tests which seek to predict potential skin irritation or corrosion. Like all the other tests in what is classically called a range-finding, tier I, or acute battery, the tests used here are both among the oldest designs and are currently undergoing the greatest degree of scrutiny and change. Currently, all the established test methods for these endpoints use the same animal model, the rabbit (almost exclusively the New Zealand White), though some other animal models have been proposed. There are also new *in vitro* test systems.

Virtually all man-made chemicals have the potential to contact the skin of people. In fact, many (e.g., cosmetics and shampoos) are intended to have skin contact. Also, the greatest number of industrial medical problems are skin conditions, indicating the large extent of dermal exposure where none is intended.

Testing is performed to evaluate the potential occurrence of two different, yet related, endpoints. The broadest application of these is evaluation of the potential to cause skin irritation, characterized by erythema (redness) and edema (swelling). Severity of irritation is measured in terms of both the degree of these two parameters and how long they persist. There are three types of irritation tests, each designed to address a different concern:

1. Primary (or acute) irritation: a localized reversible dermal response resulting from a single application of, or exposure to, a chemical without the involvement of the immune system.

2. Cumulative irritation: a reversible dermal response which results from repeated exposure to a substance (each individual exposure is not capable of causing acute primary irritation).

3. Photochemically induced irritation: a primary irritation resulting from light-induced molecular changes in the chemical to which the skin has been exposed.

Although most regulations and common practice characterize an irritation that persists 14 days past the end of exposure as other than reversible, the second endpoint of concern with dermal exposure—corrosion per se—is assessed in separate test designs. These tests start with a shorter exposure period (4 hr or less) to the material of concern and then evaluate simply whether tissue has been killed or not (i.e., if necrosis is present or not).

Irritation is generally a localized reaction resulting from either a single exposure or multiple exposures to a physical or chemical entity at the same site. It is characterized by the presence of erythema (redness), edema, and may or may not result in cell death. The observed signs are heat (caused by vessel dilation and the presence of large amounts of warm blood in the affected area), redness (due to capillary dilation), and pain (due to pressure on sensory nerves). The edema often observed is largely due to plasma, which coagulates in the injured area, precipitating a fibrous network to screen off the area, thereby permitting leukocytes to destroy exogenous materials by phagocytosis. If the severity of injury is sufficient, cell death may occur, thereby negating the possibility of cellular regeneration. Necrosis is a term often used in conjunction with cell death, and it is the degeneration of the dead cell into component molecules which approach equilibrium with surrounding tissue.

There are three major objectives to be addressed by the performance of these tests:

1. Providing regulator required baseline data: Any product now in commerce must both be labeled appropriately for shipping and be accompanied by a material safety data sheet which clearly states potential hazards associated with handling it. Department of Transportation regulations also prescribe different levels of packaging on materials found to constitute hazards as specified in the regulations. EPA (under FIFRA) also has a pesticides labeling requirement. Similar requirements

exist outside the United States. These requirements demand absolute identification of severe irritants or corrosives and adherence to the basics of test methods promulgated by the regulations. False positives (type I errors) are to be avoided in these usages.

2. Hazard assessment for accidents: For most materials, dermal exposure is not intended to occur, but it will occur in cases of accidental spillage or mishandling. Here, we need to correctly identify the hazard associated with such exposures and be equally concerned with false positives and false negatives.

3. Assessment of safety for use: The materials at issue here are the full range of products for which dermal exposure will occur in the normal course of use. These range from cosmetics and hand soaps to bleaches, laundry detergents, and paint removers. No manufacturer desires to put a product on the market which cannot be safely used and will lead to extensive liability if placed in the marketplace. Accordingly, the desire here is to accurately predict the potential hazards in humans—that is, to have neither false positives nor false negatives.

Table T-20 sets forth the current regulatorily mandated test designs which form the bases of all currently employed test procedures. Though these designs vary somewhat in detail, they are by and large comparable.

All of these methods use the same scoring scale, the Draize scale, which is presented in Table T-21. However, though the regulations prescribe these different test methods, most laboratories actually perform some modified methods. Here, two modifications (one for irritation and the other for corrosion) which reflect prior laboratory experience are recommended.

Primary Dermal Irritation Test

Rabbit Screening Procedure

1. A group of at least 8–12 New Zealand White rabbits are screened for the study.

2. All rabbits selected for the study must be in good health; any rabbit exhibiting sniffles, hair loss, loose stools, or apparent weight loss is rejected and replaced.

3. One day (at least 18 hr) prior to application of the test substance, each rabbit is prepared by clipping the hair from the back and sides using a small animal clipper. A size No. 10 blade is used to remove long hair and then a size No. 40 blade is used to remove the remaining hair.

4. Six animals with skin sites that are free from hyperemia or abrasion (due to shaving) are selected. Skin sites that are in the telogen phase (resting stage of hair growth) are used; those skin sites that are in the anagen phase (stage of active growth, indicated by the presence of a thick undercoat of hair) are not used.

Study Procedure

1. As many as four areas of skin, two on each side of the rabbit's back, can be utilized for sites of administration.

2. Separate animals are not required for an untreated control group. Each animal serves as its own control.

3. Besides the test substance, a positive control substance (a known skin irritant—1% sodium laurel sulfate in distilled water) and a negative control (untreated patch) are applied to the skin. When a vehicle is used for diluting, suspending, or moistening the test substance, a vehicle control patch is required, especially if the vehicle is known to cause any toxic dermal reactions or if there is insufficient information about the dermal effects of the vehicle.

4. The intact (free of abrasion) sites of administration are assigned a code number. Up to four sites can be used:
   A. Test substance
   B. Negative control
   C. Positive control
   D. Vehicle control (if required)

5. Application sites should be rotated from one animal to the next to ensure that the test substance

TABLE T-20

Regulatorily Mandated Test Designs for Dermal Irritation/Corrosion

Agency	Test material Solid	Test material Liquid	Exposure time (hr)	No. of rabbits	Sites per animal (intact/abraded)	At end of exposure	Scoring intervals postexposure	Note	Reference
Department of Transportation (DOT)	Not specified	Not specified	4	6	1/0	Skin washed w/ appropriate vehicle	4 and 48 hr	Endpoint is corrosion in 2 of 6 animals	DOT (1980)
Environmental Protection Agency (EPA)	Moisten	Undiluted	24	6	2/2	Skin wiped but not washed	24 and 72 hr; may continue until irritation fades or is judged irreversible	Toxic Substance Control Act test (TSCA); also FIFRA	EPA (1979)
Consumer Product Safety Commission (CPSC)	Dissolve in appropriate vehicle	Neat	24	6	1/1	Not specified	24 and 72 hr	Federal Hazardous Substances Act (FHSA)	CPSC (1980)
OECD	Moisten	Undiluted	4	3[a]	1/0	Wash with water or solvent	30–60 min, 24, 48, 72 hr or until judged irreversible	European Common Market	OECD (1981)

[a]Additional animals may be required to clarify equivocal results.

TABLE T-21
Evaluation of Skin Reactions

Skin reaction	Value
Erythema and eschar formation	
No erythema	0
Very slight erythema (barely perceptable)	1
Well-defined erythema	2
Moderate to severe erythema	3
Severe erythema (beet redness) to slight eschar formation (injuries in-depth)	4
Necrosis (death of tissue)	+N
Eschar (sloughing or scab fomation)	+E
Edema formation	
No edema	0
Very slight edema (barely perceptible)	1
Slight edema (edges of area well-defined by definite raising)	2
Moderate edema (raised approximately 1 mm)	3
Severe edema (raised more than 1 mm and extending beyond the area of exposure)	4
Total possible score for primary irritation	8

and controls are applied to each position at least once.

6. Each test or control substance is held in place with a 1 × 1-in., 12-ply surgical gauze patch. The gauze patch is applied to the appropriate skin site and secured with 1-in.-wide strips of surgical tape at the four edges, leaving the center of the gauze patch nonoccluded.

7. If the test substance is a solid or semisolid, a 0.5-g portion is weighed and placed on the gauze patch. The test substance patch is placed on the appropriate skin site and secured. The patch is subsequently moistened with 0.5 ml of physiological saline.

8. When the test substance is in flake, granule, powder, or other particulate form, the weight of the test substance that has a volume of 0.5 ml (after compacting as much as possible without crushing or altering the individual particles, such as by tapping the measuring container) is used whenever this volume weight is less than 0.5 g. When applying powders, granules, and the like, the gauze patch designated for the test sample is secured to the appropriate skin site with one of

four strips of tape at the most ventral position of the animal. With one hand, the appropriate amount of sample measuring 0.5 ml is carefully poured from a flycine weighing paper onto the gauze patch that is held in a horizontal (level) position with the other hand. The patch containing the test sample is then carefully placed into position onto the skin and the remaining three edges are secured with tape. The patch is subsequently moistened with 0.5 ml of physiological saline.

9. If the test substance is a liquid, a patch is applied and secured to the appropriate skin site. A 1-ml tuberculin syringe is used to measure and apply 0.5 ml of test substance to the patch.

10. The negative control site is covered with an untreated 12-ply surgical gauze patch (1 × 1 in).

11. The positive control substance and vehicle control substance are applied to a gauze patch in the same manner as a liquid test substance.

12. The entire trunk of the animal is covered with an impervious material (such as saran wrap) for a 24-hr period of exposure. The saran wrap is secured by wrapping several long strips of athletic adhesive tape around the trunk of the animal. The impervious material aids in maintaining the position of the patches and retards evaporation of volatile test substances.

13. An Elizabethan collar is fitted and fastened around the neck of each test animal. The collar remains in place for the 24-hr exposure period. The collars are utilized to prevent removal of wrappings and patches by the animals, while allowing the animals food and water *ad libitum.*

14. The wrapping is removed at the end of the 24-hr exposure period. The test substance skin site is wiped to remove any test substance still remaining. When colored test substances (such as dyes) are used, it may be necessary to wash the test substance from the test site with an appropriate solvent or vehicle (one that is suitable for the substance being tested). This is done to facilitate accurate evaluation for skin irritation.

15. Immediately after removal of the patches, each 1 × 1-in. test or control site is outlined with an indelible marker by dotting each of the four corners. This procedure delineates the site for identification.

Observations

1. Observations are made of the test and control skin sites 1 hr after removal of the patches (25 hr postinitiation of application). Erythema and edema are evaluated and scored on the basis of the designated values presented in Table T-21.

2. Observations are again performed 48 and 72 hr after application and scores are recorded.

3. If necrosis is present or the dermal reaction is unusual, the reaction should be described. Severe erythema should receive the maximum score (4), and +N should be used to designate the presence of necrosis and +E the presence of eschar.

4. When a test substance produces dermal irritation that persists 72 hr postapplication, daily observations of test and control sites are continued on all animals until all irritation caused by the test substance resolves or until Day 14 postapplication.

Evaluation of Results

1. A subtotal irritation value for erythema or eschar formation is determined for each rabbit by adding the values observed at 25, 48, and 72 hr postapplication.

2. A subtotal irritation value for edema formation is determined for each rabbit by adding the values observed at 25, 48, and 72 hr postapplication.

3. A total irritation score is calculated for each rabbit by adding the subtotal irritation value for erythema or eschar formation to the subtotal irritation value for edema formation.

4. The primary dermal irritation index is calculated for the test substance or control substance by dividing the sum of the total irritation scores by the number of observations (3 days × six animals = 18 observations).

5. The categories of the primary dermal irritation index (PDII) are as follows (this categorization of the dermal irritation is a modification of the original classification described by Draize *et al.*):

    PDII = 0.0   nonirritant
    >0.0–0.5   negligible irritant
    >0.5–2.0   mild irritant
    >2.0–5.0   moderate irritant
    >5.0–8.0   severe irritant

Other abnormalities, such as atonia or desquamation, should be noted and recorded.

Factors Affecting Irritation Responses and Test Outcome

The results of local tissue irritation tests are subject to considerable variability due to relatively small differences in test design or technique. Weil and Scala arranged and reported on the best known of several intralaboratory studies to clearly establish this fact. Though the methods presented previously have proven to give reproducible results in the hands of the same technicians over a period of years and contain some internal controls (the positive and vehicle controls in the PDI) against large variabilities in results or the occurrence of either false positives or negatives, it is still essential to be aware of those factors that may systematically alter test results. These factors are summarized as follows:

1. In general, any factor that increases absorption through the stratum corneum or mucous membrane will also increase the severity of an intrinsic response. Unless this factor mirrors potential exposure conditions, it may, in turn, adversely affect the relevance of test results.

2. The physical nature of solids must be carefully considered both before testing and in interpreting results. Shape (sharp edges), size (small particles may abrade the skin due to being rubbed back and forth under the occlusive wrap), and rigidity (stiff fibers or very hard particles will be physically

irritating) of solids may all enhance an irritation response.

3. Solids frequently give different results when they are tested dry than if wetted for the test. As a general rule, solids are more irritating if moistened (referring to item 1, wetting is a factor that tends to enhance absorption). Care should also be taken regarding moistening agent—some (few) batches of U.S. Pharmacopeia physiological saline (used to simulate sweat) have proven to be mildly irritating to the skin and mucous membrane on their own. Liquids other than water or saline should not be used.

4. If the treated region on potential human patients will be a compromised skin surface barrier (e.g., if it is cut or burned) some test animals should likewise have their application sites compromised. This procedure is based on the assumption that abraded skin is uniformly more sensitive to irritation. Experiments, however, have shown that this is not necessarily true; some materials produce more irritation on abraded skin, while other produce less.

5. The degree of occlusion (in fact, the tightness of the wrap over the test site) also alters percutaneous absorption and therefore irritation. One important quality control issue in the laboratory is achieving a reproducible degree of occlusion in dermal wrappings.

6. Both the age of the test animal and the application site (saddle of the back vs flank) can markedly alter test outcome. Both of these factors are also operative in humans, of course, but in dermal irritation tests the objective is to remove all such sources of variability. In general, as an animal ages, sensitivity to irritation decreases. For the dermal test, the skin on the middle of the back (other than directly over the spine) tends to be thicker (and therefore less sensitive to irritations) than that on the flanks.

7. The sex of the test animals can also alter study results because both regional skin thickness and surface blood flow vary between males and females.

8. The single most important (but also most frequently overlooked) factor that influences the results and outcome of these (and, in fact, most) acute studies is the training of the staff. In determining how test materials are prepared and applied and in how results are "read" against a subjective scale, both accuracy and precision are extremely dependent on the technicians involved. To achieve the desired results, initial training must be careful and all-inclusive. Equally as important, some form of regular refresher training must be exercised, particularly in the area of scoring of results. Use of a set of color photographic standards as a training reference tool is strongly recommended; such standards should clearly demonstrate each of the grades in the Draize dermal scale.

9. It should be recognized that the dermal irritancy test is designed with a bias to preclude false negatives and, therefore, tends to exaggerate results in relation to what would happen in humans. Findings of negligible irritancy (or even in the very low mild irritant range) should therefore be of no concern unless the product under the test is to have large-scale and prolonged dermal contact.

Problems in Testing (and Their Resolutions)

Some materials, by either their physicochemical or their toxicological natures, generate difficulties in the performance and evaluation of dermal irritation tests. The most commonly encountered of these problems are compound volatility, pigmented material, and systemic toxicity.

Compound Volatility

One is sometimes required or requested to evaluate the potential irritancy of a liquid that has a boiling point between room temperature and the body temperature of the test animal. As a result, the liquid portion of the material will evaporate off before the end of the testing period. There is no real way around the problem; one can only make clear in the report on the test that the traditional test requirements were not met, though an evaluation of potential irritant hazard was probably achieved (because the liquid phase would also have evaporated from a human that it was spilled on).

Pigmented Material

Some materials are strongly colored or discolor the skin at the application site. This makes the traditional scoring process difficult or impossible. One can try to remove the pigmentation with a solvent; if successful, the erythema can then be evaluated. If use of a solvent fails or is unacceptable, one can (wearing thin latex gloves) feel the skin to determine if there is warmth, swelling, and/or rigidity—all secondary indicators of the irritation response.

Systemic Toxicity

On rare occasions, the dermal irritation study is begun only to have the animals die very rapidly after test material is applied.

Design Alternatives and Innovations

In vivo alternative approaches to evaluating dermal toxicity are limited to one other dose site and two other species of small animals. These are the guinea pig, mouse ear, and rabbit ear tests. Gilman has previously presented a short overview of these three alternatives, but some additional information has since become available.

Guinea Pig

The response of the guinea pig has been reported as being less severe and more like that of a human, and there have been recommendations that it be the species of choice with the test being performed in the same manner as in the PDI. FIFRA guidelines, indeed, name the guinea pig as an alternative species for the PDI test. However, the rabbit is cheaper and its larger size makes multiple patching more practical than that in the guinea pig.

Mouse Ear

The ear of the albino mouse has been proposed as an alternative test system. As originally proposed by the author, the test was performed as follows:

- Ten microliters (liquid) or 10 mg (solid paste) is applied to the dorsal aspect of one ear; the contralateral ear serves as a control.

- Test material is applied topically, daily on four consecutive days.

- Dermal reactions are read on Day 5 as follows:
   0: No visible blood vessels or erythema
   2: Few blood vessels, barely visible; no erythema
   4: Main blood vessels visible on lower half of ear; slight erythema over lower third or base of ear
   6: Main blood vessels more obvious; suggestion of capillary network of tips of main vessels; slight or generalized erythema
   8: Main blood vessels extended to edge of ear; more extensive capillary network between main blood vessels; possibly internal hemorrhage; erythema more pronounced; ear may begin to fold back and lose suppleness
   10: Pronounced blood vessels and extensive capillary network evident; marked erythema; possibly "frilling" of ear margin
   12: Pronounced blood vessels and extensive capillary network extending to ear margins; severe erythema; frilling and thickening of ear margins; crusting more in evidence
   14: Pronounced blood vessels and severe erythema; obvious thickening of ear; possibly necroses; crusting may extend over whole ear surface

- Daily differences between control and treated ears for each animal are added. A correction is given for any difference between the control and treated ears initially, divided by 5 and interpreted as follows:
   0–9: Probably not irritating to human skin
   10–15: May be slightly irritating to some users
   Over 15: Likely to prove sufficiently irritating to elicit user complaints at unacceptable levels

In 1985, Patrick *et al.* utilized the mouse ear model to evaluate dermal irritants and try to distinguish mechanisms behind irritation. In 1986, Gad *et al.* published a paper in which a new method for evaluating dermal sensitization was described, but in doing so, they also presented a substantial amount of dermal irritation data arising from a mouse ear model.

The mouse is cheaper than the rabbit and appears to give results analogous to those in the rabbit. The chief drawbacks to the model appear to be custom and the existence of a large database in the rabbit model.

Rabbit Ear

Over the years, several people have proposed a dermal irritation evaluation model based on the test material being applied to the inside surface of the rabbit ear. The advantages are that this site does not have to be shaved and may not overpredict results as much. Seemingly no formal evaluation of a method based on this site has been performed and published.

The reader should also be aware that there are variety of cumulative irritancy test designs available, such as the guinea pig immersion test and the 21-consecutive-day occluded patch test in rabbits.

In Vitro Alternatives

The state of development of alternative models for dermal irritation or corrosion is improving rapidly. As was noted previously, though there have been attempts to utilize other animals as models, these have not been well received or widely adopted—nor do they seem to offer better results. There are seven categories of potential nonanimal test systems.

Physicochemical Test Methods

Analysis of the physicochemical properties of test substances, including the pH, absorption spectra, partition coefficients, and other parameters, often indicates potential dermal toxicity. According to OECD guidelines, substances with a pH <2 or >11 do not need to be tested for irritancy *in vivo* (OECD 81). The potential effects of acids and bases to produce irritancy has been well established.

Physicochemical analysis has evaluated the particular chemical properties of test substances which have been identified as key structural components contributing to penetration, irritation, or sensitization. Absence of absorption in the ultraviolet (UV) range also has been used to suggest lack of photoirritant potential. Physicochemical tests are rapid, cost-effective, easily standardized, and transferable to outside laboratories.

For penetration, a partition coefficient of the test sample provides a useful guide. The size of a chemical is also indicative of potential penetration. Many of the physicochemical properties of surfactants have been found to be potential indicators of their action on skin.

Target Macromolecular and Biochemical Systems

Test methods which utilize the analysis of biochemical reactions or changes in organized macromolecules evaluate toxicity at a subcellular level. Because of their simplicity, they can be readily standardized and transferred to outside laboratories to provide yardstick measurements for varying degrees of dermal toxicity.

One *in vitro* irritation prediction method that utilizes nonbiological, nonliving substances can be described as a biomembranebarrier–macromolecular matrix system. This method is known as the Skintex system. The Skintex system makes use of a two-compartment physicochemical model incorporating a keratin/collagen membrane barrier and an ordered macromolecular matrix. The effect of irritants on this membrane is detected by changes in the intact barrier membrane through the use of an indicator dye attached to the membrane. The dye is released following membrane alteration or disruption, which can occur when the synthetic membrane is exposed to an irritant. A specific amount of dye corresponding to the degree of irritation can be liberated and quantified spectrophotometrically. The second compartment within the system is a reagent macromolecular matrix that responds to toxic substances by producing turbidity. This second response provides an internal detection for materials which disrupt organized protein conformation after passing through the membrane barrier.

Test samples can be applied directly to the barrier membrane as liquids, solids, or emulsions and inserted into the liquid reagent. The results are directly compared to the Draize dermal irritation results.

More that 5300 test samples have been studied in the Skintex system, including petrochemicals, agrochemicals, household products, and cosmetics. The reproducibility with standard deviations of 5–8% is excellent. New protocols applicable to very low irritation test samples and alkaline products have increased the applicability of this method. Skintex validation studies resulting in an 80–90% correlation to the Draize scoring have been reported by S. C. Johnson & Son and the Food and Drug and Safety Center.

Thus far, most *in vitro* irritation methods, including Skintex, have relied heavily on the vast Draize rabbit skin database for validation. As previously discussed, the discrepancies in the information generated by the Draize system raise questions about the applicability of this information to irritation reactions in man.

A new Skintex protocol called the "human response assay" optimizes the model to predict human irritation. A collaborative study with Dr. Howard Maibach and co-workers at the University of California at San Francisco demonstrated good correlations to human response for pure chemicals with diverse mechanisms of dermal toxicity. Ongoing studies have evaluated pure chemicals, surfactants, vehicles, and fatty acids.

The Skintex test is a rapid, standardized approach with well-refined protocols and an extensive database. The results produced are contiguous with the historical *in vivo* database. However, the method cannot predict immune response, penetration, or recovery after the toxic response.

Cell Culture Techniques

In vitro cytotoxicity tests that indicate basic cell toxicity by measuring parameters such as cell viability, proliferation, membrane damage, DNA synthesis, or metabolic effects have been used as indicators of dermal toxicity.

The most commonly used approaches are the neutral red assay (cell viability and membrane damage), the Lowry (labeled proline) Coomassie blue and Kenacid blue assays (cell proliferationand total cell protein), the MTT or tetrazolium assay (mitochondrial function), and the intracellular lactate dehydrogenase activity test (cell lysis).

In the neutral red (cell viability) and total protein (cell proliferation) assays, cells are treated with various concentrations of a test substance in petri or multiwell dishes; after a period of exposure, the substance is washed out of the medium. (An analytical reagent is added in the case of protein measurements.) Neutral red is a supravital dye which accumulates in the lysosomes of viable, uninjured cells, and it can be washed out of cells which have been damaged. In the protein test, Kenacid blue is added and reacts with cellular

protein. Controlled cells are dark blue; killed cells are lighter colored. The IC_{50} (the concentration which inhibits by 50%) is determined; the test can be rapidly performed with automation. However, materials must be solubilized into the aqueous cell media for analysis. For many test materials this will require large dilutions which eliminate properties of the materials which cause irritation.

The MIT test assays mitochondrial function by measuring reduction of the yellow MTT tetrazolium salt to a blue insoluble product. It has been compared with the neutral red technique for testing the cytotoxicity of 28 test substances, including drugs, pesticides, caffeine, and ascorbic acid. With the mouse BALB/c 3T3 fibroblast cell line, for any given cell density the two assays ranked the test substances with a correlation coefficient of 0.939 on the basis of IC_{50} concentrations. The two assays did differ in sensitivity for a few test agents, suggesting that a combination of the two might be most effective.

Some cytotoxicity tests are likely to underestimate the toxicity of chemicals which are metabolically activated in the body, but this problem can be overcome by the addition of liver enzymes, preferably from a human source to eliminate species differences.

Inhibition of mitogen-stimulated thymidine incorporation in human peripheral blood mononuclear cells has been reported as a method for screening for photosensitizers. Cells from at least three volunteers were used for testing each chemical.

Microorganism Studies

An important method using a fungi is Daniels' test for phototoxicity, which utilizes the yeast *Candida albicans* as the test organism. A 1988 study compared favorably the results of this test with the results of photo-patch testing in volunteers for samples from six furocoumarin-containing plants. Many test materials which produce an erythemic response in the photoirritant test are not analyzed as positive in this test. A new test method, Solatex-pi, has demonstrated capability to predict the potential for photoirritation of materials in this class as well as that of other well-known photoirritants. Solatex-pi utilizes the two-compartment physicochemical model of Skintex to predict the interactive effects of specific chemicals and UV radiation. Solatex-pi is being validated by

Frame and the BGA (Zebet) as an *in vitro* test to predict photoirritants.

Human Tissue Equivalents

Human skin equivalents have been developed by several laboratories. One equivalent, Testskin, consists of human keratinocytes seeded onto a collagen base or collagen–glycosaminoglycan matrix containing human fibroblasts. In many respects, the epidermis which develops resembles epidermis *in vivo*. The tissue culture system survives for several weeks and may be useful in studying skin penetration. Testskin is a commercially produced skin equivalent system marketed by Organogenesis, Inc. (Cambridge, MA); it is currently being assessed for use in skin penetration studies. Several companies launched studies of Testskin in 1990 and 1991.

Marrow-Tech, Inc. (Elmsford, NY) has also developed a human skin model. Marrow-Tech's skin equivalent consists of (1) a dermal layer of fibroblasts and naturally secreted collagen and (2) an epidermal layer of keratinocytes separated by a dermal–epidermal junction. Whereas Testskin uses bovine collagen, Marrow-Tech's skin model consists solely of human tissue.

Both of these skin equivalent methods permit higher concentrations of test samples to be studied. However, dilutions are still necessary when, after the physical chemistry of the test sample, the chemical structure may be responsible for irritation. Many protocols and endpoints have been evaluated as predictive of eye or skin irritation.

Isolated Tissue Methods

Skin isolated from rats, rabbits, and humans has been monitored *in vitro* to predict penetration and irritation. The rat epidermal slice technique has been validated as a screen for corrosive substances. The electrical impedance changes as the integrity of the stratum corneum is altered. The use of this technique to predict irritancy is being investigated in the United Kingdom. Another method studies enzyme changes when a substance is applied to a slice whose lower surface is bathed in culture medium. Enzyme changes separate irritant and nonirritant chemicals.

Human cadaver skin has also been studied *in vitro*. Human skin shows a higher threshold of sensitivity than does rat skin. The excised or full-thickness slices are also studied in Fran 2 diffusion chambers to evaluate diffusion or absorption characteristics of test materials. Changes in the amount of a test material at different times and different depths are monitored and are very useful in predicting penetration rates for simple solutions and solvents.

Human Volunteer Studies

Human volunteer studies are widely used to assess skin irritation, penetration, and sensitization.

Many industries regularly conduct repeat insult patch tests on human volunteers to evaluate topical irritancy. Groups of human volunteers are patched with test substance. One to five concentrations can be tested simultaneously, which is a wide enough range to yield results relevant to the usage. Cumulative skin irritancy is measured by applying patch applications each day for 3 weeks (30). Skin irritation is usually assessed visually, but blood flow and skin temperature can be measured objectively by laser Doppler flowmetry, ultrasound Doppler, heat flow disc measurement, sensitive thermocouple devices, or noncontact infrared radiative techniques. In these tests, dose–response curves can be obtained. Skin thickness can be measured with calipers as a measure of edema formation.

Human volunteers are also used in many industries in tests for allergic sensitization by cosmetic substances and formulations. The repeat insult patch test includes an induction phase (repeat applications during 3 weeks) and a 2-week rest period (incubation phase), followed by a challenge to see if sensitization has occurred. A pilot study of 20 human volunteers can be followed by more extensive testing (80–100 subjects). Positive results at more than the 10% level in the human volunteers would suggest a major problem with the formulation. User tests with the sensitized individuals and nonreactive matched control subjects can often determine the importance of these results to end use. Such a procedure may determine whether the sensitivity is significant under normal conditions of product use. Broader tests can be carried out with 250–500 subjects.

TABLE T-22
In Vitro Dermal Irritation Test Systems

System	Endpoint	Validation data?[a]
Excised patch of perfused skin	Swelling	No
Mouse skin organ culture	Inhibition of incorporation of [³H]thymidine and [¹⁴C]leucine labels	No
Mouse skin organ culture	Leakage of LDH and GOT	Yes
Testskin: cultured surrogate skin patch	Morphological evaluation (?)	No
Cultured surrogate skin patch	Cytotoxicity	No
Human epidermal keratinocytes (HEKs)	Release of labeled arachidonic acid	Yes
Human polymorphonuclear cells	Migration and histamine release	Yes (surfactants)
Fibroblasts	Acid	
HEKs	Cytotoxicity	Yes
HEKs	Cytoxicity (MIT)	Yes
HEKs, dermal fibroblasts	Cytotoxicity	Yes
HEKs	Inflammation mediator release	No
Cultured Chinese hamster ovary (CHO) cells	Increases in β-hexosamindase levels in media	No
Cultured C_3 H10T½ and HEK cells	Lipid metabolism inhibition	No
Cultured cells BHK21/C13	Cell detachment	Yes
BHK21/C13	Growth inhibition	
Primary rat thymocytes	Increased membrane permeability	
Rat peritoneal mast cells	Inflammation mediator release	Yes (surfactants)
Hen's egg	Morphological evaluation	
Skintex; protein mixture	Protein coagulation	Yes
Structure–activity relationship (SAR) model	NA[b]	Yes
SAR model	NA	No

[a]Evaluated by comparison of predictive accuracy (in the sense used in this chapter) for a range of compounds compared with animal testing results (adapted from Gad, 1993).
[b]NA, not available.

Conclusions

Whole animal tests represent true physiological and metabolic relationships of macromolecules, cells, tissues, and organs which can evaluate the reversibility of toxic effects. However, these tests are costly, time-consuming, insensitive, and difficult to standardize and are sometimes poorly predictive of human *in vivo* response.

New *in vitro* test methods target the behavior of macromolecules, cells, tissues, and organs in well-defined methods which control experimental conditions and standardize experimentation. These tests provide more reproducible, rapid, and cost-effective results. In addition, more information at a basic mechanistic level can be obtained from these tests. Table T.22 provides a summary of current test systems.

The challenge of the 1990s will be to understand the capabilities and limitations of these methods. Combining information on new molecules obtained from structure–activity relationships with results on macromolecular alterations in Skintex that occur for undiluted molecules may provide more information on dermal toxic effects of particular chemical classes. Combining test methods can provide a greater understanding of the mechanism of toxic molecules. Test batteries evaluating cell cytotoxic responses at high dilutions and changes in macromolecules at low dilutions will be more informative than visual scoring of complex events *in vivo*.

—Shayne C. Gad

Related Topics

Analytical Toxicology
Animal Models
Delayed-Type Hypersensitivity
Eye Irritancy Testing
Organophosphates
Photoallergens
Poisoning Emergencies in Humans
Radiation Toxicology
Skin
Tissue Repair
Toxicity Testing, Alternatives

Toxicity Testing, Developmental

Introduction

Human concern with birth defects is as ancient as human awareness. Through the nineteenth century, the prevailing view was that "maternal impressions," maternal experience during the pregnancy, directly affected the newborn. Teratology, the study of monsters ("terata"), was essentially an observational "art" with perceived supernatural implications. The development of basic concepts of genetics early in the twentieth century provided a scientific basis for causation of congenital defects. The recognition that environmental insult also produced birth defects in mammals inexorably followed, such as ionizing radiation (1907), sex hormones (1917), dietary deficiencies (1933), and chemicals (1948). The supposed safety of the human conceptus was refuted by German measles (rubella) epidemics in Australia in 1941 and in the United States in 1964 that resulted in thousands of children born with cataracts, deafness, and congenital heart disease from infected pregnant mothers.

The thalidomide "epidemic" in the late 1950s and early 1960s, involving at least 8000 malformed children in 28 countries, confirmed the vulnerability of the human conceptus to environmental insult, especially in the first trimester of pregnancy. It also precipitated worldwide concern for the safety of the unborn and the role of governments to ensure testing of drugs and other chemicals in pregnant mammals.

The early term for the study of birth defects, teratology, has been supplanted by a more general term, developmental toxicology, to enable inclusion of a more diverse spectrum of adverse developmental outcomes (which may be separate and distinct in etiology or the result of a continuum of response) and to make overt the recognition that specific results of insult in one species may not be the same in other species, including humans.

Developmental toxicity may be currently defined as any structural or functional alteration, caused by envi-ronmental insult, which interferes with normal growth, differentiation, development, and/or behavior. The targets for such insult(s) include the fertilized egg or zygote prior to implantation and the establishment of the three primary germ layers, the embryo during the period of major organ formation (i.e., organogenesis), the fetus in the postembryonic period of histogenesis, and the neonate or postnatal offspring, occurring or expressed through the postnatal period until sexual maturity. The expressions of developmental toxicity encompass death, frank structural malformations, functional defects, and/or developmental delays.

Factors Affecting the Vulnerability of the Conceptus

The vulnerability of the conceptus is viewed as due to qualitative or quantitative characteristics of both structure and function: (1) It is composed of a small number of rapidly dividing undifferentiated cells with absent or limited metabolic capabilities to alter or detoxify xenobiotics, repair lesions, etc.; (2) there is a necessity for precise temporal and spatial localization of specific cell numbers and types, as well as specific cell products, for normal differentiation, including programmed cell death; (3) sensitivies of certain cell types to certain insults may be unique to specific periods of cell movement, induction, or differentiation (i.e., transient vulnerability during the period of formation of tissues or organs); and (4) the immunosurveillance system (to provide recognition of "self" and detection of xenobiotics or lesions) is absent or immature in the prenatal or perinatal individual.

A number of factors influence the teratogenic response. Genetic susceptibility varies among species. For example, aspirin is teratogenic in rodents but not in primates, imipramine is teratogenic in rabbits but not in humans, and thalidomide is teratogenic in primates but not in rodents. Differences also exist among strains. Inbred mouse strains differ radically in their response to many teratogenic agents, e.g., to cortisone-induction of cleft palate and cadmium-induced testicular and embryotoxicity. Individuals also vary in their response to teratogenic agents in outbred strains and heterogeneous human populations. The current interpretation is that teratogens act on a susceptible genetic locus or loci which may control disposition of the agent including absorption, metabolism, transport, or excretion and/

or direct susceptibility of the target tissue or organ. The teratogen therefore increases the incidence of previously existing malformations; its action must be viewed against the "background noise" of spontaneous malformation rates, which also vary among species, strains, and individuals. For example, the phocomelic (seal limb) syndrome, induced by thalidomide, occurs at a low rate spontaneously in human populations; approximately 20–80% of the human fetuses, exposed to the "appropriate" dose of thalidomide at the "appropriate" time, developed the malformations.

There is some specificity of agent on the teratological response, with acetazolamide causing perhaps the most specific lesion—right forelimb postaxial ectrodactyly (fourth and fifth digits). However, there are almost always effects on other systems derived in many cases from different primary embryonic germ layers. The gestational stage of the embryo or fetus at the time of environmental insult appears to be the most critical determining factor. The predifferentiation period, from fertilization to establishment of the three primary embryonic germ layers, has been considered refractory to teratogenic agents (although there are some exceptions, such as hypoxia, hypothermia, actinomycin, and ethylene oxide). This resistance has been explained as due to the small, omnipotent cell population of the pre- and immediately postimplantation embryo. Cell damage or death is either corrected for by the surviving cells, which regulate to produce a normal albeit small term fetus, or the cell loss is so devastating that the embryo dies. Once implantation and establishment of the primary germ layers have occurred, the major period of organogenesis begins—a period of approximately 10 days in rodents and 50 days in humans. This is the period of maximal susceptibility to teratogenic agents causing structural anomalies. Even within the organogenic period, there are differential susceptibilities of embryonic organ systems to teratogenic agents. For example, administration of an agent on Gestational Day 10 in the rat affects eye, brain, heart, and anterior axial skeletal development. The same agent, administered on Day 15, affects palate, urogenital, and posterior axial skeletal system development. These times of specific sensitivity need not correspond to the morphological appearance of the organ or organ system but rather to the time of cell biochemical commitment: the shift of cells from presumptive to determined status.

Once histogenesis has begun, defined as the differentiation of tissue-specific biochemical and morphologi-

cal characteristics, the conceptus is termed a fetus and is viewed as increasingly refractory to teratogenic agents. However, this is true only of most morphological or structural manifestations. Increasing evidence indicates susceptibility of the fetus to agents causing functional deficits that presumably have a biochemical or microstructural basis. Those systems not yet complete, especially the nervous system, are most vulnerable. For example, vitamin A, lead, and methylmercury cause neurofunctional lesions when administered during this period. In addition, chemicals such as diethylstilbestrol and ethylnitrosourea act during this period to produce a system-specific tumor after a long latency in the postnatal mature animal. However, the only exposure and therefore the initiation of the later carcinogenic event occurs *in utero,* and these agents are therefore called transplacental carcinogens.

The route and duration of administration of the agent are also critical for the development of the teratogenic anomaly. Human industrial exposure is almost always by inhalation or percutaneous absorption of fumes, dusts, aerosols, or vapors. Consumer or other end-use or accidental exposure would be by more varied routes. Experimental evaluations are most useful if they duplicate the human route of exposure for experimental animal models. First-pass organ absorption and metabolism may differ if the exposure is by inhalation to the lung or orally to digestive system and liver, although subsequent transport and organ exposure may yield equivalent metabolite patterns. Most teratology studies usually employ administration of the test compound in the feed, by oral intubation, or by injection into the pregnant animal.

Timing is important. Experimental exposure before implantation or during early organogenesis may result in interference with implantation or in early embryonic death, resulting in no term fetuses. Exposure before peak susceptibility or repeated exposure may induce activating and/or detoxifying enzymes in dam, placenta, and/or fetus. This may result in increased or decreased blood levels of the active metabolite in the dam and, therefore, altered exposure to the fetus. Conversely, these enzymes may be inhibited by accumulation of metabolite(s), again altering blood levels of parent compound and metabolite(s). Other effects of repeated or early exposure may be to alter liver or kidney function, for example, as well as to induce pathological changes in these organs that will affect quantity and quality of compound reaching the fetus.

Saturation of protein-binding sites may also occur in the dam to alter transport. All of these effects may alter disposition parameters and obscure or change any teratological effects of the agent being examined.

Dose range and schedule are also critical. Three or four dose levels are usually employed: A high dose, which is toxic to the maternal organism, perhaps lethal up to 10% of dams, is used essentially to obtain an effect and to establish target organ(s); mid-dose(s), which is embryotoxic or embryolethal and possibly teratogenic; and a low dose, which is comparable on a body weight basis to possible human exposure levels or small multiples thereof.

Categories of Teratogenic Agents

Many substances are known to cause malformations in one or more species of mammals. Almost all known human teratogens are drugs, with data generated by drug research companies adhering to U.S. FDA guidelines for reproductive testing of drugs, and there is an awareness that in our drug-permissive society, women consume an average of four drugs, both by prescription and over-the-counter administration, during pregnancy. Various texts have identified from 600 to 1200 drugs as teratogens, only 20 of which are currently documented as human teratogens.

Human teratogenic agents have been discovered initially from anecdotal observations and then more rigorously examined in epidemiological studies and confirmed with animal studies, or they have initially been identified in animal studies with subsequent confirmation by human data. Animal model researchers have suggested that any agent positive in two or more mammalian species must be considered a suspect human teratogen.

Approximately 7% of all live-born humans bear birth defects. This value may be as high as 10% if children are evaluated to age 10 years to include subtle structural or functional deficits such as minimal brain dysfunction or attention deficit disorders. More than 560,000 lives out of approximately 3 million births per year in the United States are lost through infant death, spontaneous abortion, stillbirths, and miscarriage due presumably to defective fetal development. The relative contributions to human teratogenesis have been estimated as follows: known germinal mutations, 20%; chromosomal and gene aberrations, 3–5%; environ-mental causes such as radiation, <1%; infections, 2 or 3%; maternal metabolic imbalance, 1 or 2%; drugs and environmental chemicals, 4 or 5%; contributions from maternal dietary deficiencies or excesses and combinations or interactions of drugs and environmental chemicals are unknown. The contribution from unknown sources is 65–70%. The estimated 20–25% pregnancy loss due to chromosomal aberrations may be even higher due to early losses currently diagnosed as late menstrual bleeding. Recovered tissues from spontaneous abortions prior to the thirteenth week of gestation exhibit chromosomal anomalies on the order of 560 per 1000 abortions; the value at term is 5 per 1000. Of the children born live who subsequently die in the first year of life, approximately 20% of the deaths are associated with or caused by birth defects, more than any other single factor.

One almost plaintive maxim, sometimes termed Karnofsky's law, states that almost any substance may be teratogenic if given in appropriate dose regimens to a genetically susceptible organism at a susceptible stage or stages of embryonic or fetal development.

Government Regulation

Soon after the worldwide thalidomide disaster in 1966, governmental regulation of the evaluation of test agents for developmental toxicity by formal testing guidelines and rules began when the U.S. FDA established *Guidelines for Reproductive Studies for Safety Evaluation of Drugs for Human Use*. These guidelines were promulgated "as a routine screen for the appraisal of safety of new drugs for use during pregnancy and in women of childbearing potential." Three phases or segments were proposed: Segment I, Study of Fertility and General Reproductive Performance, provides information on breeding, fertility, nidation, parturition, neonatal effects, and lactation (see Reproductive Toxicity); Segment II, Teratological Study, provides information on embryotoxicity and teratogenicity; and Segment III, Perinatal and Postnatal Study, provides information on late fetal development, labor and delivery, neonatal viability, and growth and lactation (Fig. T-8).

Segment II testing guidelines are currently followed by U.S. FDA (since 1966), U.S. EPA, Toxic Substances Control Act (TSCA) (since 1985), and Federal Insecticide, Fungicide and Rodenticide Act (FIFRA) (since 1982). International regulations also followed suit: Or-

A. SEGMENT II - Developmental Toxicity Study

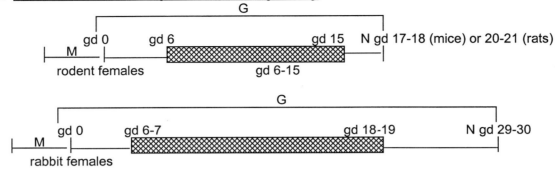

Information on: Embryotoxicity, fetotoxicity, teratogenicity

B. SEGMENT III - Perinatal and Postnatal Study

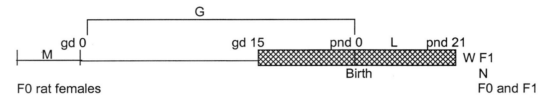

Information on: Parturition, lactation, peri- and neonatal effects

C. Developmental Neurotoxicity Study

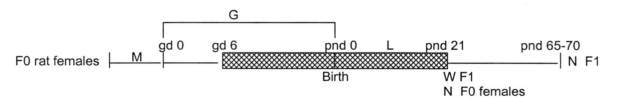

Information on: Parturition, lactation, survival and growth of offspring, landmark acquisition and behavior of offspring, structural effects on CNS

M =	Mating
G =	Gestation
L =	Lactation
W =	Wean
N =	Necropsy
gd =	gestational day
pnd =	postnatal day
▨	Direct gavage exposure to adults

FIGURE T-8.   *U.S. governmental guidelines for study designs of developmental toxicity assessments in animal models. A, Segment II—developmental toxicity study; B, segment III—perinatal and postnatal study; C, developmental neurotoxicity study.*

ganization for Economic Cooperation and Development (OECD, 1981), Great Britain (1974), Japan (1984), and the European Community (1994).

Considerations for Segment II Studies

The test animals are usually rodents and nonrodents. The rodent of choice is usually the rat and, less often, the mouse. Both species satisfy the need for a small mammalian species with known (and relatively straightforward) husbandry requirements, short pregnancy, high fertility, large numbers of offspring, a low background incidence of spontaneous malformations, and a reasonably well-known embryology. The nonrodent species is usually the rabbit. The rabbit is not a rodent; mice and rats belong to the order Rodentia and rabbits to the order Lagomorpha. The requirement for the use of rabbits is predicated on the awareness that it was the only common test mammal in use in the 1960s which responded to thalidomide and (it is hoped) would have indicated the prenatal risk to humans, and on the need to distinguish between agents with specific or unique specificity (i.e., a rodent-specific teratogen) and those with more universal effects, presumably then also a greater potential risk to human development.

Prenatal development in the Rodentia and Lagomorpha differs in significant ways from that in humans. All three have a chorioallantoic placenta; the structure differs among species. The human and rat placenta also differ functionally with different secretory patterns of placental lactogen and with the presence in primates of chorionic gonadotropin. What effect, if any, these differences have on placental transport is not fully understood. In addition, rodents and lagomorphs also form a yolk sac placenta immediately after implantation, which is the major (only) mechanism for nutrient processing and transport until Gestation Day 11 to 11.5, and persists as functional, even when the chorioallantoic placenta forms, almost to parturition. Again, what effect this has on embryo and fetal vulnerability is not yet known, although at least one teratogenic agent, trypan blue, appears to act solely on the yolk sac placenta. In multifetal pregnancies, there are differences in blood flow to left and right uterine horns and to implants at ovarian versus cervical ends of the uterine horns. Different fetuses within the same dam have been shown to be at differential risk. In addition, fetal loss is handled differently in test animals: Dead implants are not expelled in a spontaneous abortion as in single-birth mammals but are resorbed *in situ*. It is not uncommon to recover healthy, viable fetuses side by side with large numbers of resorption sites. Maternal, placental, and fetal metabolism of xenobiotics may also differ, hence the need for prior characterization, at least of the test organism's metabolic capabilities of the substance to be tested.

The placenta is both a transport and a metabolizing organ. Transport is accomplished by simple diffusion, facilitated diffusion, active transport across membranes, and by special processes such as pinocytosis, phagocytosis, and breaks in the "barrier." The placenta also contains a full complement of mixed function oxidases located in the microsomal and mitochondrial subcellular fractions capable of induction and metabolism of administered chemicals.

Metabolism in the test dam and/or fetus and its relevance to the pregnant human are critical. For example, the parent compound may be teratogenic and is metabolized to innocuous products as with diphenylhydantoin, an antiseizure drug used in the treatment of epilepsy. In contrast, the parent compound may be harmless and metabolized to the proximal teratogenic agent as with chlorcyclizine, an antihistamine metabolized *in vivo* to the active teratogen norchlorcyclizine. One of the current hypotheses concerning the mechanism of thalidomide-induced teratogenesis suggests that thalidomide is transmitted to the human fetus and metabolized to a more polar metabolite(s), the putative proximal teratogenic agent(s), which cannot cross the placenta back to the maternal organism for further metabolism and excretion. This sequence may be qualitatively or quantitatively different in the insensitive pregnant rodent. In contrast, imipramine, an antidepressant, is teratogenic in rabbits in which blood levels of the parent compound stay high. In the human, imipramine is rapidly metabolized by demethylases and is not teratogenic.

Pregnancy per se causes many physiological changes, which may alter over the duration of the pregnancy. These changes include alterations in gastrointestinal function which may affect absorption rates of chemicals in the stomach and/or intestine and ventilatory changes which may modify pulmonary uptake, absorption, and/or elimination of chemicals with a 20–30% increase in maternal oxygen consumption and greater oxygen debt after physical activity (in humans). Changes also occur in the cardiovascular system that alter hemodynamics (there is a 30–40% increase in blood volume and a 33% decrease in erythrocytes in humans) and alter body water compartments which may influence distribution and elimination of chemicals. Plasma components with roles in chemical binding, transport, and disposition also change during pregnancy. Renal elimination is normally enhanced and hepatic elimination may be modified during pregnancy, also affecting xenobiotic elimination.

The concern for and emphasis on thorough evaluation of the maternal organism are based on the need to determine, when study results are interpreted, whether the maternal toxicity per se is responsible for the observed embryo/fetal results. A number of fetal malformations in rodents and rabbits have been identified which are observed in the presence of maternal toxicity, regardless of the agent, route, or dose, with the clear implication that the maternal toxicity is the cause of the developmental toxicity, not the test agent. Mechanistic studies have also implicated the compromised status of the maternal organism as the cause for the adverse embryo/fetal outcome for many drugs. For example, elevation of endogenous corticosteroids, as a result of maternal stress irrespective of the source, or administration of exogenous corticosteroids result in cleft palate in offspring from susceptible strains of mice; hypercapnia (elevated blood CO_2) in mice has been proposed as the cause of forelimb ectrodactyly in mice exposed to acetazolamide, and bradycardia (slowed heart rate) in mice from phenytoin administration has been suggested as the cause of cleft lip/palate in offspring. Renal toxicity from mercuric chloride exposure to the mother may be the cause of hydrocephalus in the offspring. If maternal toxicity per se, including even "stress" from restraint, for example, is the cause of the developmental toxicity, then the classification of the test agent as a teratogen may be erroneous.

In addition, information on maternal toxicokinetics and metabolism of the test agent is essential to characterize the conditions under which the toxicity to dam or conceptuses is observed; these conditions include evidence of systemic exposure, blood levels of parent compound and/or metabolite(s), identification of metabolites, bioavailability, half-life, and evidence for or against bioaccumulation. This information is necessary for studies by any route to extrapolate results from one route to another, one species to another, and for human risk assessment. That is, the handling of the test agent by the test species must be characterized so that one

can say that, for a specific test agent, maternal and/or developmental toxicity occurs in the presence of the parent compound or identified metabolite(s) at specific blood levels for a specified duration, with the expectation that another species that produces the same metabolite(s) at comparable levels for comparable duration will exhibit the same or similar toxicities. In the absence of metabolic information, one cannot assess whether the test animal is an appropriate surrogate for humans for specific test agents. Histopathology of target organ(s) and organ function tests may also be appropriate.

Standard Segment II Testing Protocol

In brief, the segment II study consists of exposure of a pregnant rodent species (rats or mice) and a nonrodent species (usually rabbits) to the test agent during organogenesis, sacrifice of the maternal animals 1 or 2 days prior to the date of expected parturition, Caesarean delivery of the gravid uterus, and thorough evaluation of the fetuses by examination of external, visceral (including craniofacial), and skeletal structures.

The current guidelines call for at least 20 rodent litters and at least 12 rabbit litters per group (although current draft guidelines call for the same numbers of rabbit litters as required for rodent litters) and at least four dose groups (three agent-exposed groups and a concurrent vehicle control group). On Gestational Day (GD) 0, mated animals are placed on study and randomly assigned or assigned by a randomization procedure (stratified by body weight) into test groups.

The period of exposure of the maternal organism to the test agent is usually during the period of major organogenesis. This corresponds to GD 6 through 15 for rodents and GD 6 or 7 through 18 or 19 for rabbits (current draft guidelines call for dosing from implantation to term, with no postdosing recovery period). This period of dosing was specifically chosen to preclude efforts on implantation so there would be conceptuses to evaluate and to maximize the chances of inducing and detecting structural changes in the conceptuses. The possible effects of the test agent on the reproductive and developmental processes prior to organogenesis are evaluated in segment I or multigeneration studies. This dosing period also precludes induction of maternal metabolizing enzymes prior to the presence of implanted conceptuses and allows for a postexposure recovery period prior to scheduled sacrifice close to term. Variations in the exposure period include exposure from implantation to scheduled sacrifice with no recovery period; exposure during the entire gestational period, GD 0 to term sacrifice; or exposure beginning prior to GD 0. These latter extended exposure periods may be useful and appropriate if the test agent, or route of administration, results in slow and/or limited systemic absorption and therefore delayed attainment of steady state or maximal blood levels in the maternal organism. In these circumstances, the usual dosing period could result in the conceptuses exposed to less than maximum levels during some or most of organogenesis and the misleading conclusion of little or no developmental toxicity. Extended exposure periods may also be called for if bioaccumulation of or cumulative toxicity from parent compound or metabolites is an important aspect of known or potential human risk.

The guidelines specify the preferred route of administration as gavage (orogastric intubation) to deliver the largest possible bolus dose in order to maximize the potential of the test agent to cause maternal and developmental toxicity, i.e., "worst case scenario." Use of other routes to simulate possible human exposure situations is becoming increasingly popular and is acceptable if scientifically defensible. These alternative routes include dosed feed, dosed water, inhalation by whole body or nose-only exposure, cutaneous application, injection by intravenous, subcutaneous, intraperitoneal, or intramuscular routes, or subcutaneous insertion (for implants or for minipumps for continuous infusion).

Maternal data to be collected from segment II studies include maternal mortality; pregnancy rate; maternal body weights on GD 0 and first and last day of dosing at sacrifice (at least); sacrifice weight corrected for the weight of the gravid uterus; body weight changes through gestation, pre- and posttreatment, and during the treatment period; feed and/or water consumption; clinical observations; gravid uterine weight; and weight of other organs (absolute and relative to sacrifice body weight). Improvement would include more frequent body weights and weight change calculations during the dosing period to detect early, possibly transient, treatment-related effects and during the postexposure period to detect effects continued from the dosing period (e.g., body weights 24 hr after last dose and at least midway during the postdose period) and possible rebound and recovery. Food and/or water consumption

frequency as described previously for body weights would enhance sensitivity for detection of toxicity and aid in the interpretation of weight loss or reduced weight gain. Additional evaluations of physiological and/or biochemical status, such as more detailed behavioral evaluation, clinical pathology such as hematology, clinical chemistry, and urinalysis, may be performed on maternal animals during and after the treatment period.. These tests may duplicate those performed in other studies, but pregnant animals may respond quantitatively or qualitatively differently (vide supra), and these data will be critical in interpretation of any observed developmental toxicity.

Reproductive and embryo/fetal data to be collected from segment II studies at sacrifice include number of ovarian corpora lutea (number of eggs ovulated); number of total, nonviable (resorptions and dead fetuses) and live implantations; and calculation of pre- and postimplantation loss. For litters with live fetuses, data collected include number, sex, and body weights (by sex) of fetuses/litter, fetal crown–rump length/litter (by sex), incidence of individual external, visceral and skeletal and total malformations, and variations by fetus, by litter, and per fetus per litter (male and female fetuses differ in body weights, with males significantly heavier). These procedures thoroughly assess two of the four embryo/fetal endpoints—death and structural malformations—and also assess developmental delays, but only in terms of delays in growth such as reduced body weight, reduced crown–rump length, and delays in structural development, such as reduced ossification relative to concurrent and historical control fetuses (usually designated as variations), especially in those skeletal districts that ossify late in prenatal development. Double-staining of fetuses with alizarin red S for ossified bone and alcian blue for cartilaginous bone provides information on the status of bone not yet ossified. These techniques aid in the interpretation of a finding as a skeletal malformation or permanent skeletal variation (when there is no cartilage in a short bone or for a missing bone, so no subsequent growth, ossification, or correction would be anticipated) versus a variation of transient delay in ossification (where there is cartilage with anticipated subsequent growth, ossification, and possible correction). Current draft guidelines call for double staining of fetal preparations.

There is apparent potential for extensive remodeling of the skeletal system in the postnatal period; extra ribs become vertebral arches, and fused ribs and other skeletal malformations disappear prior to sexual maturity. This plasticity of the skeletal system may result in revision of the current classification of morphological findings in term fetuses as malformations or variations. The current definition of a malformation specifies a permanent morphological change which is incompatible with or detrimental to postnatal survival, normal growth, and development. Short ribs, extra ribs, fused ribs, alterations in sternebrae (which fuse to form the sternum), alterations in vertebral centra, and arches are currently designated malformations or variations depending on the laboratory; if these changes do not persist, their designation could change. The reverse situation is also true; that is, findings commonly designated as variations, usually delays, in term fetuses may, in fact, sometimes develop into findings designated as malformations in postnatal life. For example, a dilated renal pelvis (reduced renal papilla) may or may not be the precursor of hydronephrosis, and dilated lateral ventricles of the fetal cerebral hemispheres may or may not be the precursor to hydrocephaly in the postnatal organism. With strictly limited evaluations of term fetuses, "frozen in time," there is no way to project the postnatal consequences of the initial findings. In addition, the term evaluation is based on structure. If the lungs or kidneys, etc., are in the right location and are the right size, shape, and color under a dissecting microscope, they are designated as normal; there is no assessment of microscopic integrity or of function. Additional evaluations of term fetuses should perhaps include biochemical assessment of organ function, histological examination of structure, as well as postnatal assessment of the reversibility of detected structural lesions and of the structural and functional sequelae of the prenatal insult (see below, Standard Segment III Testing Protocol and Developmental Neurotoxicity Test).

Statistical Analyses of Maternal and Developmental Toxicity Data

As part of protocol development, the choice of statistical analyses should be made a priori although specific additional analyses may be appropriate once the data are collected. The unit of comparison is the pregnant female or the litter and not individual fetuses as only the dams are independently and randomly sorted into dose groups. The fetus is not an independent unit and cannot be randomly distributed to groups. Intralitter interactions are common for a number of parameters, e.g., fetal weight or malformation incidence. Two types of data are collected: ordinal/discrete data, which are essentially present or absent (yes or no) such as incidence of maternal deaths, abortions, early deliveries, clinical signs, and incidence of fetal malformations or variations; and continuous data, such as maternal body weights, weight changes, food and/or water consumption, organ weights (absolute or relative to body or brain weight), and fetal body weights per litter. For both kinds of data, three types of statistical analyses are performed. Tests for trends are available and appropriate to identify treatment-related changes in the direction of the data (increases or decreases), overall tests are performed for detecting significance among groups, and specific pairwise comparison tests (when the overall test is significant) to the concurrent vehicle control group values are the critical endpoint to identify statistically significant effects at a given dose relative to the concurrent vehicle control group. Continuous data are designated parametric (distributed along a bell-shaped curve) or nonparametric (skewed distribution), with different specific tests employed for the three types of statistical analyses depending on whether the data are parametric or nonparametric.

Risk Assessment

The U.S. government's approach to health assessment of agents involves four major components: hazard identification, dose–response assessment, exposure assessment, and risk characterization (Fig. T-9). This section relies heavily on the U.S. EPA guidelines for the health assessment of suspect developmental toxicants which describe how the government uses, and plans to use, developmental toxicity data as part of their "weight-of-evidence" approach to both the hazard identification and the dose–response assessment components of risk assessment.

Standard developmental toxicity studies are performed, under the appropriate governmental toxicity guidelines, for a drug early in the drug discovery period (FDA), for a pesticide prior to registration (as required by EPA's FIFRA), or for an industrial chemical (performed on a case-by-case basis under EPA's TSCA). These studies provide information on the intrinsic capacity of the test agent to cause developmental toxicity under conditions to maximize the opportunity, i.e.,

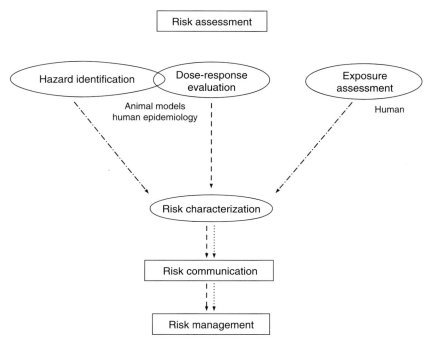

FIGURE T-9. *U.S. National Research Council risk assessment paradigm.*

hazard identification, and the dosage or dosages at which the developmental toxicity (death, malformation, delays, and/or deficits) is observed, i.e., dose–response assessment. Of the three dosage levels employed, the highest dose should result in overt maternal toxicity, including significantly reduced body weight, weight gain, and specific organ toxicity, with maternal mortality up to 10% viewed as acceptable. This dose level should characterize embryo/fetal outcome in a compromised dam/doe and should represent a "worst-case scenario" for hazard identification. However, the presence of maternal toxicity per se confounds the interpretation of observed developmental toxicity since these effects may be due to the status of the dam and not to the test agent per se (see previous discussion for maternal toxicity data).

The low dose should be a no-observable-adverse-effect level (NOAEL) for both dams and conceptuses. The NOAEL is defined as the highest dose (or exposure concentration) at which no statistically significant and/or biologically relevant adverse effects are observed in "any adequate developmental toxicity study." The middle dose may or may not result in maternal and/or developmental toxicity and should be a lowest-observable-adverse-effect level (LOAEL), defined as the lowest dose or exposure concentration at which a statistically

significant and/or biologically relevant adverse effect is observed in "any adequate developmental toxicity study." The characteristics of the NOAEL (or the LOAEL) are that (1) it is obviously experimentally derived and therefore dependent on the statistical power of the study (which is in turn dependent on the number of animals employed); (2) it is dependent on the number and sensitivity of the parameters examined; and (3) its presence implies a "threshold," i.e., a dose below which adverse effects would not be observed, again with the same experimental caveats. The attainment of a NOAEL (or LOAEL) is critical for subsequent risk assessment processes since it is used to ultimately extrapolate to human exposure limits. However, the NOAEL is not a characteristic of the population (all rats, all mice, etc.) but only of the group under test and, in a real sense, specific to the species, strain, laboratory, staff, specific time of performance, source and purity of test material, identity of any vehicle, parameters evaluated, etc. The NOAEL also does not provide information on the slope of the dose–response curve (steep or shallow), although it is obviously at the low end of the dose–response continuum. These characteristics are very important since regulators are usually extrapolating from (relatively) high dose levels in animal studies to (relatively) low exposure levels for humans, and the presence and location of the threshold is crucial to risk assessment.

Once a NOAEL (or LOAEL) is provided by the experimental data, the proposed next step by risk assessors is to define a reference dose for developmental toxicity (RfD$_{DT}$) according to the following equation:

$$\text{RfD}_{DT} = \frac{\text{NOAEL} / \text{LOAEL}}{\text{UF}}$$

where UF is uncertainty factors. The RfD$_{DT}$ is defined by the U.S. EPA as an estimate of the daily human exposure that is likely to be without appreciable risk of adverse developmental effect and is characterized by the use of NOAEL or LOAEL (if NOAEL is unavailable) of most sensitive indicators for most appropriate (if known) and/or most sensitive mammalian species. If the NOAEL is used,

$$\text{RfD}_{DT} = \frac{\text{NOAEL of most sensitive indicator}}{\begin{array}{cc} \text{interspecies} & \text{intraspecies} \\ \text{variability} \quad \times & \text{variability} \\ \text{(UF; 10)} & \text{(UF; 10)} \end{array}}$$

The RfD$_{DT}$ is assumed to be below the threshold for an increase in adverse developmental effects in humans and is used for risk characterization along with human exposure assessments.

A second use for NOAELs (or LOAELs) is in the calculation of a proposed margin of exposure (MOE) for developmental toxicity to be used in risk characterization. The MOE is defined as the ratio of the NOAEL from the most sensitive or appropriate species to the estimated human exposure level from all potential sources. If the MOE is very high relative to the estimated human exposure level, then risk to the human population would be considered low.

The proposed U.S. EPA weight-of-evidence (WOE) scheme for suspect developmental toxicants defines three levels of confidence for data used to identify developmental hazards and to assess the risk of human developmental toxicity: (1) definitive evidence for human developmental toxicity or for no apparent human developmental toxicity, (2) adequate evidence for potential human developmental toxicity or no apparent potential human developmental toxicity, and

(3) inadequate evidence for determining potential human developmental toxicity. The scheme may require scientific judgment based on experience to weigh the implications of study design, statistical analyses, and biological significance of the data.

Standard Segment III Testing Protocol

There is growing concern about postnatal sequelae to *in utero* structural and/or functional insult as well as a recognition that exposure to a developing system may result in qualitatively or quantitatively different effects than an exposure to an adult system. In brief, the segment III study consists of exposure of pregnant rats to the test agent starting at the end of the organogenesis period (GD 15), through the histogenesis period (concepti are fetuses), through parturition (birth), and through the lactational period until the offspring are weaned [Postnatal Day (PND) 21]. The offspring are "exposed" only from possible transplacental and/or translactational (via the milk) routes (Fig. T-8). There are usually three test material groups and a vehicle control group, with at least 20 litters per group; exposure is by gavage (to minimize disruption of the mother and her litter). During gestation, the dam is weighed periodically and feed consumption is measured. Dams and pups are weighed, sexed, and examined externally, and feed consumption is measured at birth (PND 0) and repeatedly during the lactation period (e.g., on PNDs 0, 4, 7, 14, and 21). Litters are culled to eight pups on PND 4. The time of acquisition of developmental landmarks is recorded, such as surface righting reflex, pinna (external ear) detachment, incisor eruption, eye opening (pups are born blind with eyes shut), auditory startle (pups are born deaf with the external ear canal closed), and mid-air righting reflex. If the pups are maintained after weaning, then vaginal patency (opening of the vaginal canal), testes descent, and/or preputial separation are monitored as well as motor activity (initial exploratory behavior as well as habituated behavior); learning and memory may also be assessed. This test provides information on the last "trimester" of pregnancy; delivery; maternal–pup interactions and behaviors, such as pup retrieval, nursing, grooming, nest building, etc.; and on pup postnatal growth and development. At weaning, the dam is sacrificed and the number of uterine implantation scars are counted to obtain information on prenatal loss; pups can be necropsied at wean or beyond, with target tissues examined histologically.

Developmental Neurotoxicity Test

The nervous system—with its long developmental phase, involving proliferation, migration, and differentiation of cells and regions at different gestational and perinatal ages, and its complexity—is one for which there is special concern for postnatal consequences of *in utero* exposure to developing systems. In response, the U.S. EPA has developed a "stand-alone" standardized developmental neurotoxicity test to assess "potential functional and morphologic hazards to the nervous system which may arise in the offspring from exposure of the mother during pregnancy and lactation" (Fig. T-8). When this study design would be employed, i.e., the "triggers" for its requirement, is still not fully established and will (and should) probably be decided on a case-by-case basis. Agents which should be candidates for developmental neurotoxicity or behavioral teratology testing include those that cause central nervous system (CNS) malformations, that are psychoactive, that are adult neurotoxicants, that are hormonally active, and that are peptides or amino acids. The last agents might be antagonists or agonists of endogenous CNS chemical signalers and could easily cross the blood–brain barrier. Such testing protocols should assess sensory and motor function, neuromotor development, learning and memory, reactivity and/or habituation, reproductive behavior, and other functions such as social or aggressive behaviors.

The study design for the developmental neurotoxicity screen as currently mandated by both TSCA and FIFRA testing guidelines is very similar. The design involves performance in rats, at least three agent-exposed groups and one vehicle control group, and at least 20 usable litters in each group. The route of administration should be "orally by intubation"; other routes of administration are acceptable, on a case-by-case basis, with appropriate justification. If the agent has been previously shown to be developmentally toxic, "the highest dose for this study shall be the highest dose which will not result in perinatal deaths or malformations sufficient to preclude a meaningful evaluation of neurotoxicity." If there are no developmental toxicity study data, "the highest dose shall result in overt maternal toxicity, with weight gain depression not to exceed 20% during gestation and lactation." The lowest dose should not result in either overt maternal or developmental neurotoxicity, while the intermediate dose(s) must be equally spaced between the highest and lowest doses. With GD 0 designated as the day of copulation, the dosing period extends from GD 6 through weaning (PND 21) or GD 6 through PND 10. Live pups are counted and weighed at birth and throughout the lactation and postwean periods. On PND 4, litters are culled to yield eight pups. Developmental landmarks assessed on all appropriate pups include age of vaginal opening and testes descent or preputial separation. Motor activity is monitored at multiple pre- and postweaning times. The period of evaluation for motor activity will include the exploratory phase and the habituation phase. Auditory startle test, including magnitude and habituation of response, and tests to evaluate learning and memory are performed at wean and at 60 days. Necropsy and histopathology requirements include perfusion of pups with fixative *in situ* and specified central and peripheral nervous system tissues examined histologically with "qualitative, semiquantitative and simple morphometric analysis." Additional animals are decapitated, the brains are removed, and regional brain weights are obtained.

This test is perceived as useful in the risk estimation process, to identify specific agents, or classes of agents, for which acceptable exposures in the adult may not be acceptable to the developing organism, to elucidate long-term consequences of pre- and perinatal exposures and results, to determine the relationship of lowest effective (or highest no effect) dose for behavioral effects versus the dose for overt or general toxicity effects, and to identify, for human exposures, those effects which may be important to monitor.

Although the developing nervous system has received the most attention from researchers and governmental regulators, there are many other systems with continuing proliferation and differentiation in the postnatal period. Evaluation of the postnatal sequelae of prenatal exposure has been done for three of these: the renal system, gastrointestinal tract, and immunosurveillance system. Transplacental carcinogenesis, expressed in the adult from late gestational *in utero* exposure, is also well documented in test animal species; diethylstilbestrol is currently the only documented transplacental carcinogen in humans.

Male-Mediated Developmental Toxicity

All of the previously described approaches focus on the maternal–placental–fetal unit as the subject of testing and the object of concern. However, increasing evidence has implicated the male as the cause of any of the classic four endpoints of developmental toxicity. Human male exposure, such as operating room personnel, to waste anesthetic gases results in increased incidences of spontaneous abortions, stillbirths, and congenital defects. Male production worker exposure to Oryzalin has been implicated in congenital heart defects in their children. The pesticide DBCP (1,2-dibromo-3-chloropropane) is a human male sterilant. Elevated caffeine consumption in men has been reported to result in spontaneous abortions, stillbirths, and premature births. In animal studies, exposure of the male to methadone, thalidomide, lead, narcotics, alcohol, and caffeine results in malformations in the offspring. Possible mechanisms of male-mediated developmental toxicity include genetic or epigenetic damage to the sperm, the presence of the agent or its metabolite(s) in the semen which may affect the conceptus directly or act on the gravid uterus, or indirect or more systemic actions on the male affecting the hormonal milieu and perhaps libido.

Developmental Toxicity Screening Protocols

Over 70,000 chemicals are listed in the TSCA registry, with 1500–2000 new chemicals added each year; 20,000 chemicals are commonly found in the workplace (NIOSH list) with <1% tested for reproductive and developmental hazard potential. It is therefore necessary and appropriate to develop fast, inexpensive, sensitive, and accurate methods to prescreen the plethora of chemicals and concentrate resources on those identified by the screening test(s) as potential human health hazards. However, the mechanisms of action of developmental toxicants appear numerous and frustratingly difficult to identify (see Mechanisms).

A number of approaches have been taken to develop screening protocols, herein arbitrarily classified into *in vivo*, *in vivo/in vitro*, and *in vitro* categories. *In vivo* screens include developmental toxicity range-finding studies, which also can be used to identify (or prioritize)

agents which produce developmental toxicity for more rigorous testing, and the so-called Chernoff–Kavlock assay. The assay employs a block design of one dose (the maternal minimally toxic dose; MTD) per chemical for a number of chemicals and a concurrent control group, with 24–50 timed-pregnant animals, usually mice, per group. Dosing is on GD 8–12, with the date of a vaginal plug being designated GD 1. The earliest version of this study design collected maternal weights at the beginning and end of the treatment period and also weight change, with dams allowed to litter. Litters were counted, sexed, weighed, and examined externally on PNDs 1 (date of birth) and 3 and discarded. This protocol does not require extensive or intensive technical training in pup visceral or skeletal examinations and assumes that the pups will be their own assay system, i.e., if the pups survive and thrive, then they do not exhibit significant toxicity at a dose which is minimally toxic to the dam (the MTD) and they do not bear malformations or variations which preclude or affect normal early postnatal growth and development. Chernoff and Kavlock set up three levels of concern: If there is pre- or postnatal mortality and/or malformations of the offspring, then the test agent has the "highest priority" for further classic developmental toxicity testing; if the pups exhibit reduced weight gain, then the agent has a "lower" priority for further testing; and if there is no evidence of developmental toxicity, pre- or postnatally, then the agent has the "lowest priority" for further testing. The block design described previously provides comparisons among the test agents in the block, all at the MTD, for relative potency with regard to developmental toxicity.

Modifications to the initial protocol include multiple dose levels, dosing during the entire period of major organogenesis, and more thorough evaluation of pups on PND 3 (including visceral and skeletal examinations) so that this protocol resembles more closely the classic segment II protocol but with a postnatal component to assess viability and growth.

One *in vivo/in vitro* screening protocol involves administration of the test agent to pregnant rodents, removal (on GD 10 after one or more daily doses to the dam), explantation and culture of embryos for 24–48 hr, and evaluation of toxicity and teratogenicity. This protocol allows for the full mammalian complement of metabolizing enzymes in the dam to act on the conceptuses *in utero* and for the full range of early expression of developmental toxicity to be detected in the explanted embryos *in vitro*. The next step is one whereby explanted rat headfold embryos are cultured for 48 hr in human, monkey, or rodent serum after the serum donor had been exposed to the test agent. This protocol utilizes serum containing whatever metabolites, etc. are produced by and transported in the blood of the donor mammal—a condition duplicating the embryonic exposure *in utero*. In a fascinating offshoot of this work, serum from women who were chronic aborters has been used in the embryo culture system to identify missing nutrients and the women were supplemented prior to and during subsequent pregnancies with some early apparent success.

There are a number of fully *in vitro* screens as well, employing mammalian, lower vertebrate, and invertebrate species. Materials used include explanted rat or mouse embryos cultured in rodent serum to which is added the test agent or known metabolites, and portions of rodents, as intact organs or as dissociated cells (e.g., limb buds, dissected midbrain cells, palatal cells, etc.), explanted and cultured in medium containing test agents and/or metabolites. When explanted embryos (or parts thereof) are exposed to the test agent in culture, they are exposed only to the added test agent since metabolic capability is minimal or absent, so this study paradigm may expose embryos to situations they would not encounter *in utero* and therefore result in false-positive or (worse) false-negative study results. Cloned totipotent stem cell lines from murine embryonal teratocarcinoma or pluripotent lines from neuroblastoma are cultured, exposed to test agents (including those which are "proteratogens" requiring metabolic activation), and the cultures scored for effects on differentiation. Both tumor lines are capable of extensive differentiation in culture; restriction of this capability is presumed indicative of potential developmental toxicity *in vivo*.

In a novel approach to examine a fundamental property of differentiating cells, cell-to-cell communication, Chinese hamster lung cells or normal embryonic palatal mesenchymal cells in culture are exposed to the test agent and evaluated for disruption of cell-to-cell communication. Cell attachment is another presumed universal cell function during development and therefore a basis for a screen. Ascites or dissociated solid tumor cells are grown in culture in the presence of the test agent and scored for attachment (or inhibition of attachment) to surfaces as a measure of potential developmental toxicity.

Explanted chick embryos at presomite or multiple somite stages or chick embryonic parts are cultured with the test agent incorporated into the culture medium and evaluated for growth and differentiation.

Amphibians are also proposed for use in screening protocols. The FETAX system (Frog Embryo Teratogenesis Assay: Xenopus) involves exposure of early *Xenopus laevis* (African clawed frog) embryos at the notochord stage and/or as late premetamorphic larvae to test agents in the water. A teratogenic index (TI) is proposed to compare relative potencies of test agents and to identify any agents which affect development at doses below which general toxicity is observed; the TI is defined as LC_{50}/ED_{50} (the concentration lethal to 50% of the animals divided by the concentration producing effects in 50% of the animals).

Drosophila melanogaster (the fruit fly) is used in two ways: Larvae are grown on feed containing the test agent, are allowed to pupate, and emerging adults are scored for viability (toxicity) and malformations from alterations in imaginal discs present in the larvae and used to form adult structures; or early primary embryonic cell cultures are grown in medium containing the test agent and are scored for differentiation of embryonic cell types.

Synchronous cultures of *Artemia* sp. (brine shrimp) in seawater or rodent or human serum have also been suggested as a screen, with scoring for survival, growth, and morphological and molecular differentiation after exposure directly to agents or to serum from agent-exposed individuals.

Hydra attenuata (a coelenterate) is the source of the "artificial embryo" assay. The adult Hydra can be dissociated and the cells pelleted by centrifugation. The cells of the pellet will sort and reaggregate by cell type and redifferentiate into an adult Hydra. The assay consists of pellets ("artificial embryos") and adult Hydra exposed to the test agent to determine the lowest effect concentration (or the highest no-effect concentration) of the developing "embryo," as measured by inhibition of redifferentiation or abnormal differentiation, and of the adult, as measured by mortality or overt damage to adult structures. An A/D ratio is calculated: that is, the ratio of the adult toxicity lowest effect (or highest no-effect) concentration to the developmental toxicity lowest effect (or highest no-effect) concentration. The developers and users of this assay suggest that an A/D ratio ≥3 indicates a unique or greater susceptibility of the developing organism relative to that of the adult and therefore a potential of the test agent for mammalian developmental toxicity. They also claim that the A/D ratio is fairly consistent across widely divergent species and therefore predictive of relative risk, although this latter claim has been contested by other workers in the field.

The consensus on screening assays appears to be that the *in vivo* protocols are appropriate and useful to prioritize chemicals for subsequent testing, to decide early in the chemical/drug development phase whether to pursue a particular formulation, to evaluate what effect changes in chemical structure have on toxicity, and to "fill in the blanks" on a chemical series, all relative to the potential for developmental toxicity, including teratogenicity. The *in vivo/in vitro* assay requires the same number of maternal animals to do fully *in vivo* studies, requires sophisticated technical procedures for culturing embryos, and provides for only a limited number of embryological endpoints due to the limitations on the length of time embryos can be maintained in culture. There does not appear to be an advantage in using these assays as screens.

The *in vitro* assays with mammalian embryos or tissues have two critical limitations. First, the metabolic capabilities of the embryo are very limited and only the embryo is cultured. Any metabolic changes to the parent compound by the maternal organism and therefore the metabolites to which the embryo would be exposed *in vivo* are totally missing in the explant system. Currently, attempts are being made to provide metabolic capability by coculturing embryos with adult liver cells or cell fractions, which are the major source of metabolism of xenobiotics to obviate the first limitation. Second, the duration of sustained normal growth and development of embryos appears very limited (24–48 hr) so that the numbers of structures differentiating and the extent of differentiation are similarly limited. A two-system approach, e.g., midbrain plus limb bud micromass culture assay, is an attempt to increase the number of systems evaluated, but it is still very limited relative to the tremendous range of systems developing which may be vulnerable. The *in vitro* assays are very useful in answering research-oriented questions since the age of the embryos (as judged by somite number or other specific morphological signposts) can be precisely controlled, identity and concentration of the test agent are precisely controlled, and early responses can be observed and characterized. They can be used to identify the proximate teratogen by exposing the explanted embryos to specific metabolites

which they cannot further transform and to elucidate mechanisms of action of known teratogens at the organ, tissue, cellular, subcellular, or molecular levels early in the toxic response, prior to cell death or demise of the embryo. The utility of nonmammalian (nonvertebrate) assays as predictors of potential mammalian developmental toxicity appears unclear at this time, although the concept of phylogenetically conserved universal processes in embryonic development is attractive and compelling.

Mechanisms

There is no mechanism fully understood for any developmental toxicant causing fetal malformations. Although in many cases the proximate teratogen is known and maternal and/or developmental toxicity is well characterized, what is not known is how the observed effects result in the malformation(s). The site(s) of action may be intranuclear, intracellular, at the cell membrane, extracellular, outside the conceptus, in the placenta, or in the maternal organism. The mode(s) of action may be general or specific, biochemical, physiological, or microstructural. It is also likely that the mechanism(s) will vary from agent to agent. The two extremes in mechanisms, from very specific to very general, may be exemplified by those proposed for 2,3,7,8-tetrachlorodibenzo-*p*-dioxin (TCDD) and for valproic acid. TCDD produces cleft palate (and hydronephrosis at higher doses) in susceptible mouse strains. The putative mechanism for the induced cleft palate is that TCDD binds to certain epidermal growth factor (EGF) receptors and prevents the normal reduction in expression of certain EGFs in the medial epithelial cells of the palatal shelves just prior to fusion. Therefore, with TCDD, abnormally high levels of certain EGFs apparently continue to stimulate proliferation and differentiation of the cells normally destined to die, and the shelves do not fuse. Valproic acid causes neural tube defects, including exencephaly (in mice) and spina bifida (in humans). It has been proposed that valproic acid and other weak acid teratogens (of which there are many) reach the mammalian embryo and lower the intracellular pH of the embryonic cells. (The embryonic intracellular pH is more basic than the maternal intracellular pH, especially early in development, and changes over time.) The specificity of the effect probably lies in the sensitivity of the target neural tube. The suggested mechanism for TCDD may explain the cleft palate but it does not explain the hydronephrosis. The suggested mechanism for valproic acid (and other teratogens which are weak acids) does not explain the specificity and susceptibility of the targets since other weak acid teratogens do not affect the neural tube and many weak acids are not teratogens. Perhaps the most important barrier to understanding the mechanism(s) of abnormal development is that we do not know enough about the mechanism(s) of normal development.

Studies performed initially on *Drosophila* embryos indicated that sequential activation of a hierarchy of regulatory genes occurs during development of multicellular organisms. These genes regulate the transcription and translation of genetic information into structures (and functions) by orchestrating a precise temporal and spatial expression of structural genes, which in turn control differentiation, i.e., establishment of cell types and organ formation. Many of these genes also appear to play a role in pattern formation during or after gastrulation in vertebrates. The mechanisms of these regulatory genes include genetic and epigenetic control. Genetic mechanisms include the role of genes in establishing the basic embryonic axes (cephalocaudal and dorsoventral), specifying specific embryonic regions, controlling the transition of cells from presumptive to determined in the establishment of the fate of diverse cell types and ultimately specifying directly the differentiated patterns of gene expression, including inter- and intracellular molecules, structure, and functions. Epigenetic mechanisms include the interactions between cells, between cell types, and between cells and the products of other cells. The genetic and epigenetic roles are linked and integrated by so-called second messengers which translate molecular signals by individual cells into commands to produce specific effects on cell growth and patterns of gene activity.

Abnormal expression of mutated genes from this regulatory class results in abnormalities in development which produce information on normal development as well as suggest a mechanism(s) of action of xenobiotics. It is clear that cell division, cell migration, and differentiation are directed by regulatory gene classes that control which genes are expressed in which tissues at which times in development. The molecular approach to identifying these fundamental controlling factors of mammalian development may be the most fruitful in the long run in elucidating mechanisms of normal and abnormal development and providing mechanisms of action of developmental toxicants.

Further Reading

Tyl, R. W. (1993). Developmental toxicology. In *General and Applied Toxicology* (B. Ballantyne, T. Marrs, and P. Turner, Eds.), Vol. 2, pp. 1021–1068. Macmillan, Basingstoke, UK.
Wolkowski-Tyl, R. (1981). Reproductive and teratogenic effects: No more thalidomides? In *The Pesticide Chemist and Modern Toxicology* (S. K. Bandal, G. J. Marco, L. Golberg, and M. L. Leng, Eds.), ACS Symposium Series 160, pp. 115–155. American Chemical Society, Washington, DC.

—Rochelle W. Tyl

Related Topics

Ames Test
Carcinogen–DNA Adduct Formation and DNA Repair
Chromosome Aberrations
Developmental Toxicology
Dominant Lethal Tests
Dose–Response Relationship
Environmental Hormone Disrupters
Epidemiology
Host-Mediated Assay
Levels of Effect in Toxicological Assessment
Molecular Toxicology
Mouse Lymphoma Assay
Mutagenesis
Reproductive System, Female
Reproductive System, Male
Risk Assessment, Human Health
Sister Chromatid Exchange
Toxicity Testing, Reproductive
Toxicology, History of

Toxicity Testing, Inhalation

Introduction

Inhalation is, in many senses, a special case of administration for toxicology as a whole and for acute toxicology in particular. As will be reviewed here, animal inhalation studies are difficult and complex to perform correctly. The complexity and cost of such studies firmly dictates that they be performed only when there are substantial opportunities for human inhalation exposure. Such opportunities occur under three broad categories. In order of decreasing occurrence or importance, these are (1) occupational exposure (from the workplace, either in the normal course of operations or during maintenance or accident situations), (2) environmental exposure (a spectacular example being Bhophal), or (3) therapeutic exposure (when the material is to be used as a therapeutic by this route).

All inhalation studies can be classified either by the pattern of exposure or by the physical nature of the contaminant. Both of these classifications are important because they dictate equipment and animal selection and details of study design.

Pattern refers to how (or how much of) a test animal is exposed to the contaminated atmosphere of interest. In practice, there are only very limited situations in which exposure to a toxicant is purely by inhalation (these cases are with therapeutics when an individual has a material administered directly into the nasal or oral cavity and with an inhalation test system, where only nose exposure is truly achieved). Rather, both in the real world and in the laboratory, inhalation exposure is accompanied by dermal and oral exposure. How concerned one is with the possible confounding effects of such other route exposures on the evaluation of biological outcome dictates selection of pattern exposure.

The three categories of exposure patterns are nose only, head only, and whole body. There are also minor routes (lung only and partial lung) that will not be discussed here and see use only in special research settings but seek to allow precise delivery of doses of test material directly to the lungs.

In "nose only," which is the least commonly used (particularly in acute tests), the test animal is situated so that only its nasal region (or, for dogs and primates, where a mask is used to administer test compound, only the mouth and nasal region) is exposed to a test atmosphere. This can be achieved by having the animal restrained with only its nose poking through an elastic barrier or with a breathing mask fitted over the nose and mouth region. There is still some small amount of oral exposure in such a system because animals will swallow any material deposited on the surface of their

mouths or "cleared" from the nasal region or lungs back into the trachea.

"Head-only" exposure, in which the entire animal is in a chamber into which a test atmosphere is introduced, is both the most common and the least complicated pattern of exposure. There are a wide variety of chamber designs available such that all common laboratory species can be exposed using this methodology. There will be extensive dermal and oral administration in animals exposed whole body (particularly oral in rodents and rabbits which carefully "preen" themselves after an exposure). For gases and vapors, of course, such considerations have minimal impact. The advantages and disadvantages associated with each of these exposure patterns are summarized in Table T-23.

The other manner of classifying inhalation studies is in terms of the exposure "media"—that is, the physical nature of the contaminant atmosphere that is being evaluated. Though several of these categories can be subdivided or defined somewhat differently, for our purposes the types of possible test media are gases, aerosols, and dusts.

Gases are generally the easiest type of exposures to perform because the contaminants in the test atmosphere are in the gaseous phase. This makes handling and manipulating the atmosphere relatively easy. There are two subcategories, however (based on the predominant physical state of test material under "ambient" conditions). These are "true" gases (which can be metered from tanks or generated simply from a highly volatile liquid) and vapors (where the test material is a liquid of low volubility). Generating vapors can be a very difficult problem.

Technically, aerosols include any liquid- or solid-phase material which forms a stable suspension in air. Using this definition and depending on the size of the particles or droplets involved, there is a wide range of subcategories and materials that can be of concern. For purposes here, however, aerosols mean only liquid-phase materials. These are generally more difficult to properly conduct an exposure of, but not as difficult as the dusts.

Dusts are solid-phase (contaminant) particles suspended into a gaseous (atmosphere) phase. They are generally the most difficult to conduct a proper study with for reasons which will become obvious. There is a subcategory of dusts (fibers) which represents an even more difficult special case.

TABLE T-23
Advantages, Disadvantages, and Considerations Associated with Patterns of Inhalation Exposure

Mode of exposure	Advantages	Disadvantages	Design consideration
Whole body	Variety and number of animals Chronic studies possible Minimum restraint Large historical database Controllable environment Minimum stress Minimum labor	Messy Multiple routes of exposure: skin, eyes, oral Variability of "dose" Cannot pulse exposure easily Poor contact between animals and investigators Capital intensive Inefficient compound usage Difficult to monitor animals during exposure	Cleaning effluent air Inert materials Losses of test material Even distribution of space Sampling Animal care Observation Noise, vibration, humidity Air temperature Safe exhaust Loading Reliability
Head only	Good for repeated exposure Limited routes of entry into animal More efficient dose delivery	Stress to animal Losses can be large Seal around neck Labor in loading/unloading	Even distribution Pressure fluctuations Sampling and losses Air temperature, humidity Animal comfort Animal restraint
Nose/mouth only	Exposure limited to mouth and respiratory tract Uses less material (efficient) Containment of material Can pulse the exposure	Stress to animal Seal about face Effort to exposure large number of animals	Pressure fluctuations Body temperature Sampling Airlocking Animals comfort Losses in plumbing/masks
Lung only	Precision of dose One route of exposure Uses less material (efficient) Can pulse the exposure	Technically difficult Anesthesia or tracheostomy Limited to small numbers Bypasses nose Artifacts in deposition and response Technically more difficult	Air humidity/temperature Stress to the animal Physiologic support
Partial lung	Precision of total dose Localization of dose Can achieve very high local doses Unexposed control tissues from same animal	Anesthesia Placement of dose Difficulty in interpretation of results Technically difficult Possible redistribution of material within lung	Stress to animal Physiologic support

Basic Steps

A technically good inhalation exposure can be broken into four major parts or basic steps. One can consider these four problems which must be solved before an actual study is undertaken. In fact, it is this aspect of inhalation which makes the proper conduct of such studies difficult and expensive. The following are the four basic steps:

1. Generation of a test atmosphere

2. Containment, mixing, and movement or test atmosphere and animals (both before and after exposure)

3. Measurement and characterization of what animals have been exposed to

4. Clean up and disposal of resulting "wastes" (gaseous, solid, and liquid)

Each of the first three steps will be discussed in further detail; the fourth step, however, is beyond the scope of this entry. First, however, some basic information on the behavior of gases (and of other phases suspended in gases) and of respiratory physiology should be reviewed.

Mechanics of Exposure

The mechanics of performing acute inhalation exposures to state-of-the-art standards can be complex in their entirety but the individual technical components of the problem are rather simple.

As a starting place, remember that there are four sequential components to an exposure system. These are generation systems, exposure chambers, systems for measuring exposure, and systems for cleaning the effluent air stream.

Generation Systems

Optimal generation systems have four major desirable features:

1. Uniform rate of sample delivery
2. Uniform character of sample delivered
3. Ability to deliver in desired range of concentrations
4. Safety of operations

For each of our types of exposure (vapor, aerosol, and dust), there are a multitude of systems available.

Definitions

- Aerosols: Fine particles, solid or liquid, in a stable gaseous suspension; maximum diameter is 1 μm; there are monodisperse and polydisperse aerosols; for monodisperse aerosols, the standard deviation of geometric diameter about the mean is less than 1.25

- Dusts: Solid particles in air—not necessarily a stable suspension

- Fumes: Agglomerates of many fine particles

- Smokes: Stable agglomerates of fine particles in a gaseous suspension

- Mists: Liquid particles $\geq$40 μm in diameter

- Fogs: Liquid particles ranging from 5–40 μm in diameter

Dust Generation

The following factors need to be considered in dust generation:

- Particle size (and size distribution)
- Particle shape
- Density of material
- Concentrations needed (necessary capacity)

Vapor Generation

Vapor generation systems are based on the principle of maximizing surface area of the liquid, temperature (within the limits of chemical stability), and airflow across the surface of the liquid as a means of increasing efficiency. The four most common generation systems utilizing these principles are tube generators, wick generators, bubble generators, and special instrument generators.

- Tube generators: The parent liquid flows along the inside surface of a tube while an airflow is passed over this surface.

- Wick generators: A liquid phase is passed up some form of porous wick while an airflow is played over it.

- Bubble generators: The airflow is passed through the liquid phase.

- Special instrument generators: A turning tube generator has the internal surface area of the tube maximized by adding ridges or sections and the tube itself, as well as the airflow passing through it, is warmed.

Aerosol Generation

Liquid aerosols only present generation problems and concerns in a couple of special cases. First, if they are extremely volatile, one may actually end up generating a vapor. Second, denser or more viscous liquids require greater energy to overcome surface tension and form droplets of the desired size. The following are four widely used aerosol generation systems:

- Spray nozzle
- Ultrasonic generation (uses sound to provide energy to disrupt liquid into droplets)
- Spinning discs
- Nebulizers

After a stream of test material is generated into an airflow, it is mixed (usually by allowing the mixture to transit a ducting system of sufficient length, with some turbulence) and then introduced into the exposure chamber system in which test animals are contained (or are to be contained).

Exposure Chambers

Technically, chamber exposures can be dynamic or static. In a static chamber exposure, there is no airflow through the system; animals are entered into a closed system that contains an atmosphere "precharged" with the desired test material. Only dynamic systems are considered state of the art, with static systems being inadequate for anything other than some minor short-term rank hazard-type assessments.

The special case of "whole body" versus head-only exposure deserves some consideration. The head-only approach is especially favored for pharmaceuticals and has the following considerations associated with it: Advantages (contrasted to whole body exposure)

- Limited oral and dermal exposure
- Less test material required
- Physiological measurements possible during exposure

Disadvantages

- Time-consuming (i.e., labor-intensive; less a factor in acute studies than in longer term)
- Stressful to animals

Model systems:

- Tubular
- Cylinders
- Rings

Measurement Systems

While animals are being exposed in a chamber, what they are being exposed to must be measured and (if appropriate) characterized. The desirable characteristics for a measurement system include:

- Accuracy and precision across the range of concentrations (or characteristics) to be measured
- Reproducibility
- Continuous measurement (failing this, as frequent intermittent measurement as possible)
- Direct measurement of the variable interest (not interference or calculation based on an indirect measurement)
- Automatic measurement at a number of discrete different locations in the exposure chamber

Achieving all of these characteristics in practice is usually impossible. Rather, a set of acceptable compromises are made. The particular problems, concerns, and instruments associated with each of the separate types of exposure (dust, aerosol, and vapor or gas) are again very different.

Dusts

Parameters affecting measurement and characterization of dusts:

- Size-dependent forces (inertia, gravity, diffusion, and electrical charge)
- Weight/density
- Shape (five basic classes)

Aerosols

Measurement of aerosol concentrations (and characterization of droplet sizes) requires two separate and independent measurements. Such measurements must be performed with sampling in such a way that the droplets in the sample are not altered in either concentration (by deposition on the walls of the sampling device) or aerodynamic size (by either evaporation or condensation). Concentration sampling is the easier of the two measurements to make and is almost always done by volumetric sampling into some form of impinger. Droplet size characterization can be done optically using microscope slides precoated with a film of magnesium oxide, with a laser particle sizer (for low to medium concentrations), or with an impactor if the liquid is not too volatile.

Gases or Vapors

For gases and vapors, there exists the possibility of doing either continuous (but limited in scope) measurements of concentration of many materials by infrared spectrophotometric or all-inclusive intermittent measurements by either gas chromatography or high-pressure liquid chromatography.

Cleanup Methods

The last step or phase in the process of properly conducting an inhalation exposure is cleaning up the airstream leaving the exposure system before releasing it into the atmosphere, and, of course, then properly disposing of the collected waste products that result from such a cleanup.

Depending on the nature of the chemical being evaluated, one or a combination of three methods may be utilized. These are filters, incinerators, or scrubbers (using either water or more exotic fluids such as sulfuric acid or potassium permanganate).

For acute studies, one has much more flexibility in applying these methods than for longer term studies in which logistics limit choices.

Filters can be fibers (e.g., high-efficiency particulate air filters and cotton), particulates (e.g., activated charcoal), or the simple expedient of (for relatively nontoxic materials) a series of cloth bags in a box (in effect, a miniature baghouse). A variant on filtration for acute studies is to pass the airstream through a glass or metal container packed with glass wool as a filtration medium.

Incineration is only used when the material being tested is a gas or volatile, is very toxic, and is subject to thermal degradation. This approach is very rarely used for acute studies.

Washing or "scrubbing" the contaminants out of an air-stream is an attractive alternative for acute studies. A liquid (usually water of an aqueous solution of sulfuric acid, sodium hydroxide, or potassium permanganate) is used as a filter medium for the contaminated air-stream. This can be done by either bubbling the gas stream through a volume of the liquid or passing the gas stream through a spray "curtain."

Dose Quantitation

Measuring (or even expressing) the dose received in an acute inhalation study is a difficult and, at best, inexact undertaking. Unlike other routes, one starts out knowing two variables: how long an animal was exposed and the concentration of material in the atmosphere. One does not know what volume of the atmosphere was inhaled by the animal or how much of the material in the inhaled volume was absorbed.

The traditional approach has been to express and evaluate exposures in terms of concentrations and times of exposure. For most acute exposures, Haber's rule generally holds. That is,

$$ct = K$$

where c is the concentration, t is the time of exposure, and K is the constant specific for material.

This is a handy relationship for comparing exposures at different concentrations or for different lengths of time, but it obviously has limits as to applicability (e.g., very high or very low concentrations).

Just expressing exposure concentrations is more complex than it appears on the surface. First of all, in a dynamic exposure situation, the initial concentration in a chamber is clearly lower than the final concentration. It takes more time for the atmosphere to equilibrate to a concentration at or very near the desired target concentration. This equilibration time can be calculated as

$$C = (w/b) \ [1\text{-evp} \ (bt/a)]$$

where C is the desired chamber concentration, w is the weight of material introduced per unit time, b is the total airflow through chamber, t is the time, and a is the chamber volume.

A second complication to expressing concentration is that, for gases and vapors, it is properly expressed as parts per million (ppm). Interconversions can be calculated with the formulas

$$mg/liter = g/m^3$$
$$mg/m^3 = (ppm)(MW)/24.5$$

where MW is the molecular weight.

One consideration in model selection is the comparability of doses received with those likely in humans. In the special case of the inhalation route, doses received must be calculated (rather than measured) in a manner somewhat specific to the animal models being employed.

Calculated inhalation dosimetry models, thought not extremely accurate, do have some utility in the cases of (1) comparing toxicity via the inhalation route with toxicity via other routes, (2) risk assessment models and calculations, and (3) interspecies calculations and extrapolations.

These calculations are performed using the formula

$$E = [RF \times TV \times C \times 60 \times T]/1000$$

where E is the total maximum possible exposure, RF is the respiratory frequency (per minute), TV is the tidal volume in ml, C is the concentration of test agent in mg/liter, and T is the daily exposure time in hours. Note that this formula can also be used to compare total doses received over different lengths of exposure. If exposure is repeated over a period of several days, the result of the previous calculation is also multiplied by the number of days of exposure.

Values to be used in this equation for the laboratory species commonly used in inhalation studies and humans are as follows:

Species	RF	TV
Rat	85.5	0.86
Mouse	109	0.18
Guinea pig	90	1.8
Rabbit	49	15.8
Human	11.7	750.0

These values will, in turn, result in hourly exposure values which can be reduced to the following:

Rat = (44,118 liters) C

Mouse = (11,772 liters) C

Guinea Pig = (9.720 liters) C

Rabbit = (46,452 liters) C

Human = (526.5 liters) C

It must be remembered in using this model that the values obtained will be the maximum average limits on the dose received. A number of factors that are not included will affect the actual values in all cases except one (item 1 below), serving to reduce the actual values of doses received. These factors include the following:

1. There will be variations in individual animals (and in the same animal at different ages, weights, and states of exercise) in RF and TV.

2. If the material is a particulate or is water insoluble, the degree of deposition and clearance (respectively) in and from the lungs will vary.

3. The degree of absorption from the lungs into the body will vary from compound to compound.

All analytical monitoring systems must accomplish two basic tasks: collection of a representative sample and accurate analysis. Samples may be collected directly from the chamber, a technique which is efficient, rapid, easily automated, and less prone to error. Extractive techniques may need to be used which, although more time-consuming and hence less efficient, are able to concentrate low-level atmospheric samples in order to increase analytical sensitivity (Table T-24).

The most commonly used analytical techniques for analyzing gases and vapors are gas chromatography and infrared spectroscopy. Aside from the need to accurately measure the test atmosphere, these readings are needed to control the intended concentrations during the test.

Direct samples may be collected in evacuated glass or metal containers, inflatable bags of polymeric (nonreactive) material, or most commonly by gas-tight syringe. These may also be taken remotely by the use of transfer lines fed directly into the analytical instrument. Extractive sampling involves passing the test atmosphere through a solvent, adsorption onto a collecting surface, or condensation onto a cold surface.

TABLE T-24
Analysis of Inhalation Chamber Concentrations

Vapors
  Gas chromatography
    Direct sampling
    Extracted samples
  Infrared spectroscopy
  Ion-selective electrodes
  UV–visible spectrophotometers
Aerosols
  Concentration
    Gravimetric
    Forward-scatter detectors
    Back-scatter detectors
    J-attenuation detector
    Quartz crystal microbalance detector
  Particle size
    Cascade impactors
    Microscopy (fiber morphology)
    Laser/Doppler type

Gas chromatography (GC) is the most versatile and frequently used analytical technique for monitoring gases and vapors. GC offers chemical separation of components for specific analysis, low detection limits, and rapid turnover of data for feedback control. Infrared spectroscopy works well since most gases and vapors give reasonably intense and unique spectra. The Miran portable gas analyzer is particularly useful for continuous monitoring. Other techniques shown to be useful include the use of ion-specific electrodes, ultraviolet–visible spectrophotometers, and scrubbing colorimeters. As is always the case, frequent calibration of analytical instruments is essential.

Aerosols present a special case in that the investigator needs to measure the mass concentration of the chemical, the chemical composition as a function of particulate size, and the particle size distribution of the aerosol. No continuous sampling instruments are available to measure both particle size and chemical concentration. Particle detection can be accomplished using both forward- and back-scatter detectors. The back-scatter described allows for noninvasive determinations over a range from 6 to 10,000 mg/m^3. An infrared light beam is projected through a 1.5-mm Lucite window into the chamber. The light back-scattered from a sensing volume of about 12 cm^2 is detected, in the same unit, containing the light mean by the outer edges of the light focusing optics. The unit is capable of monitoring without invading the chamber to remove an aliquot of the test atmosphere.

The quartz crystal mass monitor reflects the mass change on the face of the crystal when particles are drawn through an orifice and are deposited on the face by electrostatic precipitation. Mass concentration instruments can detect the collected mass by the attenuation of J radiations. The aerosol is drawn through an orifice and articles impact on a surface positioned between a source and a counter.

For particle sizing, many varieties of cascade impactors perform well, although care must be taken to avoid errors introduced by sampling (such as collection in sampling lines). The laser/Doppler-type particle size device can be used to measure aerodynamic size and low concentrations with a rapid readout. In a system described by Cook, a powerful pulsed laser using temporal analysis of back-scattered light indicates the spatial distribution of particles.

Study Types

The type of inhalation test conducted will depend on the question being addressed, but generally we look to determine what happens when biological systems are exposed infrequently or frequently and whether the exposures are to high or low concentrations. Studies can be described as acute, chronic, or subchronic depending on number of test exposures. Acute effects usually occur rapidly as a result of short-term exposures (most often single) and are of short duration. Chronic studies involve repeated exposures which may need to be continued for the duration of the animal's lifetime. These studies are generally conducted at relatively low concentrations.

Acute studies are generally conducted at a relatively high concentration and are useful in determining the approximate range of toxicity of a chemical. Acute studies can be used as a starting point in the determination of dose levels for longer term tests. The clinical signs evoked at three high exposures often allow determination of the nature of the toxic effect induced. The two most common numerical values that come from an acute study are the approximately lethal concentration (ALC) and the LC_{50}. The ALC is defined as the lowest concentration that produces death in at least one of a group of exposed animals, while the LC_{50} is the calculated concentration at which half of the exposed population would be expected to die. Generally the exposures are conducted for a single 4- or 6-hr period and the animals are observed for 14 days after treatment.

Subchronic studies generally precede lifetime studies and are conducted to determine what the target organ or organ system might be and what exposure regimen (concentration × time) is required to produce this change. For this purpose, our practice is to expose groups ($n = 10$) of male rats to three test concentrations (Table T-25). The highest concentration tested is set at one-fifth the ALC (or the LC_{50} depending on the steepness of the mortality dose–response curve) and the lower two would be $\frac{1}{15}$ and $\frac{1}{50}$. Rats are exposed 6 hr a day for 5 days, given a 2-day rest period, and are again exposed for 5 days. *In vivo* observations, including body weight measurements, are made daily. Following the 10th exposure, all rats are subjected to hematological, clinical blood chemistry, and urine analysis evaluations. Half of the rats are sacrificed at that time and complete pathological examination including histological evaluation is conducted. The remaining rats are held without additional exposures for 2 weeks and the parameters altered in rats sacrificed immediately following the 10th exposure are evaluated to determine the reversibility of the change(s).

A variation of this design uses an increasing exposure regimen which continues until severe biologic effects are observed. This provides target organ toxicity data using fewer animals (only one group is treated), but the quantitative aspects can be masked in cases in which chemical buildup in the body occurs or change occurs only after some protective function in the body has been depleted. In both of the subchronic studies, the

TABLE T-25
Design of Subchronic Inhalation Study

Test species:	Rat
Sex:	Male
Number of test groups:	3(1/5, 1/15, 1/50 ALC)
Number per group:	10
Exposures:	6 hr/day, 5 days/week, 2 weeks
Animal sacrifice:	5/group after 10th exposure
	5/group after 14-day recovery period
Parameters measured:	Growth and *in vivo* responses
	Clinical pathology
	Urine analyses
	Gross pathology with organ weights
	Microscopic pathology
	Chemical index of exposure (where possible)

importance of adequate concurrent control animals needs to be underscored.

Chronic studies are conducted to determine effects of long-term exposures at levels where acute toxicity is not obvious. Chronic exposure patterns generally follow those encountered in the workplace—animals are exposed 6 hr/day, 5 days/week for their lifetime. For environmental contaminants, continuous exposures of 23 hr/day (allowing 1 hr/day to feed the animals and clean the exposure chambers) for 7 days/week might be considered more appropriate.

In both chronic study types, exposures are designed to be as constant as possible with minimal deviation from the target or design concentrations. Investigators go to great lengths to be able to report test concentrations of x ppm with $x/100$ standard deviations. This needs to be contrasted with the work or living environment where chemical pollutant levels fluctuate greatly depending on released amounts, ventilation, meteorological conditions, and many other important factors. Investigators measuring effects of airborne chemicals in confined spaces such as a submarine would more likely choose continuous exposures, while evaluation of materials in industrial atmospheres would likely involve intermittent exposures similar to those described previously.

Chemicals irritating to the sensory apparatus of the upper respiratory tract can be identified by measuring the reflex-induced decrease in respiratory rate. The basic design uses an apparatus which translates changes on thoracic body cavity volume into tracings that can determine the depth and rate of respiration. The animal, usually a rat or mouse, is placed into a whole body tube connected to an inhalation chamber such that only the nose protrudes into the exposure area. Animals are often exposed to various concentrations of the agent and a plot of respiration rate against concentration is made. Dose–response curves and minimum effect levels can be determined, the data usually being expressed in terms of the RD_{50} (concentration required to lower the respiration rate by 50%). The method is quite simple and detects the effects of irritational concentrations where no associated pathological modifications occur.

Intratracheal instillation of materials is a popular alternative to inhalation exposure of animals. The advantages of this type of exposure include the need for very small amounts of test agent (hence, a safety feature in terms of handling and containment of the chemical), extensive chambers are not required, and the complex technical support needed to generate and maintain experimental exposure conditions is avoided. These factors make this type of study very inexpensive to conduct. Furthermore, the dose can be delivered very precisely to the respiratory tract tissues. However, dose distribution to the respiratory tract tissues does not accurately simulate an inhaled dose and, hence, does not reflect the real-life response very clearly. Inhalation of airborne toxins generally results in a relatively well-distributed dose throughout the respiratory system. Intratracheal instillation tends to lead to a less uniform deposition and to favor the lower portions of the lung due to gravimetric settling of material. Rats and hamsters were exposed to radioactive particles and the distributions following both inhalation and instillation were examined. The resulting distributions were strikingly different with instillation producing heavy deposits in the medium-sized bronchi. Instilled materials seldom reached the alveoli, whereas inhalation led to considerable deposition in the small airways. High local concentrations following instillation can lead to localized tissue damage which would not be seen following more uniform deposition. The use of this technique then is basically limited to situations in which tissue reactions [both of an acute (inflammation) and chronic (neoplasia and fibrosis) nature] to a variety of materials are to be compared side by side.

Further Reading

Gad, S. C., and Chengelis, C. P. (1981). *Acute Toxicology: Principles and Methods.* Telford, Caldwell, NJ.
Leong, B. K. (1981). *Inhalation Toxicology and Technology.* Ann Arbor Science, Ann Arbor, MI.
Salem, H. (1987). *Inhalation Toxicology.* Dekker, New York.

—*Shayne C. Gad*

Related Topics

Analytical Toxicology
LD_{50}/LC_{50}
Levels of Effect in Toxicological
    Assessment
Occupational Toxicology
Pollution, Air
Respiratory System
Toxicity, Acute

Toxicity, Chronic
Toxicity, Subchronic

Toxicity Testing, Reproductive

Introduction

It is currently estimated that approximately 15% of couples are clinically infertile (no conception after 1 year of unprotected intercourse); approximately 30% of the infertility is attributable to the male partner, 20% to the female partner, 20% to a combination of problems in both partners, and another 30% is not explained by diagnosis of adverse conditions in either partner. Once conception has occurred, up to 80% of human pregnancies may be lost, most in the first trimester. Reproductive toxicity may be defined as an adverse effect on any aspect of male or female reproductive structures or functions, on the developing offspring, or on lactation, which would interfere with the development of normal offspring through sexual maturity, in turn capable of normal reproduction. This definition includes aspects of developmental toxicity, including teratogenesis, and developmental neurotoxicity. This entry focuses on the male and female mammalian reproductive systems, chemicals which affect the status and functions of these systems, and tests currently utilized to detect such effects.

Male and Female Reproductive Systems

The reproductive system in the embryo consists of paired gonadal ridges in the dorsal midline containing all components but the gonial cells; the gonial cells differentiate during embryogenesis, external to the embryo in the yolk sac, and migrate into the embryo and into the gonadal ridges along prescribed routes. The gonial cells in transit number in the hundreds; once they arrive, they proliferate and the gonad develops into sex-specific structures.

Male System

The mammalian male reproductive system consists of the testes and associated structures: epididymides, vas deferens, accessory sex glands (seminal vesicles and prostate), and intromissive organ (the penis) (Fig. T-10). The testis consists of two major compartments: the seminiferous tubules, which contain the spermatogonial cells, which differentiate into the spermatozoa (sperm), and the Sertoli cells, which provide nutrition to the developing sperm, produce androgen-binding protein to bind the male sex hormone testosterone (and produce a glycoprotein in the fetus and neonate which suppresses female development), and which maintain the blood–testis barrier; and the interstitial compartment, which contains Leydig cells which produce testosterone and other androgens for transport within and outside the testis.

The androgens control spermatogeneis, the growth and activity of accessory sex glands, masculinization, and male behavior (see Androgens). *In utero*, androgen production by the fetal testis early in development (e.g., Weeks 4–6 in humans) is essential for sexual differentiation of the gonads triggering male development and repressing female development (along with a product of the Sertoli cells). The predominant masculinizing hormone *in utero* is dihydrotestosterone, produced from testosterone by the enzyme 5-α-reductase. The presence of testosterone during puberty in males triggers the development of male secondary sex characteristics and the initiation of spermatogenesis.

The control of testicular function is via the hypothalamus–pituitary–testis axis (Fig. T-10). In the hypothalamus of the forebrain, neuroendocrine neurons secrete gonadotrophic hormone releasing hormone (GnRH) into the anterior pituitary gland. Here GnRH stimulates release into the blood of the two gonadotrophic hormones, luteinizing hormone (LH) and follicle stimulating hormone (FSH), named for their first-discovered roles in the female reproductive system. Prolactin (PRL) is also released into the blood from the anterior pituitary under control of dopamine from the median eminence in the forebrain. FSH and LH act on the testis: FSH stimulates the Sertoli cells to enhance sperm pro-

duction, and LH acts on the Leydig cells to stimulate synthesis of testosterone, which in turn is required in high concentrations in Sertoli cells for sperm production. PRL acts to enhance the effects of LH on the testes.

Spermatogenesis (the process which produces mature sperm) in the testis begins at puberty in mammalian males, when high blood levels of FSH and LH are attained (Fig. T-11). The spermatogonial cells (diploid) at the periphery of each seminiferous tubule undergo repeated mitoses; one "daughter" cell in each cell division replaces the original spermatogonial cell, and the other daughter cell becomes a primary spermatocyte and begins the process toward production of sperm. Each primary spermatocyte undergoes meiosis I (the first reduction division) to form two secondary spermatocytes; each secondary spermatocyte undergoes meiosis II (the second reduction division) to form two spermatids. Each spermatid undergoes a differentiation process, termed spermiogenesis, to produce a mature sperm (haploid); this process involves compaction of the DNA into the headpiece of the sperm, covered by an acrosomal cap (used to penetrate the egg); formation of a midpiece with mitochondria (to fuel the swimming function of the flagellum); and formation of a flagellum ("tail") to drive the sperm. The process produces, from one primary spermatocyte, four functional sperm. The sperm are passively moved from the center (lumen) of each seminiferous tubule into the epididymis where the sperm acquire the capacity for movement and fertilization—the capacitation process. The epididymides, seminal vesicles, and prostate secrete fluids to nourish and capacitate the sperm, to provide the hydrodynamic force for ejaculation, and to neutralize the acidic environment of the female's reproductive tract. Large numbers of sperm are produced in waves, with 10,000–10 million ejaculated per time in males. The process of spermatogenesis takes approximately 5 weeks in mice, 8 weeks in rats, and 10 weeks in humans and occurs continuously from puberty until death.

Female System

The female reproductive system consists of the ovaries, oviducts (fallopian tubes), uterus, cervix, and vagina (Fig. T-10). The ovary consists of oocytes, follicles where the oocytes develop, and support cells. The follicles produce estrogen and progesterone during *in utero* development and during the reproductive period of the female.

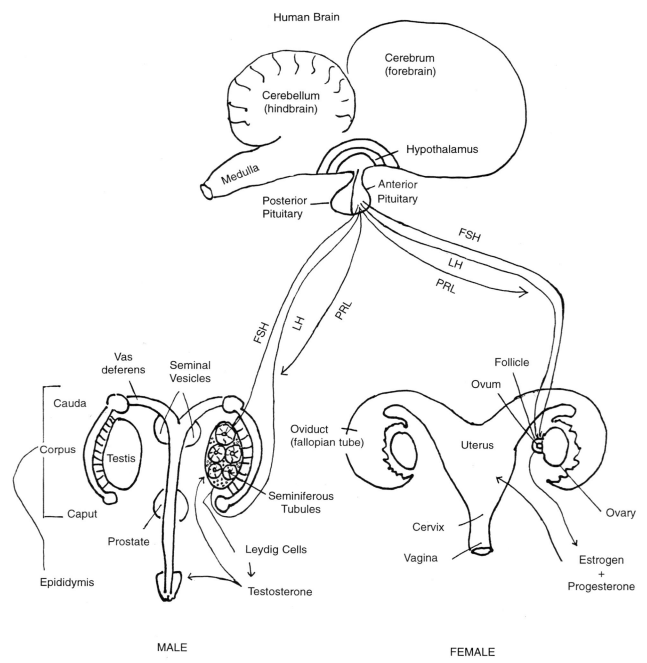

FIGURE T-10.   *Male and female reproductive systems with hormonal controls from the brain (hypothalamus) and anterior pituitary gland.*

The control of ovarian structure and function is via the hypothalamus–pituitary–ovary axis (Fig. T-10). As with the male, cells in the hypothalamus secrete GnRH in the female which acts on the anterior pituitary to release FSH and LH in a cyclical pattern (PRL is also released). FSH and LH act on the ovary; FSH stimulates the growth of follicles which in turn secrete estrogen and progesterone to prepare the uterus for implantation of the fertilized egg (zygote), and LH, in a midcycle surge, triggers the rupture of the follicle and release of the ovum (ovulation). PRL plays a role in the maintenance of the corpus luteum (the collapsed follicle after ovulation which produces estrogen and progesterone), the rupture of the follicle, and in lactation.

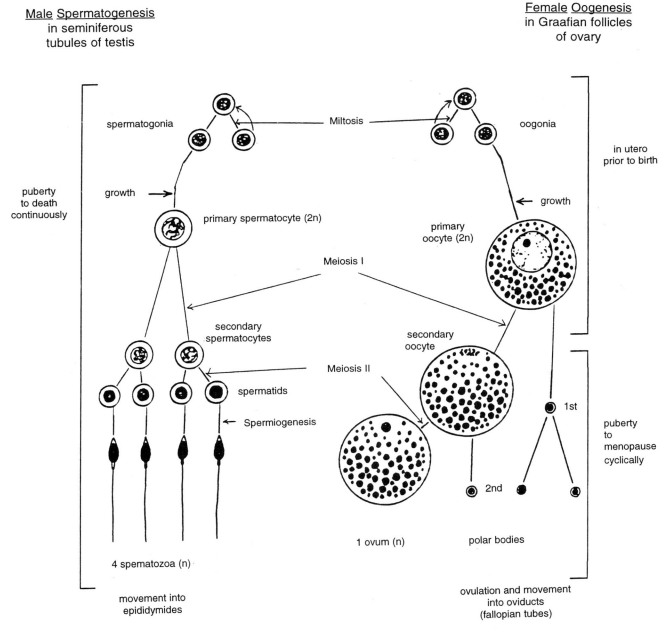

FIGURE T-11. *The process of male spermatogenesis and female oogenesis.*

Oogenesis (the process which produces mature ova) begins at puberty in mammalian females (Fig. T-11). Note that the series of mitoses of oogonial cells occurs only *in utero;* all the primary oocytes a female will have (approximately 500,000 in humans, most of which will die) are present in her ovaries prior to her birth. *In utero,* each primary oocyte proceeds through the second (of four) phases of meiosis I and waits until puberty and the onset of the cyclical release of FSH and LH.

A number of follicles begin the process of oogenesis during each cycle. In each follicle, the primary oocyte completes meiosis I to produce a large secondary oocyte and a very small first polar body. The secondary oocyte (and in some species, the first polar body) undergoes meiosis II to produce a very large ootid (haploid) and a second polar body (if the first polar body divides, it produces two secondary polar bodies). The objective of the "lopsided" division, and the functions of the layers of accessory nurse cells which surround the developing oocyte in the follicle, is to produce a very large ovum, filled with nutrients and preformed genetic blueprints to sustain the early embryo. In response to the LH surge, the most advanced follicle ruptures (the other ova die in a process termed atresia) and the mature ovum is released into the oviduct; the collapsed follicle becomes a corpus luteum with endocrine functions (producing estrogen and progesterone to sustain the uterine lining, i.e., ovulation). One or more eggs is ovulated during each cycle (up to 24 in rodents). These ovulation (estrous) cycles, and associated menstrual cycles, to shed the uterine lining if the egg or eggs are not fertilized last a few days (mouse) to several weeks (dogs, humans, and horses). Estrus, at the time of ovulation of the egg(s), corresponds to the period of female receptivity (high levels of circulating estrogens) in species which undergo periodic "heats." In humans, there are approximately 500 ova ovulated from puberty to menopause.

Fertilization and Offspring Development

The mature ovum is released from the ruptured follicle and is drawn into the oviduct. The egg is transported from the ovary to the uterus by movement of cilia at the opening of the oviduct, muscle contractions of the oviduct, and subsequent fluid movement; fertilization takes place in the oviduct. Meanwhile, the uterus, in response to preovulatory estrogen production from the follicle, increases blood flow and the uterine lining (endometrium) begins to thicken. After ovulation, the actions of progesterone (and estrogen) complete the growth of the endometrium including increased blood supply, formation of microvilli, increased synthesis of proteins, and other sources of nutrition.

Fertilization consists of penetration of the ovum by a sperm and union of the two haploid nuclei (one from each gamete); early cell divisions then begin (cleavage). The fertilized egg (zygote) continues to travel down the oviduct into the uterus and implants in the receptive uterine lining. The time from fertilization to implantation is relatively short and comparable among mammalian species: 5 (mice) to 8 (humans) days. Implantation is accomplished by the invasive destruction of uterine lining by the outermost extraembryonic cells of the conceptus (trophoblast cells) and ultimate intimate at-

tachment of these cells with the maternal uterine blood vessels; the uterine lining heals over the conceptus and the exchange of nutrients and wastes begins at the site of the future placenta. Once implantation is complete, the conceptus proper begins to differentiate into outermost (ectoderm), innermost (endoderm), and middle (mesoderm) cell layers and the major organ systems begin to form. This period of organogenesis lasts approximately 10 days in rodents and 50 days in humans. At the end of organogenesis (signaled by the closure of the secondary palate), the conceptus is termed a fetus and the fetal period of histogenesis (differentiation of cells and tissues within systems) begins. It lasts until parturition, approximately 7 days later in rodents and 7 months later in humans. Delivery occurs after approximately 19–22 days of pregnancy in rats and 270 days in humans; the perinatal period involves adjustments of the offspring to air breathing, nursing, and rapid growth, development, and learning in a gravity-based environment, aided by required postnatal care by the mother (maternal behaviors such as pup retrieval, milk production and delivery, grooming, and teaching).

Considerations for Reproductive Toxicity Evaluations

The male reproductive system is at risk during fetal development *in utero*, postnatally during puberty, and during his reproductive lifetime with targets including the processes and structures involved in primary sex differentiation (testes and accessory organs), secondary sex characteristics and sexual behaviors (e.g., libido), and performance (e.g., erection and ejaculation). The traditional endpoint of concern is production of normal numbers of normal (genetic, chromosomal, and structural) sperm.

Identification of reproductive toxicants is made from clinical workups on men in infertility clinics or undergoing drug treatments, predominantly for cancer, from epidemiologic studies on general populations (environmental exposures) or on worker populations (industrial exposures in production plants or users of chemicals as pesticides and commodity chemicals), and from animal studies (usually in rodents). Animal studies allow for more invasive examinations such as histopathology of the testes, close scrutiny of mating behavior, and mating to proven breeders. They can be used to confirm or extend initial observations in humans or to initially identify a potential reproductive toxicant. One epidemiologic study of male workers exposed to dibromochloropropane (DBCP) was apparently triggered by the men talking at work breaks and a request from the workers to OSHA for an investigation. DBCP proved to be a testicular toxicant that affected spermatogenesis in the male workers; however, data on rats exposed to DBCP with the same testicular findings were, in fact, available in the literature 15 years prior to the worker concerns. One additional unique characteristic of male reproductive toxicity is that if effects are limited to postspermatogonial cells, then the effect is transient (limited to the time when these cells become sperm and are ejaculated) and subsequent waves of spermatogenesis from the intact spermatogonial cells are not affected.

Since the definition of normal sperm includes normalcy of the haploid genetic complement as well as normal sperm structure, numbers, and functions (i.e., ability to swim in the female's reproductive tract, penetrate the ovum, and join its genetic material with the egg's haploid genetic material), analyses of sperm parameters are performed in all three categories of investigations: clinical, epidimiologic, and animal testing. Routinely, sperm numbers, motility (viability), and morphology are ascertained. The interpretation of the human worker reproductive data is usually confounded by, among many factors, job experience, exposures to multiple materials, lifestyle (e.g., smoking, drinking, "recreational" use of drugs, and hobbies), age, status of spouse, number and ages of children, socioeconomic class, educational level, and ethnic/religious factors. Exposure data for workers are usually very poor [e.g., concentrations are not precise for individuals or job descriptions/titles, and measurements are not frequent and not long term (over the job or employee working lifetime)]. Exposure data for nonworking environments (i.e., contaminated foodstuffs, water, soil, or air) may be even worse.

The female reproductive system is also at risk during fetal development *in utero*, postnatally during puberty, and during her reproductive lifetime until menopause (cessation of ovulation). The traditional endpoints of concern are ovulation of a normal ovum, fertilization, uterine status, implantation and prenatal development, parturition, and lactation involving nursing (normal amounts and constitution of milk) and other maternal behaviors.

Identification of reproductive toxicants in females is made from clinical work-ups (as with males), from epidemiologic studies in general populations, and to a lesser degree on worker populations (the industrial work-force in chemical production and end use has been traditionally male). Animal studies are also very important to initially identify an agent or to confirm and extend initial findings in women. The risk to women's reproductive status may be transient (limited to the pregnancy at risk) or permanent, if effects are to the primary oocytes, since no additional oocytes will be made during her reproductive lifetime. The confounders for male reproductive risk assessments are essentially the same for female reproductive risk assessments. The cyclical nature of the female's reproductive activity (e.g., hormone levels, ovulation, and uterine lining build-up and shedding) makes both human and animal research more difficult; synchronized populations (by estrous cyclicity) are necessary to identify most effects on reproductive structures and functions.

Categories of Reproductive Toxicants

Reproductive toxicants can be categorized by type of agents, e.g., physical agents such as ionizing radiation; pharmaceuticals such as therapeutic drugs, especially those used in treatment of cancers (which target DNA and therefore act as mutagens and/or those which target cell proliferation and therefore affect spermatogenetic and oogenetic cell divisions); recreational drugs/drugs of abuse; industrial chemicals (including pesticides, solvents, and commodity chemicals); environmental chemicals (contaminating air, water, soil, and foodstuffs); and naturally occurring toxicants such as plant toxins (e.g., mushrooms and herbs) and animal toxins (from invertebrates such as certain shellfish, spiders, and insects and from vertebrates such as certain fish, frogs, toads, and snakes).

Categorization of reproductive toxicants by function or mechanism would include agents acting as mutagens (causing changes within and between genes), clastogens (causing changes in parts of chromosomes, including chromosome breakage), cytotoxins (killing cells in general or specific cell types such as gonial cells, Leydig cells, Sertoli cells, etc.), mitotic (meiotic) poisons (interfering with cell division, e.g., by damage to the assembly/dissociation of spindle fibers which control the movement of chromosomes), agonists or antagonists of endogenous hormones (e.g., environmental antiestrogens and antiandrogens), inhibitors of hormone synthesis or degradation, and neurotoxicants (affecting central nervous system control of reproduction).

Governmental Regulations

The thalidomide disaster in the late 1950s and early 1960s resulted in over 8000 children in 28 countries with major malformations. The U.S. governmental agencies recognized that it was only extraordinary luck and Dr. Frances Kelsey of the U.S. FDA which averted huge numbers of children in the United States being affected (the manufacturers were not permitted to market the drug in the United States); there were no mandated testing procedures in place at the time which would have identified the risk. In response, in 1966, Dr. E. I. Goldenthal, Chief of the Drug Review Branch, sent a letter to all corporate medical directors establishing *Guidelines for Reproductive Studies for Safety Evaluation of Drugs for Human Use*. These guidelines were promulgated "as a routine screen for the appraisal of safety of new drugs for use during pregnancy and in women of childbearing potential." Three phases or segments were proposed. Segment I, Study of Fertility and General Reproductive Performance (Fig. T-12A) was designed to provide information on breeding, fertility, nidation (implantation), parturition, neonatal effects, and lactation in rats. It involves exposure of weanling males for at least one full spermatogenic cycle (10 weeks) and/or of adult females for at least two ovulation cycles (2 weeks) and then a mating period, with males necropsied after the mating period. Females continue exposure through gestation with one-half of the parental females and their litters necropsied during gestation to identify pregnancy rate and pre- and postimplantation loss. The other half of the females and their litters are necropsied at weaning at the end of lactation (Postnatal Day 21) to identify *in utero* losses, postnatal losses, and growth and development of the offspring. Acquisition of developmental landmarks such as surface righting, pinna (external ear) detachment, pilation, auditory startle, eye opening, incisor eruption, mid-air righting reflex, negative geotaxis, etc. is noted. The offspring can be maintained beyond weaning to ascertain time of vaginal patency, testes descent, preputial separation, and, if appropriate, motor activity, learning and memory, and mating compe-

tence. (Segment II and III studies are discussed under the Toxicity Testing, Developmental.) An additional reproductive toxicity study, used by EPA (TSCA and FIFRA) (Fig. T-12B), involves a long prebreed exposure to both sexes (designated F_0) begun after weaning, with continuing exposure during mating, gestation, and lactation (with F_0 males necropsied usually after mating), selection of offspring (designated F_1) and prebreed exposure, mating, gestation, and lactation of F_1 parents and F_2 offspring. The F_1 generation is the major focus of this protocol since it is the only generation that receives exposure from the time its members were gametes through their reproductive performance; treatment-related effects on structures and functions of the reproductive system would be discernible in the F_1 animals by this protocol. A third reproductive toxicity protocol promulgated by FDA involves a three-generation, two litter per generation study design (Fig. T-12C). It is similar to the two-generation study, except that F_0 animals, after they produce the first litters (designated F_{1a}) are rebred (within groups to different partners) to produce F_{1b} offspring. Usually the F_{1a} offspring are retained for prebreed exposure and generation of F_{2a} and F_{2b} offspring. (F_{1b} offspring are terminated at weaning, with representative animals, usually 10/sex/group, necropsied.) The F_{2a} animals are usually retained for prebreed exposure and generation of F_{3a} and F_{3b} offspring. The last breed of F_{2a} animals to produce the F_{3b} litters can be executed like previous breeds, with the offspring terminated at weaning, or the F_{2a} mothers can be necropsied on Gestational Day (GD) 20 (prior to expected parturition) and the F_{3b} fetuses evaluated for developmental toxicity (examination of external, visceral, and skeletal morphological development *in utero*; see Toxicity Testing, Developmental).

The International Conference on Harmonization of testing guidelines for reproductive toxicity has recently (1994) promulgated five study designs (Fig. T-13).

The first study design (Fig. T-13A), termed Study of Fertility and Early Embryonic Development, is similar to an FDA segment I study except that male prebreed exposure is for 4 weeks and female exposure begins at mating and extends only to GD 6 (at the time of implantation). F_0 males are sacrificed after GD 6 and F_0 females are sacrificed at midpregnancy (GD 15) or just prior to term (GD 20). It is designed to assess effects of exposure during prebreed (males), mating (both sexes), and the pre-implantation period (females) on *in utero* reproductive indices.

A. Segment I

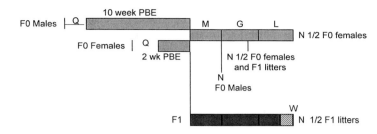

B. Reproductive Toxicity Study Design (Two-Generation, One Litter/Generation)

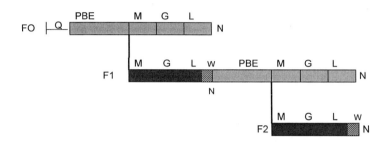

C. Reproductive Toxicity Study Design (Three-Generation, Two Litters/Generation)

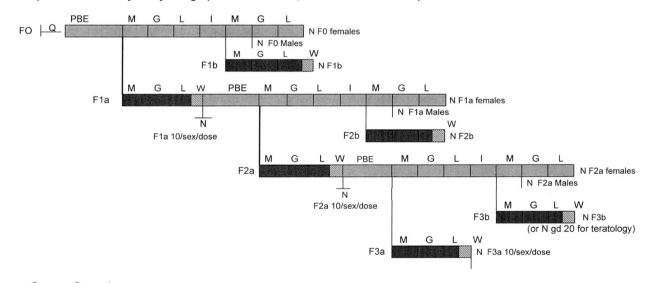

Q =	Quarantine
PBE =	Pre-Breed Exposure
M =	Mating
G =	Gestation
L =	Lactation
W =	Weaning
N =	Necropsy
I =	Interbreed Period (7-14 days)

　　Direct exposure

　　Possible indirect exposure from transplacental and/or translactational exposure

　　Both direct and possible indirect exposure (nursing pups also self-feeding) if administration of test material is by feed or water

FIGURE T-12.　U.S. government guidelines for study designs of reproductive toxicology. A, Segment I (U.S. FDA); B, two generation (U.S. EPA, TSCA, and FIFRA); C, three generation (U.S. EPA).

A.   Study of Fertility and Early Embryonic Development, (4.1.1) rodent (See Segment I)

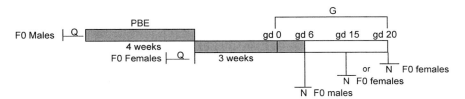

Assess: Maturation of gametes, Mating behavior, Fertility, Preimplantation, Implantation

B.   Study for Effects on Prenatal and Postnatal Development, Including Maternal Function
      (4.1.2) rodent (See Segment III)

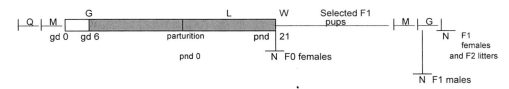

Assess: Toxicity relative to nonpregnant females, Prenatal and postnatal development of offspring,
           Growth and development of offspring, Functional deficits (behavior, maturation,
           reproduction)

C.   Study for Effects on Embryo-fetal Development (4.1.3) rodent and non-rodent (See Segment II)

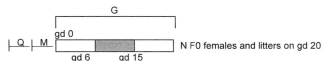

Assess: Toxicity relative to nonpregnant females, Embryofetal death, Altered growth of offspring <u>in
           utero,</u> Structural changes of offspring <u>in utero</u>

*FIGURE T-13.   International Conference on Harmonization guidelines on detection of toxicity to
reproduction for medicinal products. A, Study of fertility and early embryonic development
(4.1.1); B, study for effects on prenatal and postnatal development, including maternal function
(4.1.2); C, study for effects on embryo–fetal development (4.1.3); D, single-study design (4.2);
E, two-study design (4.3).*

The second study design (Fig. T-13B), termed Study for Effects on Prenatal and Postnatal Development, assesses exposure to the parental female from implantation (GD 6) through weaning of her litter (Postnatal Day 21) with selected offspring pups retained for mating (to produce F_2 litters). This design is similar to an FDA segment III study (see Toxicity Testing, Developmental).

The third study design (Fig. T-13C), titled Study for Effects on Embryo–Fetal Development, is essentially an FDA segment II study (see Toxicity Testing, Developmental) with exposure of the mother during organogenesis of her offspring *in utero* (GDs 6–15).

Additional study designs (Figs. T-13D and T-13E) are essentially combinations of the first two or three designs.

Obviously, specific studies are also designed to investigate a specific endpoint and/or agent.

Statistical Analyses

Statistical analyses of continuous data which distribute according to a bell-shaped curve (e.g., adult body weights, weight changes, feed consumption, pup body weights, and percentage male pups) employ parametric

D. Single Study Design (4.2) rodents (Combine 4.1.1 and 4.1.2)

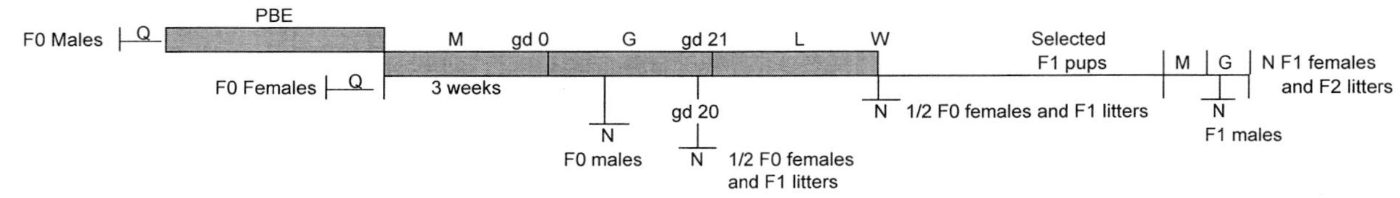

E. Two Study Design (4.3) rodents

4.1.1 with 4.1.2 1/2 F0 females and F1 litters necropsied on gd 20
1/2 F0 females and F1 litters necropsied on pnd 21
(retained selected F1 pups followed through mating and gestation of F2 litters)

Q = Quarantine
PBE = Prebreed Exposure Period
M = Mating
G = Gestation
L = Lactation
W = Wean
N = Necropsy
gd = Gestational Day
pnd = Postnatal Day
Direct exposure to adults

FIGURE T-13. (Continued).

methods, including analysis of variance, tests for homogeneity of variances, tests for dose-related trends, and pairwise comparisons to the concurrent control group values. Continuous data which do not distribute as discussed previously are examined by nonparametric methods to identify trends and pairwise comparisons. Nominal (noncontinuous) data such as reproductive indices (e.g., mating, fertility, fecundity, and incidence of adult clinical signs) and survival indices are also analyzed for trends and pairwise comparisons. The unit of analysis for all tests is the male, the female, or the litter.

Risk Assessment

The U.S. government's approach to health assessment of agents involves four major components: hazard identification, dose–response assessment, exposure assessment, and risk characterization (Fig. T-14). Animal studies usually provide hazard identification and dose–response assessment. Animal studies usually employ a route of administration which delivers a bolus dose for a worst-case scenario (i.e., gavage) or which duplicates the known or potential human route of exposure, usually dosed feed or dosed water for end-use consumer exposure, or inhalation or cutaneous routes (for industrial exposures). The highest dose in these studies which produces no effects is designated the no-observable-effect level (NOAEL) or the no-observable-adverse-effect level (NOAEL). The NOEL/NOAEL is then divided by one or more uncertainty or safety factors to obtain the "acceptable daily intake" value (ADI; U.S. EPA, FIFRA) or the reference dose (RfD; U.S. EPA, TSCA). These doses (ADI or RfD) are defined as an estimate of the daily human exposure that is likely to be without appreciable risk of adverse effect. The uncertainty factors include 10 for extrapolation from animal models to humans and 10 for the diversity of human populations (to protect the most sensitive human subpopulations) for a total of 100. One more uncertainty factor (usually 10) is commonly used to cover possible lifetime exposures in humans versus the exposure period in the test animal species, especially for ADIs, for a total of 1000 (to protect against chronic, long-term exposure in the diet from food, water, etc. contaminated with pesticide residues). Exposure assessment is then performed for human exposure from all

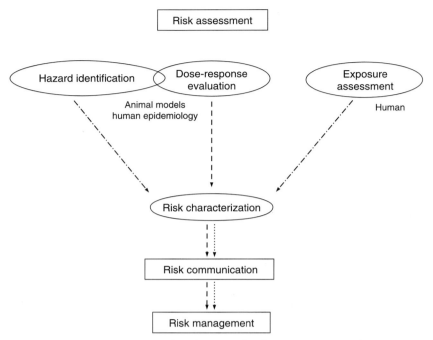

FIGURE T-14.   *U.S. National Risk Council risk assessment paradigm.*

sources. A value termed margin of exposure (MOE) is defined as the ratio of the NOAEL from the most sensitive test species to the human estimated exposure from all possible sources. If the MOE is very large, the risk to humans is perceived as low.

The U.S. EPA also employs a weight-of-evidence scheme to factor in the levels of confidence for the data and to emphasize human data over animal data.

—Rochelle W. Tyl

Related Topics

Developmental Toxicology
Dose–Response Relationship
Endocrine System
Levels of Effect in Toxicological Assessment
Neurotoxicology
Radiation
Reproductive System, Female
Reproductive System, Male
Risk Assessment, Human Health
Risk Characterization
Toxicity Testing, Developmental

Toxicology

All chemical or physical agents can affect living organisms. Some of these actions can be beneficial while others can be adverse to the well-being of the organism. "Toxicology," in its broadest sense, is the science that concerns itself with the adverse effects of chemical or physical agents on living organisms. There are two important elements in the definition of toxicology: the first is the aggressor substance (the chemical or physical agent), and the second is the target (the living organism affected). Since the target is a living organism, toxicology, as a consequence, is a biological science. The science of toxicology, however, draws heavily from the knowledge acquired in other sciences: chemistry, physics, physiology, biochemistry, pathology, pharmacology, immunology, genetics, molecular biology, mathematics, statistics, etc. Although the aggressor substance may be a physical agent (e.g., an electromagnetic field), the major concern of modern toxicology deals with chemicals (e.g., medicinal products, drugs of abuse, occupational chemicals, pesticides, industrial effluents, and hazardous wastes). Biological targets include humans as well as other species. In our egocentric fashion, we humans place much emphasis on ourselves as potential biological targets, but we must not forget that chemicals can have an important impact on other biological targets. While humans are considered a target of particular interest, other terrestrial and aquatic species are of considerable importance as potential biological targets. Toxicological problems worthy of societal concern are not limited only to those that affect human beings.

Toxicology is a very broad science. Toxicity studies are essential to the safe use of chemical substances in various aspects of our lives. In medicine, one must know the adverse effects of therapeutic agents as well as their beneficial utility. In the workplace, chemicals used as solvents, components of a process, or as intermediates must be handled safely. In agriculture, the safe use of pesticides, feed additives, or growth regulators as well as the problem of food residues are major considerations. Industrial effluents and their impact on the environment are societal preoccupations. Identification of chemically induced diseases and their prevention is an important public health undertaking. Regulatory controls essential for the safe use of chemicals require a broad and detailed understanding of toxicology.

Subdivisions of Toxicology

There are a number of subdivisions to the science of toxicology, and these vary according to the particular interests of the toxicologist concerned. No single classification system of categorization is entirely satisfactory. About 30 years ago, however, T. A. Loomis divided the science of toxicology into three major subdivisions: environmental, economic, and forensic. These subdivisions were in large part based on how humans would come in contact with potentially harmful chemicals. Generally, the scheme is still valid today.

Environmental toxicology, according to Loomis, is concerned primarily with the harmful effects of chemicals that are encountered by humans because of the presence of chemicals in the atmosphere, in the occupational setting, through recreational activities, or by ingestion as food residues. Environmental toxicology is the branch of toxicology that deals with the incidental exposure to chemicals that appear basically as contaminants of air, food, or water. This characterization of environmental toxicology is still appropriate today, although toxicologists are also interested in the impact of chemical substances on species native to various parts of the environment.

Economic toxicology, according to Loomis, deals with the potentially harmful effects of chemicals that are intentionally administered to living organisms for the purpose of achieving a specific beneficial effect. Here we find drugs developed for medicinal therapeutic purposes in human or veterinary medicine, chemicals developed for use as pesticides or insecticides, or substances designed as food additives. The term "economic" used by Loomis stems from the work of Adrian Albert, who coined the phrase "selective toxicity" to describe the use of chemicals by one species (humans) to eliminate an undesirable species, such as insects, in a specific circumstance. In this context, humans were called the "economic species" and the insect the "uneconomic species." Regardless of the terminology, economic as used by Loomis denotes that the potentially toxic chemical in question is being developed for some specific purpose, and we are interested in the undesirable effects that may accompany the beneficial effect.

Loomis categorizes forensic toxicology as the subdivision of toxicology that deals with the medical and legal aspects of the harmful effects of chemicals on humans. Therefore, here one finds those aspects of toxicology related to the diagnosis and treatment of chemical intoxications. The legal aspects of the subdivision pertain to cause–effect relationships between exposure to an aggressor agent and the adverse consequences observed in humans. We are very familiar with certain aspects of forensic toxicology, like the operation of a motor vehicle while under the influence of alcohol or the use of performance-enhancing drugs in sporting events. The detection and quantification of chemicals in biological fluids or tissues is a very important phase of forensic toxicology.

Scope and Activities of Toxicology

While Loomis' three-category scheme covers the broad use of applied toxicological information, it does little to denote the wide scope that represents the activities of toxicologists. In 1987, E. Hodgson and P. E. Levi formulated another set of characteristics in an attempt to cover the scope of the many activities encompassed by the discipline. They chose to organize toxicology into five broad categories, each with a number of sub-categories:

A. Mechanisms of toxic action: All events leading to adverse effects at the level of the organ, cell, or molecular function
   1. Biochemical toxicology (enzymes, receptors, molecular events, etc.)
   2. Behavioral toxicology (peripheral and central nervous system; endocrine system, etc.)
   3. Nutritional toxicology (influence of diet on the expression of toxicity)
   4. Carcinogenesis (chemical and biochemical events that lead to cancer)
   5. Teratogenesis (effects on embryonic and fetal developmental processes)
   6. Mutagenesis (effects on the genetic material and inheritance of these defects)
   7. Organ toxicity (effects at the level of organ function)

B. Measurement of toxicants and toxicity: These include the use of analytical chemistry, bioassays, and applied mathematics
   1. Analytical toxicology (identification and assay of toxic chemicals in biological material)
   2. Toxicity testing (use of living systems to estimate toxic effects)
   3. Toxicological pathology (branch of pathology dealing with the effects of toxic substances)
   4. Structure–activity study (relationship between chemical structure and toxicity)
   5. Biomathematics and statistics (determination of significance; risk estimates)
   6. Epidemiology (occurrence of toxicity)

C. Applied toxicology: Applications as they occur in the field
   1. Clinical toxicology (diagnosis and treatment of human poisoning)
   2. Veterinary toxicology (diagnosis and treatment of poisoning of animals)
   3. Forensic toxicology (medicolegal aspects of clinical poisonings, including analytical detection)
   4. Environmental toxicology (movement of toxicants in the environment and food chain; effects on various species)

5. Industrial toxicology (deals with the occupational environment)

D. Chemical use classes: Includes the toxicological aspects of the development of new chemicals for commercial use
   1. Agricultural chemicals (pesticides; targeted species)
   2. Clinical drugs (adverse effects of pharmaceutical agents)
   3. Drugs of abuse (chemicals taken for psychological effects that cause dependency and toxicity)
   4. Food additives (food preservatives; facilitate food processing)
   5. Industrial chemicals (solvents, degreasers, intermediates, etc.)
   6. Naturally occurring substances (includes phytotoxins, myocotoxins, and inorganic minerals)
   7. Combustion products (generated from fuels and other industrial chemicals)

E. Regulatory toxicology: Concerned with laws and regulations and their enforcement
   1. Legal aspects (governmental agencies)
   2. Risk assessment (definition of risk; risk/benefit considerations)

Early History of Toxicology

Obviously, the earliest humans gathered toxicological information through experience and trial-and-error. Animal venoms and plant poisons eventually were used for killing other animals or humans. In time, an art of poisoning developed, including the training of professionals. (Much of what follows is based on a chapter written by J. F. Borzelleca in the third edition of A. W. Hayes' *Principles and Methods of Toxicology*, 1994.) The interested reader is encouraged to consult this comprehensive work for more detail. Also, see Toxicology, History in this encyclopedia for additional insights.

Poisons, antidotes, and case histories are found in early Egyptian writings (Ebers Papyrus, circa 1500 BC); toxic agents were used by Egyptians in the administration of justice. Additional lists of poisons and antidotes appear in early Chinese (Shen Nung, circa 2700 BC) and Hindu (the Riga-Veda, circa 1500 BC) writings. The contributions of Hypocrates (circa 400 BC) and

Diocles (circa 350 BC) in ancient Greece described rational methods for the treatment of poisoning. Theophrastus (circa 350 BC) is said to be the first to recognize the adulturation of food. The Roman physician Celsus' treatise (circa 40 BC), *De Medicina*, continues Hippocratic teaching and contains a separate section on poisons and anidotes; *De Medicina* gained worldwide importance since it was the first medical work published (circa 1500 AD) after the invention of the printing press.

Important writings came from Avicenna of Persia (circa 1000) and Maimonides, court physician to Saladin and rabbi of Cairo (circa 1200). These texts exerted an enormous influence for nearly 500 years. Finally, Paracelsus (circa 1525), a Swiss physician, made the important declaration that "all things are poison . . . solely the dose determines that a thing is not a poison." This concept is the cornerstone of modern toxicology.

Toxicology was brought to other areas of human endeavor. An Italian physician, Ramazzini (circa 1700) is credited with bringing toxicology to the workplace with his works on health problems related to the occupational setting. He is considered the founder of occupational medicine. The application of analytical chemistry to food and drug safety was introduced by the works of Accum (circa 1800), *A Treatise on Adulterations to Food, and Culinary Poisons*. Finally, Orfila's (circa 1815) classic work on toxicology combined forensic and clinical toxicology with analytical chemistry; it is said to be the first book devoted entirely to toxicology. The father of experimental physiology, C. Bernard, used toxic chemicals (circa 1850) as laboratory tools to understand mechanisms involved in normal physiological processes. As such, he contributed greatly to the understanding of mechanisms of action of toxic substances.

Modern toxicology, which is more than 100 years old, is both an experimental and an applied science. It is a predictive science that has evolved remarkably since the time of Orfila, particularly in the past 60 years. Recent additions to the discipline include safety evaluation and risk assessment. In the next century, toxicology will be influenced greatly by expanding knowledge in immunology and genetics.

Further Reading

Albert, A. (1960). *Selective Toxicity*, 3rd ed. Wiley, New York.

Borzelleca, J. F. (1994). History of toxicology. In *Principles and Methods of Toxicology* (A. W. Hayes, Ed.), 3rd ed., pp. 1–17. Raven Press, New York.

Gallo, M. A., and Doull, J. (1991). History and scope of toxicology. In *Casarett and Doull's Toxicology* (M. O. Amdur, J. Doull, and C. D. Klaassen, Eds.), 4th ed., pp. 3–11. Pergamon, New York.

Hodgson, E., and Levi, P. E. (1987). *A Textbook of Modern Toxicology*. Elsevier, New York.

Loomis, T. A. (1968). *Essentials of Toxicology*. Lea & Febiger, Philadelphia.

Lu, F. C. (1991). *Basic Toxicology*, 2nd ed. Hemisphere, New York.

Ottoboni, M. A. (1984). *The Dose Makes the Poison*. Vincente, Berkeley, CA.

—*Gabriel L. Plaa*

Related Topics

Analytical Toxicology
Behavioral Toxicology
Developmental Toxicology
Ecological Toxicology
Environmental Toxicology
Forensic Toxicology
Information Resources in Toxicology
Molecular Toxicology
Neurotoxicology: Central and Peripheral
Occupational Toxicology
Radiation Toxicology
Toxicology, Education and Careers
Toxicology, History of

Toxicology, Education and Careers

Education

Individuals take formal courses in toxicology for several reasons. For example, some individuals might simply want to know more about the field of toxicology to make more informed decisions concerning the chemicals they use on a daily basis or to better evaluate reports in newspapers and magazines or on radio and television regarding potentially toxic materials. Other individuals, especially those with a basic interest in science, might desire formal training to pursue a career in toxicology. The objective of this entry is to provide more information to those individuals considering a career in toxicology and to provide guidance as to the educational training required. Much of the information presented in this entry was obtained from the book *Resource Guide for Careers in Toxicology* developed by the Society of Toxicology. A complete reference for this guide is presented in Further Reading. A copy of the *Resource Guide* can be obtained from the Society of Toxicology, and individuals serious about toxicology education are encouraged to obtain a copy.

Why Consider a Career in Toxicology?

Challenges

Chemicals are an essential component of the high standard of living we enjoy. The challenge to toxicologists is to ensure that we do not endanger our health or the environment with products or by-products of modern and comfortable living. Toxicology careers provide the excitement of science and research while contributing to the well-being of current and future generations. Few other careers offer such exciting and socially important challenges as protecting public health and the environment.

Opportunities

A wide variety of career opportunities exists in toxicology. For example, toxicologists participate in basic research using the most advanced techniques in molecular biology, chemistry, and biomedical sciences. Many toxicologists work in the chemical, pharmaceutical, and consumer products industries to test and ensure that their products and workplaces are safe. Some toxicologists work for state and federal governments to develop and enforce laws to ensure that chemicals are produced, used, and disposed of safely. Many toxicologists work in academic institutions, where they teach others about the safe use of chemicals and train future toxicologists. Clinical toxicologists are engaged in the development of new treatments and diagnostic (forensic) methods.

Attractive Salaries and Professional Advancement

The demand for well-trained toxicologists is high. Competitive salaries are available in a variety of employment sectors. Specific information on salaries for toxicologists in various workplace settings and with various levels of experience can be found in a survey published by the American College of Toxicology. A reference for this survey can be found in Further Reading.

What Do Toxicologists Do?

Research

Many toxicologists are involved in the acquisition of new knowledge of the mechanisms by which toxic substances produce their effects. Research is conducted in specialty areas in toxicology such as carcinogenesis, reproductive and developmental toxicology, neurotoxicology, immunotoxicology, and inhalation toxicology. Researchers use *in vitro* systems, laboratory animals, and occasionally humans to examine the cellular, biochemical, and molecular processes underlying toxic responses. Research opportunities exist for individuals employed in industry, academia, and government. Many commercial and nonprofit laboratories also provide interesting and challenging research opportunities.

Toxicology research can be considered basic or applied. In basic research, no immediate commercial application is expected, but the knowledge adds to our understanding of basic life processes and is eventually of great value in solving important problems. Examples of basic research are studies of how a particular enzyme involved in the detoxification of a chemical is regulated at the gene level or how a chemical affects the rate of cell division in cell culture.

In contrast, results from applied research are expected to yield direct social or commercial benefit. Examples of applied research are studies to identify new chemicals that selectively kill certain pests or studies to determine if a particular industrial process is responsible for a specific disease identified in a population of workers. Toxicologists working in applied areas also conduct studies directly related to determining whether or not a new chemical is toxic to laboratory animals and by inference toxic to people.

Product Safety Evaluation

Many industries employ toxicologists to assist in evaluating the safety of their products. For therapeutic drugs, food additives, cosmetics, agricultural chemicals, and other classes of chemicals, federal laws require that the manufacturer provide adequate testing of the product before it is released into commerce. Tests to determine if a chemical has the potential to cause cancer, birth defects, reproductive effects, neurological toxicity, or other adverse effects are commonly conducted by the manufacturer. Toxicologists involved in product safety evaluation have the responsibility to ensure that such tests are designed, conducted, and interpreted in a scientifically sound manner. Information from such studies is, in turn, reviewed by toxicologists in various regulatory agencies such as the Food and Drug Administration or the United States Environmental Protection Agency to ensure that the products will not present an unreasonable risk to human health or the environment.

Teaching

Toxicologists employed in colleges and universities are frequently involved in teaching toxicology. Because of the growing interest in the impact of chemicals on our society, most colleges and universities are developing new courses at both the undergraduate and graduate levels to provide students with a background in the science of toxicology. There are already many graduate programs in toxicology. A partial list of these programs is presented in Table T-26, and details of each of these programs are provided in the *Resource Guide to Careers in Toxicology*. Many other academic institutions that do not have a specific graduate program in toxicology employ toxicologists to participate in curriculum development and teaching in more basic programs such as chemistry and biology. Thus, opportunities exist to teach toxicology in small colleges as well as in major universities.

Public Service and Regulatory Affairs

The tremendous growth in public awareness of chemical hazards over the past two decades has resulted in passage of many laws governing the production, use, and disposal of chemicals. Many local, state, and federal agencies employ toxicologists to assist in the development and enforcement of these laws. An increasingly

TABLE T-26
Partial Listing of Academic Institutions That Support Training Programs in Toxicology at the Graduate Level (Grouped by Geographical Location)

Mid-atlantic
  Clemson University
  Duke University
  The George Washington University
  North Carolina State University
  University of Kentucky
  University of North Carolina at Chapel Hill
  University of South Carolina
  Vanderbilt University
  Virginia–Maryland Regional College of Veterinary Medicine
  West Virginia University
North central
  Indiana University School of Medicine
  Michigan State University
    Dept. of Pharmacology and Toxicology
    Institute for Environmental Toxicology
  The Ohio State University
  Purdue University
  University of Cincinnati College of Medicine
  University of Illinois at Urbana–Champaign
  University of Kansas Medical Center
  The University of Michigan
  University of Minnesota
    Department of Pharmacology
    Toxicology Program
  University of Nebraska
  University of Wisconsin–Madison
Northeast
  The Johns Hopkins University
    Department of Environmental Health Sciences
    Division of Toxicological Sciences
  Massachusetts Institute of Technology
  Northeastern University
  Rutgers University
  State University of New York at Buffalo
  The University of Connecticut
  University of Maryland
  University of Medicine and Dentistry of New Jersey
  University of Pittsburgh
  University of Rochester
  University of Toronto
Northwest
  Orgeon State University
  University of Washington
  Washington State University
South central
  Louisiana State University
    Institute for Environmental Studies
  Louisiana State University Medical Center
  Mississippi State University
  Texas A&M University
  University of Arkansas for Medical Sciences
  The University of Mississippi
    Department of Pharmacology
  The University of Mississippi Medical Center
  The University of Oklahoma
    Health Sciences Center

(continues)

TABLE T-26 (*Continued*)

University of Texas at Austin
The University of Texas
 Health Science Center of Houston
The University of Texas
 Medical Branch at Galveston
Southeast
 The University of Alabama at Birmingham
 University of Florida
 University of Georgia
  Interdepartmental Program
Southwest
 Colorado State University
 San Diego State University
 University of Arizona
  Center of Toxicology
  Department of Pharmacology and Toxicology
 University of California at Berkeley
 University of California, Davis
 University of California, Irvine
 University of California, Riverside
 University of California, San Francisco
 University of Colorado Health Sciences Center
 University of New Mexico
 University of Utah
 Utah State University

Note. From Society of Toxicology, *Resource Guide for Careers in Toxicology,* 3rd ed., Society of Toxicology, Reston, VA.

important area of toxicology is in public communication of chemical risks. Toxicologists employed by regulatory agencies may often be called on to examine the scientific basis for regulatory actions or to assist in communicating to the public why regulatory actions are or are not taken in particular situations. There are many private consulting firms with expanding expertise in toxicology that can now provide such services to local and state health departments, public utilities, and private industries. Thus, many employment opportunities in the private sector are available to the toxicologist interested in assisting public agencies and private industries in resolving many public health and environmental problems.

Clinical and Forensic Toxicology

Clinical toxicologists are health professionals concerned with disease caused by exposure to toxic agents. Generally, clinical toxicologists are physicians, pharmacologists (e.g., Pharm.D.), and veterinarians who receive specialized clinical training in toxicology. These individuals are engaged in the diagnosis and treatment of poisoned patients. Poisoning may result from acci-

dental, deliberate, environmental, or occupational exposure to a toxicant. Forensic toxicologists interact with clinical toxicologists to establish analytical chemical methods for the detection of toxic agents in tissue samples from poisoned patients. Research performed by many clinical and forensic toxicologists has led to the recognition of new chemical hazards and the development of novel therapies for poisoning. Clinical and forensic toxicologists may be found in academia, industry, and other places in which health professionals are employed.

Who Employs Toxicologists?

Industry

Chemical, pharmaceutical, and support industries are the major employers of toxicologists and account for about 37% of the toxicologists employed. Product development, product safety evaluation, and regulatory compliance generate a large job market for toxicologists. These industries often employ toxicologists trained at all levels of education, including those holding bachelor, master, and doctoral degrees. Many industries have their own research programs in product safety evaluation, whereas others may contract their work to specific research organizations.

Academia

Academic institutions account for about one-third of all employed toxicologists. The rapid growth in toxicology programs at academic institutions has generated a large market for toxicologists with doctoral-level training. Although most of these opportunities are in schools of medicine and public health in major universities, smaller colleges are beginning to employ toxicologists to teach toxicology in basic biology, chemistry, and engineering programs.

Government

The government employs about 15% of toxicologists. Although most government jobs are with federal regulatory agencies, many states employ toxicologists with a master's or doctoral degree. While most of the toxicologists employed by the federal government are involved in the development and enforcement of laws related to the toxicity of materials, a number of federal agencies employ toxicologists to conduct both basic and applied research in toxicology.

Consulting Firms

The professional service industry is a growing employer of toxicologists and currently accounts for about 8% of toxicologists. Many graduates of baccalaureate and master's programs in toxicology are finding employment with consulting firms. Individuals with doctoral training and several years of experience in applied toxicology may also find opportunities to direct projects and serve as team leaders or administrators. In the consulting field, experienced individuals provide professional guidance and advice to local public agencies, industries, and attorneys involved in problems with toxic chemicals. Consulting is a rapidly growing activity for the experienced toxicologist.

Research Institutes

Research foundations provide opportunities for research in toxicology to about 7% of the toxicologists. Numerous public and private research foundations employ toxicologists to conduct research on specific problems of industrial or public concern. Toxicologists at all levels of education might find employment with these research foundations.

Preparing for a Career in Toxicology

For those individuals who are in the midst of their college education, careful planning of undergraduate courses will enhance graduate education opportunities. For those who have already received an advanced degree such as a PhD, an MD, or a DVM in a biomedical science other than toxicology, careers can be focused toward toxicology through postdoctoral clinical or research training.

Undergraduate and Graduate Training

Planning

Depending on career aspirations, a bachelor's degree may not be sufficient for achieving career goals. Although there are some employment opportunities in toxicology for those with a bachelor's degree, the breadth of career choices and opportunities for advancement are much greater for those with postbaccalaureate degrees. Acceptance into graduate programs in toxicology generally requires a strong academic record and evidence of research and/or leadership abilities. Most graduate toxicology programs have specific prerequisites for admission. The primary requirement is a baccalaureate degree in a relevant field of study, such as biology, chemistry, environmental health, or other science-related field. Prerequisites often include advanced course work in chemistry, especially organic chemistry; at least 1 year of general biology; a year of college mathematics, usually including calculus; and general physics. Additional upper division courses in biochemistry and physiology will often increase competitive advantage for admissions. As the ability to be an effective communicator becomes an increasingly important skill for toxicologists, course work in scientific writing and public speaking is also useful. Performance on the Graduate Record Examination (GRE) is often one criterion evaluated by graduate admissions committees and the exam should be prepared for in advance. Many programs require GRE scores on both the General Test and on the Subject Test if it is given in an undergraduate major such as biology or biochemistry. The GRE should be taken at least 5 months prior to the time one plans to begin graduate study. Individual graduate programs should be consulted in advance to determine specific admission requirements.

In addition to a strong academic record, demonstration of basic laboratory research skills enhances the chance of admission. Laboratory courses in chemistry and biology are an important part of an undergraduate education and help develop research skills. Cooperative work-study programs enhance those skills by placing students in a research setting during the semester. Summer internships in a research laboratory are another approach to enhancing laboratory skills. Research internships provide interested undergraduate science majors with a stimulating research experience in toxicology. These internships are available in academic and industrial research laboratories across the country. More information on research internships in toxicology can be obtained by contacting the Society of Toxicology.

Selection of an Appropriate Toxicology Program

Identifying a graduate training program and mentor most appropriate for a particular individual requires some advance planning. First, individuals should establish a potential career plan. By considering the various subspecialties in toxicology, such as neurotoxicology, chemical carcinogenesis, teratology, inhalation toxicology, mathematical modeling, and risk assessment, a specific field of research that is of particular interest can be identified. Although such a choice early in the education process does not commit one to this direction, careful assessment helps in deciding which programs are most likely to meet various needs. Talking with toxicologists in local universities, industries, and governmental agencies is helpful in selecting a training program and in focusing the future career direction.

The admission requirements of the graduate program must be identified well in advance since these requirements must be met prior to the time of beginning the program. Requirements vary among programs and from the general requirements described previously. Details as to the specific requirements of toxicology graduate programs can be obtained by referring to the *Resource Guide to Careers in Toxicology*.

Financial Assistance

University financial assistance is often available through research and teaching assistantships, fellowships, traineeships, and grants. The National Institutes of Health, other federal institutions such as the Environmental Protection Agency, private foundations, and the Society of Toxicology are all potential sources of financial support.

Resource Guide to Careers in Toxicology

The *Resource Guide to Careers in Toxicology* contains descriptions of a large number of very diverse academic programs in toxicology located throughout the United States (Table T-26). Geographical considerations may be important to some individuals and may substantially limit the number of potential toxicology programs of interest to those individuals. Review of this document in the early stages of planning a career in toxicology is one of the most important steps that individuals can take in planning their toxicology education. The listing for each toxicology program starts with a brief general description of the university and the local environment and a description of the department where the toxicology program resides. The *Resource Guide* lists all the

faculty members who participate in the university program, including the year the faculty members obtained doctoral degrees and the institutions granting the degrees and their current area of research interest in toxicology.

Each program description outlines the prerequisites for admission into the specific graduate program. This is a very important section and provides clear direction as to the types of courses that need to be taken at the college level to be accepted into toxicology graduate programs. It is also useful to know at an early point other information required by the toxicology program. For example, most programs require official college transcripts, GRE scores, a letter of intent, and letters of recommendations. The letter of intent describes why the individual wants to be admitted into the graduate program and general career goals. Recommendations are generally from individuals who know the applicant on a professional or academic level. Examples of appropriate references are teachers, advisors, or employers.

The final section on each program description is the curriculum. The description of the curriculum includes the degrees that are offered by the particular toxicology program, the areas of specialization for the degrees, required courses, optional course work, and specific dissertation requirements. The graduate curriculum for a doctorate in toxicology often includes courses in biochemistry, physiology, anatomy, histology, pathology, pharmacology, and statistics. The academic program can also include areas such as analytical methods, carcinogenesis, mutagenesis, teratogenesis, comparative toxicology, molecular mechanisms of toxicology, and organ-specific toxicity. Some programs may also include course work in such fields as statistics, computer science, mathematical modeling, immunology, and toxicokinetics.

Because the doctorate in toxicology is a scholarly degree, the student is required to conduct a program of original research that extends over a period of 2 or more years. Part of this research requirement is the completion of a dissertation. This document is written by the student and includes an introduction or literature survey, a statement of the hypothesis underlying the dissertation, methods, results (including figures, graphs, and tables), and a discussion. By conducting original research, the student can understand and experience the application of observation and analysis to specific problems in toxicology. In keeping with the tradition of the doctorate degree, defense of the graduate research is expected. The final section of the program description provides the name and address of a person who can be contacted for more information and applications for admission and financial assistance.

Further Reading

Gad, S. C. (1996). Third triennial toxicology salary survey. *J. Am. Coll. Toxicol.* **15**, 83–89.

Society of Toxicology. *Resource Guide for Careers in Toxicology*, 3rd ed. Society of Toxicology, Reston, VA. (Available from the Society of Toxicology offices, 1767 Business Center Drive, Suite 302, Reston, VA 22090-5332: 703-438-3115; sothq@toxicology.org)

—*Michele A. Medinsky*

Related Topics

Academy of Toxicological Sciences
American Academy of Clinical Toxicology
American Board of Toxicology
American College of Medical Toxicology
American College of Toxicology
Chemical Industry Institute of Toxicology
European Society of Toxicology
Information Resources in toxicology
International Life Sciences Institute—North America
Society of Environmental Toxicology and Chemistry
Society of Toxicology
Toxicology
Toxicology, History of

Toxicology, History of

Introduction

Toxicology is concerned with the adverse effects of chemicals on living systems. When written records first began to appear, it was evident that early humans had already amassed a considerable amount of information on the deleterious effects of ingesting parts of certain plants and being bitten by certain animals (e.g., snakes), had speculated on the cause of these effects, and had developed some therapeutic measures.

Language was first expressed in written form beginning around 4500 BC in rather rudimentary ideograms and pictograms, used for inscriptions on monuments and as a means of recording information. Sumerian cuneiform and Egyptian hieroglyphics are early examples. In the three or four millennia BC, some medical works were produced which revealed excellent descriptions of various diseases, proposed explanations as to cause, and offered a wide range of treatments. Some of these writings are extant in complete form, others in fragments only. A famous one is the Ebers papyrus of about 1550 BC, which was deciphered in 1874 AD by the German Egyptologist Georg Moritz Ebers. He was able to decipher this thanks to the work of Jean-François Champollion, who had announced his deciphering of the hieratic and demotic hieroglyphics of the Rosetta stone in 1822. The Ebers papyrus, a continuous scroll about 1 ft wide and 65 ft long, now held in Berlin, is a compendium of some 800 formulas and prescriptions and reveals a remarkable familiarity with dozens of substances and materials with toxic properties. These include opium, aloes, aconite, turpentine, cumin, squill, mandrake, and many others of plant origin. Mineral substances include lead, magnesia, copper sulfate, iron, sodium chloride and carbonate, mercuric sulfide, and potassium nitrate. There are thus early associations between toxicology and medicine and also with occupational activities such as mining. Drug toxicology and industrial toxicology are among the many specialized areas of modern toxicology that can trace their origins back for three or four millennia.

The Classical Period

In the millennium of the classical period, from approximately 700 BC to 300 AD, knowledge of poisonous substances grew considerably and attempts at a classification of poisons were made. The "father of medicine," Hippocrates (ca. 460–375 BC) is most noted for his introduction of a rational approach to medicine, attempting to rid it of its superstitious trappings, and for the oath which bears his name. However, he paid some attention to medicinal preparations and named some 400 components to be used in them, including some toxic alkaloids. He also described the adverse effects of lead exposure on miners.

Socrates (469–399 BC), a famous contemporary of Hippocrates, voluntarily accepted self-immolation by drinking a cup of hemlock, as described in Plato's *Phaedra*. This was an aqueous or vinous extract of the poison hemlock, *Conium maculatum;* its lethal constituent is coniine.

Theophrastus (ca. 370–286 BC), the "father of botany," succeeded Aristotle as head of the Peripatetic school in Athens. His work on the history of plants identifies many of the poisonous plants known at that time. The use of poisons, occasionally for suicide and often for politically motivated murder, had been going on for centuries and has not completely disappeared even today. Cleopatra provides us with an example of both uses. She poisoned her younger brother, Ptolemy XIII, who was also her husband, in order to remove him from the throne of Egypt in 44 BC. Later, in 30 BC, Cleopatra, her fortunes declining, utilized the poisonous bite of an asp to commit suicide to avoid becoming a Roman slave.

The only remaining works of the prolific Pliny the Elder (23–79 AD) are the 37 books of his *Historia naturalis.* This encyclopedic but not entirely accurate work has sections on the medicinal uses of plants and curative preparations derived from the animal kingdom. He mentions that cinnabar (mercuric sulfide) refiners used "bladders" over their faces to reduce the inhalation of dust, showing an awareness of the toxic hazard of the dust. He was also cognizant of the toxic potential of contaminants in food and medications. There is a well-known story of Pliny dissolving Cleopatra's pearls in vinegar to show that it was possible to consume a million sesterces, about $35,000–40,000, at a single banquet. His scientific curiosity was his downfall because he was asphyxiated while investigating the crater of Vesuvius, which erupted in 79 AD burying Pompeii and Herculaneum. It is not clear whether he died from acute hydrogen sulfide poisoning or from the dust overload phenomenon.

Dioscorides (ca. 40–90 AD) is the author of *De Materia Medica,* which described several hundred plants and plant principles. He developed a system of classification of poisons based on their source—plant, animal, or mineral—which remained a standard for centuries. Like many other early physicians, Dioscorides was interested in the treatment of disease and is credited with

the formulation of medicated wines. Polypharmacy, the use of many ingredients in a prescription, was common, with some preparations containing 60–70 components. Dioscorides employed emetics for ingested poisons and described the use of clays to treat skin disorders such as erysipelas.

The last medical writer of significance in the classical period was Galen (130–201 AD), a Greek physician who spent much of his career in Rome as physician to a succession of emperors. Galen surpassed Pliny the Elder in his volume of writings, which comprise 130 books on medicine, 83 of which are extant, and 125 books on law, grammar, mathematics, and philosophy. His contribution to toxicology was minimal. He gave good descriptions of the anatomy and structure of the animal organs but was frequently in error when describing function. Human dissection was forbidden at this time, hence the use of animals. This proscription against human dissection had been in effect in Egypt and the process of mummification was carried out not by priests, who might have left anatomical descriptions, but by technicians of low repute.

In these ages, the practice of medicine, including especially the formulation of remedies, was surrounded by mysticism and charlatanism. People seemed to feel the need for authorities in whom they could place unthinking trust. Thus, the writings of Pliny, Galen, Aristotle, and Ptolemy, the Greek astronomer, were accepted as gospel for centuries. To criticize these authorities was tantamount to sacrilege. During the Dark Ages, scientific inquiry was stifled by the blind acceptance of early authorities, despite obvious errors in their pronouncements. In the Arab world, not dominated by Western attitudes, some of the statements of Galen were contradicted by such physicians as Avensoar (1096?–1162 AD) and Averroes (1126–1198 AD). Their contemporary, Moses ben Maimon or Maimonides (1135–1204 AD), wrote a text on poisons and their antidotes. While some of his recommendations have merit, such as drinking oily liquids like cream to delay absorption of ingested poisons, others reveal the superstitions that prevailed in those days. He advised that the most efficacious antidote for snakebite was an emerald held in the mouth or applied to the stomach.

A recapitulation of toxicologic knowledge to this point in time shows that the toxic properties of various plant, animal, and mineral substances were well-known, that some attempts at classification of poisons by source had been made, and that a great variety of polypharmacal formulations were recommended that were of doubtful efficacy if not actually harmful. In those times, appeal to ancient authorities and omnipresent superstition prevailed.

The chief use that seems to have been made of the knowledge of poisons was to eliminate rivals for power and position or to remove unwanted husbands. It has been said that poison is a woman's weapon and the record lends some support to this allegation. The list of female poisoners in history is long. Cleopatra has been already mentioned. In Rome, Locusta was Nero's specialist in poisons and, in the Middle Ages in Italy, Toffana, Hieronyma Spara, and Lucretia Borgia were active. Catherine de Medici practiced in France, as did the Marchioness de Branvilliers and the notorious La Voisine, Catherine Deshayes.

The Renaissance

The state of affairs outlined previously continued as the Roman empire declined and the European world suffered through the Dark Ages, the intervening medieval period, and gradually emerged from its scientific hiatus with the coming of the Renaissance, which began in Italy in the fourteenth century and burgeoned in the fifteenth and sixteenth centuries. A difference in the quality of toxicologic observations and speculations becomes increasingly evident. Observations become more factual and there is an increase in the objectivity of investigators as the following account will show. What today is called industrial or occupational toxicology provides the first examples of this more enlightened approach.

Ulrich Ellbog, or Ellenbog (1459–1499), a German physician, published in 1473 a leaflet warning goldsmiths about the harmful effects of inhaling the fumes and vapors of lead, mercury, and antimony compounds. He stressed the use of preventive measures—a modern emphasis.

Philippus Aureolus Paracelsus, also known as Theophrastus Bombastus von Hohenheim (1493–1541), was a Swiss physician who studied the diseases of miners. His famous work, *Von der Bergsucht und anderes Bergkrankheiten,* was published posthumously. Diseases associated with the mining, washing, and smelting of lead, arsenic, copper, iron, zinc, mercury, silver, and gold ores were described. It was Paracelsus who enunciated the well-known principle that it is the dose that makes the poison, with its implication of a threshold below which there will be no toxic manifestations. Paracelsus was contemptuous of the medical orthodoxy of the time and earned the hearty dislike of more conventional physicians. His role in the advancement of industrial toxicology is pivotal, spanning the earlier, superstition-ridden practices and ushering in a more rational and objective approach to this specialized area of toxicology. His quest for truth impressed the English poet Robert Browning who, in an early work in 1835, described his life in a lengthy poem titled "Paracelsus."

A contemporary of Paracelsus, Georg Bauer or Georgius Agricola (1494–1555), was the author of *De Re Metallica,* which also did not appear until after the author's death. Agricola described the diseases prevalent among miners, both in the mines and above ground where smelting of the ores of lead, iron, tin, antimony, and bismuth was performed. His descriptions of diseases, while not too precise, probably correspond to tuberculosis and carcinoma of the lung and silicosis. It is interesting to note that Herbert Hoover, the 31st president of the United States, published with his wife an English translation of *De Re Metallica* in 1912.

An important figure of this time was Francis Bacon, Viscount St. Albans (1561–1628), author of *The Advancement of Learning* (1605) and the *Novum Organum* (1620). This new instrument of thinking was the inductive method of reasoning which characterizes modern sciences and rejects the a priori approach of the early scholastics. He warned against formulating theories on insufficient data and emphasized the need for thorough investigation of phenomena. The application of his precepts are to be found in subsequent investigations.

Many of the writers considered to this point were concerned with the toxic hazards that accompanied industrial activities, especially the mining and processing of ores. However, other toxic threats had been recognized in England where coal was used for the heating of houses and other buildings. As early as the thirteenth century, complaints about the smoke and odor from coal burning were recorded. John Evelyn (1620–1706), the diarist, wrote his *Fumifugium or the Inconveniencie of the Aer and Smoak of London Dissipated* in 1661. Although there was no exact knowledge of the active constituents of the smoke, its adverse effects on health were recognized. It was not until the air pollution disaster of 1952 in London, when some

4000 excess deaths were recorded, that fully preventive actions were taken.

Toxicology is a multidisciplinary science and its progress has often depended on advances in such fields as anatomy, pathology, chemistry, and biochemistry. The errors in Galen's statements on anatomy were largely rectified by Andreas Vesalius (1514–1564), who personally carried out dissections on human cadavers. His great work, *De Humanis Corporis Fabrica*, appeared in 1543. The erroneous ideas about the circulation of the blood were discarded when William Harvey (1578–1657) published his *Exercitatio Anatomica de Motu Cordis et Sanguinio* in 1628. However, advances in other sciences were not so early. It might be noted that the famous English chemist, Joseph Priestly (1733–1804), who studied the effects on animals of various gases such as nitric oxide, sulfur dioxide, and ammonia, died a staunch phlogistonist (one who believes in phlogiston, a nonexistent chemical once thought to be released during combustion). Quantitative analytical chemistry developed in the eighteenth and nineteenth centuries and led to the introduction of tests for the detection of arsenic, the perennial favorite of poisoners, and for the detection of several metals. Even the first organic compound to be synthesized, urea, had to await the skill of Friedrich Wöhler (1800–1882), who prepared it from cyanic acid in 1828.

Another important contributor to the field of industrial toxicology was the Italian Bernardino Ramazzini (1633–1714), dubbed the "father of occupational medicine." His interest in the field was aroused by observing a poor laborer who had the unenviable job of cleaning cesspools, spending hours each day in these noisome pits. Ramazzini noted his red and rheumy eyes and learned from him that those who did this work suffered from photophobia and could ultimately go blind. This would appear to be the result of chronic exposure to low levels of hydrogen sulfide, which is a mucous membrane irritant. During his career, Ramazzini studied the diseases seen in miners, masons, metal workers, potters, painters, and many other trades. He wrote a score of books of which the most noteworthy was *De Morbis Artificum Diatriba,* published in 1700. He advocated various therapeutic measures for the treatment of occupational diseases rather than controlling them by the utilization of preventive methods. He suggested that physicians should add one more question to those that Hippocrates had advised when examining a patient—that is, What is your occupation?

The English surgeon, Percival Pott (1713–1788), was the first to demonstrate clearly a cause–effect relationship between an occupation and the induction of cancer, namely, the scrotal cancer suffered by chimney sweeps described in his *Chirugical Works* (1775). Pott considered that soot was the agent responsible, a view strongly supported by Benjamin Bell (1794). However, many years would pass before polyaromatic hydrocarbons, particularly benz(*a*)pyrene, would be identified.

Mattieu J. B. Orfila (1787–1853) has been called the "father of modern toxicology." He was a Spanish physician whose main work was performed in Paris. He conducted numerous quantitative toxicological studies on experimental animals, relating the dose of a toxic agent to the biological effects that ensued. This aspect of his work was described in the celebrated text, *Toxicologie Générale,* which appeared in 1814/1815. Orfila also made major contributions to the special area of forensic toxicology. He applied the methods of analytical chemistry to tissues from autopsy material and exhumed bodies to detect the presence of poisons so that accidental or deliberate poisoning now became detectable. His investigations in this field were recorded in *Leçons de Médecine Légale* (1823) and *Traité des Exhumations Juridiques* (1830). Orfila also critically examined the procedures used at this time in the treatment of poisoning, many of which were ineffective. Many of his own recommendations concerning the elimination of poisons from the body and the use of artificial respiration remain valid today. Orfila recognized toxicology as a discipline in its own right and his writing stimulated others in the nineteenth century, such as Thackrah (1831), Christison (1845), and Kobert (1893), to produce textbooks of toxicology.

Charles Turner Thackrah (1795–1833) deserves special mention. He was a physician who practiced in Leeds, England, where the worst features of industrial activities which followed the Industrial Revolution were in evidence. Thackrah's patients were the workers in over 100 different trades being carried on in Leeds and its vicinity. Thackrah published a book in 1831 which gave in detail his observations on chronic lead poisoning in painters and potters, the incidence of lung diseases, especially tuberculosis, in miners and grinders, and other occupational diseases. He also wrote about the appalling conditions under which children worked and the postural defects that followed long hours of work in unnatural positions of the body. A series of acts of Parliament in 1802, 1819, 1825, and 1833 were passed to correct some of the worst abuses but their provisions were poorly enforced. However, the Act of 1833 resulted in the establishment of the Factory Inspectorate and was the forerunner of today's laws designed to control toxic hazards in the workplace, which are to be found in every modern industrial society. Much credit for this development must go to Thackrah, whose book is considered to be superior to Ramazzini's in its comprehensiveness, detailed clinical observations, and constructive proposals for improvements.

The Scope of Modern Toxicology

The history of toxicology up to the early nineteenth century has been recounted in terms of the contributions of gifted individuals, but by the middle and latter half of the 1800s, the day of the single gifted individual began to give way to team efforts in which, for instance, a professor with a group of students would study toxicological problems. Monographs, such as that of H. Eulenberg in 1865 on toxic gases, began to appear the year, incidentally, when F. A. Kekulé had his inspiration about the cyclic structure of the benzene molecule. By the end of the nineteenth century and into the twentieth century to the present time, the number of toxicologists and their productivity has increased almost exponentially. The toxicological characteristics of every class of chemical were explored and dozens of textbooks, monographs, and journals appeared. Special areas of toxicology have developed in addition to those already noted (i.e., industrial, drug, air pollution, and forensic toxicology). It is too early to assess who, among the many thousands of contributors to toxicological progress world-wide, will be recognized by history as having made seminal discoveries and advances, but a survey of the topics that have concerned toxicologists in this century may be undertaken.

Initially, many studies were of an acute nature—that is, the response of test animals to single graded doses of the toxic agent at levels in the lethal range were examined. Questions as to the species to employ and the route of administration arose. Small rodents, rats and mice, were frequently employed; guinea pigs, hamsters, and rabbits were also utilized. Larger species, dogs and non-human primates, especially macaques, are used in many studies, and even larger animals (e.g., pigs, sheep, and donkeys) have been employed in special studies. Much effort has gone into improving the quality and uniformity of experimental animals and today pure-bred strains and even genetically altered species are available. The investigator can administer the toxicant by any one of several routes: orally, by injection into veins and arteries, into the peritoneal cavity, intrathecally, into muscle, by inhalation or instillation, or by application to the skin and mucous membranes. The choice of route usually depends on the route of entry in the body that is encountered in real-life situations. Thus, a food additive will be administered orally since food is ingested.

Since there are many circumstances in which the subject is exposed daily to a toxicant, types of repeated-dose studies were introduced. These are called subacute, indicating that the dose is less than that used in an acute study, or subchronic, implying a duration of study shorter than that employed in a chronic investigation (discussed later). In a repeated-dose trial, nonlethal doses of a toxicant are given daily for periods as short as 1 or 2 weeks and occasionally for as long as 3–6 months. From such studies, some information on the cumulative action of the agent may be gained.

Because of the concern about the exposure of subjects in occupations and in urban environments where quite low concentrations of toxicants may be encountered for many years, toxicologists conduct chronic trials in animals. If the study is performed with rats or mice, the administration will continue daily for 2 years or even somewhat longer, all or nearly all of the life span of the animal. If larger species are employed, administration may be continued much longer. Some studies in monkeys have gone on for 10 years.

In evaluating the effects of the toxicant in the assays described previously, a great many variables may be examined. Changes in body weight, food consumption, and the general behavior of the animals are recorded. Hematological and various clinical biochemical indices may be included. Organ or system functional tests may be conducted. It is usual at termination of the trial to sacrifice the animals, perform a gross necropsy, and take tissues for microscopic examination using histological methods. To some extent, the design and conduct of tests have led to the development of standardized protocols, but these may be modified because of some special anticipated action of the toxicant or because of a particular interest of the investigator. In many cases, investigations in applied toxicology are conducted under contract in consulting laboratories;

there are a number of these in North America, Japan, several European countries, and elsewhere.

Despite the wide range of effects that may be detected by the battery of tests listed previously, toxicologists have found it necessary to add to the list as new effects and new concerns have assumed importance. Some substances when applied to the skin or inhaled may elicit a sensitization response. This has led to the development of a special area in toxicology, immunotoxicology. The thalidomide tragedy of 1961 led to the addition of experimental designs to detect teratogenic potential and these were broadened to include other reproductive effects. The specific action of substances like tri-*o*-cresyl phosphate and *n*-hexane on the nervous system has opened up the field of neurotoxicity. Many compounds, such as carbon monoxide and certain mercurials, are noted for their effects on behavior and this has stimulated the development of behavioral toxicology. The growing recognition that many toxicants exert their effects by interfering with the molecular mechanisms that control gene expression has elucidated certain aspects of toxic action. Thus, toxicology has entered this frontier area of research already productively exploited in modern biochemistry and pharmacology.

One large area of toxicological endeavor is connected with the induction of cancer by chemical agents. Conventional cancer bioassays involve chronic exposure of animals with particular attention being directed to the overt appearance of lumps and nodules and to the histological observations on tissues. In order to establish statistically the significance of the incidence of cancerous lesions, large numbers of animals are needed in each test group. These whole animal studies are expensive and time-consuming so efforts have been made to gain evidence of carcinogenic potential by short-term tests in simpler systems. Microbial cultures and plant and animal tissue preparations are available which detect mutagenic and genotoxic effects and thus provide indications of the capacity to induce cancer. The opposition of some people to the use of animals in toxicology with its attendant suffering, however minimal, and sacrificing at the end has added further impetus to the search for non-animal test systems, not only for cancer bioassay but also for other biological endpoints. Many such systems are under active development and validation.

A principal objective of toxicological investigations is to be able to extrapolate accurately the findings to humans. It is a process fraught with difficulty. The prediction of qualitative effects can be made with some confidence, especially when the same effect is elicited in several laboratory species. Thus, if a chlorinated hydrocarbon causes liver damage in rats, dogs, and monkeys, it is highly likely that the compound will prove to be hepatotoxic in humans. The problems arise when an attempt is made to quantify the response. Translation of the administered dose to the human on a weight or body surface basis has not proven to be very reliable. It is believed that knowledge of the amount or concentration at the site of action, the target organ or tissue, would permit a better quantitative prediction. To this end, toxicokinetic studies are performed. The absorption, distribution, and metabolism of the toxic compound are taken into account so that the conditions that exist at the target organ can be ascertained. The use of physiologically based pharmacokinetic models has resulted in improvement in the quantitative aspects of prediction. With increased confidence in extrapolating the results of animal studies to humans, attention has been directed to risk assessment in which mathematical models are used to estimate the proportion of a human population exposed to a given toxic agent that may experience the toxic effect associated with the agent. Of particular concern is the possible induction of cancer in humans from carcinogenic chemicals. Such substances are found in industrial operations, in polluted air, water, and soil, and in food treated with additives or pesticides; these substances may be a concomitant of certain features of life-style, for example, tobacco smoking.

Ever since the first writings on toxic substances began to appear, there has been a continuing interest in finding remedies or treatments for poisoning. Lacking an understanding of the basic mode of action of toxicants at the biochemical level, early physicians proposed formulations that were either useless or, if they had any merit, their discovery was largely fortuitous. In this century, a rational approach to experimental therapeutics has developed as the fundamental mechanism of action of various poisons has been elucidated. Notable success has been achieved in the treatment of poisoning caused by arsenic, carbon monoxide, cyanides, and other toxicants.

The widespread interest in toxicological hazards in the twentieth century has resulted in a spate of publications, the establishment of institutes devoted to pure and applied toxicological investigations, and the formation of national societies of toxicology in more than two dozen developed and emerging countries on all continents. Preparatory discussions in Helsinki, Finland, in 1975 culminated in the creation of the International Union of Toxicology (IUTOX), in Toronto, Canada, in 1979. IUTOX meets in various cities around the world every 3 years. Today national governments have instituted laws and regulations designed to identify and control toxic hazards in the environment. Permissible levels and limits for the concentration of toxic substances have been established for industrial chemicals and food contaminants and for the maintenance of air and water quality.

Toxicology today is a science in and of itself, but it draws without reservation from the physical and biological sciences and mathematics to serve its purposes. Training in toxicology as a recognized discipline is offered today at both the undergraduate and graduate levels in many universities throughout the world. Although new chemicals produced by humans will require toxicological examination, enough of the basic knowledge of the toxic properties of most known substances is at hand. In the future, efforts will be directed toward increasing the public's awareness and knowledge of toxic hazards and, as always, continuing vigilance in enforcing the regulations to control these hazards.

Further Reading

Adams, F. (Trans.) (1939). *The Genuine Works of Hippocrates*. Williams & Wilkins, Baltimore, MD.

Amdur, M. O., Doull, J., and Klaasen, C. D. (Eds.) (1991). *Casarret and Doull's Toxicology*, 4th ed. Pergamon, New York.

Barnard, C. (Trans.) (1932). Verhütung and Rath für giftige Dampffe der Metal (1473). U. Ellbog (Ellenbog). *Lancet.*

Beeson, B. B. (1930). Orfila—Pioneer toxicologist. *Ann. Med. Hist.* **2**, 68–70.

Bryan, C. P. (1930). *The Papyrus Ebers*. Geoffrey Bales, London.

Christison, R. (1845). *A Treatise on Poisons*, 4th ed. Barrington & Howell, Philadelphia.

Clendening, L. (1942). *Source Book of Medical History*. Dover, New York.

Debus, A. G. (1977). *The Chemical Philosophy: Paracelsian Science and Medicine in the Sixteenth and Seventeenth Centuries*. Science History, New York.

Dierbach, J. H. (1969). *Die Arzneimittel des Hippokrates, oder Versuch einer systematischen Aufzahlung der in allen hippokratischen Schriften vorkommenden Medikamenten*. H. Olms, Hildesheim

Dubois, K. P., and Geiling, E. M. K. (1959). *Textbook of Toxicology*. Oxford Univ. Press, New York.

Eulenberg, H. (1865). *Die Lehre von den schädlichen und giftigen Gasen*. Friedrich Vieweg und Sohn, Braunschweig.

Evelyn, J. (1959). *Fumifugium, or the Inconveniencie of the Aer and Smoak of London Dissipated. The Diary of John Evelyn* (E. S. Baer, Ed.). Oxford Univ. Press, London.

Fabricius, C. (1972). *Galens Exzerpte aus alteren Pharmakologen*. De Gruyter, New York.

Gilman, A. G., Rall, T. W., Nies, A. S., and Taylor, P. (Eds.) (1990). *The Pharmacological Basis of Therapeutics*, 8th ed. Pergamon, New York.

Gunther, R. T. (1934). *The Greek Herbal of Dioscorides*. Oxford Univ. Press, New York.

Haber, F. (1924). *Fünf Vortrage aus den Jahren 1920 to 1923*. Springer, Berlin.

Hawes, L., Clarke, W., and Collins, R. (Eds.) (1775). *The Chirurgical Works of Percival Pott. F. R. S.*, London.

Hayes, A. W. (Ed.) (1989). *Principles and Methods of Toxicology*, 2nd ed. Raven Press, New York.

Hodgson, E., Mailman, R. B., and Chambers, J. E. (1988). *Dictionary of Toxicology*. Van Nostrand Reinhold, New York.

Kobert, R. (1893). *Lehrbuch der Intoxikationen*. Enke, Stuttgart.

Lambert, S. W., and Goodwin, G. M. (1929). *Minutemen of Life: The Story of the Great Leaders in Medicine from Hippocrates Down to the Present Day*. Grossett & Dunlap, New York.

Levey, M. (1966). Medieval arabic toxicology. *Trans. Am. Philos. Soc.* **56** (7).

Levey, M. (1973). *Early Arabic Pharmacology: An Introduction Based on Ancient and Medieval Sources*. Brill, Leiden.

Lewin, L. (1920). *Die Gifte in der Weltgesichte. Toxikologische allgemeinverständliche Untersuchungen der historischen Quellen*. Springer, Berlin.

Lewin, L. (1929). *Gifte und Vergiftungen*. Stilke, Berlin.

Lloyd, J. U. (1921). *Origin and History of All the Pharmacopeial Vegetable Drugs, Chemicals and Preparations with Bibliography*. Caxton Press, Cincinnati, OH.

Munter, S. (Ed.) (1966). *Treatise on Poisons and Their Antidotes*, Vol. II of the Medical Writings of Moses Maimonides. Lippincott, Philadelphia.

Orfila, M. J. B. (1814/1815). *Traité des Poisons Tirés des Rènges Minéral, Végétal et Animal, ou Toxicologie Générale Considerée sous les Rapports de la Physiologie, de la pathologie et de la Médecine Légale*. Crochard, Paris.

Orfila, M. J. B. (1818). *Secours à Donner aux Personnes Empoisonées et Asphyxiées*. Feugeroy, Paris.

Orfila, M. J. B. (1821). *Leçons de Médecine Légale*. Bechet Jeune, Paris.

Orfila, M. J. B., and Lesueuer, O. (1831). *Traité des Exhumations Juridiques*. Bechet Jeune, Paris.

Ottoboni, M. A. (1991). *The Dose Makes the Poison: A Plain Language Guide to Toxicology*. Van Nostrand Reinhold, New York.

Pagel, W. (1958). *Paracelsus: An Introduction to Philosophical Medicine in the Era of the Renaissance.* Karger, New York.

Paracelsus (Theophrastus ex Hohenheim Eremita) (1567). *Von der Bergsucht und anderes Bergkrankheiten.* Dillingen, Germany.

Pollak, K. (1963). *Die Junger des Hippokrates; der Weg des Arztes durch sechs Jahrtausende.* Econ-Verlag, Dusseldorf.

Ramazzini, B. (W. C. Wright, Trans.) (1940). *De Morbis Artificium Diatriba* (1713). Univ. of Chicago Press, Chicago.

Riddle, J. M. (1985). *Dioscorides on Pharmacy and Medicine.* Univ. of Texas Press, Austin.

Thompson, C. J. S. (1931). *Poisons and Poisoners. With Historical Accounts of Some Famous Mysteries in Ancient and Modern Times.* Shaylor, London.

Walton, J., Beeson, P. B., and Scott, R. B. (Eds.) (1986). *The Oxford Companion to Medicine.* Oxford Univ. Press, London.

Williams, R. T. (1959). *Detoxication Mechanisms*, 2nd ed. Chapman & Hall, London.

—*Harold MacFarland*

Related Topics

Analytical Toxicology
Behavioral Toxicology
Developmental Toxicology
Ecological Toxicology
Environmental Toxicology
Forensic Toxicology
International Union of Toxicology
Metals
Molecular Toxicology
Neurotoxicology: Central and Peripheral
Occupational Toxicology
Pollution, Air
Radiation Toxicology
Toxicology
Toxicology, Education and Careers

Toxic Substances Control Act

◆ TITLE: TSCA
◆ AGENCY: U.S. EPA

◆ YEAR PASSED: 1976
◆ GROUPS REGULATED: Manufacturers and marketers of chemicals in the United States

Synopsis of Law

The Toxic Substance Control Act (TSCA) represents Congress' most ambitious effort to control the hazards of chemicals in commercial production. TSCA covers all chemical substances manufactured or processed in or imported into the United States, except for substances already regulated under other laws. A chemical substance is defined broadly as "any organic or inorganic substance of a particular molecular identity."

TSCA gives U.S. EPA three main powers. The agency is empowered to restrict, including banning the manufacture, processing, distribution, use, or disposal of a chemical substance when there is a reasonable basis to conclude any such activity poses an "unreasonable risk of injury to health or environment." In determining whether a chemical substance presents an unreasonable risk, the agency is instructed (TSCA Section 6) to consider:

> The effects of such substance or mixture on the health and the magnitude of the exposure of human beings to such substance or mixture; the effects of such substance or mixture on the environment and the magnitude of the exposure of the environment to such substance or mixture; the benefits of such substance for various uses and the availability of substitutes; and the reasonably ascertainable economic consequences of the rule, after consideration of the effect on the national economy, small business, technological innovation, the environment, and public health.

U.S. EPA also must consider any rule's positive impact on the development and use of substitutes as well as its negative impact on manufacturers or processors of the chemical and weigh the economic savings to society resulting from reduction of the risk.

TSCA's trade-off approach to regulation has proved a major challenge to U.S. EPA. The agency has not used its authority often under TSCA Section 6. In 1989, it issued a comprehensive rule prohibiting the future manufacture, importation, processing, and distribution of almost all products containing asbestos. Despite the documented hazards of asbestos, a court overturned the rule; the ban failed to satisfy the statutory requirement that U.S. EPA promulgate the "least burdensome" regulation required to protect the environment (*Corrosion Proof Fittings v. Environmental Protection*

Agency, 1991). The ban continues to govern products that were not in manufacture as of mid-1989, but U.S. EPA has not attempted to issue a new rule regulating existing asbestos-containing products.

If U.S. EPA suspects that a chemical may pose an unreasonable risk but lacks sufficient data to take action, TSCA empowers the agency to require testing to develop the necessary data. Similarly, U.S. EPA may order testing if the chemical will be produced in substantial quantities that may result in significant human exposure whose effects cannot be predicted on the basis of existing data. In either case, U.S. EPA must consider the "relative costs of the various test protocols and methodologies" and the "reasonably foreseeable availability of the facilities and personnel" needed to perform the tests (TSCA Section 4).

Finally, to enable U.S. EPA to evaluate chemicals before humans are exposed, TSCA requires the manufacturer of a new chemical substance to notify the agency 90 days prior to production or distribution [TSCA Section 5(a)(1)]. The manufacturer's or distributor's notice must include any health effects data it possesses. However, U.S. EPA is not empowered to require that manufacturers routinely conduct testing of all new chemicals to permit an evaluation of their risks; Congress declined to confer that kind of pre-market approval authority that the U.S. FDA exercises for drugs and food additives and U.S. EPA exercises for pesticides. This has proven to be a major limitation on the value of the premarket notification process. At least 50% of submissions present no safety data and 90% have at most an Ames mutagenicity study and a lethality study.

U.S. Congress enacted TSCA in 1976 to give U.S. EPA broad control over chemicals not regulated by other statutes, largely industrial chemicals. Under TSCA, importers and manufacturers of chemicals have a wide variety of requirements for reporting and conduct. TSCA gives U.S. EPA the authority to require manufacturers and importers of chemicals to test chemicals, to control the way chemicals are manufactured or used, and to report certain activities to U.S. EPA.

The law is divided into a number of sections. Section 5 is largely devoted to new chemicals, while other sections of TSCA deal with existing chemicals. The main sections of the law are listed in Table T-27.

Currently, U.S. EPA's office of Pollution Prevention and Toxic Substances enforces TSCA. U.S. EPA is divided into a number of offices, each of which has re-

TABLE T-27
Major Sections of the Toxic Substances Control Act

Section no.	Subject	40 CFR reference
4	Authority to require chemical testing	Parts 790–799
5	New chemicals	Part 720
	Premanufacture notification (PMN) exemptions	Part 723
5 (a)	Significant new use rules	Part 721
6, 7	Existing chemicals control	Part 750
8 (a)	Chemical use reporting	Parts 704, 712
8 (b)	Inventory reporting rules	Part 710
8 (c)	Adverse reactions	Part 717
8 (d)	Health and safety data reporting	Part 716
8 (e)	Substantial risk reporting	—
12	Export rules	Part 707
13	Import rules	Part 707

sponsibility for laws associated with different aspects of the environment: air pollution, water pollution, waste, pesticides, and toxic substances. The organization of these offices has changed from time to time. Agency scientists and staff in these offices belong to various divisions which usually have responsibility for one law. There is very little interchange between the offices and divisions which administer different laws. It is not unusual for representatives of industry who deal with various types of products or problems to find that scientists at U.S. EPA who are responsible for pesticides, for example, use a whole set of different tools and standards than the scientists who are responsible for industrial chemicals.

A toxicology staff needs to play a key role in the TSCA compliance efforts of an industrial chemical manufacturing or processing company. In addition, TSCA compliance requires the involvement of other regulatory affairs specialists who will monitor new products and procedures, as well as new TSCA regulations, in order to keep a company in compliance.

One key to compliance is keeping up with new information on TSCA. In addition to new regulations, policy statements and guidance documents are frequently published and made available from U.S. EPA. One key publication is called the *Chemicals in Progress Bulletin*. This is published quarterly by U.S. EPA and subscriptions are free. This publication keeps abreast of new projects of the agency and contains frequent summaries of regulatory and compliance activities related to TSCA.

TSCA is somewhat of a misnomer; the law gives U.S. EPA control over all chemicals (at least those that are not exempt due to the authority of other laws, such as pesticides and food additives), not just toxic substances. Actually, the term toxic is not defined in TSCA or in any of U.S. EPA's regulations. When TSCA was enacted, one of U.S. EPA's first responsibilities was to require reporting from industry to identify the chemicals which were currently being manufactured or imported in the United States. U.S. EPA "grandfathered" these chemicals by entering the chemicals reported into a list called the *TSCA Inventory*. After the reporting period became closed, all chemicals not on the inventory were then considered "new" under TSCA and therefore reportable under Section 5.

New Chemicals

Before a new chemical under TSCA can be manufactured or imported in the United States, Section 5 of TSCA requires the submission of a Pre-Manufacture Notification (PMN). A PMN must be submitted to U.S. EPA at least 90 days before manufacture or import. In the PMN, information must be included on the identity, manufacture, use, exposure, and disposal of the chemical. The PMN must also include reports of any testing which indicate the chemical substances' possible impact on health or the environment.

A PMN must be submitted on a special form. A reprint of the 10-page form used for many years can be found in the *Code of Federal Regulations* (CFR) Title 40, Part 720, Appendix A. Recently, U.S. EPA expanded this form, making it several pages longer. As of this writing in 1993, U.S. EPA plans to further expand the form. Copies of the latest form can be obtained from U.S. EPA at an address listed in 40 CFR. An instruction manual for submitting PMNs can also be obtained from the same address.

After submitting a PMN, it is not unusual for the submitter to hear very little from U.S. EPA. Within a few weeks, the submitter usually receives the letter from U.S. EPA indicating the file number for the PMN and the date that the 90-day review period will expire. U.S. EPA will notify the submitter, however, if additional time is needed for review of the PMN. U.S. EPA is not required to inform the submitter that the PMN is approved, or more precisely under the law, that the notice review period has expired. If the submitter does

not hear from U.S. EPA by the last day of the review period, the submitter can begin manufacture or import.

After beginning manufacture or import, the submitter must send U.S. EPA a letter known as the Notice of Commencement (NOC) within 30 days. When U.S. EPA receives the NOC, the chemical substance is added to the *TSCA Inventory*, and anyone, including competitors of the PMN submitter, can then begin manufacture or import of the chemical substance.

Regulatory affairs staff of chemical companies need to be familiar with many regulatory nuances needed to comply with PMN and *TSCA Inventory* requirements. These are listed in Part 720 of 40 CFR, as well as in the discussion in *Federal Register* documents where these regulations were published and a myriad of other official and semi-official guidance published by U.S. EPA. Some discuss market exemptions, polymer exemptions, research and development chemicals, mixtures, and rules to maintain confidentiality of information. These requirements are quite complex, and the regulatory affairs staff of chemical companies need to stay current with new and changing requirements.

—*Shayne C. Gad*

Related Topics

Clean Air Act
Clean Water Act
Comprehensive Environmental Response,
    Compensation, and Environmental Toxicology
Liability Act
National Environmental Policy Act
Toxic Torts

Toxic Torts

Introduction

Acute risks, carcinogenicity, causation, chronic exposures, enterprise liability, epidemiology, latency period, medical monitoring, risk assessment, teratogen, toxicity, ultrahazardous—the language of toxic tort law would indeed have been foreign to an attorney practicing 50 years ago. The jargon, in fact, would hardly have been recognizable as belonging to the legal profession. By the 1990s, however, most personal injury attorneys, if not most of the general public, had some familiarity with the terminology.

One of the most challenging and complex specialties in which a trial attorney can practice today is a somewhat ill-defined branch of tort law called "toxic torts." The area provides the legal practitioner an opportunity to become immersed in a diversity of technical disciplines such as toxicology, physiology, dosimetry, epidemiology, risk assessment, and many other medical and scientific specialties and subspecialties, unparalleled by any other area of the law. The request to act as a consulting or testifying expert in a toxic tort case likewise affords the scientist or physician a unique forum in which to communicate to the public those "truths" discerned through scientific research or, perhaps more often, to describe the extent to which consensus has yet to be attained.

The result of the legal process whereby a claim is made that one's conduct or product resulted in compensable harm may be that (1) an individual is awarded a monetary sum, i.e., damages; (2) a product is removed from the market; (3) further scientific research is directed to be funded by the tortfeasor, as the wrongdoer is called; (4) medical monitoring or surveillance is ordered; (5) the tortfeasor is ordered to modify or cease its offending conduct; or any combination of these types of remedies. The actor sought to be held responsible in the toxic tort setting may be an individual, a corporation, an entire industry, or the local, state, or federal government. Aside from technical complexity and the concern of many in the plaintiff's bar and in the judiciary that remedies are far too elusive, the hallmark of the toxic tort case which has generated the most commentary in the legal community is its weighty transaction costs. The costs associated with the legal process are heavy and exact a toll on courts and administrative forums as they struggle with the resource demands such complex litigation imposes; likewise, litigants must bear severe financial, and often emotional, costs as well, while they seek to resolve the issues in a system notorious for its inefficiencies.

Exposure to any of a myriad of agents inside or outside the workplace, including dioxin-containing Agent Orange, asbestos, benzene, electromagnetic radiation, ionizing radiation, solvents, and the like, or the use of consumer products or pharmaceuticals, including HIV-contaminated blood products, Halcion, the Dalkon Shield, tobacco, cellular telephones, computer terminals, diethylstilbestrol (DES), ear drops, heart valves, herbicides, electric blankets, insecticides, breast implants, and lead-based paint, has given rise to lawsuits. The list will continue to expand as our society increases its dependence on technological innovation if the legal system continues to offer some prospect of obtaining redress, albeit costly and inefficient, for those who believe they have been injured by a toxic agent.

Some discussion has been devoted in law journals, treatises, and reported court decisions to defining the toxic tort. It is far easier to describe the typical features of the toxic tort than it is to arrive at a satisfactory definition. *Black's Law Dictionary* (p. 1489, 6th ed., 1990) defines a tort as "a civil wrong or injury . . . for which the court will provide a remedy in the form of an action for damages," or

> A legal wrong committed upon the person or property independent of contract. It may be either (1) a direct invasion of some legal right of the individual; (2) the infraction of some public duty by which special damage accrues to the individual; (3) the violation of some private obligation by which like damage accrues to the individual.

The toxic tort is a civil action including a demand for damages for personal injury or property damage arising from alleged exposure to a toxic substance, emission, or product. The probable outcome is more likely a settlement than an award. Many different factual settings may give rise to the so-called toxic tort. Environmental contamination resulting from pollutant discharges may form the basis for a toxic tort suit. Suits have been brought in connection with incidents or releases which occurred at such places as Three Mile Island, Love Canal, and Times Beach. A worker's compensation case in which an employee alleges that his or her injury or disease "arose out of and in the course and scope of employment" from occupational exposures may properly be characterized as a "toxic tort-type" action. What have been traditionally viewed as product liability actions also resemble the toxic tort.

Practical Considerations

Toxic torts are characterized by a vast array of complex practical and policy problems that baffle litigants, their attorneys, and the courts. Causation will be a proof problem in almost all toxic tort cases. That the plaintiff must show (1) exposure to a toxin, (2) that the toxin caused the injury, and (3) that an injury, in fact, occurred is a simple formula which belies the immense complexities which typify the kind of evidence required to establish causation. The two general standards for legal cause, "but for" and "substantial factor," must be met "by a preponderance of the evidence," i.e., >50% likelihood, or "more probable than not." Scientific and medical issues will invariably require the use of expert witnesses. Commercially available computer databases, such as Toxline, SciSearch, and Medline, are one tool useful in identifying potential experts, determining their standing in the scientific community, and evaluating their qualifications and, in particular, the applicability of their research to the scientific issues of a case. Even the most sophisticated, astute fact finder in a toxic tort case—judge, jury, or administrative law judge— will likely be challenged by its technical, scientific, and medical complexity. The need to simplify the issues without sacrificing technical accuracy will be an overriding practical concern for counsel and expert witnesses alike, as will the need to obtain sufficient time on the court's calendar to educate regarding the scientific concepts necessary to understand the expert testimony.

These often highly publicized lawsuits typically involve intractable pretrial maneuvers, extensive motions practice, lengthy discovery, and voluminous technical, medical, and scientific evidence. More so than perhaps any other type of litigation, the toxic tort case will require the attorney to place special importance on case management from the outset. One of the first of such tasks which must be accomplished in all but the most straightforward of cases is to assemble a "litigation team." Members of the team will assume responsibility for aspects of the case including factual development, legal research and briefing, identification of expert witnesses, and, in many cases, public relations. Even worker's compensation cases are likely to attract the attention of the media where they involve a worker's alleged exposure to occupational toxins. Toxic tort cases are but one product of a highly industrialized, litigious society concerned about not only health problems which have already been manifested but also potential future health consequences; the "fear of cancer" or "increased risk" claim is becoming increasingly common.

Bases for Toxic Tort Actions

Liability for property damage or personal injury from exposure to hazardous materials or substances is primarily rooted in judge-made, common law theories, although an increasing number of federal statutes provide various specific bases of recovery for groups of similarly situated individuals. Such statutes may seek to lessen the burden on plaintiffs of establishing "causation," compared to traditional tort requirements, by creating a presumption which places the burden on the defendant to prove that he or she did not cause the plaintiff's injury once the plaintiff makes a threshold showing, in effect requiring that the defendant prove a negative, or by requiring that the court accept animal data as evidence of causation, where human epidemiologic data are nonexistent. One statutory device which has been employed with varying degrees of success is to mandate the establishment of an administrative scheme which provides a fixed amount of compensation to individuals who meet certain criteria. For example, see Veterans' Dioxin & Radiation Exposure Compensation Standards Act of 1984, 38 U.S.C. § 1154 note (1992); National Swine Flu Immunization Program of 1976, 42 U.S.C. §201 note, 247b (1991); Black Lung Benefits Act of 1972, Pub. L. 92–303, 86 Stat. 150 (codified as amended in scattered sections of 26, 29, and 30 U.S.C.); and H. R. 3743, 103d Cong., 2d Sess. (1994), the Radiation Experimentation Compensation Act of 1994—"a bill to provide for payments to individuals who were the subjects of radiation experiments conducted by the Federal Government."

Common law causes of action which have been employed in the toxic tort area include public nuisance, private nuisance, negligence (including negligence per se, where there is an alleged violation of a statutory standard), trespass, strict liability (for an abnormally dangerous activity or defective or unreasonably dangerous product), fraud, intentional or negligent infliction of emotional distress or fear, express or implied warranties, and assault and battery. Causes of action of recent derivation include market share liability, concert of action, enterprise liability, alternative liability, and successor and creditor liability.

In addition, toxic substances are heavily regulated by state and federal governments. These statutes may play a significant role in a toxic tort suit. Statutory standards may establish the minimum duty owed by manufacturers, transporters, or users to third parties or employers to employees. Violation of an applicable standard may constitute negligence per se or, alternatively, evidence of negligence. See *O'Conner v. Commonwealth Edison Co.,* 13 F.3d 1090, 1104–1105 (7th Cir. 1994), *cert. denied,* 512 U.S. 1222 (1994), holding that federal safety standards for nuclear power plants constitute the appropriate standard of care in state tort actions.

The statutes which regulate environmental quality may provide a remedy through citizen suit provisions for injunctive relief or federal funds for remediation.

Common Law

"A nuisance may be merely a right thing in the wrong place, like a pig in the parlor instead of the barnyard" (*Euclid v. Ambler Realty Co.,* 272 U.S. 365, 388, 1926). A cause of action sounding in public nuisance must allege harm, injury, inconvenience, or annoyance arising out of the invasion of a public interest. Actual harm need not be shown. A public nuisance has been defined as "an unreasonable interference with a right common to the general public" [*Restatement (Second) of Torts* 2d § 821B, 1979]. The restatement notes that analysis of whether an interference is unreasonable or not should include the following:

> (a) Whether the conduct involves a significant interference with the public health, the public safety, the public peace, the public comfort or the public convenience, or (b) whether the conduct is proscribed by a statute, ordinance or administrative regulation, or (c) whether the conduct is of a continuing nature or has produced a permanent or long-lasting effect, and, as the actor knows or has reason to know, has a significant effect upon the public right.

The goal of this legal theory is the protection of community rights. It is generally agreed that a private individual does not have an action for damages on a public nuisance theory unless he or she can establish that he or she suffered special damage different from that sustained by other members of the general public. Where an injunction is the remedy sought, this requirement may be less stringently imposed in some courts.

A private nuisance is defined as a "nontrespassory invasion of another's interest in the private use and enjoyment of land." The invasion may be caused by conduct that is intentional, where the defendant acts for the purpose of causing it, knows that it is resulting

from his or her actions, or knows it to be substantially certain to result from his or her conduct, or unintentional, as where the defendant's conduct is negligent, reckless, or ultrahazardous. See *Morgan v. High Penn Oil Co.,* 238 N.C. 185, 77 S.E.2d 682 (1953) and *Good Fund Ltd.—1972 v. Church,* 540 F. Supp. 519, 533 (D. Colo. 1982), *rev'd sub nom. McKay v. United States,* 703 F.2d 464 (10th Cir. 1983).

Negligence began to emerge as a separate cause of action for unintentional torts in the early 1800s, coinciding with the industrial revolution in England. The textbook elements of the tort of negligence are (1) duty, an obligation recognized by the law, requiring the actor to conform to a certain standard of conduct, for the protection of others against unreasonable risks of harm; (2) breach of duty, or the failure to comply with the standard of care; (3) proximate or legal cause, a reasonably close causal connection between the conduct and the resulting injury—this includes both cause in fact and certain legal limitations on the extent to which the law will recognize "cause"; and (4) actual loss or damage to the interests of another. See generally *Restatement (Second) of Torts* 2d § 281–328D (1965). *Sterling v. Velsicol Chemical Corp.,* 855 F.2d 1188, 1198–1201 (6th Cir. 1988) provides an excellent discussion of proximate cause in the context of a toxic tort case involving environmental exposures from a chemical waste burial site. The court noted that the requirement that plaintiffs prove bodily injury to a "reasonable medical certainty" is of "particular importance when dealing with injuries or diseases of a type that may inflict society at random, often with no known specific origin" (855 F.2d 1188, 1200). This requirement is crucial under traditional tort analysis where a plaintiff alleges that exposure to some potentially toxic agent caused his or her cancer.

Trespass is an intentional tort where the plaintiff's possessory interest is intentionally invaded by the defendant, causing harm. See *Martin v. Reynolds Metals Co.,* 221 Or. 86, 342 P.2d 790 (1959), *cert. denied,* 362 U.S. 918 (1960), in which fluoride gases and particulates from the defendant's plant were dispersed on plaintiff's land rendering it unfit for livestock, and *Good Fund, Ltd.—1972 v. Church, supra* at 532–533, in which plaintiffs alleged that radioactive material from defendant United States' nuclear weapons plant deposited on plaintiff developer's land constituted trespass.

A cause of action alleging that defendant's conduct was unreasonably dangerous or ultrahazardous is a strict liability theory frequently relied on in a toxic tort suit or products liability action. In such an action, the plaintiff need not prove fault, but rather must show, to varying degrees depending on the factual setting and the court, (1) that the defendant's conduct posed a high degree of risk of harm, (2) likelihood that the harm will be great, (3) the defendant cannot eliminate the risk by exercising reasonable care, (4) the activity or conduct is not of common usage, (5) the activity was carried out in an inappropriate place, and (6) the extent to which the value of the activity or conduct is outweighed by its dangerousness. There is much overlap between products liability theory and practice and toxic torts. In a product liability action, the plaintiff need not prove fault but must prove that a defect in the product rendered the product unreasonably dangerous, the defect was present when it left the manufacturer's control, and the injury complained of was proximately caused by the defect. For example, see *In re Swine Flu Products Liability Litigation*, 764 F.2d 637 (9th Cir. 1985); *In re Swine Flu Immunization Products Liability Litigation*, 508 F. Supp. 897 (D. Colo. 1981), *aff'd sub nom. Lima v. United States*, 708 F.2d 502 (10th Cir. 1983); *Smith v. Paslode Corp.*, 799 F. Supp. 960 (E. D. Mo. 1992), *aff'd in part and rev'd in part*, 7 F.3d 116 (5th Cir. 1993) (HIV-contaminated blood products); *Anguiano v. E. I. DuPont de Nemours Co.*, 808 F. Supp. 719 (D. Ariz. 1992), *aff'd*, 44 F.3d 806 (9th Cir. 1995) (polytetrafluoroethylene temporomandibular joint prostheses); *Quillen v. International Playtex, Inc.*, 789 F.2d 1041 (4th Cir. 1986) (toxic shock syndrome associated with tampon usage); and *Lee v. Baxter Healthcare Corp.*, 721 F. Supp. 89, *aff'd*, 898 F.2d 146 (4th Cir. 1990) (defective breast implant).

Fraud is another common law cause of action which is sometimes alleged in the complaint in a toxic torts case. The plaintiff must show (1) a misrepresentation of fact, (2) that defendant had knowledge of the falsity, (3) that the defendant intended to induce the plaintiff to act in reliance on the factual misrepresentation, (4) plaintiff's justifiable reliance on the misrepresentation, and (5) damage or loss as a result of the plaintiff's reliance.

The cause of action for intentional or negligent infliction of emotional or mental distress, or fear, arises frequently in the toxic tort setting as a "cancerphobia" claim. For example, see *Cantrell v. GAF Corp.*, 999 F.2d 1007 (6th Cir. 1993), which notes the distinction between cancerphobia and increased risk of cancer. Phobia is a claimed present injury consisting of mental anxiety and distress over contracting cancer, whereas increased risk is the potential physical predisposition to develop cancer in the future. In the oft-cited case of *Ferrara v. Galluchio*, 5 N.Y.2d 16, 152 N.E.2d 249, 176 N.Y.S.2d 996 (1958), cancerphobia was determined to be actionable as the result of negligent X-ray treatment where the circumstances provided a substantial guarantee of genuineness of the fear. Other illustrative cases include *Sterling v. Velsicol Chemical Corp.*, *supra* (cancerphobia is a specific type of mental anguish); *Hagerty v. L & L Marine Services, Inc.*, 797 F.2d 256 (5th Cir. 1986) (plaintiff developed cancerphobia after incident where he was drenched with toxic and carcinogenic chemicals; court held that the "impact requirement" had been met, supporting recovery for emotional distress damages); and *Jackson v. Johns-Manville Sales Corp.*, 781 F.2d 394 (5th Cir. 1986), *cert. denied*, 478 U.S. 1022 (1986) (former shipyard worker brought strict liability action against asbestos products manufacturer for failure to warn; court permitted recovery for reasonable medical probability of contracting cancer in the future and mental distress from knowledge of increased risk). Judicial recognition of these causes of action is a rather recent development. Some 50 years ago, the *Restatement of Torts* took the position that one's interest in freedom from emotional or mental distress was not of sufficient importance to require others to refrain from conduct intentionally designed to cause such distress upon pain of adverse legal consequences. The interest in emotional and mental tranquillity was simply one for which the law formerly provided no protection. Early cases required the plaintiff to show some bodily harm as a result of severe emotional distress, intentionally caused, as a prerequisite to recovery. Later, in many cases the physical harm requirement was abandoned. See Gregory G. Sarno, Annotation, *Infliction of Emotional Distress: Toxic Exposure*, 6 A.L.R. 5th 162 (1992) for a discussion of cases in which recovery was sought for intentional or negligent infliction of emotional distress from exposures to certain toxins.

Express warranties can be said to result generally from express statements, representations, or affirmations of fact about the characteristics of goods sold, whereas implied warranties arise simply as a consequence of a sale by a merchant. The original purpose of these theories was to protect purchasers from economic and commercial losses and not to protect from dangerous products. This area of the law expanded well beyond the original goal, however, to the point where strict liability would eventually make the seller an insurer of the safety of his or her product, even though he or she exercised reasonable care and even when there was no privity of contract between himself/herself and the purchaser. Liability became grounded on the notion that the seller ought to be obligated to absorb the costs attributable to injuries caused by a defective, dangerous product without proof of negligence, manifestation of intent to guarantee, privity of contract, and even where there were clear contractual disclaimers of liability. The policy grounds for imposing strict liability on the supplier of a defective product are explained in W. Page Keeton *et al.*, *Prosser and Keeton on The Law of Torts* § 95A, 96–98 (5th ed. 1984). *City of Philadelphia v. Lead Industries Ass'n*, 994 F.2d 112 (3d Cir. 1993) is an example of a suit alleging breach of warranty. The city and the housing authority sued manufacturers of lead pigment and a trade association to recover costs of the city's lead paint abatement program. The complaint included allegations of negligence, strict liability, fraud, and misrepresentation.

Assault and battery are two intentional tort causes of action which have been employed in the toxic tort arena. Assault is an intentional, unlawful offer to touch another person in such circumstances which create in the mind of the other person a well-founded fear of an imminent battery, coupled with an apparent present ability to effectuate the attempt. Alternatively, an assault is an act intended to put another person in reasonable apprehension of an immediate battery, which succeeds in causing such apprehension. The defendant must have been in a position to carry out the threat immediately and he or she must have taken some affirmative action to do so. Battery is an intentional harmful or offensive contact with another person. A defendant may be liable for battery where he or she acts intending to cause such contact or an imminent apprehension of such a contact, and the harmful contact directly or indirectly results.

These "traditional" tort theories of liability will form the basis for recovery in some toxic tort cases, but in many, novel problems involving proof, causation, scientific uncertainties, multiple tortfeasors, latent injuries, "subclinical" injuries, statutes of limitation, and the disproportionately high cost of litigation in relation to the amount of damages which might be expected to be obtained will cause the plaintiff to go home empty-handed. The societal goal of creating an economic incentive for manufacturers and suppliers to reduce injuries will sometimes also go unrealized. These problems have driven litigants, creative attorneys (for both plaintiffs and defendants), and the courts to develop new theories of recovery, unprecedented methods for managing cases that must be tried, and original ways of resolving cases in lieu of trial.

The market share liability theory has been employed in cases in which the plaintiff can show causation between a mass-produced defective product and injury but is unable to trace the product to a specific manufacturer because of a long latency period between use or exposure and the injury or some other characteristic of the product itself. This theory places the burden of proof of causation on the parties that gained commercially from the creation of the risk. Market share liability was first applied in a case involving DES, a synthetic estrogen [see *Sindell v. Abbott Laboratories*, 26 Cal.3d 588, 163 Cal. Rptr. 132, 607 P.2d 924, *cert. denied*, 449 U.S. 912(1980)]. The injury here was manifested not in the user herself, but in the next generation. Under this theory, the amount of defendant's liability is proportional to the percentage of the market held by the defendant at the time the drug was marketed. The cases of *Martin v. Abbott Laboratories*, 102 Wash.2d 581, 689 P.2d 368 (1984) and *Collins v. Eli Lilly & Co.*, 116 Wis. 166, 342 N.W.2d 37 (1984), *cert. denied*, 469 U.S. (1984) also discuss this theory.

The concert of action theory of liability is used to overcome the problem of tying a particular manufacturer or seller to the plaintiff's injuries. Under this theory, the plaintiff seeks to establish liability by showing that his or her injuries are caused by "concerted action" within the industry. Concert of action is discussed in *Shackil v. Lederle Laboratories*, 116 N.J. 155, 561 A.2d 511 (1989), where the plaintiff developed chronic encephalopathy after receiving a DPT immunization.

The case of *Abel v. Eli Lilly and Co.*, 418 Mich. 311, 343 N.W.2d 164 (1984) describes the theory of "alternative liability." This theory, aptly named, has been asserted in circumstances in which all defendants who could have caused the injury in question are before the court and the plaintiff shows that all of them acted negligently in a way that could have caused the injury, but no proof is available to identify the actual actor whose conduct, in fact, caused the injury.

Enterprise liability, or liability imposed on an entire industry where the number of defendants is deemed to be manageable, is examined in *Hall v. E. I. Dupont de Nemours & Co.*, 345 F. Supp. 353 (E.D.N.Y. 1972).

Successor and creditor liability was first discussed in the case of *Ray v. Alad Corp.*, 19 Cal.3d 22, 136 Cal. Rptr. 574, 560 P.2d 3 (1977). The crux of this recent theory is whether a successor corporation continues to manufacture essentially the same product line as its predecessor. Under this theory, where the original manufacturer of an injury-causing product is purchased by another company, the purchasing company may be liable for the toxic torts of the original manufacturer under the "product line" exception to the general rule that where a company sells all of its assets to another company, the latter is not liable for the liabilities of the transferor company, including its tort liabilities. The theory may be especially appurtenant in toxic cases in which a long latency period before injuries are manifested increases the possibility that the original manufacturer will have been sold.

Statutory Law

As noted previously, federal statutes, and state statutes patterned after their federal counterparts or which implement federal regulations under joint state–federal programs, may play a role in a toxic tort suit. For example, see E. T. Tsai, Annotation, *Liability Under Federal Employers' Liability Act* (45 U.S.C. § 51 et seq.) *for Industrial or Occupational Disease or Poisoning*, 30 A.L.R.3d 735 (1993). The major federal environmental statutes are outlined in the following sections.

Major Federal Environmental Statutes

Comprehensive Environmental Response and Compensation Liability Act

The Comprehensive Environmental Response and Compensation Liability Act (CERCLA), also known as the "Superfund Act", 42 U.S.C. § § 9601–9657, provides for the cleanup of hazardous waste disposal sites and authorizes the U.S. Environmental Protection Agency (U.S. EPA) to pay for cleanup costs and to recover such costs from generators and disposers determined to be "responsible parties." Injunctive relief is also available under the statute.

Resource Conservation and Recovery Act

The Resource Conservation and Recovery Act (RCRA), 42 U.S.C. § § 6901–6987, regulates the current and future disposal of hazardous waste at operating waste disposal sites, regulates the transport and disposal of hazardous waste, and requires that manifests be maintained to document the movement of such waste. The manifest may be a useful document in establishing causation in a toxic tort case. Citizen suits for injunctive relief are available under RCRA.

Toxic Substances Control Act

The Toxic Substances Control Act (TSCA), 15 U.S.C. § § 2601–2629, regulates the manufacture and distribution of new chemicals and requires testing and special handling of potentially injurious chemicals. Under TSCA, the U.S. EPA may seize or ban the use of certain chemicals. Citizen suits are available, but there is no private right of action for compensatory damages for injuries caused by toxic exposures.

Federal Insecticide, Fungicide, and Rodenticide Act

The Federal Insecticide, Fungicide, and Rodenticide Act (FIFRA), 7 U.S.C. § § 136–136y, requires the registration of pesticides used in the United States. Under FIFRA, the EPA makes a cost/benefit determination before a pesticide can be used commercially. Registration of a pesticide does not mean that it has been determined to be "safe." The act also imposes labeling requirements, including warnings of potentially hazardous properties, and restricts the methods of a pesticide's use.

Clean Water Act

The Clean Water Act, 33 U.S.C. § § 1251–1376, makes it illegal to discharge any pollutant into waters of the United States, unless the discharge is permitted. A National Pollutant Discharge Elimination System permit allows a discharge within limits set by the permit based on national standards for the specific pollutant in question. The act generally subjects the more toxic pollutants to greater control than the less toxic ones.

Safe Drinking Water Act

The Safe Drinking Water Act, 42 U.S.C. § § 300F–300J-10, applies to public drinking water systems, primarily from groundwater sources. The act calls for the setting of standards for drinking water pollutants, directs the EPA to implement an underground injection control program, and includes citizen suit provisions.

Clean Air Act

The Clean Air Act, 42 U.S.C. § § 7401–7642, regulates emissions of airborne pollutants. Under this act, the EPA establishes national air quality standards and the states prepare State Implementation Plans which include standards for specific sources of pollutants. Citizen suits are allowed, but there is no provision for compensatory damages.

Worker's Compensation

Worker's compensation cases involving alleged exposures to toxic materials feature some of the same difficulties and challenges as tort actions based on negligence, strict liability, or any of the other theories noted herein. In fact, the majority of toxic chemical exposure cases are believed to have arisen in the worker's compensation arena. Notable toxic tort-type compensation cases include *Dow Chemical Co. v. Gabel*, 746 P.2d 1357 (Colo. App. 1987) (affirming Industrial Commission award of death benefits to widow up until the time of remarriage for death of husband from brain cancer, determined to have been caused by exposure to radiation at nuclear weapons plant); *Krumback v. Dow Chemical Co.*, 676 P.2d 1215 (Colo. App. 1983) (claim for worker's compensation death benefits for colon cancer alleged to have been caused by radiation exposure); *Dow Chemical Co. v. Downing*, 843 P.2d 122 (Colo. App. 1992), *cert. denied*, 1993 Colo. LEXIS 18 (Jan. 19, 1993) (compensation allowed for death from esophageal cancer alleged to have been caused by radiation exposure acting on preexisting, nonoccupational condition known as "Barrett's esophagus"); and *Meyer v. Rockwell International Corp.*, No. 91CA0633, (Colo. App. May 28, 1992), *cert. denied*, 1992 Colo. LEXIS 1045 (Nov. 16, 1992) (compensation denied for lung cancer alleged to have been caused by occupational radiation exposure where deceased employee had substantial smoking history).

The employee who desires to bring a toxic tort suit against his or her employer more than likely will face the statutory bar of exclusive remedy. For example, see *Silkwood v. Kerr-McGee Corp.*, 464 U.S. 238 (1984), *on remand*, 769 F.2d 1451, 1458 (10th Cir. 1985). The reason a negligence suit against the employer is generally barred derives from the origins of the worker's compensation system. Worker's compensation, which awards no-fault benefits to workers injured by conduct arising out of and in the course and scope of the employment, is the product of a social reform movement that sought to provide the injured worker a speedy remedy for occupational injuries in exchange for relinquishment of the right to sue the employer. It was assumed that in a tort action, the worker would find a heavier burden of proof and a less certain chance of recovery. The compensation system generally awards an injured worker somewhat less than regular wages as an incentive to avoid injury and return to work sooner following an injury. Suits against third parties, e.g., manufacturers and sellers of chemicals or equipment used in the workplace, are generally not barred by worker's compensation statutes.

The exclusive remedy doctrine has come under attack by plaintiffs because it produces smaller damage awards where larger awards are believed to have more of a deterrent value that could produce safer workplaces. The doctrine, however, is generally not a bar to an intentional tort action against the employer and will usually apply equally to claims for accidental injuries and occupational diseases. Some circumstances may give rise to applicability of the "dual capacity" doctrine and a negligence suit may be allowed against the employer in its capacity as a producer of a defective product. Suits against vendors, suppliers, and contractors are likely to be the only means of recovery in tort for workplace exposures. Plaintiffs may face insurmountable difficulties in alleging an intentional tort with regard to chemical exposure or the inadequate communication of health and safety hazard information. Such conduct will probably fall squarely within the scope of the applicable worker's compensation statute.

History of the Toxic Tort: Notable Cases

There is a substantial body of legal commentary on the subject of the tension between two predominant approaches to dealing with toxic tort cases. The first approach treats toxic torts as involving merely another application of traditional principles of tort law. The second emerging perspective shows a tendency to develop special rules and principles because of the perceived unique nature of the toxic tort case. As noted previously, the latter view has sometimes resulted in the enactment of legislative compensation schemes providing specified remedies for groups of people seeking recompense for alleged injuries arising out of similar circumstances. The two approaches as applied by the courts are best illustrated by reviewing a sample of archetypal cases. This entry will only briefly explore the major developments in the field of toxic torts in the context of Agent Orange, asbestos, environmental exposure, and radiation cases.

Agent Orange

The first occurrence of the term toxic tort in a reported court decision was in an Agent Orange decision by Judge George Pratt in 1979: *In re Agent Orange Products Liability Litigation,* 506 F.Supp. 737 (E.D.N.Y. 1979), *rev'd,* 635 F.2d 987 (2d Cir. 1980), *cert. denied,* 454 U.S. 1128 (1981). The term caught on and became widely used in both legal and popular publications. Today, toxic tort is nearly a household word. In this Agent Orange case, Judge Pratt employed a "case management plan" to deal with what he perceived to be the unique procedural problems of the so-called "mass tort." The plan dealt with issues such as class certification, the application of state statutes of limitation, trials of common issues, and the need for limitations on discovery. Judge Pratt appreciated firsthand that this case was not the conventional tort case for many reasons: (1) Plaintiffs and defendants numbered in the thousands; (2) defendants had differing levels of responsibility; (3) there were different state laws to be applied which governed standards of conduct, rules of causation, and damages; (4) the complexity of the causation issues differentiated the case from a "mass disaster" case, such as a plane crash or hotel fire; (5) plaintiffs suffered different injuries; (6) exposure pathways differed among plaintiffs; (7) each plaintiff theoretically exhibited different contributing causal factors other than dioxin exposure; (8) there were difficult legal issues involving the application of strict liability and enterprise liability theories, latency of the injuries, and "genetic injuries"; (9) many of those who claimed to have been exposed had not yet manifested any discernible harm; (10) there were a multitude of scientific and medical issues which were unresolved because inade-

quate data would not permit scientifically sound conclusions; (11) a major issue was whether legally permissible conclusions could be drawn based on data which were insufficient for scientific purposes; (12) the government's involvement was uncertain; (13) there were many potential procedural options for resolving the litigation; and (14) the litigation exposed widely conflicting public policies pertaining to substantive and procedural matters. Judge Pratt was not alone in being acutely aware of the fact that more than money was at stake. He acknowledged the poignancy of the circumstances and noted that apparently both sides believed that, in fairness, the government should assume responsibility, where most plaintiffs had limited financial resources and severe health problems, despite the contrary result that existing law might dictate. A multimillion dollar settlement was later approved by Judge Jack Weinstein as the most expeditious way of resolving the extremely complex factual and legal issues posed by this case.

The Agent Orange litigation was fraught with many factual problems, not the least of which for the plaintiffs was the fact that the evidence was largely inconclusive as to whether a causal relationship could be supported by existing studies and data. The magnitude of the case is reflected in the number of scientists who worked on the case (hundreds), the number of motions heard by the court (hundreds), the number of depositions taken by the parties (hundreds), the number of pages of documents exchanged in discovery or filed with the court (millions), the number of cases encompassed by the litigation (approximately 600 separate cases), and the number of plaintiffs (approximately 15,000). A major problem was that many of the diseases from which various plaintiffs suffered could well have been caused by some etiologic agent other than dioxin. This case is a good illustration of why many litigants, courts, and legal scholars perceive there to be a need for innovative rules of causation or, alternatively, a legislative remedy.

Subsequent orders issued by Judge Weinstein are instructive on the immensity of the problems one very perceptive federal judge saw in the management of mass toxic tort cases. Judge Weinstein viewed animal data and industrial accident data as unreliable and, therefore, inadmissible, and epidemiologic studies as the only kind of evidence of any probative value on the issue of causation. He saw a great need for judicial screening of expert witnesses and called for judges to be vigilant in their "gatekeeping" role regarding expert testimony. The propriety and importance of this role was reaffirmed by the Supreme Court in the much-heralded decision of *Daubert v. Merrell Dow Pharmaceuticals, Inc.,* 509 U.S. 579 (1993). Judge Weinstein saw many inefficiencies in toxic tort litigation, including the high ratio of transaction costs to recoveries, and inequities in that some plaintiffs get sizable awards and some "equally deserving" get nothing. Questioning the deterrent value of a toxic tort case, he expressed the view that better health protection is achieved through government and private-sector testing, control, and regulation and a preference for government-mandated or privately funded compensation programs. In a later opinion dealing with attorney fees, Judge Weinstein cautioned that federal courts should not actively encourage the filing of "such dubious actions in the future" [*In re Agent Orange Products Liability Litigation,* 818 F.2d 226, 236 (2d Cir. 1987)].

Asbestos

Asbestos litigation represents the largest mass toxic tort in the United States. To date, more than 200,000 cases have been filed and approximately 250,000 additional cases are expected to be filed [*Malcolm v. National Gypsum Co.,* 995 F.2d 346, 348 (2d Cir. 1993)]. Although there are reported decisions involving asbestos dating back to the 1950s, major litigation with industrywide impact began with the frequently cited case of *Borel v. Fibreboard Paper Products Corp.,* 493 F.2d 1076 (5th Cir. 1973), *cert. denied,* 419 U.S. 869 (1974). Of the subsequent cases to plague the asbestos industry, among the most notable is *Jackson v. Johns-Manville Sales Corp.,* 727 F.2d 506 (5th Cir. 1984), ("*Jackson I*"), *vacated and question certified,* 750 F.2d 1314 (5th Cir. 1985) (en banc), ("*Jackson II*"), *cert. declined,* 469 So.2d 99 (Miss. 1985), *decision after cert.,* 781 F.2d 394 (5th Cir. 1986) (en banc), ("*Jackson III*"), *cert. denied,* 478 U.S. 1002 (1988). One of the most perplexing issues to be decided was whether a plaintiff could recover for "mere" increased risk of developing cancer in the future. In *Jackson I,* the court held that a claim for cancer accrues at the time of actual physical manifestation of the disease (727 F.2d at 520). Strict liability was held to be a valid basis for a claim in *Jackson II, supra.* In *Jackson III,* the court held that state law governs, rather than federal common law, opining; however, short of a legislative remedy, guid-

ance from the Supreme Court was needed to assure uniformity and equity in the resolution of the unique problems posed by mass asbestos litigation. The Fifth Circuit sought to certify issues pertaining to punitive damages and the accrual of cancer claims, but the state court declined to offer guidance. The Fifth Circuit ultimately held, in *Jackson III,* that applicable state law would allow the recovery of punitive damages, damages for the reasonable probability of contracting cancer in the future, and damages for mental distress and fear.

The Chief Judge, with four other Circuit Judges, filed a dissent in *Jackson III* that was scathing and yet not unsympathetic of the majority. The dissenters stated:

> The learned, lengthy opinion of the majority is the wrong response by the wrong court. . . . The court responds to innovative lawyers pressing for immediate financial recovery for individual clients (and themselves). We do not fault the lawyers. That counsel wish to continue the present mode of case-by-case adjudication is all too understandable. . . . They are not required to anticipate the impact of today's judgment on the host of tomorrow's claimants they do not represent but who deserve to share in the finite proceeds which a limited group of defendants can provide to compensate mass tort victims. . . . [T]he court is frustrated by lack of congressional action. A number of legislative solutions has been proposed for the problems we must confront today and tomorrow throughout America because of yesterday's production and use of asbestos. None has been enacted. Clearly the powers of Congress to tax and regulate give that forum the interstate reach and flexibility needed to allocate the relatively scarce resources that must be available to present and future claimants to achieve the greatest good for society. Yet, Congress can refuse to act while the court cannot abstain from resolving a case presented. . . . This seminal case concerns much more than James Leroy Jackson's individual claim against some companies that may have furnished an unsafe product to a shipyard in which he worked. We know better.
>
> Our dockets tell us so. . . . Tens of thousands of similarly situated claimants are already seeking relief against the same defendants, and a legion of other potential plaintiffs stand in the wings, awaiting the predictable manifestations of their identical exposure.
>
> We are not passing a milepost along a known path leading to a chosen goal. Instead, the court, without a goal, chooses a new path which will compel the way of all litigants who come later.
>
> The *wrong court* gives this wrong response. . . . We say so because a national problem has been adjudicated as though it were a state problem.

The problems we face in this litigation are ones of national public policy. To respond to such policy is beyond the ability of this diversity-based court. Such policy cannot be subject to the whims of individual states because matters of national public policy have nationwide application. Since Congress has not provided a solution, the Supreme Court of the United States should have been asked to provide one. (781 F.2d at 415-416)

Ironically, the majority would not disagree with much of what the dissenters said. In response to the dissenters, they justified their decision by stating:

Congress' silence in the face of a desperate need for federal legislation in the field of asbestos litigation does not authorize the federal judiciary to assume for itself the responsibility for formulating what essentially are legislative solutions. Displacement of state law is primarily a decision for Congress, and Congress has yet to act. (781 F.2d at 415)

Theories of negligence, failure to warn, breach of implied warranty, fraudulent concealment, conspiracy, and strict liability have been typically relied on in asbestos suits. Other noteworthy cases include *Jenkins v. Raymark Industries, Inc.,* 782 F.2d 468 (5th Cir. 1986), in which the Fifth Circuit upheld the certification of a plaintiff class to resolve common questions, in particular the viability of the "state-of-the-art" defense, and *Burns v. Jaquays Mining Corp.,* 156 Ariz. 375, 752 P.2d 28 (Ct. App. 1987). In *Burns,* claims for fear of contracting future disease were disallowed in the absence of manifestations of bodily injury. Medical monitoring, however, was allowed, through the use of a court-supervised fund rather than an award outright to individual plaintiffs, and "subclinical" asbestos-related injuries were held insufficient to constitute actual loss or damage to support a cause of action. In contrast to DES, which was manufactured by many companies pursuant to only one formula, asbestos products exhibit different toxicities depending on the percentage of asbestos contained in the product, the type of asbestos fiber, and characteristics of the product itself. Thus, application of even the newer theories of liability, such as market share liability, was problematic.

Toxic tort problems have even invaded the bankruptcy arena. The Bankruptcy Court recognized the need to protect the interests of current and future victims, creditors, and shareholders in *In re Johns-Manville Sales Corp.,* 78 B.R. 407 (Bankr. S.D.N.Y.

1987). In *In re UNR Industries, Inc.,* 46 B.R. 671 (Bankr. N.D.Ill. 1985), the court stated that the Bankruptcy Code was flexible enough to allow the fashioning of relief for asbestos claimants where potential suits exceeding hundreds of thousands in number drove the debtor to seek bankruptcy protection.

Environmental Exposure

Many environmental exposure cases surfaced in the 1980s alleging personal and property injuries from exposure to toxic substances in the ambient environment. Statutory and common law issues frequently overlap in this area. *Ayers v. Township of Jackson,* 106 N.J. 557, 525 A.2d 287 (1987), is one of the most significant of the earlier environmental exposure cases and arose out of massive groundwater contamination of a drinking water supply from landfill seepage. The New Jersey Court determined that damages for emotional distress were not recoverable under the applicable Tort Claims Act but allowed damages for disruption of quality of life and for medical surveillance based on increased risk of future harm. The court declined, however, to allow recovery for increased risk where the risks were speculative and unquantified. With respect to medical monitoring expenses, the court considered the policy implications and concluded that the award of such expenses was warranted where reliable expert testimony established the extent of exposure, toxicity of the chemicals in question, the gravity of the disease, an increased risk of disease among the exposed population, and a value to early diagnosis. Medical monitoring is also given an extended discussion in *In re Paoli R. R. Yard PCB Litigation,* 916 F.2d 829, 849–852 (3d Cir. 1990), *cert. denied,* 499 U.S. 961 (1991).

In a prophetic statement, the *Ayers* court opined: "We note the difficulty that both law and science experience in attempting to deal with the emerging complexities of industrialized society and the consequent implications for human health" (525 A.2d at 298). In a theme to be echoed repeatedly in toxic tort cases which have continued to confound the wisest and most innovative of judges, the court concluded that common law tort doctrines are inadequate and statutory compensation schemes are critically needed.

The issue of whether recovery should be permitted for increased risk is one which has received considerable attention in other environmental cases as well. In *Anderson v. W.R. Grace & Co.,* 628 F. Supp. 1219

(D. Mass. 1986), *aff'd in part, rev'd in part,* 862 F.2d 910 (1st Cir. 1988), which involved chemical contamination near Woburn, Massachusetts, the court expressed concern that permitting recovery for enhanced risk of contracting some future affliction would produce a windfall to the plaintiff who remains healthy. In *Sterling v. Velsicol Chemical Corp.,* 647 F. Supp. 303 (W.D. Tenn. 1986), *rev'd in part,* 855 F.2d 1188 (6th Cir. 1988), Judge Guy upheld damage awards for increased risk where expert evidence demonstrated to a degree of reasonable medical certainty that the disease would occur in the future:

> Where the basis for awarding damages is the potential risk of susceptibility to future disease, the predicted future disease must be medically reasonably certain to follow from the existing present injury. While it is unnecessary that the medical evidence conclusively establish with absolute certainty that the future disease or condition will occur, mere conjecture or even possibility does not justify the court awarding damages for a future disability which may never materialize. . . . Therefore, the mere increased risk of a future disease or condition resulting from an initial injury is not compensable. (855 F.2d at 1204)

Although both common law tort principles and statutes tend to play a role in environmental exposure cases, there are vast differences especially in the area of remedies. The case of *Artesian Water Co. v. New Castle County,* 659 F. Supp. 1269 (D. Del. 1987), *aff'd,* 851 F.2d 643 (3d Cir. 1988), illustrates the point. In this suit by a water company against a county government to recover costs incurred as a result of the release of hazardous substances from a county landfill, as "response costs" under CERCLA, the court observed that unlike an action for damages in tort, Congress did not intend for CERCLA to be a general vehicle for toxic tort actions or to provide whole relief for injured parties.

Class actions under Rule 23 of the Federal Rules of Civil Procedure have been a widely used device for controlling mass toxic tort litigation not only in environmental exposure cases but also in other types of toxic torts. Representative cases discussing the class action approach include *In re Agent Orange Products Liability Litigation,* 506 F. Supp. 737 (E.D.N.Y. 1979), *rev'd,* 635 F.2d 987 (2d Cir. 1980), *cert. denied,* 454 U.S. 1128 (1981), and *In re Three Mile Island Litigation,* 87 F.R.D. 433 (M.D. Pa. 1980). Problems frequently arise where individual causation and damage issues abound and where there is a possibility of multiple classes, which may be defined by economic harm, medical monitoring, punitive damages, and nuisance abatement considerations, for example.

Radiation

The United States' nuclear weapons program has been a fertile ground for major toxic tort litigation. *Allen v. United States,* 588 F. Supp. 247 (D. Utah 1984), *rev'd,* 816 F.2d 1417 (10th Cir. 1987), *cert. denied,* 484 U.S. 1004 (1988), is a prime case in point. Multiple actions were brought against the United States under the Federal Tort Claims Act to recover for injuries and deaths allegedly resulting from radioactive fallout from atmospheric nuclear weapons testing in Nevada in the 1950s and 1960s. In a 225-page opinion which included "mini-primers" on the basic principles of radiation, nuclear physics, and health physics, Judge Bruce Jenkins entered final judgment against the government on 9 of 24 "bellwether" cases selected from nearly 1200 consolidated claims. Difficult as it was, the judge concluded that familiarity with the applicable basic scientific principles was absolutely essential to knowledgeable application of the law. One of the fundamental problems he acknowledged was the non-specific nature of the alleged injury. Cancers which occur in radiation-exposed populations are indistinguishable from cancers among the unexposed.

As has been seen in other mass toxic tort litigation, the emotional plight of the plaintiffs weighed heavily on Judge Jenkins. He wrote:

> This case is concerned with atoms, with government, with people, with legal relationships, and with social values.
> This case is concerned with what reasonable men in positions of decision-making in the United States government between 1951 and 1963 knew or should have known about the fundamental nature of matter.
> It is concerned with the duty, if any, that the United States government had to tell its people, particularly those in proximity to the experiment site, what it knew or should have known about the dangers to them from the government's experiments with nuclear fission conducted above ground in the brushlands of Nevada during those critical years.
> This case is concerned with the perception and the apprehension of its political leaders of international dangers threatening the United States from 1951 to 1963. It is concerned with high level determinations as to what to do about them and whether such determinations legally excuse the United States from being

answerable to a comparatively few members of its population for injuries allegedly resulting from open air nuclear experiments conducted in response to such perceived dangers.

It is concerned with the method and quantum of proof of the cause in fact of claimed biological injuries. It is concerned with the passage of time, the attendant diminishment of memory, the availability of contemporary information about open air atomic testing and the application of a statute of repose.

It is concerned with what plaintiffs–laymen, not experts–knew or should have known about the biological consequences that could result from open air nuclear tests and when each plaintiff knew or should have known of such consequences.

It is ultimately concerned with who in fairness should bear the cost in dollars of injury to those persons whose injury is demonstrated to have been caused more likely than not by nation-state conducted open air nuclear events. (588 F. Supp. at 257)

The 10th Circuit Court of Appeals reversed on the ground that the government was immune from suit because its conduct fell within the scope of the discretionary function exception of the Federal Tort Claims Act, 28 U.S.C. 2680(a).

Johnston v. United States, 597 F. Supp. 374 (D. Kan. 1984), involved allegations by employees at an aircraft instrument plant that their cancers were caused by exposure to luminous dials and instrument parts procured from government surplus sources which contained minute quantities of radium-226. The reported opinion of Judge Patrick Kelly, like the opinion in *Allen,* is noteworthy for its primer on radiation health effects and dosimetry. Judge Kelly held that the United States was immune under the discretionary function exception of the Federal Tort Claims Act.

Other notable radiation litigation includes *In re Consolidated United States Atmospheric Testing Litigation,* 616 F. Supp. 759 (N.D. Cal. 1985), 820 F.2d 982 (9th Cir. 1987), *cert. denied,* 485 U.S. 905 (1988), in which actions were brought against the United States and it contractors for alleged injuries arising out of exposures to radiation resulting from the bombing of Hiroshima and atmospheric testing of nuclear weapons, and *Prescott v. United States,* 523 F. Supp. 918 (D. Nev. 1981), *aff'd,* 731 F.2d 1388 (9th Cir. 1984), 724 F. Supp. 792 (D. Nev. 1989), *aff'd,* 959 F.2d 793 (9th Cir. 1992), *amended,* 973 F.2d 696 (9th Cir. 1992), 858 F. Supp. 1461 (D. Nev. 1994), in which workers at the Nevada Test Site sought to recover for alleged radiation injuries.

Role of the Expert Witness: The "Junk Science" Issue

One of the most obvious features of the toxic tort case is the expert witness. It is unlikely that such a case has been decided by a judge or jury anywhere without the benefit of expert opinion. Much has been written of late, however, about the "helpfulness" of opinions offered by putative "experts" with dubious credentials, methodologies outside of the so-called "mainstream," and conclusions never before revealed in any peer-reviewed journal. The origins of "the science of things that aren't so" are explored by Peter W. Huber in his book which popularized the expression "junk science," *Galileo's Revenge: Junk Science in the Courtroom* (1991, pp. 24–35). The book was apparently so widely read that one unsuccessful litigant, in a suit against the United States for damages arising out of an automobile accident with a vehicle being pursued by the Border Patrol, argued on appeal that the trial judge had been improperly influenced by the book and, as a result, refused to hear testimony proffered by the plaintiff from an accident reconstruction or traffic safety expert [*Stuart v. United States,* 1994 U.S. App. LEXIS 3462, at *6 (9th Cir. March 1, 1994), *amended,* 23 F.3d 1483 (9th Cir. 1994)]. The 9th Circuit upheld the trial judge's determination that the expert testimony in question would have provided little help and was therefore properly excluded.

Some 20 reported federal court decisions have used the phrase, although the one Supreme Court case characterized as the court's first foray into the "junk science issue" avoided the term altogether. *Daubert v. Merrell Dow Pharmaceuticals, Inc., supra,* involved a claim against a pharmaceutical company for birth defects allegedly caused by the mother's ingestion of the anti-nausea drug, Bendectin. Merrell Dow moved for summary judgment based on an expert affidavit to the effect that there were no human epidemiologic studies among 30 that had been done finding an association between Bendectin and birth defects. (Despite this, the company had removed Bendectin from the market in 1983 in the face of some 2000 lawsuits.) Plaintiffs countered with expert affidavits based on animal studies, pharmacological studies showing similarities in the chemical structure of Bendectin and known teratogens, and an unpublished, non-peer-reviewed reanalysis of the epidemiologic studies prepared specifically for purposes of the litigation. The district court excluded the plaintiffs' evidence on the grounds that the opinions had not been subjected to peer scrutiny and had not attained general acceptance in the applicable scientific field. The animal data and chemical studies, it held, were insufficient to establish causation in the face of contrary epidemiologic studies. The 9th Circuit affirmed, citing the "*Frye* rule," which required that expert opinion be based on reliable technique and methodology generally accepted in the scientific community. This "consensus science" approach had been the standard for admission of scientific opinion in nearly all of the federal courts and in most state courts, even after adoption of Rule 702 of the Federal Rules of Evidence. Rule 702 provides:

> If scientific, technical, or other specialized knowledge will assist the trier of fact to understand the evidence or to determine a fact in issue a witness qualified as an expert by knowledge, skill, experience, training, or education, may testify thereto in the form of an opinion or otherwise.

The Supreme Court vacated and remanded, concluding that *Frye* had been superseded by the Federal Rules of Evidence and general acceptance in the scientific community is not a prerequisite to the admission of scientific opinion. Under Rule 702, the trial judge must assure that an expert's opinion rests on a reliable, relevant scientific foundation. The court cautioned that trial judges must become gatekeepers and exercise more control over experts than lay witnesses because of the greater likelihood of expert testimony to confuse, mislead, and persuade. In evaluating an expert's methodology, the trial court should consider (1) whether the theory or technique can be or has been tested, (2) whether the basis for the opinion has withstood peer review, (3) the known or potential rate of error, and (4) whether the bases for the opinion, rather than the conclusions reached, are generally accepted in the scientific community.

The opinion generated a flurry of interest. Some commentators believed that the court was prodding the trial courts to exercise greater discretion in admitting expert testimony that would not have been admitted under the *Frye* rule, which was thought to be more restrictive. Others commented that *Daubert* would do much to ensure that only reliable, legitimate, consensus opinion becomes the basis for decisions in toxic tort litigation [see Special Report, Daubert: What's Next? *Toxics Law Reporter* (BNA)8(9), Part II, (1993, Summer/Fall,)]. The court itself stated its disapproval of the wholesale exclusion of novel scientific evidence that it believed occurred under *Frye's* general acceptance test. In a partial dissent, Chief Justice Rehnquist said,

> I do not doubt that rule 702 confides to the judge some gatekeeping responsibility in deciding questions of the admissibility of proffered expert testimony. But I do not think it imposes on them either the obligations or the authority to become amateur scientists in order to perform that role. (509 U.S. 600-601)

Whether *Daubert* results in the novel scientific opinion being given more credence or the closer scrutiny of theoretical evidence with more frequent exclusion of pseudo-scientific opinions which have no valid scientific basis remains an issue to be answered as its guidelines are fleshed out in the federal trial courts and state courts with evidentiary rules based on the Federal Rules.

Toxic Torts: Prospects for the Future

The law of toxic torts is in a state of flux, as is science by definition. New theories of liability are being tested continually and creative remedies urged. Traditional legal rules are being strained in the process. Proponents of the new theories have argued that the old tort rules no longer fit the problems claimed to be caused by new products, new chemicals, and new toxins. Areas of scientific uncertainty are not shrinking either. The tension created by the juxtaposition of scientific uncertainty and unsettled law will afflict toxic tort litigants indefinitely. The most difficult obstacle for plaintiffs will continue to be causation. Latency between exposure and manifestation of injury is perhaps the single scientific factor most responsible for the complex, time-consuming, and expensive nature of the toxic tort. A concomitant problem is the inability of epidemiologic studies to detect in populations effects from chronic, low-level exposures to many toxins and there are inherent problems in applying causal associations observed in such studies to individuals. Toxic tort cases are being filed with increasing regularity, perhaps even disproportionately to the number of potentially toxic substances being developed and put into commerce. Professional and societal challenges in science and law can be expected to expand dramatically as our society becomes technologically more complex than one could ever have envisioned even a half-century ago.

Further Reading

Annotations

Annotation, *"Concert of Activity," "Alternative Liability," "Enterprise Liability," or Similar Theory as Basis for Imposing Liability Upon One or More Manufacturers of Defective Uniform Product, in Absence of Identification of Manufacturer of Precise Unit or Batch Causing Injury,* 22 A.L.R.4th 183 (1983).

David A. Klein, Annotation, *Reliability of Scientific Technique and Its Acceptance within Scientific Community as Affecting Admissibility at Federal Trial of Expert Testimony as to Result of Test or Study Based on Such Technique Modern Cases,* 105 A.L.R. Fed. 299 (1991).

David Carl Minneman, Annotation, *Future Disease or Condition, or Anxiety Relating Thereto, as Element of Recovery,* 50 A.L.R.4th 13 (1986).

Gregory G. Sarno, Annotation, *Infliction of Emotional Distress: Toxic Exposure,* 6 A.L.R.5th 162 (1992).

Allan L. Schwartz, Annotation, *Recovery of Damages for Expense of Medical Monitoring to Detect or Prevent Future Disease or Condition,* 17 A.L.R.5th 327 (1994).

Law Journals

Ayala, F. J., and Black, B. (1993). The nature of science: A primer for the legal consumer of scientific information. *Sci. Cts.* 1, 1.

Benjamin, D. M. (1993). Elements of causation in toxic tort litigation; science and law must agree. *J. Legal Med.* 14, 153.

Bernstein, D. E. (1990). Out of the frying pan and into the fire: The expert witness problem in toxic tort litigation. *Rev. Litig.* 10, 117.

Black, B., and Hollander, D. H., Jr. (1993). Unravelling causation: Back to the basics. *U. Balt. J. Environ. L.* 3, 1.

Black, B., and Lilienfeld, D. E. (1984). Epidemiologic proof in toxic tort litigation. *Fordham L. Rev.* 52, 732.

Blomquist, R. F. (1992). American toxic tort law: An historical background 1979–1987. *Pace Environ. L. Rev.* 10, 85.

Brennan, T. A. (1988). Causal chains and statistical links: The role of scientific uncertainty in hazardous substance litigation. *Cornell L. Rev.,* 73, 469.

Chesebro, K. J. (1993). Galileo's retort: Peter Huber's junk scholarship. *Am. U. L. Rev.* 42, 1637.

Gold, S. (1986). Note, Causation in toxic torts: Burdens of proof, standards of persuasion, and statistical evidence. *Yale L. J.* 96, 376.

Poulter, S. R. (1992). Science and toxic torts: Is there a rational solution to the problem of causation? *High Tech. L. J.* 7, 189.

Trauberman, J. (1983). Statutory reform of "toxic torts": Relieving legal, scientific, and economic burdens on the chemical victim. *Harvard Environ. L. Rev.* 7, 177.

Weinstein, J. B. (1985). The role of the court in toxic tort litigation. *Geo. L. J.* 73, 1389.

Weinstein, J. B., and Hershenov, E. B. (1991). The effect of equity on mass tort law. *U. Illinois L. Rev.* 1991, 269.

Scientific and Medical Journals

Brannigan, V. M., Bier, V. M., and Berg, C. (1992). Risk, statistical inference, and the law of evidence: The use of epidemiological data in toxic tort cases. *Risk Anal.* 3, 343–351.

Byinton, S. J. (1990). Risk assessment: Legal judgments and products liability, or the nexus between science and the law. *Sci Total Environ.* 99, 245–253.

Henderson, T. W. (1990). Toxic tort litigation: Medical and scientific principles in causation. *Am. J. Epidemiol.* 132, 569.

Johnson, R. H. (1989). Radiation litigation in retrospect. *Radiat. Res.* 117, 173–177.

Jose, D. E. (1989). Radiation litigation: Future issues. *Radiat. Res.* 117, 181–184.

Muscat, J. E., and Huncharek, M. S. (1989). Causation and disease: Biomedical science in toxic tort litigation. *J. Occup. Med.* 31, 997–1001.

Treatises and Loose-Leaf Services

Bernstein, D. E., *et al.* (Eds.) (1993). *Phantom Risk: Scientific Inference and the Law.* Cambridge, Ma. MIT Press.

Dore, M. (1987). *Law of Toxic Torts.* Deerfield, Il. Clark Boardman Collaghan.

Frumer, L. R., *et al.* (Eds.) (1988). *Personal Injury: Actions, Defenses, Damages.* N.Y., N.Y. Matthew Bender.

Houts, M., *et al.* (1993). *Courtroom Toxicology.* N.Y., N.Y. Matthew Bender.

Huber, P. W. (1991). *Galileo's Revenge: Junk Science in the Courtroom.* Basic Books/Harper Collins.

Madden, M. S. (1993). *Toxic Torts Deskbook.* Boca Raton, Fl. Lewis Publishers.

O'Reilly, J. T. (1992). *Toxic Torts Practice Guide,* 2nd ed. Colorado Springs, Co., Shepard's/McGraw Hill.

Searcy, M. (1987). *A Guide to Toxic Torts.* N.Y., N.Y. Matthew Bender.

Toxics Law Reporter: A Weekly Review of Toxic Torts, Hazardous Waste, and Insurance Litigation (BNA). Washington, D. C. Bureau of National Affairs.

—Michele A. Reynolds

Related Topics

Carcinogenesis
Clean Air Act
Clean Water Act
Comprehensive Environmental Response, Compensation, and Liability Act
Developmental Toxicology
Dose–Response Relationship
Epidemiology
Environmental Toxicology
Exposure Assessment
Exposure Criteria
Federal Insecticide, Fungicide, and Rodenticide Act
Food, Drug and Cosmetic Act
Forensic Toxicology
Hazard Identification
Information Resources in Toxicology
Medical Surveillance
Occupational Safety and Health Act
Occupational Toxicology
Pollution Prevention Act
Psychological Indices of Toxicity
Radiation Toxicology
Resource Conservation and Recovery Act
Risk Assessment, Ecological
Risk Assessment, Human Health
Safe Drinking Water Act
State Regulation of Consumer Products
Toxic Substances Control Act
Uncertainty Analysis

Triadimefon

- CAS: 43121-43-3
- SYNONYMS: 1-(4-Chlorophenoxy)-3,3-dimethyl-1-(1*H*-1,2,4-triazol-1-yl)-2-butanone; Amiral; Azocene; BAY 6681 F; BAY-MEB 6447; MEB 6447; Bayleton; Triadimefone
- CHEMICAL CLASS: Triazole fungicide
- CHEMICAL STRUCTURE:

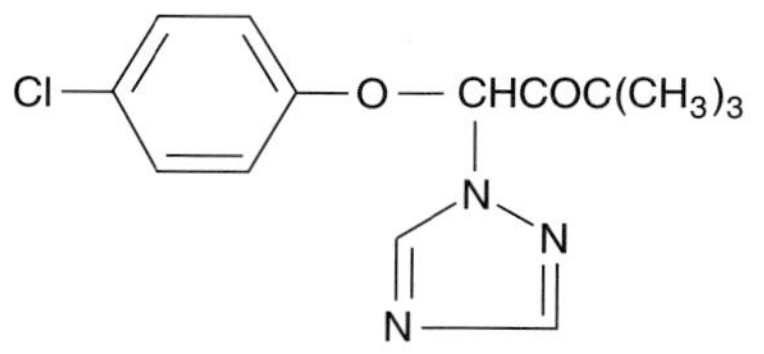

Use

Triadimefon is an agricultural fungicide used to protect various fruits and grains.

Exposure Pathways

Exposure to triadimefon may occur by the oral, dermal, and inhalation routes.

Toxicokinetics

Triadimefon may be absorbed through oral, inhalation, or dermal routes of exposure. One study demonstrated $\geq$50% percutaneous absorption of triadimefon in rats, with an elimination half-life of 29–54 hr. The major mammalian metabolite of triadimefon is triadimenol. Metabolic products are excreted via the urine and feces.

Mechanism of Toxicity

Triadimefon has effects on behavior that are similar to those of psychomotor stimulants like amphetamine. Triadimefon substituted completely for the effects of methylphenidate in one study. The mechanism of action of triadimefon may involve the interaction of its azole nitrogen with the heme iron. The affinity of triadimefon for cytochrome P45014DM is extremely high compared with usual nitrogenous ligands. Another study suggested that triadimefon acts by releasing and/or blocking reuptake of dopamine and serotonin, thereby altering their actions.

Human Toxicity

Human exposure to triadimefon has been associated with contact dermatitis.

Clinical Management

Toxicity is unlikely following acute exposure to triadimefon and any treatment is symptomatic.

Animal Toxicity

In both mice and rats, triadimefon produces hyperactivity similar to that seen following administration of compounds with catecholaminergic activity (e.g., *d*-amphetamine). Prominent effects of triadimefon exposure in rats (100 and 300 mg/kg) included increased arousal and stereotypies involving repetitive sniffing, head bobbing, pacing, and self-mutilation. Dose-related handling-induced convulsions, changes in reflex and sensory reactivity, hypothermia, and body weight loss were also significant findings. Doses of 30, 75, and 150 mg/kg triadimefon increased figure-eight maze activity, whereas 300 mg/kg decreased activity. The effects of triadimefon on operant performance were similar to those seen following *d*-amphetamine and were attenuated by pretreatment with chlorpromazine (0.5 mg/kg). Female rats appeared to be somewhat more sensitive than males. Recovery was evident in some measures on the day after dosing, but the effects of high doses ($\geq$100 mg/kg) were typically prolonged several days. The oral LD_{50} for the rat ranges from 350 to 1000 mg/kg. The inhalation LC_{50} is $\geq$450 mg/m^3/4 hours, and the dermal value is $\geq$2–5 g/kg.

—*Todd A. Bartow*

Related Topic

Pesticides

Trichloroethanes

◆ SYNONYMS: 1,1,1-Trichloroethane, methyl chloroform (CAS 71-55-6); 1,1,2-trichloroethane, vinyl trichloride, β-trichloroethane (CAS 79-00-5)

◆ CHEMICAL CLASS: Chlorinated hydrocarbon

◆ CHEMICAL STRUCTURE:

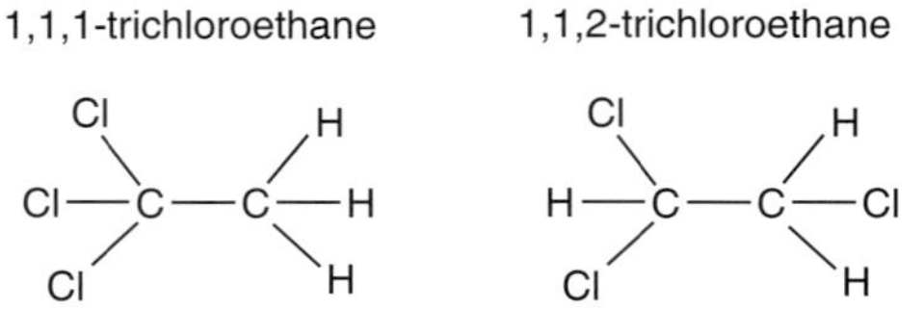

Uses

Trichloroethane is a solvent used for cleaning and degreasing. It is a substance of abuse in glue sniffing. It is also used in organic synthesis.

Exposure Pathways

Inhalation and dermal and mucous membrane contact are possible routes of exposure.

Toxicokinetics

The lungs excrete most of the dose unchanged. The biological half-life is 8.7 hr.

Mechanism of Toxicity

Trichloroethane is absorbed through the lungs, gastrointestinal tract, and skin (in higher concentrations). It blocks the release of adrenal catecholamines. It also defats and disrupts membranes.

Human Toxicity

Trichloroethane can irritate the eyes and mucous membranes. It is a central nervous system (CNS) depressant. The ACGIH TLV is 350 ppm for 1,1,1-trichloroethane and 10 ppm for 1,1,2-trichloroethane.

Clinical Management

Respiratory support and cardiac monitoring should be provided. Hypotension should be treated.

Animal Toxicity

Trichloroethane is a mutagen and primary irritant (due to defatting). The reported inhalation LC_{50}s for 1,1,1-trichloroethane are 18,000 ppm/3 hr in rats and 13,500 ppm/10 hr in mice. The reported inhalation LC_{50} for 1,1,2-trichloroethane is 2000 ppm in rats. The reported oral LD_{50} for 1,1,2-trichloroethane is 100–200 mg/kg in rats.

—*Shayne C. Gad and Jayne E. Ash*

Related Topic

Pollution, Water

Trichloroethylene

◆ CAS: 79-01-6

◆ SYNONYMS: Trichloroethene; 1,1,2-trichloroethene; TCE; TRI; trichlor

◆ CHEMICAL CLASS: Chlorinated olefinic hydrocarbon

Uses

Manufactured by chlorination of acetylene or other two carbon hydrocarbons or by dehydrohalogenation of tetrachloroethane, trichloroethylene may contain one of numerous additives for stabilization. This chlorinated solvent has been used extensively as a degreasing and metal cleansing agent, as a heat exchange liquid, as a diluent in paints and adhesives, in textile processing and in the manufacture of organic chemicals and pharmaceuticals, as a cleaning solvent, for dyeing of polyester textiles, as an extractant for caffeine from coffee, in typewriter correction fluid, and as a solvent for insecticides and other chemicals. It is a volatile liquid whose vapor is heavier than air. The vapor has a chloroform-like odor detectable at around 21 ppm in air.

Exposure Pathways

The two main sources of human exposure to trichloroethylene are the environment and the workplace, although home use of products containing trichloroethylene is not uncommon. Background levels of trichloroethylene can be found in the outdoor air we breathe (30–460 parts per trillion) and in many lakes, streams, and underground water used as sources of tap water for homes and businesses. An important source of environmental release of trichloroethylene is evaporation into the atmosphere from work performed to remove grease from metal (degreasing).

Toxicokinetics

Absorption of trichloroethylene following inhalation exposure in humans is characterized by an initial rate of trichloroethylene uptake that is quite high. Retention of inhaled trichloroethylene has been measured at up to 75% of the amount inhaled. Absorption of trichloroethylene following oral exposure in both humans and animals is rapid and extensive. In animal studies absorption from the gastrointestinal tract has been measured at 91–98%, and peak trichloroethylene blood levels are attained within a matter of hours.

Trichloroethylene is extensively metabolized (40–75% of the retained dose) in humans to trichloroethanol, its glucuronide, and trichloroacetic acid. Saturation of metabolism has not been demonstrated in humans up to an exposure concentration of 300 ppm. Mathematical models predict, however, that saturation of metabolism is possible at trichloroethylene concentrations previously used for anesthesia (i.e., 2000 ppm). Although the liver is the primary site of trichloroethylene metabolism in animals, there is evidence for extrahepatic metabolism of trichloroethylene in the kidneys and lungs.

The distribution of trichloroethylene in rats following exposure to 200 ppm, 6 hr/day for 5 days was studied. Seventeen hours after exposure on Day 4, there were relatively high levels of trichloroethylene in the perirenal fat (0.23 nmol/g) and in the blood (0.35 nmol/g) and virtually no trichloroethylene in the other tissues. Following exposure on Day 5, tissue levels in brain, lungs, liver, fat, and blood reached a steady state within 2 or 3 hr.

In humans, about 11% of trichloroethylene is eliminated through the lungs, whereas more than 50% of the absorbed dose is metabolized and excreted in the urine as trichloroacetic acid (TCA) and other metabolites. Elimination is relatively slow in humans, with TCA being detected in the urine of exposed individuals up to 12 days postexposure suggesting a cumulative process, probably related to storage in fatty tissue. The biological half-life of urinary metabolites of trichloroethylene in humans is about 41 hr.

Mechanism of Toxicity

Extended exposure (e.g., occupational exposure) to a chlorinated solvent like trichloroethylene typically results in signs of central nervous system (CNS) disturbance and hepatotoxicity. Administration of this chemical to mice induces neoplasms in the liver, as is typical of virtually all the chlorinated hydrocarbons. Trichloroethylene is readily converted to trichloroacetic acid, which acts as a peroxisome proliferator, and hepatic neoplasms in mice may arise through this mechanism. Glutathione adducts of trichloroethylene are thought to be converted to the reactive metabolite in the kidneys through the action of β-lyase. These processes may account for nephrotoxicity exhibited in rats.

Human Toxicity

Trichloroethylene is readily absorbed through ingestion, inhalation, and dermal exposure, the latter producing a defatting effect if contact is prolonged, resulting in erythema and vesication followed by exfoliation. The liquid solvent is also an eye irritant producing pain and irritation but apparently no permanent injury. Exposure to high vapor concentrations of trichloroethylene has been reported to result in irritation of the mucous membranes of the eyes, nose, and throat, conjunctivitis, rhinitis, and pharyngitis. Exposures exceeding 100 ppm in air are reported to result in restlessness, peripheral neuritis, impaired concentration, irritability, euphoria, lightheadedness, dizziness, depression, reversible trigeminal degeneration, psychic disturbances, cranial nerve deafness, alterations in electrical patterns in the brain, bronchoconstriction, fatal cardiac arrhythmias, and renal and hepatic damage. In combination with alcohol, trichloroethylene exposure produces a vasodilation which has been described as "degreasers flush." Optic neuritis, hallucinations, and gastrointestinal changes have been reported after ingestion of trichloroethylene. These symptoms are often accompanied by nausea and vomiting, as well as major cardiovascular effects including hypotension, conduction defects, myocardial injury, and cardiac arrhythmias. The latter has been reported to be the cause of death in some individuals who have been exposed to high levels of trichloroethylene, but usually death is preceded by coma and subsequent hepatic or renal failure. The estimated dose in humans to cause death is reported to be 3–5 ml/kg. The ACGIH TLV for trichloroethylene is 50 ppm, while OSHA's PEL is 100 ppm and NIOSH's recommended exposure limit is 25 ppm.

There is limited evidence in humans for the carcinogenicity of trichloroethylene in humans. There are reports of increased risks of multiple myeloma, childhood leukemia, non-Hodgkins lymphoma, and cancer of the biliary passages, but the studies are limited. Limited evidence also exists supporting a relationship between trichloroethylene exposure and autoimmune disease including systemic lupus erythematosis and scleroderma. There is sufficient evidence, however, in experimental animals for the carcinogenicity of trichloroethylene; therefore, trichloroethylene has been classified as being a probable human carcinogen.

Animal Toxicity

Acute toxicity data indicate that trichloroethylene is relatively nontoxic by the inhalation and oral routes. In mice, LC_{50}s ranged from 7480 to 49,000 ppm,

whereas in rats the range was 12,500–26,300 ppm. By the ingestion route, acute LD_{50}s in dogs, cats, rats, mice, and rabbits ranged from approximately 2000 to 8000 mg/kg. Inhalation and oral studies indicate that the bone marrow, CNS, liver, and kidneys are the principal targets of trichloroethylene in animals. Effects on the liver and kidney include enlargement with hepatic and biochemical and/or histological alterations. Other reported effects include indication of impaired heme biosynthesis and other hematological alterations in rats exposed by inhalation and immunosuppression in orally exposed mice.

Inhalation studies with mice and rats indicate that trichloroethylene is a developmental toxicant. Fetotoxicity is expressed mainly as skeletal ossification anomalies and other effects consistent with delayed maturation. Oral studies with rats and mice showed no trichloroethylene-related effects on fertility or other indicators of reproductive performance. No definitive teratogenic effects have been reported regarding exposure to trichloroethylene.

Chronic inhalation exposure to trichloroethylene produced lung and liver tumors and leukemia in mice and Leydig cell tumors in rats. Chronic oral exposure to trichloroethylene produced increased incidences of hepatocellular carcinomas in mice and marginally significant increased incidences of renal adenocarcinomas in rats.

Mutagenic responses generally occurred with metabolic activation only, suggesting the involvement of metabolites of trichloroethylene. The mutagenic potential of pure trichloroethylene is unclear; however, the limited information available suggests that trichloroethylene would be a weak mutagen. Positive findings showing frameshift and base pair mutations in *Saccharomyces cerevisiae* and reverse mutations in *Escherichia coli* K12 have been reported.

—*R. A. Parent, T. R. Kline,*
and D. E. Sharp

Related Topics

Kidney
Pollution, Water
Sensory Organs
Skin

Tricyclic Antidepressants

♦ REPRESENTATIVE COMPOUNDS: Imipramine; amitriptyline; doxepin; desipramine; nortriptyline

♦ CHEMICAL CLASS: The tricyclic antidepressants are a group of drugs that have a three-ring molecular core and share a similar pharmacologic effect.

♦ CHEMICAL STRUCTURE:

$CHCH_2CH_2N(CH_3)_2$
Amitroptyline

$CH_2CH_2CH_2NHCH_3$
Desipramine

Uses

Tricyclic antidepressants are used to treat depression. They are also used for treatment of enuresis in children, chronic pain syndromes, the fibromyalgia syndrome, and chronic headaches.

Exposure Pathways

Ingestion is the most common route of exposure. Several tricyclic antidepressants are also available in injectable form.

Toxicokinetics

The tricyclic antidepressants are well absorbed following oral ingestion. Large ingestions may be more slowly absorbed. There is extensive first-pass metabolism that limits oral bioavailabilty.

The tricyclic antidepressants are highly lipid soluble and bind extensively to tissue and plasma proteins. The volume of distribution ranges from 10 to 50 liters/kg.

The tricyclic antidepressants are extensively metabolized by the liver and partially enterohepatically recirculated. They undergo demethylation, hydroxylation, and glucuronide conjugation. The demethylated metabolites of the tertiary amine tricyclic antidepressants are pharmacologically active. Drugs that induce hepatic microsomal enzymes speed the metabolism of tricyclic

antidepressants. The half-life of various tricyclic antidepressants ranges from 10 to 50 hr. Less than 5% of the drugs appear unchanged in the urine.

Mechanism of Toxicity

The toxicity of tricyclic antidepressants is related to their anticholinergic, quinidine-like, and α_1 blocking properties. The inhibition of reuptake of biogenic amines in the central nervous system (CNS) probably also plays a role in CNS toxicity. Early CNS toxicity may be due to anticholinergic toxicity, while other central effects of the tricyclic antidepressants probably contribute to more severe toxicity. The anticholinergic effects may produce sinus tachycardia. The tricyclic antidepressants block fast sodium channels in cardiac tissue in a fashion similar to quinidine. This leads to altered conduction, slowing of both depolarization and repolarization, and decreased inotropy. This can lead to ventricular arrhythmias, bradyarrhythmias, and asystole. Decreased inotropy, peripheral α_1 blockade, and bradycardia can all contribute to hypotension and shock.

Human Toxicity: Acute

Early signs of tricyclic antidepressant toxicity are due to anticholinergic effects and include tachycardia, mydriasis, dry mouth, low-grade fever, diminished bowel sounds, CNS excitation, delirium, or drowsiness. More serious toxicity is manifest by coma, respiratory depression, seizures, and cardiovascular toxicity including conduction disturbances, hypotension, ventricular arrhythmias, and asystole. Seizures may cause hyperthermia, rhabdomyolysis, and metabolic acidosis. Death usually results from cardiovascular collapse or intractable seizures. Clinical deterioration can be rapid and catastrophic with tricyclic antidepressant overdose. The typical therapeutic dose of a tricyclic antidepressant is 2–4 mg/kg/day. Doses of 15–20 mg/kg are potentially lethal. Therapeutic drug levels for most tricyclic antidepressants range from 100 to 260 μg/ml. Toxicity may be seen at levels only modestly elevated, although severe symptomatology is usually associated with levels >1000 μg/ml. However, drug levels are not useful in predicting toxicity, complications, or patient management. EKG changes include sinus tachycardia, a rightward deviation of the terminal vector of the frontal plane QRS complex to >120°, QT and PR prolongation, intraventricular conduction disturbances with a prolongation of the QRS duration, and T wave changes. The risk of seizures and cardiac arrhythmias has been reported to be increased with a QRS duration >0.10 and >0.16 sec, respectively. In distinction to the tricyclic antidepressants, newer cyclic antidepressants such as the bicyclics and dibenzoxazepines are less cardiotoxic but associated with an increased risk of seizures.

Clinical Management

If the patient is seen early postingestion, up to several hours, gastric decontamination by lavage should be performed. Because of the risk of rapid CNS depression, ipecac should be avoided and airway protection by endotracheal intubation should beconsidered. Activated charcoal should be given. Flumazenil and physostigmine should be avoided. Sodium bicarbonate administration is beneficial in treating cardiac toxicity and hypotension. It is not clear if the effects are a consequence of alkalinization or sodium administration. With signs of impaired conduction (QRS >0.10–0.12 sec), ventricular arrhythmias, or hypotension, alkalinize serum to pH 7.45–7.55. Sodium bicarbonate should be used by bolus injection followed by continuous infusion to maintain target pH. Ventricular arrhythmias unresponsive to alkalinization should be treated by standard ACLS methods, avoiding the class Ia antiarrhythmics (quinidine, procainamide, and disopyramide). Phenytoin has been used in this setting. Hypotension may be multifactorial and treatment should include volume resuscitation if not contraindicated, serum alkalinization, and pressor support if needed. Central hemodynamic monitoring may be useful in this setting. Seizures should be treated in the usual fashion. For uncontrolled seizures, paralysis may be indicated as the associated acidosis and hyperthermia may aggravate cardiac toxicity. Hemodialysis and hemoperfusion are not effective treatment modalities. Extracorporeal membrane oxygenation has been used to treat severe cardiac toxicity unresponsive to other therapy. Only patients free of any signs of toxicity during the first 6 hr, with the exception of a resolved tachycardia, can be considered medically clear at that time.

—Michael J. Hodgman

Trihalomethanes

- ◆ REPRESENTATIVE COMPOUND: Chloroform (CAS 67-66-3; see Chloroform)
- ◆ SYNONYMS: Carbonyl chloride; chloroformyl chloride
- ◆ OTHER CHEMICAL COMPOUNDS: Bromoform; dichlorobromomethane; dibromochloromethane (see Bromoform)
- ◆ CHEMICAL STRUCTURE:

$$\text{Cl}-\underset{\underset{\text{Cl}}{\overset{\overset{\text{H}}{|}}{|}}{\text{C}}-\text{Cl}$$

Use

Trihalomethanes are by-products of the chlorination process.

Exposure Pathways

Ingestion, inhalation, and dermal contact are possible routes of exposure.

Toxicokinetics

Trihalomethanes are absorbed readily then distributed primarily to stomach, liver, and kidneys. Elimination is primarily by the lungs. Chloroform undergoes more conjugation than other trihalomethanes (see Phosgene for related toxicokinetics).

Mechanism of Toxicity

Chloroform inhibits the function of kidney tubulars. It increases nitrogen in blood urea, renal concentrating ability, and glomerular filtration rate. It also increases the metallothionein concentration and reduces the level of cytochrome P450. P450 oxidation also contributes to metabolic release of carbon monoxide. Glutathione is a cofactor of this process. Toxic intermediates such as phosgene, a conjugate by-product of chloroform, may bind covalently with proteins and lipids, contributing to toxicity.

Human Toxicity

Chloroform is an irritant to skin and mucous membranes. It produces liver and kidney damage and central nervous system depression. Carcinogenic properties are suspected but not confirmed. The fatal dose is estimated to be 10 ml.

Clinical Management

Respiratory therapy should be administered. If ingested, emesis should be induced. Blood pressure should be maintained and normal urinary output. A high carbohydrate diet can assist in restoring normal liver function. Epinephrine should not be used.

Animal Toxicity

Chloroform has proven carcinogenic in the liver, kidneys, and/or intestines of rodents. It inhibits kidney function and increases chromosomal aberrations.

—*Shayne C. Gad and Jayne E. Ash*

Related Topics

Metallothionein
Neurotoxicology: Central and Peripheral
Pollution, Water

Trinitrotoluene

- ◆ CAS: 118-96-7
- ◆ PREFERRED NAME: TNT
- ◆ SYNONYMS: 2,4,6-Trinitrotoluene; methyltrinitrobenzene
- ◆ CHEMICAL CLASS: Explosives
- ◆ CHEMICAL STRUCTURE:

Uses

Trinitrotoluene is used as a high explosive. It is an intermediate in the production of dyes and photographic chemicals.

Exposure Pathways

Ingestion, inhalation, and dermal contact are possible routes of exposure.

Toxicokinetics

Trinitrotoluene is readily absorbed through skin. It is excreted in urine more than in feces; some is found in bile. Residual retention of trinitrotoluene is found in the body at 17 days.

Mechanism of Toxicity

Trinitrotoluene increases UDP-glucuronsyltransferase in the liver and kidneys. It increases renal epoxide hydrolase activity.

Human Toxicity

Trinitrotoluene vapors are toxic. Toxic exposure may cause weakness, anemia, headaches, liver, or central nervous system damage. Chronic exposure may cause cataracts, cyanosis, jaundice, or hepatitis. The ACGIH TLV is 0.5 mg/m^3 of air. Trinitrotoluene is a possible human carcinogen.

Clinical Management

Methylene blue should be administered with oxygen therapy.

Animal Toxicity

Trinitrotoluene is a mutagen. The oral LD$_5$0s are 795 mg/kg in rats and 660 mg/kg in mice.

—*Shayne C. Gad and Jayne E. Ash*

Related Topic

Pollution, Water

Uncertainty Analysis

Uncertainty analysis provides an evaluation of the key parameters that contribute to the uncertainty (e.g., variability and imprecision) involved in performing a risk assessment. It provides information that enables decision makers to better understand the strengths, weaknesses, and assumptions inherent in the assessment and evaluates the conclusions of the risk assessment accordingly. The results of the uncertainty analysis should serve to help the risk manager better understand the implication of the conclusions derived based on the risk assessment and to support scientifically based and economically feasible hazardous waste management decisions.

Uncertainty analysis has become a popular method for assessing the uncertainty associated with risk estimates calculated for specific receptors under designated routes of exposures. Two types of uncertainty analysis are typically performed in risk assessment: qualitative and quantitative. Qualitative uncertainty analysis literally describes and lists the parameters or assumptions likely to produce the highest uncertainties with an attached statement quantifying whether the risks have been over- or underpredicted. Although the qualitative analysis is simple to perform, it generally addresses the overall uncertainty of the assessment in vague and general terms and yields very imprecise results. Quantitative uncertainty analysis, on the other hand, is the mathematical investigation of the uncertainty in the risk output performed by varying the input parameters such that the relative contribution of each parameter is determined. This method has numerous advantages over the qualitative analysis since (1) it provides precise results that are for the most part reproducible and consistent; (2) with the advent of many commercially available software, the analysis is straightforward and easy to perform; and (3) it does not require mathematics beyond that which is commonly used in risk assessments.

In most risk assessments, calculations are based on deterministic values (i.e., all of the variables are treated as known constants). Many of these values are actually estimates of either an average, high-end, or conservative worst-case condition used in place of a range of values that would better characterize a population or condition. For example, one value may be used to represent the body weight of all the individuals in a population that is composed of people of different ages, sexes, and sizes. Clearly, the deterministic values commonly used in risk assessments represent only a portion of the overall available data.

The current practice in risk assessment of using multiple point estimates that are either high end or worst case to calculate risk often results in a compounding of conservatism intended to significantly overestimate risk. The uncertainty associated with each parameter commonly used to calculate risk usually has not been fully characterized. Major sources of uncertainty associated with risk assessment include:

- Incomplete information on the potential adverse health effects that can be caused by a chemi-

cal in the species, at the dose, and over the anticipated length of time the exposure might occur

- Natural variability (e.g., uncertainty about the
range of sensitivity in the population of interest)

- Measurement and sampling errors (i.e., errors
resulting from direct measurement due to instrument and observation variations)

- Bias (i.e., difference between the estimated value
and true value)

- Judgmental error (i.e., estimated value based on
professional opinion)

- Randomness

- Unpredictability (i.e., systems that exhibit extreme sensitivity)

- Disagreement (i.e., differences in opinion based
on expert opinion)

- Approximations (i.e., simplification of the
real world).

The uncertainty associated with certain exposure parameters may decrease once probability distributions
are used that describes the parameter of interest (e.g.,
body weight) in place of point estimates as inputs into
the calculations. This methodology is described as stochastic modeling which utilizes the full range of data
available by selecting random variables from a defined
probability distribution. There are three modeling
methods used to propagate uncertainty: analytical
methods using mathematical statistics, the delta
method, and Monte Carlo analysis. The analytical and
delta methods are only used for analysis with limited
complexity and thus are not discussed further. It should
be noted that although increasing the complexity of
the uncertainty model may initially improve its accuracy (i.e., consistent and reproducible results), uncertainty within the analysis increases with complexity as
less characterized parameters are included in the model.
The Monte Carlo method is a well-established approach used in characterizing uncertainty that can be
used to incorporate ranges of data (distributions) into
calculations. Monte Carlo involves choosing values

from a random selection scheme drawn from probability density functions based on a range of data that
characterize the parameter of interest. Monte Carlo
analysis can be selectively used to generate input parameters and mixed with point estimates, as appropriate, to calculate risk.

The use of this method has become increasingly popular as the availability of commercial software that
allows the probability distribution functions to be input
directly into a computer spreadsheet have become
available. Once the probability distribution is incorporated into a spreadsheet cell, each time the spreadsheet
is recalculated, a new value for the random variable is
selected from the distribution and used in the calculations. The key to appropriately using this method is to
run the entire simulation (choosing random samples
from each distribution) hundreds to thousands of times.
After each selection, a new representative parameter is
generated and can be used to base an estimate of exposure or risk. When the method is used to calculate
risk or exposure, the results of all risk estimates are
summarized in a histogram which provides risk assessors a full possible distribution of risk based on probability.

The Monte Carlo analysis is performed using the
following steps:

- Standard spreadsheet calculations are entered
for all chemicals and pathways to be modeled
following the methods used for deterministic calculations. For each of the random variables, discrete or continuous probability density functions
are placed in the appropriate cells.

- Any correlations among the exposure parameters must be identified. For example, body weight
and skin surface area would be positively correlated such that when a high body weight is selected a corresponding high skin surface area
should likewise be used. It is important to identify
these correlations so that individual simulations
avoid selecting values of two different random
variables that are not representative of a individual (i.e., high body weight and low skin surface
area). All other variables are assumed to be independent and are not correlated with any other
parameter selected.

- The simulation should be run thousands of
times in order to fully sample from each distribu-

tion. The summaries of the risk estimates can include statistical tables and histograms of resulting risks and intermediate calculations.

The most difficult aspect of performing a Monte Carlo analysis is estimating the probability distribution underlying many of the variables used. Because it is not immediately obvious what distribution best characterizes the exposure parameters for a particular population, the risk assessor must carefully evaluate the available data and choose the appropriate distribution based on the level of information known. A sensitivity analysis may provide additional insight into the distribution selected by indicating the significance of the parameter in affecting the conclusions. There are several general rules used in uncertainty analysis in determining the most unbiased distribution for a specific parameter. The data should initially be tested using a robust goodness-of-fit test (i.e., chi-square test, *W* test, or Kolmogorov–Smirnov test) to determine if the data are normally or lognormally distributed. If only a range of values are known for a variable, a uniform distribution is the least biased assumption. If the range and mode of values are known, a triangular distribution could be used although it may result in values being selected more from the extremes than would be expected. A beta distribution can be selected when estimates of the mean, lower bound, and upper bound are available. If the data cannot be adequately described by a standard distribution, the empirical data may be "bootstrapped" into the simulation in which the model randomly selects individual data points from the data provided. It should be noted that the use of full distribution functions (e.g., lognormal or normal) is not entirely accurate since it is impossible to have mass values beyond physical plausibility within the extreme tails of these distributions although these values may be selected during the analysis (i.e., body weights that are negative, zero, or infinite). Users of packaged software have the flexibility of describing the exposure distributions in terms of parametric and nonparametric functions including cumulative percentiles, bootstrapped values, and moments to limit the values selected from a distribution to within the physical realm.

Uncertainty analysis should account for and characterize the variability inherent in most data sets. In certain cases (e.g., use of Monte Carlo analysis), it is used to more accurately represent a parameter (e.g., body weight) that influences the calculation of risk. Quantitative analysis provides information that enhances understanding and implications of the risk assessment. In the case of human health risk assessment, the risk assessor attempts to quantify the likelihood an individual in a population will develop cancer or other adverse effect due to contact with a chemical at a particular dose level over a specified exposure period. Uncertainty analysis allows the risk assessor to more accurately account for the differences in the population being evaluated (e.g., body weight, exposure duration, ingestion rates, and other exposure parameters) that potentially impact the overall estimate of risk and the conclusions that can be made base on the assessment. It does not, however, address the uncertainty or validity of the methodologies used to develop the parameter distributions or test the underlying uncertainty model itself.

—*Virginia Lau*

Related Topics

Hazard Identification
Risk Assessment, Ecological
Risk Assessment, Human Health
Risk Characterization
Risk Communication
Risk Management
Sensitivity Analysis

Uranium (U)

- CAS: 7440-61-1
- SELECTED COMPOUNDS: Uranium dioxide, UO_2 (CAS: 1344-57-6); uranium tetrafluoride, UF_4 (CAS: 10049-14-6)
- CHEMICAL CLASS: Metals

Uses

By far the most important use of uranium is in nuclear reactors and weapons. A limited amount of uranium is used in glass, ceramics, and specialty chemicals.

Exposure Pathways

Inhalation of radon daughters (isotopes of lead, bismuth, and polonium) is a primary exposure pathway in uranium mining operations. Radon daughters attach to dust particles, which are inhaled by workers.

In addition to inhalation, ingestion and dermal contact are potential exposure pathways for uranium and its compounds. Uranium is ubiquitous; small amounts are ingested with a variety of foods (e.g., vegetables and cereals). In areas near uranium mines, drinking water becomes a source. Some water-soluble uranium compounds are absorbed through the skin.

Toxicokinetics

The soluble uranium compound (uranyl ion) is easily absorbed from the gastrointestinal tract. Most absorbed uranium is bound to carbonate; the remainder is bound to blood proteins or red blood cells. Uranium binds to phosphate moieties in proteins, nucleotides, and bone tissue. In the latter case, uranium substitutes for calcium. Approximately 25% of absorbed uranium may be fixed in the bone; 60% is rapidly excreted in the urine.

Mechanism of Toxicity

Although soluble uranium compounds are absorbed, there is apparently no information on the effect of uranium on enzymes. Much of the literature on the toxicity of uranium is related to inhalation of radon daughters.

Human Toxicity

The toxicity of uranium is primarily related to inhalation of radon daughters. Lung cancer and chronic lung diseases are associated with radon exposure. Exposure to uranium combined with exposure to cigarette smoke appears to have an additive effect in producing lung cancer. It appears that uranium per se is not carcinogenic.

Uranium tetrafluoride is a highly corrosive, radioactive poison. It is very corrosive to the lungs because, upon hydrolysis, hydrogen fluoride is formed. Uranium oxides are less toxic. The carbonate complexes, which are the common carriers in the blood, produce systemic toxicity in the form of renal damage and renal failure.

The ACGIH TLV-TWA is 0.2 mg/m³ for elemental uranium and soluble and insoluble compounds of uranium.

Clinical Management

Calcium-EDTA (the calcium disodium salt of ethylenediaminetetraacetic acid) is recommended for acute uranium poisoning; BAL (British Anti-Lewisite; 2,3-dimercaptopropanol) is not effective as an antidote.

Animal Toxicity

Uranium is a nephrotoxic agent when administered to experimental rodents; it acts selectively on the proximal tubular cells. It is not known to be teratogenic, mutagenic, or carcinogenic as the element apart from radon daughters.

—*Arthur Furst and Shirley B. Radding*

Related Topics

Metals
Radiation
Radon

Urethane

- CAS: 51-79-6
- PREFERRED NAME: Ethyl ester carbamic acid
- SYNONYMS: Ethyl carbamate; ethylurethan; urethan
- CHEMICAL CLASS: Carbamates
- MOLECULAR FORMULA: $C_3H_7NO_2$
- CHEMICAL STRUCTURE:

$$H_2N \text{---} CO \text{---} OC_2H_5$$

Uses

Urethane is used as a solvent for various organic materials, pesticides, and fumigants. It is also used as a chemical intermediate in the production of cross-linking agents in textiles and in the pharmaceutical industry. Urethane was once used as an anesthetic in veterinary medicine. Its veterinary use was discontinued in 1948 when its carcinogenic properties were revealed.

Exposure Pathways

Industrial workers can be exposed to urethane in the workplace. Urethane is unintentionally formed during the manufacture of certain consumer beverages. It has been found predominantly in bourbons, sherries, fruit brandies, whiskeys, and wines.

Toxicokinetics

Urethane can be excreted in the urine unchanged (0.5–1.7% of dose). Urethane can also be metabolized by the liver cytochrome P450 system. The urinary metabolites of urethane include *N*-hydroxy urethane, acetyl-*N*-hydroxyurethane, ethyl mercapturic acid, and *N*-acetyl-*S*-ethoxy carbonylcysteine.

Mechanism of Toxicity

Urethane is activated in the liver into a carcinogenic metabolite. The activation of urethane by cytochrome P450 involves two sequential reactions. Urethane is dehydrogenated to vinyl carbamate followed by epoxidation to form vinyl carbamate epoxide. The former is believed to be the ultimate carcinogenic metabolite of urethane.

Human Toxicity

Toxic effects from urethane exposure include vomiting, coma, hemorrhages, kidney, and liver injury. No data exist that links human exposure to urethane and cancer. Nonetheless, given its animal carcinogenicity, IARC has classified urethane as a possible human carcinogen.

Clinical Management

There is no specific treatment for urethane toxicity. Supportive and symptomatic treatment is recommended.

Animal Toxicity

Developmental defects have been produced in offspring of rats and hamsters treated *in utero* with urethane. Some of the malformations noted included eye, skeletal, and neuronal tube defects and cardiac malformations.

Chronic and acute administration of urethane through the oral, inhalation, subcutaneous, and intraperitoneal routes has produced cancer in mice, rats, and hamsters. Perinatal and newborn animals appear to be more susceptible to the carcinogenic properties of urethane. Urethane exposure in laboratory animals has produced an increased incidence of spontaneous lung adenomas in susceptible mice strains.

—*Heriberto Robles*

Related Topic

Carbamate Pesticides

Valproic Acid

- CAS: 99-66-1
- SYNONYMS:
  Valproic acid—2-propylpentanoic acid; dipropylacetic acid (*n*-DPA); 2-propylvaleric acid; Depakene
  (Semi)sodium valproate (CAS: 76584-70-8)—divalproex sodium; sodium hydrogen bis(2-propylpentanoate); Depakote
  Other proprietary names—Epilim; Convulex; Depatene; Depakin; Depakine; Deprakine
- PHARMACEUTICAL CLASS: Synthesized simple branched-chain carboxylic acid that is chemically unrelated to other anticonvulsants
- CHEMICAL STRUCTURE:

$$CH_3—CH_2—CH_2\underset{CH_3—CH_2—CH_2}{\overset{}{\diagdown}}CH—C\overset{O}{\underset{OH}{\diagup}}$$

Uses

Valproic acid is used therapeutically as an anticonvulsant. Valproic acid and valproate are used in a variety of absence and generalized seizure disorders.

Exposure Pathway

Toxicity results from acute or chronic ingestion of tablets, capsules, or elixir.

Toxicokinetics

Peak plasma levels occur 1–4 hr after a single dose. Absorption is delayed but not diminished in the presence of food. Bioavailability appears to be complete. The majority of a dose undergoes hepatic glucuronidation or oxidation. At least two metabolites, 2-propyl-2-pentenoic acid and 2-propyl-4-pentenoic acid, have anticonvulsant activity. Biotransformation can be enhanced by enzyme-inducing drugs (e.g., phenobarbital and phenytoin), but there is no apparent autoinduction. The apparent volume of distribution is 0.2 or 0.3 liters/kg (but approximately 1 liter/kg for the free, unbound portion), with high concentrations found in areas containing γ-aminobutyric acid (GABA). Plasma protein binding is 90% at therapeutic concentrations but decreases as plasma levels increase. The therapeutic level in plasma is 50–100 μg/ml.

Less than 3% of a dose is excreted unchanged in the urine or through the feces. The elimination half-life from plasma is 10–15 hr when valproic acid is used alone, but interaction with other anticonvulsant drugs can reduce the half-life to 4–10 hr.

Mechanism of Toxicity

The anticonvulsant properties of valproic acid (and/or its metabolites) are likely attributable to enhancement of GABA activity. Valproic acid may also inhibit platelet aggregation.

385

Human Toxicity: Acute

In most cases, overdoses with valproic acid are relatively well tolerated. Most patients will have nausea and/or vomiting, and a mild degree of lethargy. Miosis and confusion can also occur. Rare cases of seizures, coma, cerebral edema, hypotension, and cardiorespiratory arrest have been reported. The incidence of these effects is unknown, but they are likely dose related. Significant depression of consciousness has been associated with ingestions exceeding 200 mg/kg; recovery has occurred following ingestion of 25 g. A fatality was reported in a 15-year-old with a plasma concentration of 1914 μg/ml; cardiorespiratory arrest occurred 20 hr postingestion. Transient elevation of hepatic transaminases, acute pancreatitis, and hyperammonemia have been noted after acute overdose.

Human Toxicity: Chronic

Hepatotoxicity is of most concern. During the first few months of therapy, transient elevation of hepatic transaminases occurs in an average 11% (up to 40%) of patients. Fulminant hepatic failure will develop in 1 in 5000–10,000 patients. In these cases there is hepatic necrosis, steatosis, and a Reye's syndrome-like illness. Fatal hepatic injury is most likely in children less than 2 years old and in those patients on multiple-drug therapy.

Clinical Management

The majority of patients with acute overdose have a benign course and need supportive care alone. The gut should be decontaminated with oral doses of activated charcoal. Asymptomatic children with accidental ingestion may receive syrup of ipecac (within 1 hr of ingestion) to induce emesis. Measures to enhance elimination are not justified despite testimonials from case reports. Patients requiring treatment in the emergency department should be tested for valproic acid plasma concentration, complete blood count, liver function, and perhaps for the presence of other anticonvulsant drugs. Valproic acid therapy should be discontinued in patients with elevated hepatic enzymes or serum ammonia. There is no known antidote; however, one case report describes a positive response to naloxone in a child with a serum level of 185 μg/ml.

—*S. Rutherfoord Rose*

Vanadium

- ◆ SELECTED COMPOUND: Vanadium pentoxide (CAS: 1314-62-1)
- ◆ MOLECULAR FORMULA: O_5V_2
- ◆ CHEMICAL CLASS: Metals

Uses

Vanadium is used as an alloying addition to steel, iron, titanium, copper, and aluminum, with the primary use in the steel industry. Vanadium is also used as a target material for X-rays, as a catalyst for the production of synthetic rubbers, plastics, and chemicals, and in ceramics.

Exposure Pathways

Exposure may occur through inhalation (dust or fume), ingestion (water, food, and soil), or dermal contact (dust, soil, and solid).

Toxicokinetics

In humans, 0.1–1% of orally administered vanadium is absorbed through the gut. Lung and gut absorption increases with the solubility of the vanadium compound. Vanadium pentoxide is nearly 100% absorbed by inhalation. Vanadium is not absorbed through the skin. When absorbed, 60% of the vanadium is excreted by the kidneys within 24 hr of administration.

Mechanism of Toxicity

When inhaled, vanadium is toxic to alveolar macrophages and therefore may impair pulmonary resistance to infection and clearance of particulate matter. An increase in inflammatory cells of the nasal mucosa has been observed in workers exposed to vanadium. The toxicity of absorbed vanadium is attributed to its ability to inhibit enzyme systems such as monoamine oxidase, ATPase, tyrosinase, choline esterase, and cholesterol synthetase. Vanadate (VO_3^-) mimics the action of insulin in target tissues and is a potential inhibitor of the sodium pump.

Human Toxicity

In general, vanadium has a very low oral and dermal toxicity and a moderately low toxicity by the inhalation route. The toxicity of vanadium increases with its valence state, with vanadium pentoxide being the most toxic of the vanadium compounds. Vanadium fumes are more toxic than vanadium dust. Acute inhalation exposure has resulted in lung irritation, coughing, wheezing, chest pain, nose bleeds, atrophic rhinitis, pharyngitis, epistaxis, tracheitis, asthma-like diseases, irritation of the eyes, and a metallic taste in the mouth. Symptoms generally disappear within 2 weeks of exposure. Two to 10^4 mg/m^3 has resulted in mild to moderate respiratory effects and no systemic effects in humans. Acute oral exposures result in abdominal cramping, diarrhea, black stools, and a greenish-black coating on the tongue. Skin exposure may result in dermatitis, allergic skin lesions, and a green discoloration of the skin. Systemic symptoms of exposure to vanadium are extremely rare but could include peripheral vasoconstriction of the lungs, spleen, kidneys, and intestines. A fatal dose may result in central nervous system depression with tremors, headache, and tinnitus. Chronic exposure to vanadium may result in arrhythmias and bradycardia. There are no data on the carcinogenicity of vanadium in humans.

Clinical Management

Irrigate exposed skin and eyes with copious amounts of tepid water (with soap for exposed skin). After inhalation exposures, move to fresh air and monitor for respiratory distress. Administer 100% humidified supplemental oxygen with assisted ventilation as required. If coughing or breathing difficulties are noted, the patient should be evaluated for irritation or bronchitis, including chest X-rays and determination of blood gasses. For ingestion exposures, emesis may be indicated for recent, substantial ingestion. Activated charcoal may be considered, depending on the form of vanadium ingested. Chelation is not usually indicated since systemic effects are rare.

Animal Toxicity

Inhalation of vanadium in animals results in lung irritation, coughing, wheezing, chest pain, atrophic rhinitis, and conjunctivitis. Pulmonary edema has been observed in animals after exposure to some vanadium compounds. The acute oral toxicity of vanadium is low: In mice, 1000 mg/kg causes catarrhal gastritis. Acute oral exposure in rats results in distress, hemorrhagic exudate from the nose, diarrhea, hindlimb paral-

ysis, labored respiration, convulsions, organ congestion, fatty degeneration of the liver and kidney, focal hemorrhage of the lung and adrenal cortex, and death. Chronic inhalation and oral exposure to vanadium in laboratory animals has resulted in kidney and liver changes, decreased body weight, diarrhea, decreased erythrocyte count and hemoglobin levels, and increased reticulocyte count in peripheral blood. Chronic oral exposure to vanadium has caused an increase in minor birth defects and fetal death in pregnant rats. Vanadium has not been found to cause mutagenic, carcinogenic, teratogenic, or reproductive effects in short-term studies.

—Janice M. McKee

Related Topic

Metals

Vanillin

- ◆ CAS: 121-33-5
- ◆ SYNONYMS: *M*-anisaldehyde; 4-hydroxyl; vanillic aldehyde; lioxin; 3-methoxy-4-hydroxybenzaldehyde
- ◆ CHEMICAL CLASS: Essential oil
- ◆ MOLECULAR FORMULA: $C_8H_8O_3$

Uses

Vanillin is used in flavorings, perfumes, and pharmaceuticals; it is a source of L-dopa.

Exposure Pathways

Exposure may occur via dermal contact, inhalation, or ingestion.

Human Toxicity

Vanillin is a weak human sensitizer.

Animal Toxicity

Vanillin is a weak dermal sensitizer in guinea pigs and mice. Rat LD$_{50}$s are 1580 mg/kg (oral) and 1160 mg/

kg (intraperitoneal). The mouse LD$_{50}$ is 475 mg/kg (intraperitoneal). Oral ingestion at high doses causes hyperpnea, muscular weakness, dyspnea, collapse, and circulatory failure.

—*Shayne C. Gad*

Related Topics

Limonene
Sensitivity Analysis

Veterinary Toxicology

This entry focuses on understanding and managing chemically induced disorders in domestic animals. Approximately 10% of veterinary practice is devoted to the diagnosis and treatment of poisonings. The various animals treated range from small domestic animals (i.e., cats and dogs) to food-producing animals (i.e., dairy cattle, beef cattle, and swine), to horses, pet birds, zoo animals, and, occasionally, wild game (i.e., rabbits and fish).

Species Differences

Small animals react to chemicals more or less the same way as do humans because the species are all monogastrics. Ruminants (i.e., cattle and sheep), however, react differently than the monogastrics; they have evolved a unique digestive tract structure and microbial flora which play a major role in the fermentation of the forage ingesta. The ruminant's microflora are usually capable of metabolizing toxic chemicals. For example, cattle are more susceptible to nitrate poisoning than are horses, whereas dogs and cats are very resistant to nitrate poisoning. Cattle are very susceptible to nitrate poisoning because their digestive tract microbes will convert nitrates to the proximate toxic metabolite, nitrite.

Dogs, because of their relatively small gastrointestinal microbial population, are resistant to nitrate poisoning. The horse may succumb to nitrate poisoning because of microorganisms in the cecum in its posterior digestive tract. However, by the time nitrate reaches the cecum, more than 70% will have been absorbed; little will be available for biotransformation into the toxic nitrite ion. Horses, therefore, require threefold higher nitrate concentrations to be poisoned than do cattle.

Physiological differences among species will markedly alter the susceptibility to toxicants. Birds are more sensitive to toxic vapors and gases than mammals. Canaries have been used in mines to test for the presence of poisonous gases because their elaborate respiratory system causes them to succumb to lower concentrations of toxic gases than would endanger humans.

Biochemical differences also contribute to differential susceptibility between and within species. Cats are more susceptible to acetaminophen poisoning than other domestic animals. The cat's glucuronyl transferase activity for conjugating acetaminophen is much lower than that of other domestic species, and feline hemoglobin is more susceptible to oxidation than that of other animals. Therefore, cats given what would be considered a therapeutic dose of acetaminophen in humans die of methemoglobinemia. Biochemical differences are also found within the same species. For example, the Boston terrier is much more susceptible to copper poisoning than other species of dogs. Most biochemical differences are of genetic origin.

Adequate comprehension of the variance in toxicity from chemicals in the domesticated species requires an understanding of the anatomy, physiology, and biochemistry of the affected animals. Other general factors that affect the toxicity of chemicals must also be considered when dealing with clinical toxicities in domestic animals. These factors include the animal's age, sex, health, nutritional status, environment, and concurrent exposure to other chemicals.

The effects of these and other factors in modifying the outcome of poisoning can be of vital significance in determining its outcome and also point to appropriate management options. There is a vast amount of literature in this area.

Common Toxicoses in Food-Producing Animals

Food-producing species are cattle, swine, and small ruminants. Swine differ from other animals in this category in that they have a simple stomach (monogastric), whereas the other animals have a compound stomach. Most of the toxicants affecting other animals also affect food-producing animals, but some toxicants are peculiar to or predominantly seen in food-producing animals. Toxicoses frequently encountered in ruminants include nonprotein nitrogen toxicoses; copper, lead, or arsenic poisoning; mycotoxicoses; nitrite poisoning; plant poisoning; and algae poisoning. In swine, salt poisoning, mycotoxicoses, organic arsenicals, plant poisoning, and gases generated in swine confinement operations are often involved in toxic episodes.

Nonprotein Nitrogen Compounds

Nonprotein nitrogenous sources include urea, biuret, and ammoniated feeds. These compounds are cheap sources of the nitrogen required by the animals for protein synthesis. Nonprotein nitrogen poisoning is a common problem and is often seen in animals not gradually introduced to diets containing these compounds. It is an acute fatal condition characterized by bloating, intense abdominal pain, ammonia breath, frequent urination, and frenzy. Often several animals are affected.

In ruminant animals, the rumen microflora normally convert urea to ammonia, and the ammonia is rapidly utilized by the liver for protein synthesis. However, in cases of excess ammonia production, the blood ammonia concentration builds up to toxic levels very fast and induces central nervous system (CNS) derangement. Therefore, in addition to the gastrointestinal signs, the animals will show fulminating CNS signs. Treatment of the condition involves giving a weak acid such as vinegar and plenty of cold water orally. The rationale for giving cold water and acetic acid is to slow down the action of urease, the enzyme responsible for converting urea to ammonia, which requires high temperature and pH for optimal function. The cold water lowers the temperature and the acetic acid lowers the pH. Infusions of calcium and magnesium solutions are administered to alleviate tetany.

Another source of nonprotein nitrogen (urea) poisoning in ruminants is the accidental ingestion of nitrogen-based fertilizers such as ammonium phosphate. Occasionally, cattle break into drums or bags of fertilizers containing these nitrogen-based compounds. The prognosis is grave in most cases if several animals are affected. If only a few valuable animals are affected, a rumenotomy can be performed. Although small ruminants (e.g., sheep and goats) have the same anatomical predisposition to suffer from nonprotein nitrogen poisoning, they are rarely involved, probably because they are not usually fed rations containing these compounds.

Nitrate–Nitrite

Excessive exposure of ruminant animals to nitrates causes nitrite toxicity, an acute rapidly fatal disease. The most common source of nitrates in ruminants is through consumption of forage that was grown on heavily fertilized fields and has accumulated high levels of nitrates. All common animal feeds, such as sorghum, alfalfa, and milo, can accumulate excessive amounts of nitrates. Another common source of dietary nitrates is contaminated drinking water. Nitrates are highly water soluble and underground water can become contaminated from heavily fertilized fields. Runoff from fertilized fields is another source of contamination in surface pools and ponds.

Nitrates are reduced to nitrites by rumen microflora. In normal circumstances the nitrite ion is rapidly utilized for ammonia synthesis, but in cases of excessive acute intake of nitrate, the rapidly formed nitrite ion is absorbed into the bloodstream. In blood the nitrite ion reacts with hemoglobin to form methemoglobin. Methemoglobin is incapable of oxygen transport, and the animal compensates for the anoxia by increasing its respiratory rate. Therefore, affected animals will be hyperventilating, have brownish blood/mucous membranes, and be weak. Chronic intake of nitrates has been reported to cause reproductive problems such as abortion, but experimental results regarding this claim are currently inconclusive. Besides reacting with methemoglobin, the nitrite ion also replaces iodine in the thyroid gland, thereby interfering with the function of the thyroid hormone.

Treatment of nitrate/nitrite poisoning involves intravenous infusion of 1% methylene blue at a rate of 1.5 mg/kg body weight and withdrawal of the offending feed.

Copper–Molybdenum

Sheep are more susceptible to copper poisoning than are cattle, but cattle are more sensitive to molybdenum poisoning than are sheep. The *in vivo* relationship between copper and molybdenum is well understood. Excess copper induces molybdenum deficiency and vice versa. The most frequent cause of copper poisoning in sheep is uninformed farmers feeding sheep feed meant for cattle. Copper is an essential element for cattle and is usually added to their feeds; however, molybdenum is not considered essential and is therefore not added. Cattle feeds therefore have high copper concentrations and no molybdenum; feeding this ration to sheep upsets the normal 6 : 1 copper:molybdenum ratio *in vivo*.

Copper toxicity in sheep is an acute condition that develops after a chronic copper intake. During the chronic phase copper is stored in the liver until a critical concentration is reached. Stressful conditions, such as transportation or insufficient feed or water intake, will trigger a massive hepatic release of copper and cause a hemolytic crisis. Affected sheep have hemoglobinuria, are weak, and die acutely. The massive release of hemoglobin can block the renal tubules, inducing renal failure. The prognosis is poor for animals already showing clinical signs. Chelation therapy using D-penicillamine is recommended for the exposed animals not showing clinical signs.

In cattle, molybdenosis is characterized by a foamy diarrhea which may be bloody. Affected animals also have depigmented hair. Molybdenosis is a subacute to chronic condition and occurs when the copper:molybdenum ratio is 2 : 1 or less. The condition shows geographical distribution and occurs in areas deficient in copper or having an excess of molybdenum (e.g., parts of California, Oregon, Nevada, and Florida). Treatment of this condition involves copper supplementation in the feed.

Lead

Despite awareness regarding the dangers of lead poisoning in humans and domestic species, it is surprising that lead poisoning is the most frequently encountered toxicity in food-producing animals. Lead poisoning is more commonly seen in cattle than in other food-producing species. Young animals are mostly affected because of their curiosity and because they are indiscriminate in their feeding habits. There are several sources of lead in cattle. Discarded junk batteries and leaded water pipes are the most common sources. Quite often uninformed owners will discard or store old batteries in farm environments and cattle will chew on them. Discarded leaded pipes, especially those used around oil wells, are a common source of lead poisoning.

Lead interferes with heme synthesis and causes renal and CNS lesions in food-producing animals. Affected animals are initially anorectic. They may then become belligerent, blind, and have periodic seizures at the terminal stages of the poisoning. Once the CNS signs have set in, the prognosis is grave but treatment with chelating agents may be of value. Chelating agents include calcium disodium-EDTA (calcium disodium salt of ethylenediaminetetraacetic acid) and 2,3-dimercapto-1-propanesulfonic acid.

Arsenic

Arsenic poisoning is second to lead as the most frequently reported heavy metal toxicant in food-producing animals. Arsenic is present in the environment in two forms: inorganic and organic arsenicals. Inorganic arsenic is often incorporated into pesticides, which are the most common sources of arsenic poisoning in cattle. Inorganic arsenicals are also used as herbicides and cattle sometimes are exposed by eating grass clippings from recently sprayed forage.

Inorganic arsenic poisoning is a rapidly developing and fatal disease. Affected animals show severe gastrointestinal irritation without CNS involvement. They have severe abdominal pain and hemorrhagic diarrhea and are depressed. Usually these signs appear 24–36 hr after exposure to the inorganic arsenic.

Phenylarsonic arsenicals are less toxic to mammals than the inorganic arsenicals. Phenylarsonic compounds are usually incorporated in swine (and poultry) feed for disease control and to improve weight gain. Examples of these compounds include arsenilic acid, 3-nitroarsenilic acid, and 4-nitroarsenilic acid. Organic arsenicals are also available as trivalent and pentavalent compounds, and the trivalent forms are more toxic than the pentavalent compounds.

These phenylarsonic compounds are peripheral nervous system toxicants. They cause demyelination of peripheral nerve fibers leading to ataxia and paralysis of hindquarters. The condition occurs frequently in swine kept on feed containing 1000 ppm arsenic for at least 3–10 days or 250 ppm arsenic for 20–40 days. Therefore, unlike inorganic arsenic poisoning, which is an acute form of the disease, poisoning by phenylarsonic compounds is an insidious condition. In addition, organic arsenic is commonly involved in swine toxicities because of its incorporation in swine feeds, whereas inorganic arsenic poisoning is more commonly seen in cattle.

Treatment of inorganic arsenic poisoning involves decontamination procedures and use of the antidote BAL (British Anti-Lewisite compound; 2,3-dimercapto-propanol). Use of demulcent to coat the gastrointestinal tract and the use of antibiotics is also recommended. Organic arsenic poisoning treatment involves only withdrawal of the feed involved, with recovery occurring in 3–5 days. Severely affected pigs should be culled.

Selenium

Selenium poisoning is a regional problem occurring in areas where the selenium content in soil is high. Selenium is then absorbed and concentrated by selenium-accumulating plants such as the *Astragalus* species. Cattle, sheep, goats, and swine are exposed by consuming these plants.

Acute selenium poisoning occurs when animals consume plants containing more than 10,000 ppm. This is characterized by sudden death or labored breathing, abnormal movement and posture, frequent urination, diarrhea, and death. Because plants containing high selenium concentrations are unpalatable, they are rarely consumed by animals. Therefore, acute selenium poisoning is rare. However, chronic selenium poisoning is relatively common. Chronic consumption of plants containing as low as 50 ppm can cause chronic selenium poisoning. Affected animals are anorexic, have impaired vision, wander, salivate excessively, are emaciated and lame, and lose hair.

Removal of animals from pastures whose forages contain high selenium is the recommended cure but may be unsuccessful if the condition has persisted for several days or more.

Mycotoxins

Some of the mycotoxins of veterinary interest are aflatoxins, deoxynivalenol (DON), diacetoxyscirpenol (DAS), *T*-2, zearalenone, ochratoxins, and fumonisin B_1. Mycotoxins are especially a common problem in warm climates where high temperatures and relative humidity support fungal growth and favor mycotoxin production. All food-producing animals are susceptible and clinical signs will depend on the mycotoxins involved. Rarely is only one mycotoxin involved because several species of fungi (e.g., *Fusarium*, *Penicillium*, and *Aspergillus*) coexist and often produce more than one type of mycotoxin.

The common sources of aflatoxins for food-producing animals include corn and oats. When aflatoxins are ingested in parts-per-million quantities, acute death can occur. The affected animals show severe gastrointestinal pain and hemorrhage. Aflatoxins are severe hepatotoxicants; therefore, hepatomegaly and jaundice may be observed in severe subacute cases. Quite often, however, aflatoxin poisoning is an insidious condition due to the chronic intake of parts-per-billion aflatoxin concentrations over a prolonged period of time. Clinical signs include poor weight gain, decreased milk production, and poor reproductive performance, including abortions. Virtually every organ function is affected by aflatoxins. The immune system of the affected animals is also impaired, and animals may more easily succumb to infectious diseases.

Toxicity due to T-2 has been reported in North America and other parts of the world, including Germany, Hungary, France, and South Africa. It is less common than aflatoxin toxicity. T-2 mycotoxins act by interfering with the blood clotting mechanism. Affected animals have gastrointestinal bleeding and will pass blood-stained feces. The animals will perform poorly (i.e., have low weight gain, decreased milk production, and decreased food intake). T-2 is also an immunosuppressant. All food-producing animals are susceptible to T-2 mycotoxicosis.

Zearalenone is an estrogenic mycotoxin that usually causes toxicity in swine that consume contaminated corn. Prepubertal swine are mostly affected. Affected females show swelling of the vulva and excessive straining, which may cause vaginal prolapses. In male animals, zearalenone will cause decreased libido. There is no effective treatment apart from withdrawing the feed containing the mycotoxin.

Other mycotoxins, including DAS, DON, and ochratoxin, are not of major economic importance although they can be toxic to food-producing animals. DAS causes necrosis and erosion of the oral mucous membranes. Consequently, affected animals exhibit feed refusal and have impaired growth. DON (also called "vomitoxin") induces vomiting and feed refusal in swine. Ochratoxins cause renal problems, including hydronephrosis, especially in swine.

Ergot poisoning is occasionally encountered in livestock fed grain screenings contaminated with *Claviceps purpurea*. The active constituents are ergotoxin and ergotamine, which are vasoactive compounds. These compounds cause vasoconstriction of the peripheral vessels, especially those of extremities, causing necrosis and gangrene of hooves, ears, and tails. Abortions and agalactia have been reported in cattle fed ergot-contaminated feed. Therapy consists of discontinuation of the source of the toxicant and antibiotic therapy to prevent secondary bacterial infections in the necrotic tissues.

Fumonisin B_1 is produced by *Fusarium moniliforme*, a fungus that predominantly grows worldwide on corn. Fumonisin B_1 causes pulmonary edema and respiratory distress in swine. Numerous deaths have been reported in swine fed fumonisin-contaminated corn screenings.

The most practical treatment for mycotoxicoses consists of withdrawal of the contaminated feed from the herd and supportive care for the affected animals.

Blue-Green Algae

Blue-green algae poisoning occurs in late summer and early fall when the algae forms a scum on top of ponds or other stagnant waters. Because of husbandry practices, cattle are most frequently involved. Blue-green algae poisoning has been reported in North America and Britain. Algae of genus *Anabaena* are most frequently involved.

There are two distinct syndromes in blue-green algae poisoning: the neurotoxic effects and the hepatotoxic syndrome. The neurotoxic disorder is peracute, and cattle drinking water containing the neurotoxic principle Anatoxin A can die within a few minutes and usually are found quite close to the pond or water (algae) source. On the other hand, the hepatotoxic principle causes a less acute type of poisoning characterized by lethargy and jaundice. Death may occur 2 or 3 days after drinking contaminated water.

Because of the peracute nature of the blue-green algae-induced neurological syndrome, there is hardly time for treatment and the prognosis is universally grave. Treatment of animals affected with the liver syndrome of blue-green algae poisoning involves appropriate supportive therapy.

Toxic Gases

Toxic gases are of primary concern in closed animal housing, especially in swine operations. Because of the intensive swine confinement operations with buildings designed to save on energy, toxic gases can accumulate in swine houses and result in serious health conse-

quences to animals and caretakers in cases of ventilation failure. These toxic gases are generated from the decomposition of urine and feces, respiratory excretion, and the operation of fuel-burning heaters.

The most important gases are ammonia, hydrogen sulfide, carbon monoxide, and methane. A number of vapors which represent the odors of manure decomposition, such as organic acids, amines, amides, alcohols, carbonyls, and sulfides, are also produced. Respirable particles which may be loaded with endotoxins are also a major health problem in swine confinement operations.

Ammonia is highly soluble in water and will react with the mucous membranes of the eyes and respiratory passages. At 100 ppm or greater ammonia concentrations, toxicosis will produce excessive tearing, shallow breathing, and clear or purulent nasal discharge. The irritation of the respiratory tract epithelium leads to bronchoconstriction and shallow breathing.

Hydrogen sulfide poisoning is responsible for more animal deaths than any other gas. At concentrations of 250 ppm and above, hydrogen sulfide causes irritation of the eyes and respiratory tract and pulmonary edema. Hydrogen sulfide concentrations above 500 ppm cause strong nervous system stimulation and acute death. In order to prevent hydrogen sulfide poisoning, manure pits should not be agitated when pigs are on the premises, and proper ventilation should be in place.

Carbon monoxide is produced by incomplete combustion of hydrocarbon fuels. Poisoning by carbon monoxide is caused by operating improperly vented space heaters or furnaces in poorly ventilated buildings. Carbon monoxide binds to hemoglobin forming carboxyhemoglobin, thereby reducing hemoglobin's oxygen carrying capacity and subsequently causing hypoxia. Concentrations of carbon monoxide >250 ppm cause hyperventilation, respiratory distress, and stillbirths.

Methane is a flammable and colorless gas produced from organic wastes through bacterial action. It serves to displace oxygen in respirable air, thus producing oxygen starvation if present in high concentrations.

Nitrogen dioxide is a very poisonous gas that is responsible for causing silo fillers disease in humans. The gas is also toxic to animals. Nitrogen dioxide is produced during the first few weeks after silage has been cut and put into the silo. The highest nitrogen dioxide concentrations are reached during the first 48 hr after filling the silo. Nitrogen dioxide dissolves in water to form nitric acid, which is very corrosive to the respiratory tract epithelium and the lungs. Nitrogen dioxide concentrations as low as 4 or 5 ppm can cause respiratory system disturbances.

Exposure to sulfur dioxide concentrations of 5 ppm or greater causes irritation and salivation in swine. The gas is soluble in water, forming the more toxic sulfuric acid. It is the sulfuric acid that causes eye and nasal irritation, and in severe cases it produces hemorrhage and emphysema of the lungs.

The effect of these toxicants singly or in combination is to produce a hypofunctional respiratory system. Affected animals are also predisposed to respiratory tract infections. The end result is significantly retarded performance and productivity decreases in the affected animals. It is therefore important to ensure that proper animal housing is provided with adequate ventilation in all seasons of the year to provide animals with a healthy breathing and a highly productive environment.

Toxic Plants

Plant poisoning is very common in areas where open grazing is practiced, such as in the Great Plains of the United States, where plant poisoning is widely reported during spring and fall grazing seasons. The wide range of toxic plants and their variations in growth environments produce risks that can affect almost all body systems, depending on the plant consumed, its level of maturity, and the soil and environmental characteristics in which it is growing.

Body systems and organs most prominently affected by plants include the digestive tract, the liver, kidneys, and nervous system, the heart and blood, the skin, and the reproductive tract and its functioning. It is important to realize, however, that toxic plants rarely affect only one body system or organ and thus may generate a complex pattern of effects in any one poisoned animal. Toxicity of a given plant can vary widely depending on the prevailing natural conditions. It is therefore not surprising that a given toxic plant may be toxic under certain conditions (e.g., during stressful drought conditions) but safe during other times.

Sodium Chloride (Salt)

Salt poisoning is frequently encountered in swine operations but can also occur in feedlot cattle. The causes

of this condition are twofold. Most commonly, the pigs will be on a ration containing a recommended concentration of sodium chloride, but management failures or changes can favor conditions which cause salt poisoning to occur. These poor management conditions include the sudden absence of water, which can be caused by frozen water in winter or the breakdown of water supplies. The other possibility is the accidental addition of excessive amounts of salt or sodium-containing materials to the ration.

Salt poisoning has also been reported in swine operations even when the management situation is appropriate; the only change was that the animals had been moved into a new housing facility, as occurs with weaning. In those situations, the animals are not used to the watering facilities in the new buildings and they do not know how to obtain the water; thus, they go without water while continuing to feed on the normal salt-containing ration.

Clinically, salt poisoning is a neurological disorder, and the syndrome is rather acute. Affected pigs will spin on their hindquarters and fall down convulsing. The pigs will also show a characteristic rhythmic pattern of seizures which occurs cyclicly every 3–5 min. Many pigs are usually affected at the same time. The condition is corrected by the provision of adequate but restricted amounts of water made available gradually.

Common Toxicoses of Poultry

Even chickens, ducks, and turkeys are affected by poisonings. There is also much concern and interest in the toxicities seen in wild birds, especially those kept in zoos, as well those kept as pet birds in households. This discussion will emphasize the toxicoses encountered in poultry.

Drugs and Medications

Sulfonamides have been used as coccidiostats in poultry for several decades. Although sulfonamides possess inhibitory action against coccidiosis and other pathogenic agents, they can be toxic and have particularly been shown to be so to poultry. In poultry, sulfonamide toxicity is characterized by blood dyscrasia and renal and liver dysfunctions. Feeding chickens a mash containing as low as 0.2% sulphonamides for 2 weeks is toxic.

Clinically affected birds have ruffled feathers, are depressed, pale and icteric, and have poor weight gain and a prolonged bleeding time. In laying birds, sulfonamides cause a marked decrease in egg production, thin rough shells, and depigmentation of brown eggs. The temperature of affected birds is often elevated. At postmortem, hemorrhages are found in the skin, muscles (especially those of thighs and breast), and in the internal organs. Once these effects are noticed, the concentration of sulphonamides in the ration should be evaluated and the feed involved withdrawn.

Other chemotherapeutic agents sometimes involved in poisoning poultry are the other coccidiostats, such as nicarbazine, zaolene, and nitrophenide, and the ionophore monensin. As little as 0.006% nicarbazine in the diet causes mottled yolks, and at 0.02% there is depressed rate of growth and reduced feed efficiency. Feeding 0.025% nicarbazine to day-old chicks for 1 week results in the chicks becoming dull, listless, weak, and ataxic.

Feeding zaolene at twice the recommended level of 0.025% will cause nervous signs and depress growth and feed efficiency. The nervous signs include stiff neck, staggering, and falling over when the birds are excited.

Nitrophenide possesses marked electrostatic properties and, therefore, sticks to the walls of a feed mixer. The last bits of feed in the feed mixer will normally contain a high concentration of nitrophenide and that elevated concentration has caused disturbances in posture and locomotion, retarded growth, and increased mortality in chickens. Postural disturbances include a tilted position of the head, tremor of the neck, and difficulty in maintaining the righting reflex.

In general poultry are more resistant to monensin toxicity than other species, but there have been reports of monensin toxicity in turkeys accidentally fed rations containing 250 ppm monensin. There is a big difference in susceptibility to monensin poisoning among various species of poultry. Chickens and turkeys less than 2 weeks old are more resistant than older birds, but keets (young guinea fowl) seem more susceptible than their adults and the young of other species. For example, diets of 200 ppm monensin are not toxic for poults, whereas 100 ppm is toxic for keets.

Cresol

Cresol was a commonly used disinfectant in poultry houses but has been gradually withdrawn and replaced by less toxic disinfectants. Nevertheless, in some regions and countries cresol is still being used. Cresol poisoning in the chicken usually occurs at 3–6 weeks of age. Affected chicks are depressed and have a tendency to huddle. There are respiratory problems such as rales, gasping, and wheezing. With prolonged cresol exposure some chicks will develop edema of the abdomen.

Sodium Chloride (Salt)

All poultry and pigeons are susceptible to salt poisoning. Young birds are more susceptible than adults. Although both acute and chronic forms of salt poisoning can occur, the chronic form is more commonly encountered and is due to prolonged ingestion of feed containing high salt content. Levels of 0.5% and above in drinking water or 5–10% in feed cause death in baby chicks.

Signs of salt poisoning in poultry include anorexia, thirst, dyspnea, opisthotonos, convulsions, and ataxia. Increased water consumption may be the most significant early indicator of hazardous exposure to salt in poultry.

Insecticides

Chlorinated hydrocarbon insecticides and organophosphate compounds are used regularly around poultry houses to control external parasites. Commonly used organochlorine insecticides include chlordane, dieldrin, DDT, heptachlor, and lindane. Occasionally, birds get exposed by gaining access to sprayed grounds such as golf courses.

Chlordane causes chicks to chirp nervously, rest on their hocks, and lie on their sides. The birds then become hyperexcitable as the condition progresses. In adult birds there is reduced food consumption, decreased body weight, and a fall in egg production.

Consumption of seeds dressed with dieldrin has been a source of exposure in wild birds. Affected birds are listless and have coordination problems while lighting. Severely poisoned cases have nervous signs characterized by lateral movements of the head and tremors of the head and neck. Birds die in violent convulsions.

DDT toxicity in chickens is characterized by hyperexcitability and fine tremors in severe cases. Moderate cases are characterized by loss of weight, molting, and reduced egg production.

Lindane in the form of a dust is frequently used around chickens. Adult chickens poisoned by lindane stop eating, manifest opisthotonos and flapping of wings, have clonic muscle spasms, and die in a coma.

The organophosphate compounds commonly involved include diazinon, malathion, and parathion. Diazinon is applied to chicken premises, but this compound is very toxic to ducklings. When used at rates recommended for chickens, 100% mortality has resulted from use of this compound on 1- or 2-week-old ducklings. Experimental studies suggest that goslings are three times more sensitive than ducks, chickens, and turkeys. Poisoned birds are unable to stand, salivate profusely, and manifest tremors of the head and neck. Brain cholinesterase levels in birds that die of organophosphate poisoning are on the average 69% less than controls.

Other organophosphate compounds commonly used on chicken premises include dichlorvos, malathion, and parathion. Birds poisoned by these compounds manifest signs similar to those seen in diazinon poisoning. Other effects that may be seen include depression, ataxia, and reluctance to move, paralysis and lacrimation, gasping for breath, and development of diarrhea, crop stasis, and dyspnea.

In general ducks are more sensitive to organophosphate poisoning than are chickens, and care should be exercised when using these products on premises holding ducks.

The carbamate insecticide sevin is a widely used poultry insecticide. This compound is relatively safe, but deaths have been reported in turkey poults kept on premises where the product has been excessively applied at 10 times the recommended rate. The clinical signs are similar to those caused by organophosphate insecticides.

Heavy Metals

Lead poisoning is not as common in domestic poultry as in wild birds, but it is the most common toxicity reported in the avian species. Lead shot has caused losses in waterfowl populations throughout North America. All birds are susceptible to lead poisoning, but most losses are reported in waterfowl because their feeding habits predispose them to the ingestion of lead pellets from shotguns and other sources.

Characteristic signs of lead poisoning are related to CNS derangement, such as ataxia, depression, paralysis of the wings, and convulsions. In some cases the birds presented are anemic, emaciated, regurgitating, and weak. Green diarrhea has often been reported in lead–affected birds.

Yellow phosphorus is a highly toxic element that is still used as a rodenticide. Poultry and wild birds can be intoxicated by consumption of bait intended for rodents. Firework fragments also are a common source of poisoning in free-ranging birds. Affected birds are depressed and anorectic, have increased water consumption, and manifest diarrhea, ataxia, paralysis, coma, and death.

Rodenticides

In addition to the metal yellow phosphorus being used as a rodenticide, other rodenticides are potentially toxic to poultry and other birds. The clinical signs caused by these rodenticides in birds are similar to those observed in other animals.

Birds occasionally consume baits containing anticoagulant rodenticides. The more potent second-generation rodenticide-containing baits, such as brodifacoum, are especially dangerous to birds. These coumarin anticoagulants act by interfering with vitamin K utilization, causing bleeding because of depletion of vitamin K-dependent clotting factors. Poisoned birds bleed from their nares and subcutaneously and have oral petechiations. Quite often the birds are also weak and depressed from the resulting anemia or may be found dead due to stress superimposed on the anemic condition.

Of special interest are secondary intoxications due to free-ranging birds consuming carrions of animals that died of rodenticide poisoning. Strychnine and sodium monofluoroacetate are other rodent control compounds that are commonly involved because they cause acute death in the primary victims and are thus present in high concentrations in carrions.

Strychnine-poisoned birds show clinical effects within 2 hr of ingesting the product. The birds become apprehensive and nervous and have violent tetanic convulsions which cause them to become exhausted and to die of hypoxia. Sodium monofluoroacetate causes overstimulation of the CNS and myocardial depression. Cardiac failure is the cause of death and occurs within 1 hr of consuming the product or contaminated carcass.

Mycotoxins

Mycotoxicoses are common problems for the poultry industry in warm moist climates and in developing countries in the tropics. Aflatoxins are the most commonly involved mycotoxins. Poultry are normally exposed by consumption of contaminated feed, especially corn. Some developing countries lack the resources to adequately screen contaminated corn. In other instances poultry feed is made from the poor-quality (and contaminated) corn that has been rejected for human consumption.

Aflatoxicosis in poultry can be either acute or chronic in nature, depending on the exposure dose. Ducklings are more susceptible to aflatoxin than are turkeys, pheasant, or chickens. In acute cases, affected birds become lethargic, their wings droop, and they manifest nervous signs such as opisthotonos; they die with their legs rigidly extended backward. Chronic dietary consumption of 2.5 ppm aflatoxin causes a significant drop in weight gain and egg production.

Perhaps more important is the increased susceptibility of the affected flock to infection because chronic consumption of aflatoxin-containing feed lowers the immunity of the birds. Aflatoxicosis is therefore a disease of serious economic consequences to the poultry industry in developing countries both through lowered productivity and because of death of affected birds.

Ergot poisoning has been reported in areas where rye is commonly used as poultry feed. In acute ergot poisoning the birds' combs are cold, wilted, and cyanotic. The animals are weak, thirsty, and have diarrhea. In severe cases the birds go into convulsions, become paralyzed, and die. Ochratoxins have been reported to cause renal toxicity in poultry.

Common Toxicoses in Dogs and Cats

Dogs and cats are commonly poisoned by pesticides, herbicides, household products such as antifreeze, and drugs such as acetaminophen applied by humans to their pets. By far the most common toxicities in these small animals involve various insecticides and the overzealous use of these products by owners attempting to control fleas and ticks on their pets.

Insecticides

The insecticides most commonly involved in poisoning dogs and cats are the organophosphates and carbamates, pyrethroids, chlorinated hydrocarbons, and diethyltoluamide (DEET). The organophosphate and carbamate insecticides have a common mode of action, which is the inhibition of acetylcholinesterase. Acetylcholinesterase is the enzyme that breaks down acetylcholine, a neurotransmitter in autonomic ganglia and at cholinergic nerve endings. The inhibition of acetylcholinesterase by organophosphate and carbamate compounds causes acetylcholine to accumulate at nerve synapses and to produce persistent firing of cholinergic nerve fibers. Affected animals are overexcited and have increased respiratory rates, excessive salivation, and muscle tremors.

Treatment of animals poisoned by organophosphate compounds involves the administration of atropine and prolidoxime. Cases involving carbamates may be treated with only atropine because of the rapid biological detoxification of carbamates. The organophosphate and carbamate compounds have a relatively high acute toxicity compared to chlorinated hydrocarbons but have a lower residual activity. As such, organophosphate compounds have largely replaced the chlorinated hydrocarbons for insecticide use because of environmental concerns.

The chlorinated hydrocarbons were among the first synthetic insecticide compounds to be used but have fallen into disfavor because of their persistence in the environment. Typical examples of chlorinated hydrocarbon insecticides are DDT, lindane, and toxaphene. The toxicity of these compounds in small animals is characterized by severe CNS effects, including ataxia and convulsions. Small animals usually get poisoned by being accidentally sprayed or by drinking chlorinated hydrocarbon insecticide concentrates intended for spraying on crops. Although most of these insecticides are banned or their use highly restricted in Western countries, they are still widely applied in developing countries. Thus, chlorinated hydrocarbon insecticide poisonings still frequently occur in the developing countries.

Another group of insecticides commonly involved in small animal poisonings are plant product derivatives—pyrethrins and their synthetic congeners, the pyrethroids. These products are currently enjoying a resurgence because of their selective insecticidal properties and absence of environmental persistence. These compounds are mainly metabolized in the body by liver glucuronidation. The cat is the most sensitive domesticated animal to pyrethrin toxicity because of the low activity of the glucuronide conjugating system in this species. Young cats, less than 6 weeks of age, are the most sensitive.

Pyrethroid compounds formulated with the insect repellant DEET were responsible for numerous deaths in cats and dogs in the past decade. Pyrethroids interfere with sodium channels in nerves causing them to fire repetitively. Clinical signs of pyrethroid poisoning in small animals include ataxia, excitement, and muscle fasciculations and tremors. There is no antidote for pyrethrin poisoning, but symptomatic treatment, such as decontamination procedures and sedation, usually results in full recovery.

Rodenticides

Rodenticide poisoning is commonly seen in all small animals. Rodenticides are widely used around farm-houses to control rodents, such as rats and mice, which destroy property and farm produce. Several classes of rodenticides are currently in use, including the anti-coagulant rodenticides (warfarin and its second-generation cousin brodifacoum), zinc phosphide, strychnine, compound 1080, and arsenicals. Small animals get poisoned by either consuming baits directly or through consumption of the carrion of animals that have died of rodenticide poisoning. The clinical signs seen will vary with the compound involved and, in the majority of cases, dogs (because of their indiscriminate eating habits) are involved.

Strychnine and anticoagulant rodenticides are the most frequently reported offenders. Strychnine poisoning in dogs is a rapidly developing syndrome character-ized by tonic–clonic seizures. These signs result from strychnine competitively blocking the inhibitory neurons in the nervous system. The animals start showing clinical effects within 20 min to 1 hr of ingesting strych-nine and, if the animal has ingested a sufficient amount, death from anoxia occurs acutely. Anoxia results from paralysis of the respiratory muscles. Treatment of strychnine poisoning is symptomatic and involves general decontamination procedures, use of sedatives such as phenobarbital and diazepam, maintenance of ade-quate urine output, and respiratory support. The seda-tives control the seizures and allow the vital muscles to relax and maintain their life-saving functioning.

The anticoagulant rodenticides have been in use for many decades. Because of the long time required for them to take effect, some strains of rats have become genetically resistant to the so-called first-generation an-ticoagulant rodenticides such as warfarin. This has led to the introduction of second-generation rodenticides such as brodifacoum. Unlike the first-generation roden-ticides, which took at least 48–72 hr to take effect, the second-generation rodenticides act relatively acutely, and clinical signs can be evident within several hours and have a long residual action.

These anticoagulant rodenticides act by inhibiting vitamin K-dependent blood coagulation factors (II, VII, IX, and X), by decreasing prothrombin synthesis, and by directly damaging blood capillaries. Animals poi-soned by the anticoagulant rodenticides are weak, have swollen joints because of bleeding into the joint cavities, may hemorrhage from the nostrils, and may pass blood-stained feces. Treatment of anticoagulant rodenticide poisoning involves whole blood transfusions (if the bleeding and resulting anemia is severe) and vitamin K_1 injections for several days. Early intervention requires general decontamination procedures to limit further rodenticide absorption, especially in the case of expo-sure to second-generation rodenticides, followed by prolonged vitamin K_1 therapy.

The toxicity of zinc phosphide results from the phos-phine gas which is produced by acid hydrolysis of the pesticide in the stomach. Animals with partially filled stomachs are more sensitive to zinc phosphide poison-ing than those with empty stomachs because of the greater acid secretion precipitated by the presence of food. The generated phosphine gas is absorbed systemi-cally and exerts its effects in the lungs. Poisoned animals exhibit respiratory difficulties because of the buildup of fluids in the lungs. The cause of death is respiratory failure. Supportive therapy, including respiratory sup-port, is recommended in cases of zinc phosphide poi-soning, but the prognosis is poor because no effective antidote is available.

Compound 1080 (sodium monofluoroacetate) is a very lethal toxicant which acts by blocking the tricar-boxylic acid cycle, thereby depriving vital cells of en-ergy. Fluoroacetate is metabolized to fluorocitrate, which inhibits mitochondrial aconitase. This blocks adenosine triphosphate production. Affected animals are initially uneasy, then they become excitable and will run in various directions in a frenzy, and finally they will fall into seizures and die of anoxia. There is no antidote and, once clinical signs develop, poisoned animals will almost always die within a few hours.

Cholecalciferol is a rodenticide that has been introduced relatively recently and that has been reported to be frequently involved in poisonings of dogs. The compound alters calcium homeostasis by promoting calcium absorption from the gut and also by mobilizing calcium from bone for tissue deposition. Consequently, poisoned animals have increased levels of blood calcium. The calcium is then subsequently deposited in soft tissues like the kidneys, digestive tract mucosa, lungs, heart, liver, and muscle. Mineralization of soft tissues interferes with normal function of these organs.

Clinically, the animals do not show signs until 24–48 or more hours after ingestion of the bait. The affected animals are depressed, have reduced urine production, and the urine is of low specific gravity. Severely poisoned animals have hematemesis, azotemia, and cardiac arrhythmias. Animals with renal impairment are more susceptible to cholecalciferol poisoning than those with normal renal function. Cholecalciferol poisoning requires protracted treatment, which may require as long as 3 weeks in severe intoxications. Appropriate treatment consists of fluid therapy to assist the kidneys in removing the excess calcium, corticosteroids to minimize inflammation, and calcitonin to enhance calcium resorption into the bone.

Several other rodenticides can cause poisoning in small animals but do so less frequently because these rodenticides are used less often. Red squill and thallium have been used as rodenticides for many years. Red squill acts as a cardiotoxicant and causes death by cardiac arrest. It also produces convulsions and paralysis. Thallium is a general systemic toxicant. It has a high affinity for sulfhydryl groups throughout the body. Thallium causes cracking at the corners of lips and also causes hair loss. α-Naphthyl-thio-urea causes death by inducing lung edema and subsequently leading to anoxia. White phosphorus is a hepatorenal toxicant. Animals poisoned by white phosphorus have severe abdominal pain, hepatomegaly, and signs of hepatic insufficiency, such as prolonged bleeding and hypoglycemia.

Cases of rodenticide poisoning in small animals should always be regarded as emergencies. General decontamination procedures, such as inducing vomiting with hydrogen peroxide or apomorphine, the use of activated charcoal to bind the unabsorbed toxicant(s), and/or enterogastric lavage, should almost always be employed to minimize absorption and the resulting hazard from the toxicant(s).

Herbicides

Herbicides are not often involved in small animal toxicity despite their frequent use around farms and the continual possibility of exposure. However, toxicity in dogs from consumption of herbicide concentrates during mixing is occasionally reported. The triazine herbicides act by inhibiting plant photosynthesis and are generally safe products for mammals. The LD_{50} of these compounds is at least 1900 mg/kg body weight in the rat. Therefore, toxicity in dogs can only practically occur following the ingestion of large volumes of concentrates. In experimental situations, triazine herbicide-poisoned dogs can become either excited or depressed, develop motor incoordination, and may proceed to have clonic–tonic spasms.

Some inorganic arsenic compounds are also used as herbicides. These inorganic arsenicals are general protoplasmic poisons and are therefore hazardous to both plant and animal life. Affected dogs almost always vomit, have severe abdominal pain, and develop bloody diarrhea. The vomitus may contain mucous shreds and blood from erosion of the gastric and intestinal epithelium.

Paraquat, although restricted from use in Western countries, is a highly toxic herbicide that is still readily available in developing tropical countries. Upon intake, paraquat is rapidly metabolized in the liver and the lungs with the production of secondary oxygen radicals. It is these radicals which cause injury to tissues and especially do so to the lungs. Poisoned animals die in several days of anoxia.

Unlike other animals, the dog appears sensitive to chlorphenoxy herbicides, such as 2,4-D; the dog's oral LD_{50} to 2,4-D is 100 mg/kg body weight. Ventricular fibrillation is the cause of death in severely poisoned dogs. Ingestion of sublethal doses induces stiff extremities, ataxia, myotonia, paralysis, coma, and subnormal temperatures.

Chlorates are herbicides that are often used along roadsides. They are rapidly metabolized in the liver to the chlorate ion, which induces methemoglobinemia in both cats and dogs. Cats, however, because of the greater susceptibility of their hemoglobin molecule to oxidation, are more susceptible to chlorate poisoning than dogs.

Organophosphate herbicides (e.g., glyphosate and merphos) are weak cholinesterase inhibitors and are moderately toxic to dogs and cats. Carbamate herbicides are not inhibitors of acetylcholinesterase but are also moderately toxic to dogs. The LD_{50} of most of the carbamate herbicides is at least 5000 g/kg body weight.

Household Chemicals

Antifreeze is the household product most commonly involved in small animal poisonings. The active ingredient in antifreeze is ethylene glycol. The characteristic sweet taste of this compound makes it attractive to small animals. Ethylene glycol is metabolized in the liver by alcohol dehydrogenase to glycolic acid and then to oxalate. Glycolic acid contributes to the acidosis which is characteristic of ethylene glycol poisoning. The oxalic acid binds calcium from blood to form calcium oxalate, which is filtered by the glomerulus into renal tubules where it precipitates into crystals which cause blockage of the tubules. Consequently, severely affected animals have renal failure characterized by anuria and uremia. The binding of blood calcium to the oxalic acid produces hypocalcemia which, if severe, can also cause death.

Ethylene glycol poisoning is treated by giving ethanol if the animal is presented within 4 hr of suspected ingestion and by giving quantities of fluids containing sodium bicarbonate to facilitate flushing the calcium oxalate crystals from the kidneys and also to correct the acid-base imbalance. Alcohol dehydrogenase, the enzyme which reduces ethanol to acetic acid and water, prefers ethanol to ethylene glycol and, in the presence of both substrates will metabolize ethanol, leaves ethylene glycol to be excreted unchanged in urine. 4-Methylpyrazole is newly available for effective antidotal use in ethylene glycol poisoning of dogs.

Household products, such as sink cleaners, dishwashing detergents, and toilet cleaners, are also common causes of poisonings in small animals. The majority of cleaning detergents are corrosive compounds which contain strong alkali, acids, or phenolic compounds. These compounds act as contact poisons, causing coagulative necrosis of the tissues they come in contact with. Following ingestion of these products, the dog or cat will vomit, have severe abdominal pain, and may develop diarrhea. The animal's vomitus and feces may be bloody. Consuming animals may show other signs depending on the specific ingredients of the ingested products. For example, products containing phenolic derivatives will cause acidosis and hepatotoxicity.

In general, treatment following ingestion of household products is symptomatic and involves the administration of adsorbents such as activated charcoal, gastrointestinal protectants such as Peptobismol, and the correction of any systemic disturbances (such as acidosis) which may accompany the poisoning. Animals should also be provided abundant glucose and a high protein diet.

Garbage

Garbage poisoning is a frequently encountered problem in small animals. This condition is also referred to as enterotoxicosis or endotoxemia, depending on whether poisoning is due to bacterial infection or due to bacterial endotoxins. Dogs and cats not well fed and/or not closely supervised will eat garbage when allowed to roam. Cats may also be affected, but only rarely so, because they are discriminate eaters. The bacteria most commonly involved are coliforms, staphylococci, salmonellae, and occasionally *Clostridium botulinum*. Enterotoxemia-affected animals often develop a bacteremia after eating infected carrion. Clinical signs normally appear at least 24–48 hr after ingestion of the infected material.

The condition is characterized by anorexia, vomiting, severe abdominal pain, fever, and a bloody diarrhea. The endotoxemia poisoning is due to bacterial endotoxins which are normally present in bacterial cell walls. The clinical signs are generally indistinguishable from those of enterotoxemia, except that in the latter there is no bacteremia; this can be evaluated by performing a blood culture. Although rare in occurrence, botulism is a rapidly developing fatal disease resulting from ingesting bones contaminated with *C. botulinum*. In small animals the disease is characterized by an ascending paralysis. At first there is muscle weakness and incoordination in the hindlimbs. As the paralysis progresses anteriorly, dyspnea and convulsions develop.

Garbage poisoning is rarely a severe condition in small animals because the animals usually vomit and thereby reduce the amount of toxicant ingested. However, in severe cases veterinary attention will be required. If the cat or dog is presented early after ingestion, general decontamination procedures should be performed to reduce absorption. Antiinflammatory corticosteroids and antibiotics should be given. Further treatment involves appropriate supportive therapy.

Heavy Metals

Lead and arsenic are the heavy metals most frequently seen in small animal poisonings. Lead poisoning is more commonly reported in the dog than in the cat, but both species are susceptible. The sources of canine lead poisoning include ingested leaded objects, such as lead weights, and paint chips from old houses being renovated. The clinical signs of lead poisoning in the dog involve primarily the CNS. Dogs may also be presented having abdominal pain and diarrhea in addition to the CNS involvement. Lead poisoning is a chronic disease in dogs, but the overt CNS signs may appear suddenly. Lead poisoning also causes blood dyscrasias characterized by reticulocytosis and anemia. Similar clinical signs are elicited in the cat. Treatment consists of giving chelating agents, such as calcium-EDTA, BAL, dimercaptosuccinic acid, or d-penicillamine.

Arsenic is the active ingredient in some insecticides, in rodenticides, and in some industrially used herbicides. Inorganic and aliphatic organic arsenic compounds are rapidly absorbed from the gut, skin, and lungs and are more toxic than the cyclic organic arsenicals used as feed additives. Trivalent arsenic is the proximate toxicant of the pesticide arsenicals which reacts with sulfhydryl groups of proteins throughout the body. It is, therefore, a general poison, inhibiting the sulfhydryl-containing enzymes it comes in contact with. The clinical signs of inorganic arsenic poisoning in dogs include anorexia, severe abdominal pain, bloody diarrhea, and hair loss, as discussed previously for herbicides. Treatment includes thorough decontamination, chelation therapy with BAL, and aggressive supportive therapy.

Toxic Plants and Mushrooms

Although one would not expect dogs and cats to commonly eat plants, plant poisoning is surprisingly often reported in these species. Because of their exploratory behavior, puppies and kittens are most often involved. Boredom and unfamiliarity due to change of environment are some of the predisposing factors that lead to plant ingestions by dogs and cats. Poisonous ornamental plants (e.g., philodendron and rhododendron) and plants growing around fences (e.g., cassia and yew) are often involved. The range of potentially poisonous plants is vast, and the clinical signs are diffuse and similar to those reported for food-producing animals.

Occasionally, dogs or cats will eat poisonous mushrooms or be fed poisonous mushrooms by uninformed owners. *Amanita muscaria* and *A. pantherima* are acutely toxic and induce signs within 15–30 min of ingestion. These two mushroom species cause nervous signs which include salivation, pupillary constriction, muscular spasms, drowsiness or excitement, and, in severe intoxications, coma and death. Ibotenic acid and muscimol are the active chemical components. However, consumption of *A. phaloides*, *A. virosa*, and *A. verna* produces gastrointestinal signs which become evident 6–12 hr postingestion. These effects include violent vomition, muscle cramps, diarrhea, and dehydration. These latter mushrooms also cause hepatic damage that becomes apparent 3–5 days after ingestion. Phalloidin and α- and β-amanitin are the poisonous principles in this group of fungi.

Common Toxicoses in Horses

In comparison to food-producing animals and cats and dogs, horses are less frequently poisoned. The most commonly encountered equine toxicoses are caused by pesticides, snake bites, arsenic, selenium, monensin, cantharidin, and mycotoxins. Most plants that are hazardous to food-producing animals are also toxic to horses, but horses are less frequently affected since owners usually assure the availability of good feed. Horses are very sensitive to monensin and cantharidin poisonings.

Insecticides

The pesticides most frequently responsible for equine poisonings are the organophosphate, carbamate, and chlorinated hydrocarbon insecticides. Both the organophosphates and the carbamates are acetylcholinesterase inhibitors and present clinical pictures similar to those seen in food-producing animals. Affected horses salivate and sweat profusely and have muscle incoordination and ataxia. The chlorinated hydrocarbons are strong CNS stimulants; affected horses become hyperalert, then excited, and, in severe cases, develop convulsions. In almost all instances, the mode of horses being exposed to pesticides is topical.

Monensin

Horses are highly susceptible to monensin poisoning in comparison with the other domesticated animals.

Monensin is an ionophore normally added to cattle and poultry feed to provide growth stimulation by enhancing the intestinal absorption of calcium and sodium. Horses are easily poisoned by accidentally consuming cattle or poultry feed containing the recommended amounts of monensin for those species. Affected horses can die suddenly without any other signs. Monensin affects the cardiac and skeletal muscles, and acute cardiac failure is the cause of death.

Blister Beetles

Cantharidin is an irritant toxic agent present in blister beetles. Only a few of the several species of blister beetles contain cantharidin. Blister beetles are abundant in mid-summer and late summer when alfalfa hay containing the beetles is harvested in the central plains of the United States. Horses are poisoned by eating the alfalfa hay containing crushed swarms of blister beetles. Affected horses develop severe colic, abdominal pain, and blood-tinged urine. They will kick at their bellies and roll on the ground; severely affected horses may die of shock. Although recommended treatment involves the use of pain killers, such as banamine hydrochloride, and large volumes of intravenous fluids, there is no effective therapy for affected horses.

Heavy Metals

Lead poisoning in horses is characterized by neurological effects. Affected horses will be either depressed or excited. Colic and diarrhea are also seen. Because of laryngeal nerve paralysis, horses poisoned by lead also present with difficult respirations and a roaring syndrome. Abortions may also occur.

Arsenic poisoning in horses is usually caused by the consumption of foliage which has recently been sprayed with arsenic herbicides. The condition is acute and characterized by intense colic and hemorrhagic diarrhea. As in other animals, inorganic arsenic poisoning does not affect the nervous system, which helps differentiate this poisoning from organophosphate or carbamate poisonings.

Selenium is an essential element but is toxic when excessive quantities are ingested. Exposure of horses is usually through consumption of seleniferous (accumulator or indicator) plants (e.g., *Astragalus* spp.). Exposure to high quantities of selenium over a short time causes diarrhea (which is often foul smelling and contains air bubbles), neurological and cardiovascular effects, and respiratory difficulty. Death in these horses is due to respiratory failure. Chronic exposure to low levels of excessive selenium is characterized by hoof abnormalities at the coronary bands and by discoloration and loss of hair. The hoof deformities are painful and cause lameness.

Toxic Plants

The plant poisonings commonly encountered in horses are those which cause gastrointestinal problems, liver damage, primary or secondary nervous system involvement, and sudden death. Plants such as castor bean and oleander cause colic and diarrhea. Oleander also causes cardiac toxicity. Prolonged ingestion of some plants for several weeks can lead to liver damage and hepatic cirrhosis. This commonly occurs with the hepatotoxic plants *Amsinckia*, *Senecio*, and *Crotolaria*. Liver damage compromises the ability of the horse to detoxify ammonia which accumulates *in vivo*, leading to CNS derangement.

Plants which can cause CNS stimulation include larkspur, locoweed, lupine, water hemlock, and fitweed. Common plants which produce CNS depression are black locust, bracken fern, horsetail, milkweed, and white snake root. Like ruminants, horses will avoid eating toxic plants because they are usually not palatable. Therefore, consumption of poisonous plants will most often occur during drought conditions or following overgrazing when the animals lack suitable pasture.

Sudden death in horses can be caused by consumption of cyanide-containing plants, such as sorghum. The cyanide ion forms a complex with cytochrome oxidase. This prevents electron transport and the utilization of oxygen by body tissues. As a consequence, the circulating blood is well oxygenated and is bright cherry-red in color. This condition is an emergency, and treatment requires the prompt intravenous administration of both sodium nitrite and sodium thiosulfate.

Horses, like other monogastrics, are more resistant to plants capable of causing nitrate–nitrite poisoning than are ruminants. However, horses can modestly reduce nitrates to nitrites in their cecums, but it requires about three times as much nitrate to produce the same toxic effect in horses as in ruminants.

Mycotoxins

Contaminated grains (corn, wheat, and milo) are the sources of mycotoxin exposure for horses. The most commonly involved mycotoxins are aflatoxins, T-2, and fumonisin B$_1$. Aflatoxins will cause nonspecific effects, such as a poor thriving condition, hemorrhages, and abortions. T-2 is a trichothecene mycotoxin which causes prolonged bleeding times and digestive tract inflammation in affected horses.

A specific mycotoxin which uniquely affects horses is fumonisin B$_1$, which is produced by the fungus *F. moniliforme;* it has been responsible for causing the condition called equine leukoencephalomalacia. Horses receive the fumonisin by consumption of molded corn. Affected horses become anoretic and depressed after consuming the infested grain for only a few days. The condition then progresses, with the animals becoming blind, walking aimlessly, head pressing, being unable to swallow, and dying in a coma 1–4 days after the initial onset of signs.

Conclusions

The broad discipline of veterinary toxicology has been presented using brief accounts of the common toxicities in different animal species to draw the attention of the reader to similarities and differences in their reactions to toxicants. Because some animals are more sensitive than others receiving the same toxicant, the diagnosis of some poisonings may require the help of toxicologists within the veterinary profession.

This entry was not intended as a detailed reference for diagnosis and treatment of animal poisonings, nor was it meant to be all-inclusive. Rather, it presents the commonly encountered toxicoses in veterinary medicine.

It should be clear that all animals are susceptible to some toxicants and that some toxicants are toxic to all animals (including humans). It is therefore important to be cautious when handling and using chemicals around animals; also, a clean environment must be provided for all animals. Domestic animals particularly are subject to the whims of their owners for hazard-free environments. Animals should be fed well-balanced quality food from reputable sources, and suspect feed should be either avoided or carefully examined for potential toxicants before being given to animals.

It is also vitally important to remember that all chemicals become poisons if the exposure rate is sufficiently high. Therefore, even useful and recommended compounds used routinely around animals (e.g., growth promoters) can be life-threatening if used excessively or if given to species for which they were not intended. The susceptibility of sheep to copper-containing cattle feed or the high risk to horses when fed poultry feeds containing monensin are cases in point.

Further Reading

Beasley, V. R. (Ed.) (1990). *Vet. Clin. North Am. Small Anim. Pract. Toxicol. Selected Pesticides Drugs Chem.* **20.**

Blood, D. C., and Radostits, O. M. (1989). *Veterinary Medicine,* pp. 1241–1249. Bailliere Tindall, London.:

Cheeke, P. R., and Shull, L. R. (Eds.) (1985). *Natural Toxicants in Feeds and Poisonous Plants,* pp. 393–476. AVI Press, New York.

Dunn, H. W., and Leman, A. D. (Eds.) (1975). *Diseases of Swine,* 4th ed., pp. 854–860. Iowa State Univ. Press, Ames.

Hays, W. J., Jr., and Laws, E. R., Jr. (Eds.) (1991). *Handbook of Pesticide Toxicology,* pp. 39–105. Academic Press, San Diego.

Kingsbury, J. M. (1964). *Poisonous Plants of the United States and Canada.* Prentice Hall, Englewood Cliffs, NJ.

Osweiler, G. D., Carson, T. L., Buck, W. B., and Van Gelder, G. A. (Eds.) (1985). *Clinical and Diagnostic Veterinary Toxicology,* 3rd ed. Kendall Hunt, Dubuque, IA.

Purchase, I. F. H. (Ed.) (1974). *Mycotoxins.* Elsevier, Amsterdam.

Robinson, N. E. (Ed.) (1987). *Current Therapy in Equine Medicine 2.* Saunders, Philadelphia.

Spoerke, D. G., Jr., and Smolinske, S. C. (1990). *Toxicity of Houseplants.* CRC Press, Boca Raton, FL.

—Frederick W. Oehme and Wilson K. Rumbeiha

Related Topics

Aflatoxin
Algae
Ammonia
Arsenic
Brodifacoum
Carbamate Pesticides
Carbon Monoxide
Castor Bean
Copper

Coumarins
DDT
DEET
Dichlorvos
Dieldrin
Ethylene Glycol
Hydrogen Sulfide
Lead
Lindane
Malathion
Methane
Molybdenum
Mushrooms
Mycotoxins
Nitrites
Oleander
Organochlorine Insecticides
Organophosphates
Paraquat
Parathion
Pyrethrin/Pyrethroids
Selenium
Sodium
Strychnine
Sulfur Dioxide
Thallium
Warfarin

Vinyl Chloride

- CAS: 75-01-4
- SYNONYMS: VC; chloroethene; chloroethylene
- CHEMICAL CLASS: Vinyl monomers; halogenated hydrocarbon
- CHEMICAL STRUCTURE:

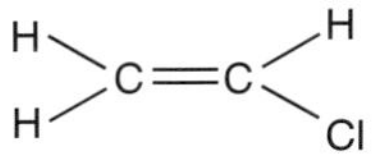

$$H_2C{=}CHCl$$

Uses

Vinyl chloride is used in plastic manufacturing, as a refrigerant, and in adhesives. It is the seventeenth highest volume chemical produced in United States (1991).

Exposure Pathways

Dermal exposure and inhalation are possible exposure routes.

Toxicokinetics

Vinyl chloride is absorbed through the respiratory system to blood and organs. Over 40% of inhaled vinyl chloride is retained in the lung. It is activated to toxic species by P450 metabolism.

Mechanism of Toxicity

Vinyl chloride is a cross-linking agent; it interferes with metabolic pathways.

Human Toxicity

"Vinyl chloride" disease results if exposure exceeds the allowable limits of concentration. It is narcotic in high concentrations and causes liver and kidney damage. Dermal contact may cause frostbite. The ACGIH TLV is 5 ppm. The OSHA PEL is 1 ppm for an 8-hr period of exposure. The human lethal dose is 20,000 ppm. Vinyl chloride is a known carcinogen.

Clinical Management

The affected individual should be removed from exposure.

Animal Toxicity

In animals, vinyl chloride is a carcinogen (liver angiosarcoma), teratogen, mutagen, and reproductive toxicant. The oral LD_{50} in rats is 500 mg/kg. The inhalation LC_{50} in rats is 18 pph/15 min. The inhalation LC_{50} in mice is 20 pph/hr.

—*Jayne E. Ash and Shayne C. Gad*

Related Topics

Carcinogenesis
Dichloroethylene, 1,1-
Indoor Air Pollution
Liver
Occupational Toxicology
Polymers

Reproductive System, Female
Respiratory Tract

Vitamin A

- ♦ CAS: 68-26-8
- ♦ SYNONYMS: Retinol; retinyl esters; antiinfective vitamin; 3,7-dimethyl-9-(2,6,6-trimethyl-1-cyclohexen-1-yl)-2,4,6,8-nonaretraen-1-ol
- ♦ PHARMACEUTICAL CLASS: Fat-soluble vitamin, derived from retinol
- ♦ CHEMICAL STRUCTURE:

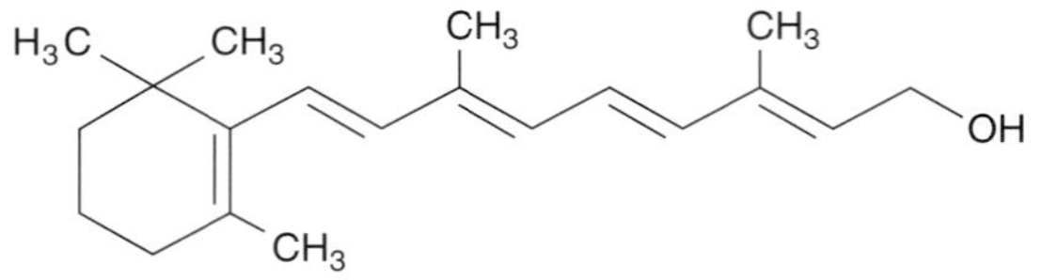

Uses

Vitamin A is used as a dietary supplement and for treatment of deficiency syndromes. It is not an endogenously produced vitamin; thus, it must be provided through dietary or vitamin supplement sources. Vitamin A is essential for normal vision in dim light. Furthermore, it is needed for regulation of all growth and development, for maintaining mucous membrane integrity and for the reproductive process.

Exposure Pathways

Ingestion is the most common route of exposure. Available forms include capsules and intramuscular solutions. Animal livers are rich in vitamin A.

Toxicokinetics

Vitamin A is readily absorbed from the intestine as retinyl esters. Peak serum levels are reached 4 hr after a therapeutic dose. The vitamin is distributed to the general circulation via the lymph and thoracic ducts. Ninety percent of vitamin A is stored in the liver, from which it is mobilized as the free alcohol, retinol. Ninety-five percent is carried bound to plasma proteins, the retinol-binding protein. Vitamin A undergoes hepatic metabolism as a first-order process. Vitamin A is excreted via the feces and urine. β-Carotene is converted to retinol in the wall of the small intestine. Retinol can be converted into retinoic acid and excreted into the bile and feces. The elimination half-life is approximately 9 hr.

Mechanism of Toxicity

The exact mechanism leading to toxicity is not known. Both acute and chronic toxicity may occur. Daily vitamin A requirements range from 1500 international units (IU) to 4500 IU for children, 5000 IU for adults, and 8000 IU for pregnant women.

Human Toxicity

Acute toxicity is uncommon in adults. Toxicity is more frequently seen with chronic ingestion of high doses of 30,000–50,000 IU/day. Children have developed acute toxicity following ingestion of 300,000 IU, but more frequently hypervitaminosis A in children develops with chronic ingestion of >10 times the recommended daily allowance for weeks to months. Malnutrition and individual tolerance may also be factors in predisposition to toxicity. Signs and symptoms of toxicity include vomiting, anorexia, agitation, fatigue, double vision, headache, bone pain, alopecia, skin lesions, increased intracranial pressure, and papilloedema.

Clinical Management

In massive acute overdose, decontamination is advised. If the ingestion is recent (<30 min) and the patient is asymptomatic, syrup of ipecac is indicated. Activated charcoal may be used to adsorb the vitamin. Plasma vitamin A levels can aid in diagnosis but are not clinically useful in treatment. Upon discovery of a potential overdose, exposure to vitamin A should be immediately discontinued. Young children should be monitored for symptoms of increased intracranial pressure; elevated intracranial pressure should be treated with mannitol, dexamethasone, and hyperventilation as needed. Vital signs and fluid and electrolyte status should be monitored closely. In general, vitamin A toxicity often resolves itself spontaneously within days to weeks following withdrawal of vitamin A. There are no known cases of vitamin A toxicity associated with β-carotene ingestion.

—*Anne E. Bryan*

Related Topics

Ascorbic Acid
Developmental Toxicology
Folic Acid
Iron
Niacin
Pyridoxine
Riboflavin
Thiamine
Vitamin D
Vitamin E

Vitamin D

- CAS: 50-14-6
- SYNONYMS: Ergocalciferol (D$_2$); cholecalciferol (D$_3$); α-calcidol; calcitriol; dihydrotachysterol (DHT)
- PHARMACEUTICAL CLASS: Fat-soluble vitamin
- CHEMICAL STRUCTURE: (Vitamin D$_2$)

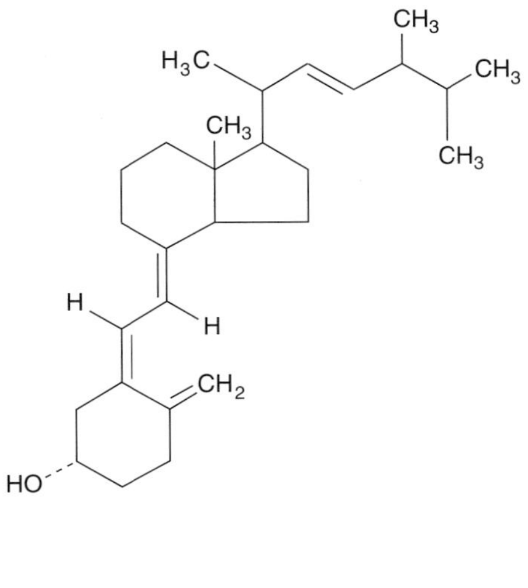

Use

Vitamin D is a dietary supplement used for the prevention/treatment of deficiency syndromes. It is the only vitamin synthesized by the conversion of 7-dehydrocholesterol to chlolecalciferol by exposure to sunlight or shortwave ultraviolet light.

Exposure Pathway

Ingestion of oral dosage forms is the most common route of exposure in both acute and chronic overdosage.

Toxicokinetics

Vitamin D is readily absorbed from the gastrointestinal tract. Cholecalciferol is metabolized in the liver to 25-hydroxycholecalciferol and then to 1-α-25-dihydroxycalciferol in the kidney. This mobilizes stores of calcium from the bone matrix to the plasma. Cholecalciferol is stored in adipose and muscle tissue. The metabolites of vitamin D compounds are excreted primarily in bile and feces.

Mechanism of Toxicity

Excess vitamin D results in hypercalcemia and hypercalciuria, due to increased calcium absorption, bone demineralization, and hyperphosphatemia.

Human Toxicity

Acute toxicity is rarely reported. Infants have reportedly tolerated up to 60,000–100,000 international units per kilogram without ill effect. Chronic toxicity occurs after the recommended daily allowance is excessively exceeded for weeks to months. Common symptoms of toxicity include nausea, flatulence, and diarrhea. Other nonspecific symptoms reported include muscle weakness, fatigue, and bone pain.

Renal failure may also occur due to precipitation of calcium in the kidneys. Vitamin D serum levels may be useful, as well as serum calcium, and alkaline phosphatase levels.

Clinical Management

Exposure to all forms of vitamin D should be stopped. Treatment should be supportive and symptomatic. Hypercalcemia treatment should include a low-calcium diet and prednisone as necessary.

Animal Toxicity

No reports of animal toxicity from vitamin D supplements could be found. However vitamin D has proven fatal to animals when they were exposed to a vitamin D-containing rodenticide.

—Anne E. Bryan

Related Topics

Ascorbic Acid
Folic Acid
Iron
Niacin
Pyridoxine
Riboflavin
Thiamine
Vitamin A
Vitamin E

Vitamin E

- CAS: 59-02-9
- SYNONYMS: Antisterility vitamin; almefrol; α-tocopherol
- PHARMACEUTICAL CLASS: Fat-soluble vitamin
- MOLECULAR FORMULA: $C_{29}H_{5}00_{2}$
- CHEMICAL STRUCTURE:

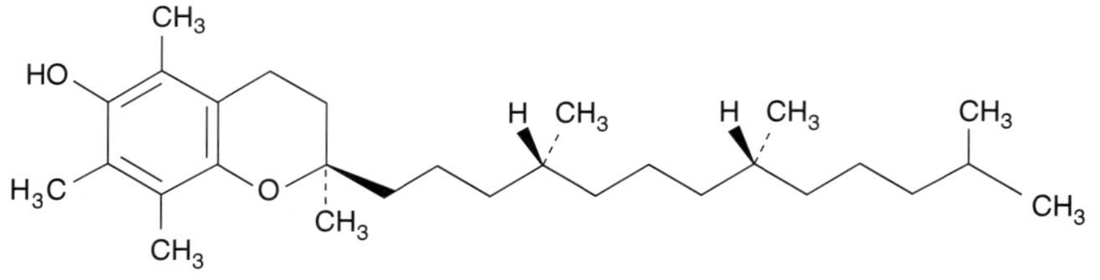

Uses

Vitamin E is used as a dietary supplement and for the treatment of deficiency syndromes.

Exposure Pathway

Ingestion as a supplemental vitamin is the most common route of exposure. Available forms include tablets, capsules, and intramuscular solutions.

Toxicokinetics

Vitamin E is absorbed in the gastrointestinal tract. Bile is necessary for absorption. Vitamin E is metabolized in the liver. The major metabolites of vitamin E are the glucuronides of tocopheronic acid. Vitamin E is distributed to all tissues. Lipid tissues store the vitamin for prolonged periods of time. Up to 30% is excreted in the urine. The half-life after intramuscular injection is approximately 45 hr.

Mechanism of Toxicity

The exact mechanism of toxicity is unknown.

Human Toxicity

The toxidrome of acute and chronic toxicity is not well defined. Subjective symptoms include nausea, vomiting, flatulence, fatigue, muscle weakness, headaches, and blurred vision. Controversy exists over whether excessive vitamin E may cause liver and renal damage. The plasma concentration levels for vitamin E vary among individuals.

Clinical Management

Since there is not any evidence that acute overdosage represents a medical emergency, decontamination is not advised. Once chronic toxicity is suspected, discontinuation of vitamin usage and supportive/symptomatic therapy is recommended.

—Anne E. Bryan

Related Topics

Ascorbic Acid
Folic Acid
Iron
Liver
Niacin
Pyridoxine
Riboflavin
Thiamine
Vitamin A
Vitamin D

VX

- CAS: 50782-69-9; 51848-47-6; 53800-40-1; 70938-84-0

- SYNONYMS: Phosphonothioic acid; methyl-*S*-(2-bis(1-methylphosphonothioate; *S*-2-diisopropylaminoethyl-*O*-ethyl methylphosphonothioate; *S*-2(2-diisopropylaminoethyl)-*O*-ethyl methylphosphonothioate; *O*-ethyl-*S*-(2-diisopropylaminoethyl) methylthiolphosphonoate; TX60; nerve gas; nerve agent

- CHEMICAL CLASS: Persistent anticholinesterase compound or sulfonated organophosphorus (OP) nerve agent

- MOLECULAR FORMULA: $C_{11}H_{26}NO_2PS$

- CHEMICAL STRUCTURE:

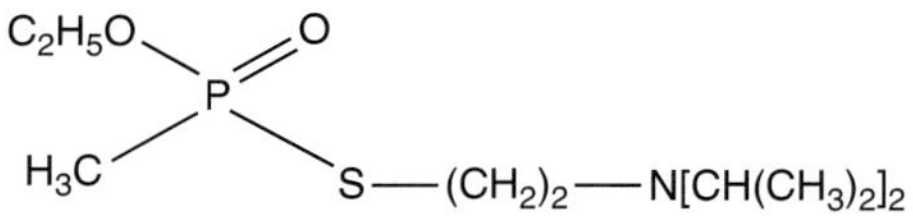

Use

VX is a nerve agent used in chemical warfare

Exposure Pathways

Casualties are caused both by inhalation and by dermal contact. Since VX is an oily liquid with a low volatility, liquid droplets on the skin do not evaporate quickly, thus facilitating effective percutaneous absorption. In addition to inhalation and percutaneous exposure, casualties can also be caused by ocular exposure, ingestion, and injection.

Toxicokinetics

VX is absorbed through the skin and respiratory system. Because it is nonvolatile, it may remain in place for weeks after dispersion and cause casualties. Thus, it is classified as a persistent agent. Although VX does not pose a major inhalation hazard in usual circumstances, by the inhalation route it is estimated to be twice as toxic as sarin. It is hydrolyzed by the enzyme OP hydrolase.

Mechanism of Toxicity

VX and the other nerve agents are irreversible organophosphorus cholinesterase inhibitors. They inhibit the enzymes butyrylcholinesterase in the plasma, the acetylcholinesterase on the red blood cell, and the acetylcholinesterase at cholinergic receptor sites in tissues. These three enzymes are not identical. Even the two acetylcholinesterases have slightly different properties, although they have a high affinity for acetylcholine. The blood enzymes reflect tissue enzyme activity. Following acute nerve agent exposure, the red blood cell enzyme activity most closely reflects tissue enzyme activity. However, during recovery, the plasma enzyme activity more closely parallels tissue enzyme activity.

Following nerve agent exposure, inhibition of the tissue enzyme blocks its ability to hydrolyze the neurotransmitter acetylcholine at the cholinergic receptor sites. Thus, acetylcholine accumulates and continues to stimulate the affected organ. The clinical effects of nerve agent exposure are caused by excess acetylcholine.

The binding of the nerve agent to the enzyme is considered irreversible unless removed by therapy. The accumulation of acetylcholine in the peripheral nervous system and central nervous system (CNS) leads to depression of the respiratory center in the brain, followed by peripheral neuromuscular blockade causing respiratory depression and death.

The pharmacologic and toxicologic effects of the nerve agents are dependent on their stability, rates of absorption by the various routes of exposure, distribution, ability to cross the blood–brain barrier, rate of reaction and selectivity with the enzyme at specific foci, and their behavior at the active site on the enzyme.

Red blood cell enzyme activity returns at the rate of red blood cell turnover, which is about 1% per day. Tissue and plasma activities return with synthesis of new enzymes. The rate of return of these enzymes are not identical. However, the nerve agent can be removed from the enzymes. This removal is called reactivation, which can be accomplished therapeutically by the use of oximes prior to aging. Aging is the biochemical process by which the agent–enzyme complex becomes refractory to oxime reactivation. The toxicity of nerve agents may include direct action on nicotinic acetylcholine receptors (skeletal muscle and ganglia) as well as on muscarinic acetylcholine receptors and the CNS.

Recently, investigations have focused on OP nerve agent poisoning secondary to acetylcholine effects. These include the effects of nerve agents on γ-aminobutyric acid neurons and cyclic nucleotides. In addition changes in brain neurotransmitters such as dopamine, serotonin, noradrenaline as well as acetylcholine following inhibition of brain cholinesterase activity have been reported. These changes may be due in part to a compensatory mechanism in response to overstimulation of the cholinergic system or could result from direct action of nerve agent on the enzymes responsible for noncholinergic neurotransmission.

Human Toxicity

Following inhalation of VX, the median lethal dosage (LCt_{50}) in man has been estimated to be 30 mg-mi/m^3 at a respiratory minute volume (RMV) of 15 liters/min. Following percutaneous exposure, the LD_{50} was estimated to be 0.315 mg/kg or 10 mg/70 kg in man. The intravenous LD_{50} was estimated to be 0.008 m/kg or 0.56 mg/70-kg man and the intramuscular LD_{50} 0.012 mg/kg or 0.84 mg/70 kg man.

The doses that are potentially life-threatening may be only slightly larger than those producing minimal effects. The ECt_{50} for miosis from ocular vapor exposure was estimated to be <0.09 mg-min/m^3, and the ECt_{50} for runny nose is also estimated to be <0.09 mg/min/m^3. For severe incapacitation for vapor inhalation the ICt_{50} was estimated to be 25 mg-min/m^3, while the LCt_{50} was estimated to be 30 mg-min/m^3.

The permissible airborne exposure concentration for VX for an 8-hr workday of a 40-hr workweek is an 8-hr TWA of 0.00001 mg/m^3.

These signs and symptoms occur within minutes or hours following exposures. The signs and symptoms following vapor exposure include miosis and visual disturbances, headache and pressure sensation, runny nose and nasal congestion, salivation, tightness in the chest, nausea, vomiting, giddiness, anxiety, difficulty in thinking, difficulty sleeping, nightmares, muscle twitching, tremors, weakness, abdominal cramps, diarrhea, and involuntary urination and defecation. These signs and symptoms may progress to convulsions and respiratory failure. After liquid exposure on the skin, the initial effects are nausea, vomiting, and diarrhea, followed by muscular weakness, seizure, and apnea.

Clinical Management

Management of nerve agent intoxication consists of decontamination, ventilation, administration of antidotes, and supportive therapy.

The three therapeutic drugs for treatment of nerve agent intoxication are atropine, pralidoxime chloride, and diazepam. Atropine, a cholinergic blocking or anticholinergic drug, is effective in blocking the effects of excess acetylcholine at peripheral muscarinic sites. The usual dose is 2 mg, which may be repeated at 3- to 5-min intervals intravenously (iv) or intramuscularly (im). Pralidoxime chloride (Protopam chloride; 2-PAM CL) is an oxime used to break the agent–enzyme bond and restore the normal activity of the enzyme. This is most apparent in organs with nicotinic receptors. Abnormal activity and normal strength returns to skeletal muscles, but no decrease in secretions is seen following oxime treatment. The usual dose is 1000 mg (iv or im). This may be repeated two or three times at hourly intervals (intravenous or intramuscular). Diazepam, an anticonvulsant drug, is used to decrease convulsive activity and reduce brain damage that may occur from prolonged seizure activity. It is suggested that all three of these drugs be administered at the onset of severe effects from nerve agent exposure, whether or not seizures occur. The usual dose of diazepam is 10 mg (im).

Miosis, pain, dim vision, and nausea can be relieved by topical atropine in the eye.

Supportive therapy may include ventilation via an endotracheal airway if possible and suctioning of excess secretions in the airways.

Animal Toxicity

Small doses of nerve agents in animals can produce tolerance in addition to their classical cholinergic effects. In rats, acute administration of nerve agents in subconvulsive doses produced tumors and hindlimb abduction. In animals nerve agents can also cause effects in behavior, analgesia, as well as cardiac effects.

The cause of death is attributed to anoxia resulting from a combination of central respiratory paralysis, severe bronchoconstriction, and weakness or paralysis of the accessory muscles for respiration.

Signs of nerve agent toxicity vary in rapidity of onset, severity, and duration of exposure. These are dependent on the specific agent, route of exposure, and dose. At the higher doses, convulsions and seizures are indication of CNS toxicity.

Following nerve agent exposure, animals exhibit hypothermia resulting from the cholinergic activation of the hypothalamic thermoregulatory center. In addition, plasma levels of pituitary, gonadal, thyroid, and adrenal hormones are increased during organophosphate intoxication.

The available animal toxicity data are presented as follows:

VX animal toxicity
  Subcutaneous LD_{50} (μg/kg)
    Rat 12
    Mouse 22
    Rabbit 14
    Guinea pig 8400 mg/kg

Intraperitoneal LD_{50} (μg/kg)
  Mouse 50
  Rabbit 66

Intramuscular LD_{50} (μg/kg)
  Chicken 30

Intravenous LD_{50} (μgkg)
  Cat 5

—*Harry Salem and Frederick R. Sidell*

(The views of the authors do not purport to reflect the position of the U.S. Department of Defense. The use of trade names does not constitute official endorsement or approval of the use of such commercial products.)

Related Topics

Acetylcholine
Anticholinergics
Atropine
Cholinesterase Inhibition
Mustard Gas
Nerve Agents
Neurotoxicology
Nitrogen Mustards
Organophosphates
Sarin
Soman
Sulfur Mustard
Tabun

Warfarin

- CAS: 81-81-2
- Synonym: Coumarin
- Pharmaceutical Class: A synthetic derivative of 4-hydroxycoumarin, the hemorrhagic component of sweet clover
- Chemical Structure:

$$\text{Structure with } ONa \text{ and } CH_2COCH_3 \text{ groups}$$

Uses
Warfarin is used therapeutically as an anticoagulant. It is also used as a rodenticide.

Exposure Pathways
Ingestion is the most common route of exposure. Warfarin is also absorbed transdermally and by inhalation. It is available in oral and injectable forms. Warfarin rodenticides are typically 0.025–0.050% warfarin by weight.

Toxicokinetics
Warfarin is rapidly and nearly completely absorbed by the oral route. Peak plasma levels typically occur within 2–8 hr. Warfarin is highly protein bound, 97–99+%. The volume of distribution approximates 0.15 liters/kg. Warfarin is extensively metabolized by hepatic microsomal enzymes. The primary metabolites are 6- and 7-hydroxy warfarin via oxidation and several warfarin alcohols via reduction. The warfarin alcohols retain weak anticoagulant activity. The metabolites undergo enterohepatic circulation. Approximately 85% of warfarin appears in the urine as metabolites. Less than 1 or 2% appears in the urine unchanged. Warfarin metabolites are also excreted in the stool. The plasma half-life varies widely, from 10 to 80 hr; it is typically 36–44 hr. The duration of clinical effects can significantly exceed the half-life of warfarin. (Note: There are many drug interactions with warfarin; the reader is referred to a standard pharmacology text for further details.)

Mechanism of Toxicity
Warfarin interferes with the hepatic production of a number of proteins involved in hemostasis. These include the coagulation factors II (prothrombin), VII, IX, and X and also proteins C and S, important modulators of coagulation. Vitamin K_1 is a cofactor for the carboxylation of specific glutamic acid groups in these proteins. During carboxylation, vitamin K_1 is oxidized to vitamin K_1 2, 3-epoxide. The regeneration of vitamin K_1 by vitamin K_1 epoxide reductase is antagonized by warfarin. As a result, dysfunctional decarboxy coagulation factors are produced and overall synthesis may be reduced. This leads to impaired coagulation.

Human Toxicity: Acute
Depletion of preformed circulating coagulation factors must occur before any effect by warfarin is apparent.

411

Factor VII has the shortest half-life and factor II the longest of the vitamin K-dependent coagulation factors. The prothrombin time (PT) may begin to increase by 24 hr and be maximal 36–72 hr postingestion. Significant toxicity from single-dose ingestion is uncommon; most instances of toxicity are the result of repeated ingestion over time. The most frequent sites of bleeding are mucocutaneous, genitourinary, and gastrointestinal, although bleeding can occur virtually anywhere. The more serious events include massive hemorrhage with shock, intracranial bleeding and stroke, and pericardial tamponade. Plasma warfarin levels are not routinely done. The effect of warfarin is best followed by the PT. Specific assays of factor activity can be measured although this is not usually necessary.

Human Toxicity: Chronic

Effects other than those outlined previously have not been described. Warfarin use has been associated with birth defects and spontaneous abortion.

Clinical Management

For acute single ingestion, activated charcoal should be administered. A gastric emptying procedure may be beneficial if performed soon afterward. Induced emesis should be avoided in the anticoagulated patient. The PT should be monitored for at least the first 48 hr for signs of toxicity. Extreme caution should be used with any invasive procedure in the anticoagulated patient. The airway should be protected if compromised. Volume resuscitation should be provided as indicated by clinical status. With active, uncontrolled, or life-threatening hemorrhage, fresh frozen plasma should be administered to provide preformed clotting factors (at least 4 to 6 units will be necessary in an adult).

Vitamin K_1 (phytonadione) is a specific antidote for warfarin toxicity. Pharmacologic doses of vitamin K_1 antagonize the inhibitory effect of warfarin on clotting factor production. The dose and route of vitamin K_1 administration depends on the clinical setting. For rapid reversal, 5–25 mg should be administered intravenously no faster than 1 mg/min; in children use 0.6 mg/kg. Clinical effects may be seen within hours. The response and duration of a single dose of vitamin K_1 is variable and dependent on the severity of the toxicity. The half-life of vitamin K_1 is less than 4 hr and repeat doses may be necessary. In less acute settings, vitamin K_1 may be given subcutaneously or orally. The PT should be monitored to follow toxicity and response to treatment. [Note: Anaphylaxis has been reported with intravenous vitamin K_1. Vitamin K_3 (menadione) is not effective therapy. In patients therapeutically anticoagulated, rapid reversal can be dangerous and should be done with caution.]

Animal Toxicity

Mammals vary in their sensitivity to warfarin. Horses are resistant to the coumarins and cats are more sensitive than dogs. Signs of toxicity in animals include depression, anorexia, weakness, vomiting, diarrhea, bleeding, and dyspnea. Toxic effects can be monitored by measurement of the PT or one-stage prothrombin time. Treatment is as for humans. The recommended dose of vitamin K_1 for dogs and cats is 0.25–1 mg/kg/d for 5–14 days.

—*Michael J. Hodgman*

Related Topic

Coumarins

Wisteria

- SYNONYMS: *Wisteria sinensis*; kidney bean tree; China kidney bean tree
- DESCRIPTION: Wisteria is a woody vine or climbing shrub of North America and Eastern Asia. The pink, white, or blue fragrant flowers bloom in clusters. The seeds are contained in pea-shaped flat pods.

Exposure Pathways

Ingestion of seeds, pods, and flowers is the primary route of exposure. Symptoms have developed following exposure to the smoke of this plant when burned.

Mechanism of Toxicity

Although toxins are identified (wistarine and lectin), clear information about their behavior does not exist.

As a saponin-containing compound, it is classified as a gastrointestinal irritant.

Human Toxicity

All parts of the wisteria plant are considered toxic, especially the pods and seeds. Although serious poisonings are not common, exposures to as few as two seeds have been known to result in serious effects. Symptoms include oral burning, stomach pain, diarrhea, and vomiting. The ingestion of large amounts may result in effects severe enough to produce hypovolemic shock. Symptoms usually resolve within 24–48 hr. The mitogenic and blood clotting effects of lectins are not seen in toxic exposures. Exposure to smoke from the burning of this plant is known to cause headaches.

Clinical Management

Initial treatment with gastric lavage or emesis is indicated. Support with fluid replacement and antiemetics may be indicated.

—*Regina Wiechelt*

Xylene

- CAS: 1330-20-7; 95-47-6; 108-38-3; 106-42-3 (*o*-, *m*-, *p*-isomers)
- SYNONYMS: Dimethyl benzene; *ksylen* (Polish); methyltoluene; NCI C55232; UN1307 (DOT); violet 3; *xiloli* (Italian); *xylenen* (Dutch); xylol; *xylole* (German); RCRA Waste No. U239
- DESCRIPTION: Xylene is essentially a benzene ring with two methyl substitutions. There are three isomers of xylene: *ortho*-, *meta*-, and *para*-xylene. Most commercially available xylenes and the xylene component of gasoline are mixtures of these three isomers. Xylene compounds are lighter than water and only slightly soluble.
- CHEMICAL CLASS: Aromatic hydrocarbon

Uses

Xylene is derived from petroleum and/or coal tar distillation, and is also a component of naphtha, asphalt, and gasoline (6% in high-octane gasoline). Xylenes are volatile, colorless, flammable liquids that are used as thinners, solvents for inks, rubbers, gums, resins, adhesives, and lacquers, as paint removers, and as intermediates in the production of plasticizer (phthalic acid and anhydride) and polyester fibers. Xylenes are also extensively used as intermediates in the manufacture of perfumes, dyes, insecticides, and pharmaceuticals. They easily dissolve hydrophobic compounds, especially fats, oils, and waxes.

Exposure Pathways

Because xylene is fairly volatile, exposure for humans would occur principally by inhalation and would most likely occur near the principal sources discussed below, such as emissions from chemical plants and refineries, gas pumps, painting or refinishing operations, and automobiles (e.g., in tunnels). Dermal exposure may also be significant, especially in an industrial setting, where skin may be exposed for long periods of time. Oral exposure is the least probable route and would occur primarily as a result of accidental poisoning or suicide.

Automobile and industrial emissions contribute the majority of xylene found in the atmosphere. Concentrations are lowest in remote areas (average concentrations are <0.5 ppb) and highest in urban areas (detected levels ranging from 0.5 to 21 ppb). Xylene is also found in plants and is present in their combustion products (i.e., forest fire smoke and tobacco smoke). Xylene has also been detected in surface water and treated wastewater effluents, with average levels generally below 1 µg/liter. It has been detected in 3% of ground-water and 6% of surface water supplies sampled in a survey sponsored by the U.S. EPA. It is a typical ground-water contaminant where gasoline releases have occurred. Xylene is readily biodegradable and will not bioconcentrate to a great degree.

Toxicokinetics

Xylene is primarily absorbed through the mucous membranes and pulmonary system. In experimental subjects, about 60% of airborne xylene is actually absorbed from the lung into the bloodstream. Xylene is also readily absorbed from the gastrointestinal tract and through broken or intact skin. Once absorbed, xylene distributes to many tissues in the body, especially lipid-rich organs, although this occurs to a lesser extent than for benzene. Chemical alteration of xylene occurs in the liver and the lung, where the compound is changed to more water-soluble metabolites (corresponding *o*-, *m*-, and *p*-toluic acids and/or methylhippuric acid) so it can be easily excreted in the urine. In animals it has been shown that metabolism is qualitatively different in the lung versus liver. Greater than 95% of absorbed xylene is excreted as a water-soluble metabolite, with the remaining fraction being exhaled unchanged. Excretion appears to occur rapidly; animal studies indicate complete clearance of the compound in 24 hr. Xylene will also cross the placenta and enter fetal tissue.

Mechanism of Toxicity

Although an exact mechanism for xylene toxicity has not been determined, it is known that the primary toxic effect of xylene is dysfunction of the brain and central nervous system (CNS; narcosis). The main function of neurons is to conduct electrochemical signals to one, several, or thousands of other cells. The normal physiology of these neurons is, in turn, largely dependent on the integrity of the neuron cell membrane, which polarizes and depolarizes during the transmission of these electric signals. Thus, the most probable mechanism of toxicity is the unique sensitivity of the cell membranes of neurons to the solvent property of xylene, which disrupts the normal transmission of nerve impulses.

Human Toxicity

Xylene is an irritant to the eyes and gastrointestinal tract. Direct contact with the skin is also irritating and will cause defatting, which may lead to dryness, cracking, blistering, or dermatitis. Xylene appears to be more acutely toxic than other structural analogs, such as benzene or toluene. CNS depression, a typical effect seen in solvent exposures, is the primary toxic effect seen following exposure to xylene. At high air concentrations, xylene may cause the following acute signs in humans: a flushing and reddening of the skin, a feeling of increased body heat, disturbed vision, dizziness,

tremors, salivation, cardiac stress, CNS depression, and confusion. Very high exposures may cause anorexia, nausea, vomiting, and abdominal pain; continued exposure may lead to coma and death, which appears to be due to cardiac fibrillation and/or lung congestion and hemorrhage. Females are reported to be more susceptible to the effects of xylene than males.

Effects from chronic exposure to xylene are similar to those from acute exposure but are systemically more severe. Repeated, prolonged exposure to xylene may result in conjunctivitis of the eye and dryness of the nose and throat. Repeated exposure of the skin will cause dryness, flaking, and/or dermatitis. Inhalation may cause CNS effects, such as excitation, then depression characterized by signs such as paresthesia, tremors, apprehension, impaired memory, weakness, nervous irritation, vertigo, headache, anorexia, nausea, and flatulence. Clinical findings may include moderate but reversible changes such as bone marrow hyperplasia, liver enlargement, and kidney nephrosis.

The current occupational exposure limit recommended by ACGIH (TLV) and enforced by the U.S government (OSHA PEL) for exposure to xylene via inhalation is 100 ppm (434 mg/m^3).

Clinical Management

Persons who have been overcome by xylene fumes or gases should be removed from the area of exposure to fresh air. Should breathing become labored or shallow, medical intervention (e.g., artificial respiration) may be necessary. Following accidental or intentional ingestion, vomiting should not be induced; stomach lavage should be initiated as soon as possible. Liquid xylene spills on exposed skin should be immediately dried with an absorbent towel; next, the affected area should be washed with soap and water. In cases of eye exposure, the eyes should be irrigated immediately.

Animal Toxicity

Acute CNS effects in animals, such as exaggerated visual disturbances, are similar to those in humans. The median lethal oral dose in rats is approximately 4.3 g/kg.

The results of early subchronic and chronic studies using xylene in laboratory animals are biased because much of the effects of the solvent were found to be caused by toxic impurities such as benzene. Later studies concur that xylene does cause a significant change in blood-forming elements and chemistry. The NTP recently conducted a study of rats and mice in which mixed xylenes were given orally. The only effect seen was decreased body weight in both sexes. This occurred at doses of 1000 mg/kg/day (5 days per week for 13 weeks). No effects were seen at the next lowest dose (500 mg/kg/day). No carcinogenic effects were apparent. The U.S. EPA has interpreted the weight of evidence for the carcinogenicity of xylene as D (i.e., insufficient evidence to classify human carcinogenicity).

In reproductive studies, effects on the fetus have been seen at oral doses higher that the no-effect level discussed previously; these effects were usually associated with concurrent maternal toxicity. In an inhalation study conducted by Biodynamics, pregnant female rats were exposed to mixed xylenes 6 hr per day for 190 days. Toxicity to the fetus was apparent in the group exposed to 500 ppm. Rat pups born to dams exposed to 500 or 250 ppm displayed reduced weight of ovaries,

but the effect was transient. Developmental toxicity was seen in another inhalation study in rats.

The no-effect level from the NTP study has been used to calculate a safe oral dose for xylene in humans of 2 mg/kg/day. Under the Safe Drinking Water Act, the maximum contaminant level (MCL) is the standard criteria for drinking water and the maximum contaminant level goal (MCLG) is the goal. The proposed MCL and MCLG for mixed xylenes is 10 mg/liter.

—*Stephen Clough*

Related Topics

Benzene
Indoor Air Pollution
Neurotoxicology
Toluene

Yew

- SYNONYMS: *Taxus baccata*, *Taxus cuspidata*, *Taxus brevifolia*; *Taxus canadensis*, Taxaceae family; ground hemlock; English yew; western yew; American yew

- DESCRIPTION: Yews are evergreen shrubs or trees with alternate branchlets and reddish brown, thin, scaled bark. The leaves are flat and needle-like, approximately 1 in. long, with green uppers and a yellowish underside. Leaves grow in opposite pairs along the stems. Only the female plants bear fleshy, scarlet red, cup-shaped fruits (arils) that have a single, hard, dark green to black seed. The yew species are native to Europe, Asia, and North America. They are common as ornamental hedging and ground covers. The yew is often brought indoors at Christmas and used as decoration.

Uses

Aqueous extracts of the yew have been used for years in Native American folk medicine for the cardiotonic, expectorant, antispasmodic, diuretic, and antiseptic properties. Experiments are being conducted on the potential for some extracts to possess central nervous system (CNS) depressant, analgesic, antipyretic, cytotoxic, and antileukemic properties.

Exposure Pathway

Ingestion of any part of the plant is the common route of exposure. The aril is not toxic. The hard seed and leaves are toxic and have the potential to release taxine. The plant parts are toxic whether green or dry.

Toxicokinetics

Taxine can be absorbed orally or by injection. Inhalation absorption is unlikely because it is not highly volatile. The onset of symptoms may be within 1 hr or delayed for several hours. Systemic symptoms are expected within 1–3 hr.

Mechanism of Toxicity

The main toxins of the yew species are the alkaloids taxine A and taxine B, which are present in all parts of the shrub except the fleshy part of the berry. These compounds are capable of causing symptoms similar to digitalis poisoning including hypotension, bradycardia, and depressed myocardial contractility and conduction delay (see Digitalis Glycosides). The mechanism appears to involve a block of the distal part of the conduction tissue of the heart, which can result in fatal arrhythmias. Atrioventricular conduction is particularly susceptible to yew alkaloids.

Human Toxicity: Acute

Serious poisoning is rare. Most cases of yew berry ingestions result in no symptoms. Ingestion of three to six intact berries without treatment would be reasonable with potential for only mild drowsiness and mild gastrointestinal symptoms within 1–3 hr. Persons who ingest other parts of the plant or multiple berries should have gastric decontamination performed.

Symptoms initially expected after ingestions of leaves or a chewed seed are dizziness, dry mouth, and mydriasis. These symptoms are followed by nausea, vomiting, and abdominal pain. A rash may appear, and facial pallor and cyanosis or reddish discoloration of the lips may occur. This is followed by generalized muscle weakness and drowsiness leading to coma. Seizures are also possible.

The primary action of these alkaloids is bradycardia and various other life-threatening arrhythmias with hypotension and decreased respiratory function. Death is due to cardiac and/or respiratory failure. Anaphylactoid reactions have been reported from chewing yew needles. If the seeds are ingested and not masticated, there is a likelihood they will pass without releasing the taxine.

Severe contact dermatitis can result from cutting yew wood. Taxines are water soluble, so drinking teas or water in which leaves are soaking is potentially dangerous.

Human Toxicity: Chronic

Chronic ingestion of the yew species has revealed liver and kidney fatty degeneration on autopsy.

Clinical Management

Basic and advanced life-support measures should be utilized as needed. Induction of emesis is not recommended. Persons who have ingested more than three to six intact berries or leaves and/or chewed or broken pits should be sent to an emergency room immediately. Lavage can be performed prior to administering activated charcoal. Whole bowel irrigation should be considered if large amounts are ingested. Observation and cardiac monitoring for a minimum of 4 hr is recommended. Those who remain asymptomatic or have only mild gastrointestinal symptoms may be discharged.

There are no antidotes. No laboratory test identifies taxine specifically.

Serious ingestions require cardiac monitoring in an intensive-care setting. Hypotension may be resistant to dopamine and dobutamine. Norepinephrine can also be used. Bradycardia can be treated with atropine and a temporary pacemaker as needed. Treatment of premature ventricular contractions and ventricular tachycardia may include lidocaine, procainamide, propranolol, phenytoin, and disopyramide.

Animal Toxicity

In animals, ingestion of large amounts of any part of the yew often causes sudden death without previous symptomatology or signs of struggle. Survival after yew poisoning is uncommon. Smaller ingestions would be expected to cause gastroenteritis. Clinical signs in a herd of 35 yew-poisoned cattle included lethargy, recumbency, dyspnea, jugular pulsation and distension, and death. Most cattle died within 4 hr. EKG changes and seizures were noted in dogs. Toxicity symptoms in a horse included weak pulse, ataxia, lower lip and tail limp, leg muscle trembling, respiratory grunt, collapse, seizures, and death within 15 min.

Induction of emesis should not be attempted. Lavage may be used if possible. Activated charcoal and a cathartic should be administered. Life support and respiratory function should be maintained as needed. Diagnosis of yew poisoning is based on the presence of yew plant in the gut on necropsy.

Lethal toxic doses reported in specific animal species are as follows:

Horse: 2 g leaves/kg body weight or 100–200 g

Sheep: 10 g/kg body weight or 100–200 g

Ruminants: 0.30–0.7 g/kg body weight

Dog: 30 g

Swine: 3 g leaves/kg body weight or 75 g

Fowl: 30 g

Oxen: 10 g leaves/kg body weight or 500 g

Goats: 12 g leaves/kg body weight

Surprisingly, deer are able to eat the foliage of *Taxus cuspidata* and apparently suffer no harm.

—*Lanita B. Myers*

Related Topic

Cardiovascular System

Yohimbine

♦ CAS: 146-48-5

♦ SYNONYMS: Aphrodine; coryine; YoYo; quebrachine

- PHARMACEUTICAL CLASS: α_2-Adrenoreceptor antagonist
- MOLECULAR FORMULA: $C_{21}H_{26}N_2O_3HCl$

Uses
Yohimbine has been on the USDA's unsafe herb list since March 1977. It is only approved for use in veterinary medicine, as a reversal agent for xylazine. Illicit uses include use as an aphrodisiac and mild hallucinogen.

Exposure Pathways
Ingestion is the most common route of exposure. The substance is extracted from the bark of the *Corynanthe yohimbe*. This tree is found in western Africa. The substance may also be smoked. It is available in an intravenous form for veterinary purposes.

Toxicokinetics
Oral absorption is rapid, with an absorption half-life of 7–11 min. Peak plasma levels occur at 45–60 min. The volume of distribution is 2.24 ± 1 to 1.25 liters/kg. Yohimbine is excreted via the kidneys. Less than 1% of the unchanged drug was recovered in the urine after 24 hr. Yohimbine is rapidly eliminated from the plasma with the half-life less than 1 hr.

Mechanism of Toxicity
Yohimbine is an α_2-adrenergic antagonist. It may also react with α_1-adrenoceptors. Yohimbine can increase sympathetic outflow and enhance the release of norepinephrine. These actions result in symptoms. A two or three time increase in plasma norepinephrine has been reported after intravenous doses of 0.016–0.125 mg/kg.

Human Toxicity
Although overdoses are rare, oral doses of 15–20 mg have produced hypertension. Doses of 0.1 mg/kg may produce stimulant effects. Other toxic manifestations are tachycardia, diaphoresis, mydriasis, salivation, nausea, vomiting, and facial flushing. Neurological signs include dizziness, anxiety, "squeezing headache," and incoordination.

Clinical Management
Basic and advanced life-support measures should be performed as needed. Treatment is focused on decreasing hypertension and anxiety. Gastric decontamination may be beneficial if performed within the first hours of ingestion. Clonidine may be effective to reverse the α-adrenergic antagonism. Diazepam is useful to decrease anxiety.

Animal Toxicity
Yohimbine has U.S. FDA approval to reverse the effects of xylazine in dogs. Toxic effects are similar to those observed in humans. Hypertension, tachycardia, central nervous system stimulation, and antidiuresis may occur.

—*Denise A. Kuspis*

Zinc (Zn)

- CAS: 7440-66-6
- SELECTED COMPOUNDS: Zinc chloride, $ZnCl_2$ (CAS: 7646-85-7); zinc chromate, $ZnCrO_4$ (CAS: 13530-65-9); zinc oxide, ZnO (CAS: 1314-13-2)
- CHEMICAL CLASS: Metals

Uses

Zinc is an essential trace element and is commonly ingested as a nutritional supplement. Divalent zinc is one of the most important of the micronutrients. More than 100 enzymes are zinc dependent; for example, carboxypeptidase, carbonic anhydrase (which is responsible for the exchange of carbonic acid in the blood and the exhalation of carbon dioxide), and the alcohol dehydrogenase (which metabolizes alcohol). Deficiency of zinc, especially in newborns, results in impaired growth, loss of hair, skin eruptions, and often impaired or delayed sexual maturation. Many medical problems are also associated with zinc deficiencies (e.g., ulcerative colitis, chronic renal disease, and anemia).

Commercially, zinc is used in galvanized iron and in various alloys (e.g., brass and bronze). It is also used in dry cell batteries, electrical fuses, fungicides, and construction materials (e.g., roofing and gutters). Zinc chloride is used in electroplating, soldering fluxes, burnishing and polishing compounds for steel, and in antiseptic and deodorant solutions. Zinc chloride is used as yellow pigment. Zinc oxide is used in ointments, rubber, and paints (for white pigments).

Exposure Pathways

Ingestion and inhalation of zinc are possible exposure pathways. Zinc is readily absorbed by most plants and, hence, is found in most foods (especially grains, nuts, legumes, meats, poultry, and most seafood). The concentration of zinc in drinking water depends on the composition of water pipes and vessels. Inhalation is a significant exposure pathway in industrial areas, where zinc levels in air are high.

Toxicokinetics

Up to 30% of ingested zinc is absorbed from the small intestine; however, a homeostatic mechanism controls the absorption. Nutritional status also influences zinc absorption; deficiency of pyridoxine or tryptophan somewhat inhibits absorption. Zinc is carried by the blood proteins, albumin and β-2-macroglobulin. Zinc induces a zinc metallothionein, the form in which it is bound to the liver and other tissues. The pancreas is high in zinc, and in males the prostate gland contains the greatest store of zinc. Zinc is excreted in the feces.

Mechanism of Toxicity

Excessive zinc interferes with iron and copper metabolism; the latter leads to copper-deficiency anemia. Salts of strong mineral acids are corrosive to skin and intestine.

Human Toxicity

It is difficult to ingest too much zinc from foodstuffs. Consumption of beverages stored in galvanized containers or pipes, use of zinc utensils, or ingestion of too many zinc supplements can result in nausea, cramps, vomiting, and diarrhea. Since the zinc:copper ratio is important, intake of too much zinc can lead to symptoms of copper deficiency. However, patients have taken 10 times the recommended daily allowance for zinc with no adverse reaction.

In the industrial setting, inhalation of fumes of zinc, zinc oxide, or zinc chloride leads to pulmonary edema and metal fume fever. Onset occurs within 4–6 hr and may be delayed up to 8 hr. Symptoms include chills alternating with fever, sweating, and weakness, which can last from 24 to 48 hr. Chronic inhalation of zinc compounds can lead to liver damage, which can be fatal.

The ACGIH TLV-TWA for zinc chloride (as fume) is 1 mg/m^3. The TLV-TWA for zinc oxide (as fume) is 5 mg/m^3.

Zinc salts (e.g., zinc chloride and zinc sulfate) are corrosive to the skin and the gastrointestinal tract and can cause acute tubular necrosis and interstitial nephritis.

Clinical Management

Clinical management is supportive. Chelating agents such as BAL (British Anti-Lewisite; 2,3-dimercaptopropanol) or D-penicillamine are not effective.

Animal Toxicity

Zinc is not carcinogenic; however, testicular tumors were induced by direct injection of zinc chloride into the testes of experimental animals (copper chloride produced the same effect).

—*Arthur Furst and Shirley B. Radding*

Related Topics

Metallothionein
Metals

Glossary

In order that the *Encyclopedia of Toxicology* may be useful to as wide a readership as possible, a Glossary of Key Terms has been provided by the publisher. For the purpose of the article text itself, it is important to use the established technical vocabulary of the science of toxicology, in the interest of accuracy, brevity, and consistency.

However, it is possible that some of these technical terms will not be entirely familiar to the nonprofessional readers of this Encyclopedia. Therefore, in the interest of greater understanding for those readers—and also for the possible benefit of professional readers consulting material outside their own area of expertise—the Glossary defines a selected group of several hundred terms. These terms occur frequently within a variety of articles in the Encyclopedia and thus can be said to represent a core vocabulary of the field of toxicology. The definitions are presented in a concise, accessible format, based on the use of the term in the context of the Encyclopedia.

absorption   the act or fact of absorbing; the process by which a chemical substance crosses the various membrane barriers of the body before it enters the bloodstream. The main sites of entry for the process of absorption are the gastrointestinal tract, the lungs, and the skin. Compare AD-SORPTION.

abuse   the intentional ingestion or injection of a substance of known toxicity, with the aim of producing some desired mental or physical effect that is other than the accepted therapeutic or industrial use of the product. Thus, **drug** or **substance (of) abuse**.

acaricide   a toxic substance that is used to kill mites (family Acaridiae).

ACGIH   American Conference of Governmental Industrial Hygienists, an organization devoted to the administrative and technical aspects of worker health and safety. ACGIH publishes a widely used set of guidelines for acceptable exposure to potentially toxic chemical compounds.

acidosis   a disturbance of the natural acid-base balance of the body (pH 7.4), so that there is excessive acidity in the blood and tissues.

active transport   the direct participation of a cell to provide a specialized carrier molecule, a protein, and the expenditure of cellular energy for the movement of a substance against a concentration gradient and usually across cell membranes. It plays an important role in the elimination of toxic chemicals by the liver and the kidneys.

acute toxicity   a toxic reaction that occurs in a relatively short period of time, following exposure to a single, typically large dose of the toxic substance. Compare CHRONIC TOXICITY.

ADI   acceptable daily intake, the level of a substance that is established as safe to be ingested on a daily basis over a lifetime without significant adverse effect.

adrenotoxic   harmful to the adrenal glands.

adsorbent   a substance used to adsorb toxic material; e.g., charcoal.

adsorption   the act or fact of adsorbing; the taking up of the molecules of a gas or liquid on the surface of another substance; contrasted with ABSORPTION in that an adsorbed material does not actually penetrate the inner structure of the other substance.

adynamia   an abnormal condition of weakness; a severe lack of energy or power.

agonist   a general term for any substance that produces a specific physiological response at cell receptors.

agranulocytosis   an abnormal condition of the blood characterized by severely low levels of granulocytes (certain white blood cells).

albuminaria   the presence of albumin in the urine.

alkalosis   an abnormal condition characterized by an excessive alkaline (base) level in the blood, or a deficient acid level, resulting in a pH level above 7.44.

alopecia   an abnormal lack or loss of hair.

alveoli   the plural of **alveolus,** one of the numerous tiny air cells of the lungs in which the exchange of oxygen and carbon dioxide takes place. Thus, **alveolar.**

Ames test   a widely used means of establishing the mutagenic qualities of a substance; i.e., its potential for causing genetic damage, based on the effect of the substance on a certain type of bacteria. [Developed by Bruce *Ames,* U.S. biochemist.]

analgesic   a substance that serves to relieve or lessen pain.

anemia   a condition in which the level of hemoglobin in the blood is abnormally low and there is a decrease in the number of red blood cells. Thus, **anemic.**

angina (pectoris)   a sharp, suffocating pain in the chest, typically caused by a lack of oxygen supply to the myocardium.

animal model   an animal used to predict or assess the effect that a toxic substance would have on humans.

anorexia   an abnormal loss or lack of appetite.

anoxia   an abnormal lack of oxygen.

antagonist   an agent or substance that reduces or counteracts the effect of another substance on cell receptors; e.g., selenium is an antagonist of arsenic.

antipyretic   a substance that serves to reduce fever.

anuria   an extreme inability to urinate, causing a harmful buildup of waste products in the bloodstream because of the failure of the kidneys to excrete them.

aplasia   a failure of cell, tissue, or organ development. Thus, **aplastic.**

aplastic anemia   a failure or reduction of the ability of the bone marrow to generate blood cells.

apnea   a temporary, involuntary halt in breathing.

arrhythmia   any variation from the normal rhythm of the heart.

asphyxia   a severe condition characterized by a lack of oxygen and an excess of carbon dioxide in the blood and tissues, leading to loss of consciousness and, if uncorrected, death.

asphyxiate   to cause to enter a state of asphyxia (suffocation). Thus, **asphyxiation.**

asymptomatic   lacking symptoms; showing or causing no symptoms.

asystole   a stoppage of the heart, or the absence of a heartbeat.

ataxia   a failure or irregularity of muscle control or muscle action.

atoxic   not toxic; not caused by or associated with a toxin.

atropinization   the therapeutic administration of atropine, $C_{17}H_{23}NO_3$, a common antidote for certain toxic substances.

azotemia   excessive levels of urea or other nitrogen compounds in the blood.

BAF   bioaccumulation factor, a measure of the tendency of an environmental toxin to accumulate in living organisms.

bioaccumulation   the process by which living organisms accumulate chemical substances within the body, either directly from the environment or from dietary sources.

bioconcentration   the process by which living organisms accumulate chemical substances directly from environmental sources rather than through dietary sources.

biologically effective dose   the amount of a substance that is sufficient to bring about some significant physiological change in the affected organism; specifically, the level of exposure to a toxic substance that is required to produce a harmful effect.

biomagnification   a phenomenon in which a certain toxic material will tend to transfer at a greater rate than the overall transfer rate for the substance in which it is contained. Thus, for example, the concentration of pesticides in a given species of predatory bird will be at a (much) higher level than in the fish that this bird eats as food, and similarly the level will be higher in the fish than in the surrounding water.

biomarker   an early change or reaction in an organism due to exposure to some chemical or agent, that can be used to predict a later and more serious toxic effect in this individual or in the larger population.

biotoxin   a toxic substance that is a product of the cells or secretions of a living organism.

biotransformation   the change of matter from one form to another within the body; specifically, the process in which fat-soluble, non-nutrient chemicals are converted by enzymatic reactions to products that are more water-soluble, thus allowing the excretion of toxic matter that would otherwise be poorly excretable.

blood-brain barrier   the barrier of capillary walls and surrounding cells that tends to prevent the passage of various substances between the blood and the brain cells.

bradycardia   an abnormally slow heart rate, usually defined as below 60 beats per minute.

carcinogen   a cancer-causing agent or substance. Thus, **carcinogenic.**

cardiomyopathy   disease of the heart muscle.

cardiotoxic   harmful to the heart.

cholestasis a lack of normal flow of liver bile.

cholinergic relating to or involving choline (acetylcholine), $C_7H_{17}O_3N$.

chronic toxicity a toxic effect that becomes evident after repeated or continued exposure over an extended period of time. Compare ACUTE TOXICITY.

circumoral relating to or affecting the area around the mouth.

clastogen a physical or chemical agent that causes chromosome damage.

CNS central nervous system, a collective term for the brain and spinal cord.

coagulopathy any disorder or abnormality of blood coagulation.

coingestant one of two or more substances taken into the body at the same time.

conjugate of chemical substances, to join by conjugation.

conjugation the joining together of two chemical compounds to produce another compound; specifically, the combination of a toxic substance in the body with some other substance to form a detoxified product.

contraindicated describing a treatment or procedure that might ordinarily be applied for a given type of toxicity, but that is specifically to be avoided in this particular situation, because of other factors.

corrosive tending to destroy or dissolve tissue. Thus, **corrosion.**

cumulative of a toxin, tending not to be eliminated from the body but rather to accumulate in certain organs over an extended time, even many years.

cutaneous relating to or involving the skin.

cyanosis an abnormal bluish or purplish appearance of the skin, caused by an excess of reduced (deoxygenated) hemoglobin in the blood.

cytotoxic harmful or destructive to cells. Thus, **cytotoxicity.**

de novo coming from something new; specifically, indicating the synthesis of a molecule from very simple precursors rather than from alterations in an already complex structure.

depressant a substance or agent that tends to reduce the activity of a system of the body; e.g., the central nervous system.

dermal relating to or involving the skin.

dermis one of two major structural components of the skin, along with the epidermis. The epidermis rests upon and is anchored onto a much thicker base of connective and fatty tissues forming the dermis. Once a chemical has crossed the epidermis and gained access to the dermis, diffusion into the bloodstream occurs rapidly.

dermotoxic harmful or destructive to skin cells.

detoxification an elimination, neutralizing, or weakening of the harmful effects of a toxic substance. Thus, **detoxify.**

dialysis the diffusion of blood across a semipermeable membrane to remove toxic materials and maintain a proper balance of fluid and blood electrolytes; used especially in reference to kidney function. See also HEMODIALYSIS.

diaphoresis excessive sweating or perspiration.

diffusion the process by which a substance spreads out or scatters over a large area. In **facilitated diffusion** the transfer of molecules across membranes is assisted by carrier proteins embedded in the membrane layer. In **passive diffusion** the passage of molecules does not involve this participation of carrier proteins.

distribution the process by which absorbed chemicals are transported to the various organs of the body where they will produce an effect, be stored, or be eliminated.

diuresis an increase in the formation and excretion of urine.

dose-response relationship the fundamental principle that as the amount of a given substance changes, its effect on an exposed organism will change as well; thus the same substance may produce no noticeable effect in a minimal dosage, a therapeutic effect in a moderate dosage, and a toxic effect in a larger dosage.

dyscrasia a general term denoting the presence of abnormal materials in the blood.

dysphagia difficulty in swallowing or an inability to swallow.

dysphoria an abnormal and ongoing feeling of restlessness, unease, or depression.

dyspnea a condition of difficult or labored breathing.

dysrhythmia a disturbance of the normal rhythms of the body; e.g., heart rate or breathing.

dystonic involving or exhibiting abnormal muscle tone or tension; e.g., stiffness of the neck or tongue.

dysuria pain or difficulty in urinating.

ED_{50} median effective dose; a commonly used measure of the beneficial effect of a substance, expressed as the amount of the substance that will produce the desired therapeutic result in 50% of a group of test organisms under specified conditions. Compare LD_{50}.

edema a swelling resulting from the accumulation of excess fluid in the tissues of the body.

EKG electrocardiogram, a graphic record of the electric activity of the heart.

embryotoxic harmful to the developing embryo.

emesis the fact of expelling matter from the stomach out through the mouth; vomiting.

endobiotic found within the organism; naturally occurring in an organism.

endocytosis   a specialized form of transport by which very large molecules and insoluble materials are engulfed by invagination of the absorptive cell membrane, forming intracellular vesicles.

endogenous   internal; found or occurring within an organism.

endorphin   any of various peptide substances produced by the pituitary gland and the brain, acting on the nervous system to reduce pain.

enervation   a lack of nervous energy; weakness or lassitude.

enterotoxic   harmful to the gastrointestinal tract.

envenomation   exposure of the body to venom, or the effect produced by this.

epidermis   one of two major structural components of the skin, along with the dermis. The epidermis is external to the dermis and is formed of several layers of cells, with the outermost layer consisting of dried dead cells forming the stratum corneum. Chemicals may move through the various cell layers of the epidermis by passive diffusion, more slowly through the stratum corneum, but more rapidly through the inner layers of live epidermal cells (stratum granulosum, stratum spinosum, and stratum germinativum).

equivocal   not producing conclusive results; not definitive.

erythema   an abnormal redness or inflammation of the skin caused by the congestion of surface capillaries.

erythrocyte   red blood cell; the major cellular component of the circulating blood, acting chiefly to transport oxygen.

etiology   the known cause or origin of a disease.

euphoria   a temporary sense of extreme physical and mental well-being, that is abnormal in its intensity, unwarranted by its supposed cause, and inappropriate for the relevant circumstances. Thus, **euphoric.**

excretion   the process by which chemical substances are eliminated from the body, chiefly through the urine and feces and the exhalation of air, and also to some extent through milk, sweat, saliva, tears, and hair.

excitant   a substance or agent that tends to arouse or stimulate the activity of a system of the body; e.g., the central nervous system. Thus, **excitatory.**

exogenous   external to an organism; found or occurring outside the organism.

extrapyramidal   relating to the pyramidal tracts of the brain that are associated with certain body movements.

facilitated diffusion   see DIFFUSION.

fasciculation   an uncoordinated twitching of a single, localized group of muscles, visible through the skin.

fibrillation   an involuntary, recurring contraction of muscle or nerve fibers, especially the abnormal random contraction of a chamber of the heart.

fibrosis   an abnormal spread of fibrous connective tissue over or in place of normal organ tissue.

first-past effect   a process in which chemicals that reach the bloodstream by absorption through the gastrointestinal tract move directly to the liver, where they undergo metabolic biotransformation to more or less active chemical forms, even before they gain access to the various tissues of the body.

FOB   functional observational battery, an array of measures of unlearned or instinctive behaviors in laboratory animals to evaluate the behavioral changes produced in the animal by a given chemical.

fugacity   literally, fleeing; a mathematical expression of the tendency of a molecule or substance to leave one particular phase (e.g., water) and move to a different phase (e.g., the air, soil, the body of an organism), or to move from one part of the same phase to another.

gastrointestinal (GI) tract or system   a collective term for the stomach, intestines, and other body parts forming a digestive pathway from the mouth to the anus.

gavage   the forced ingestion of a substance through a feeding tube.

genotoxic   causing genetic damage; harmful to the genome. Thus, **genotoxin, genotoxicity.**

GRAS   generally recognized as safe, a description of a substance that is adjudged not to have toxic properties if used or ingested in the intended manner and dosage.

gynecomastia   excessive development of the mammary glands in a male.

half-life   the time required for a given chemical reaction to affect half of the reactants involved; specifically, the length of time required for a living body to eliminate or neutralize 50% of a toxic substance introduced within it.

hematemesis   the vomiting of blood.

hematoma   a localized mass of blood, usually fully or partially clotted, found in an organ, tissue, or space external to the blood vessels.

hematopoiesis   the production in the bone marrow of the formed elements of the blood; i.e., red blood cells, white blood cells, platelets.

hematuria   the presence of blood in the urine.

hemodialysis   a procedure in which toxic substances are removed from the blood by means of a special apparatus through which the blood is shunted for filtration and then returned to the patient's body.

hemoglobin   a complex protein-iron compound in the blood that functions to transport oxygen from the lungs to tissues and carbon dioxide from tissues to the lungs.

hemolytic   acting to destroy red blood cells. Thus, **hemolysis.**

hemoperfusion the process of filtering toxic substances from the blood by means of an external absorbent device, such as activated charcoal, rather than by means of a dialysis machine.

hemoptysis the coughing or spitting up of blood.

hemotoxic harmful to blood cells.

hepat-, hepato- a prefix referring to the liver or to agents affecting the liver.

hepatic relating to or affecting the liver.

hepatoma a malignant tumor of the liver.

herbicide a substance used to kill unwanted plants.

homeostatic relating to or describing the intrinsic control mechanisms and processes that, under normal conditions, tend to maintain the body in a stable, balanced state. Thus, **homeostasis.**

hydrolyze of a compound, to decompose or otherwise become altered by a reaction with water.

hydrophobic literally, water-fearing; describing a substance that has a tendency not to absorb or combine with water.

hydroxylate to introduce a hydroxyl group, OH⁻, into a chemical compound. Thus, **hydroxylation.**

hypercalcemia an abnormally high level of calcium in the blood.

hyperemia the excessive presence of blood in an area of the body.

hyperglycemia a condition in which the level of glucose, the major blood sugar, is abnormally high.

hyperkalemia an abnormally high concentration of potassium in the blood.

hyperplasia the abnormal enlargement of a body part, due to an increase in the number of its cells rather than in the size of the cells.

hyperpyrexia an extremely high body temperature; a high fever.

hypertension abnormally high arterial blood pressure.

hypertrophy the abnormal enlargement of a body part, due to an increase in the size of its cells rather than in the number of cells.

hypogeusia a loss or weakening of the sense of taste.

hypoglycemia a condition in which the level of glucose, the major blood sugar, is abnormally low.

hypokalemia an abnormally low level of potassium in the blood.

hypotension abnormally low arterial blood pressure.

hypoxia the presence of an abnormally low level of oxygen in the blood or tissues.

IARC International Agency for Research on Cancer, a department of the World Health Organization that evaluates and defines the potential capacity of various chemicals to cause cancer in humans.

ileus an obstruction of the intestines.

immunosuppressive tending to cause the suppression of the immune function in an organism; inhibiting the normal immune response.

immunotoxic harmful to the immune system; suppressing normal immune function.

ingestion the act of feeding or swallowing.

initiator in a toxic process, the substance that actually produces the original toxic effect, as opposed to another substance (promoter) that subsequently enhances the effect. Thus, **initiation.**

intramuscular within a muscle or muscles; entering the body by means of injection or insertion into a muscle.

intraperitoneal within the peritoneum; entering the body via the peritoneum (the membrane lining the walls of the abdominal cavity).

intravenous entering the body by means of injection or insertion into a vein.

intubation the insertion of a tube in the body for therapeutic purposes.

in utero within the uterus; relating to or affecting the developing fetus.

in vitro literally, "within a glass;" in a test tube or other artificial setting, as opposed to the natural state of a living organism.

in vivo literally, "in life;" within a living organism, as opposed to an external setting; or, of a procedure or study, involving a living organism.

ipecac a syrup prepared from dried roots of the tropical shrub *Cephaelis,* especially *C. ipecacuanha* or *C. acuminata;* a common treatment to induce vomiting in cases of oral poisoning. [From a native South American word meaning "to vomit."]

ischemia a deficiency of oxygenation in a body part, caused by obstruction or constriction of blood vessels.

Itai-Itai a disease resulting from the ingestion of shellfish or other food contaminated with cadmium, characterized by bone and kidney disorders and also associated with chromosomal damage. [Literally, "Ouch-Ouch" disease, from the skeletal pain experienced by sufferers of this condition.]

Kwashiorkor a disease characterized by lack of proper nutrition, found primarily in children. [From a local term used for this disease in Africa.]

lacrimation the shedding of tears.

lacrimator a substance that is highly irritating to the eyes.

lavage the washing out of an organ or body part as a medical procedure.

LD lethal dose, the amount of a toxic substance that will cause death in a given subject.

LC$_{50}$ median lethal concentration; the concentration level of a toxic substance that is sufficient to have a lethal effect in 50% of the subject group of organisms under specified conditions.

LD$_{50}$ median lethal dose; a commonly used measure of the toxicity of a substance, expressed as the amount of the substance that will produce a lethal result in 50% of a group of test organisms under specified conditions.

leukopenia   an abnormal reduction in the number of leukocytes (white blood cells). Also, **leucopenia.**

lipid   any of a large group of fatty and fatlike substances including fatty acids, neutral fats, waxes, and steroids.

lipophilic   literally, fat-loving; describing a substance that has a tendency to combine with or dissolve in fats and oils.

LOE   lowest observed (observable) effect, a measure of the point of exposure to a substance at which some evidence of biological affect becomes apparent.

lysin   a specific toxin that destroys or damages cells.

lysis   the destruction or disintegration of a cell by a specific toxin.

MAC   maximum allowable concentration, a measure used in standards of workplace safety that represents the largest possible airborne concentration that will not produce adverse effects.

median effective dose   see ED$_{50}$.

median lethal concentration   see LC$_{50}$.

median lethal dose   see LD$_{50}$.

metabolic   having to do with metabolism; involving the natural chemical and physical processes of an organism.

metabolism   in general, the sum of all the chemical and physical processes within a living organism; specifically, a reduction or elimination of the toxic properties of a substance by means of such processes.

metabolite   any substance produced or changed by metabolism.

metabolize   of a substance, to change by metabolism; especially, to change to a less toxic state.

methemoglobin   a compound formed from hemoglobin, normally present in the blood in small amounts but not functioning as an oxygen carrier; the presence of toxic agents can increase the amount of methemoglobin in the blood to abnormally high levels.

methemoglobinemia   the appearance of excessive methemoglobin in the blood.

methylate   to introduce a methyl group, CH$_3{}^-$, into a chemical compound. Thus, **methylation.**

microvilli   the plural of **microvillus,** a small fingerlike projection found on the surface of certain cells.

minimum lethal dose   the minimal amount of a substance that is sufficient to cause death under specified conditions.

miosis   an abnormal contraction of the pupil of the eye.

miscible   able to be mixed; mixing with another substance.

morphologic(al)   having to do with or involving the form and structure of the body or a body part.

MRL   maximum residue level, the highest concentration of a substance that is considered to without significant hazard for human health if ingested in foods.

MTD   maximum tolerated (tolerable) dose, a measure used in tests of chronic toxicity that represents the largest possible dose that will not produce either death or a 10% decrease in body weight in the subject animal.

multifactorial   of a disease or pathological state, caused by the interplay of various agents and conditions rather than by a single factor.

muscarine   a toxic alkaloid substance, C$_8$H$_{19}$NO$_3$, found naturally in certain mushrooms; it has a highly characteristic toxic effect.

muscarinic   resulting from or relating to the toxic effects of muscarine. Thus, **muscarinism.**

mutagen   an agent or effect that produces an abnormal rate of genetic mutation. Thus, **mutagenesis, mutagenic.**

myalgia   a general term for pain in the muscles.

mycotoxin   a toxic substance produced by a fungus.

mydriasis   an abnormal dilation of the pupil of the eye.

myocardial   relating to or affecting the myocardium (heart wall).

myocardial infarction   a lack of blood flow to the myocardium, usually caused by the blockage of a coronary artery; in popular use, a heart attack.

myocardium   the middle and thickest layer of the heart wall, composed of cardiac muscle cells.

NAE   no adverse effect, a measure of the maximum exposure to a toxic substance that is possible before the onset of an adverse reaction to it.

narcosis   a condition of extreme depression of the central nervous system, manifested as a stupor or a lack of sensibility.

necrosis   the localized death of cells or tissue.

neonate   a newborn baby.

nephritis   an inflammation of the kidney.

nephrotoxic   harmful to the cells of the kidney. Thus, **nephrotoxicity.**

neurotoxic   harmful to the nerve tissue.

NIOSH   National Institute of Occupational Safety and Health, an organization that establishes recommended exposure limits for hazardous substances or conditions in the workplace.

NOE   no observed (observable) effect, a measure of the maximum exposure to a substance that is possible before the onset of an observable reaction to it, either beneficial or harmful.

nonthreshold describing a substance that is assumed to carry some risk of harmful effect even at the lowest levels, and that thus cannot be assigned a threshold point below which exposure is acceptable.

nystagmus an involuntary, unnatural movement of the eyeballs.

ocular relating to or affecting the eyes.

oliguria an abnormally low level of urination; failure to pass sufficient urine.

OSHA Occupational Health and Safety Act, a U.S. government measure establishing workplace standards for toxic materials and harmful physical agents.

ossification the formation or production of bone.

ototoxic harmful to the sense of hearing.

PAH polynuclear aromatic hydrocarbon, one of a major class of unsaturated hydrocarbon compounds characterized by the presence of variable numbers of rings; formed in the incomplete burning of fossil fuels and vegetable matter. Many PAHs are identified as toxic and particularly as carcinogenic.

pancytopenia an abnormal condition marked by a deficiency of all cellular elements in the blood.

parenteral entering the body by some means other than the digestive system; e.g., by injection.

paresthesia an abnormal, continued sensation of burning or tingling; i.e., the feeling of "pins and needles."

passive diffusion see DIFFUSION.

pathognomic specifically distinctive or characteristic of a given pathological condition, thus providing a basis for diagnosing the condition.

pathway a term for the means by which a toxic substance enters the body.

pediculicide a substance used to kill certain lice (family Pediculidae).

perfusion the passing of a fluid, e.g., the blood, into or through some site in the body.

pesticide any substance used to kill unwanted organisms, such as rodents, insects, or plants.

photophobia an abnormal sensitivity or aversion to light, especially by the eyes.

photosensitivity an abnormal sensitivity to light, especially by the skin. Thus, **photosensitive.**

piloerection an abnormal rising of the hair or fur.

polyuria the passage of an excessive amount of urine.

predisposed more likely to be affected by a disease or condition than the statistical probability for the general population.

promoter in a toxic process, a substance that subsequently enhances the original toxic effect, as opposed to the substance (initiator) that produced the effect. Thus, **promotion.**

prophylactic preventive; tending to prevent disease or adverse conditions.

ptosis the failure of the eye to open normally; drooping eyelid.

pulmonary relating to or affecting the lungs.

rales abnormal bubbling or cracking sounds produced during breathing.

refractory not responsive; resistant to therapy or treatment.

REL recommended exposure limit, a measure of the maximum exposure to a toxic substance that is acceptable in the workplace.

renal relating to or affecting the kidneys.

rhabdomyolysis disintegration or dissolution of muscle.

rubifacient tending to redden the skin.

sarcoma a malignant tumor made up of tissue that is comparable to the embryonic connective tissue, consisting of closely packed cells embedded in a fibrillar or homogenous substance.

scabicide a substance used to kill a certain mite, *Sarcoptes scabiei,* that is the cause of the disease scabies.

sick building syndrome (SBS) a term for the modern phenomenon that certain specific buildings, and especially mechanically ventilated office buildings and factories, will tend to produce a similar set of health complaints from various occupants; e.g., respiratory ailments, nausea, headaches, and irritation of the eyes, nose, and throat.

SLUDGE an acronym for an array of conditions often occurring together, typically in the presence of excessive acetylcholine stimulation; i.e., Salivation; Lacrimation (tears); Urination; Diarrhea; Gastrointestinal cramping; Emesis (vomiting).

stem cell a parent or progenitor cell that is capable of giving rise to increasingly more developed or differentiated cells, as in the formation of the elements of the blood.

stertorous describing heavy, labored breathing.

stratum corneum the outermost layer of the epidermis, consisting of dried dead cells that are rich in a filament-forming protein called keratin. This layer represents the major structural component of the barrier to passage of chemicals through the skin.

subcutaneous beneath the skin; occurring or passing under the skin.

suggestive indicating a possible toxic effect, but not definitive because of significant conflicting data.

supportive describing a course of treatment for toxicity that focuses on monitoring and promoting the life functions involved in the patient's own capacity for counteracting the toxic substance.

sympathomimetic describing a substance or agent that mimics the stimulative effects of the sympathetic nervous system.

symptomatic describing a course of treatment that focuses on alleviating the symptoms produced by a toxic substance, rather than on counteracting or removing the substance.

symptomatology the appearance of symptoms of a disease, or the study of these symptoms.

syncope a temporary loss of consciousness; fainting.

tachycardia an abnormally rapid heart rate, usually defined as above 100 beats per minute.

target organ an organ that is most likely to be the affected site of a toxic substance that has entered the body.

teratogenic causing physical defects in the fetus; preventing normal fetal development.

teratology the scientific study of the causes, mechanisms, and manifestations of abnormal fetal development.

therapeutic dose or **level** an amount of a substance established as beneficial in therapy but potentially toxic in a larger amount.

thermogenesis the production of heat in the body by physiological processes.

threshold a term for the point at which exposure to a potentially toxic substance becomes sufficient in amount to actually cause an adverse effect.

thrombocytopenia an abnormally low level of platelets in the blood, resulting in inadequate or slowed blood clotting.

thrombocytosis an abnormally high level of platelets in the blood.

tinnitus a persistent ringing, buzzing, or other such noise perceived in the ears, in the absence of any actual sound to produce it.

TLV threshold limit value, a widely used set of guidelines for occupational health, describing the maximum level of safe exposure for various chemical compounds and other environmental conditions.

toxicity the fact of being toxic (harmful or poisonous), or the extent to which a substance is toxic.

toxicokinetics the physical movement of a toxic substance within the body; e.g., its absorption from an external source or its elimination from the body.

transcutaneous crossing the skin; entering the body through the skin.

transplacental crossing the placenta; transferred from mother to fetus via the placenta.

tumorigenic causing or producing tumors; promoting abnormal proliferation of cells. Thus, **tumorigenicity.**

vasculotoxic harmful to the blood vessels.

vasoconstrictor a substance or agent that causes the blood vessels to become narrower. Thus, **vasoconstriction.**

vasodilator a substance or agent that causes the blood vessels to become wider. Thus, **vasodilation, vasodilatory.**

vasopressor a substance or agent that causes the blood pressure to become higher.

venotoxic harmful to the veins.

venous having to do with or affecting the veins.

vertigo an illusion that oneself or one's external world is spinning about; broadly, dizziness.

vesicant causing blisters or burning.

villi the plural of **villus,** a small protrusion or extension from a body surface, especially from a mucous membrane.

xenobiotic foreign to the body; not normally found within the body in a given species.

xenoestrogen literally, foreign estrogen; a term for a substance that is foreign to the body but that has properties similar to the natural estrogens (sex hormones), thus producing unwanted effects such as increased risk of hormonally induced cancers.

Contributors

Melissa Adams
  College of Pharmacy
  Northeast Louisiana University
  Monroe, Louisiana
Felix Adatsi
  East Lansing Crime Laboratory
  Michigan State Police
  Lansing, Michigan
Arvind K. Agarwal
  John Jay College
  City University of New York
  New York, New York
Jayne E. Ash
  Gad Consulting Services
  Raleigh, North Carolina
Kevin N. Baer
  College of Pharmacy
  Northeast Louisiana University
  Monroe, Louisiana
C. Cephas Barton
  College of Pharmacy
  Northeast Louisiana University
  Monroe, Louisiana
Todd A. Bartow
  College of Pharmacy
  Northeast Louisiana University
  Monroe, Louisiana
Janet E. Bauman
  Pittsburgh Poison Center (retired)
  Pittsburgh, Pennsylvania
Patricia J. Beattie
  Chemical Risk Management
  General Motors Corporation
  Detroit, Michigan

Edward A. Belongia
  Marshfield Medical Research Foundation
  Marshfield, Wisconsin
John R. Bend
  University of Western Ontario
  London, Ontario, Canada
Rhonda S. Berger
  Environmental Consulting and Technology
  Royal Oak, Michigan
David E. Bice
  Senior Scientist
  Lovelace Respiratory Research Institute
  Albuquerque, New Mexico
Benny L. Blaylock
  College of Pharmacy
  Northeast Louisiana University
  Monroe, Louisiana
Gary P. Bond
  Nalco/Exxon Energy Chemicals
  Sugar Land, Texas
Douglas J. Borys
  Director, Central Texas Poison Center
  Scott and White Memorial Hospital
  Temple, Texas
Michael J. Brabec
  Eastern Michigan University
  Ypsilanti, Michigan
Jules Brodeur
  Faculty of Medicine
  University of Montreal
  Montreal, Quebec, Canada
Antone L. Brooks
  Section Manager
  Batelle Pacific Northwest Laboratories
  Richland, Washington

Anne E. Bryan
Pittsburgh Poison Center
Children's Hospital of Pittsburgh
Pittsburgh, Pennsylvania
Richard J. Bull
Health Division
Battelle Pacific Northwest Laboratories
Richland, Washington
Philip J. Bushnell
Neurotoxicology Division
National Health and Environmental Effects
Research Laboratory
United States Environmental Protection Agency
Research Triangle Park, North Carolina
Zhengwei Cai
Division of Newborn Medicine
University of Mississippi Medical Center
Jackson, Mississippi
Michael P. Carver
Wyeth-Ayerst Research
Kendall Park, New Jersey
Finis L. Cavender
Information Ventures, Inc.
Durham, North Carolina
Tamal Kumar Chakraborti
Environmental Health Sciences
Baltimore, Maryland
Sanjay Chanda
North Carolina State University
Raleigh, North Carolina
Sushmita M. Chanda
Laboratory of Toxicology
NIEHS
Research Triangle Park, North Carolina
C. P. Chengelis
WIL Research Laboratories
Ashland, Ohio
Karen Chou
Institute for Environmental Toxicology
Michigan State University
East Lansing, Michigan
Stephen Clough
National Council for Air and Stream Improvement
Medford, Massachusetts
Daniel J. Cobaugh
Finger Lakes Regional Poison Center
University of Rochester Medical Center
Rochester, New York
Deborah A. Cory-Slechta
University of Rochester Medical Center
Rochester, New York

Abraham Dalu
Division of Biochemical Toxicology
National Center for Toxicological Research
Food and Drug Administration
Jefferson, Arkansas
John H. Davies
Bonnie S. Dean
Pittsburgh Poison Center
Children's Hospital of Pittsburgh
Pittsburgh, Pennsylvania
Alexis Desmoulièr
Institut Pasteur de Lyon
Lyon, France
John Doull
University of Kansas Medical Center
Kansas City, Kansas
Jeffrey H. Driver
risksciences.com, L.L.C.
Arlington, Virginia
Donald J. Ecobichon
McGill University
Montreal, Quebec, Canada
Marion Ehrich
Virginia-Maryland Regional College of Veterinary
Medicine
Blacksburg, Virginia
Janet Everitt
Upjohn Co. Research Laboratories
Kalamazoo, Michigan
Gary W. Everson
California Poison Control System
Fresno, California
M. Joseph Fedoruk
University of California, Irvine
Irvine, California
Charles Feigley
University of South Carolina School of Public
Health
Columbia, South Carolina
Paul W. Ferguson
College of Pharmacy
Northeast Louisiana University
Monroe, Louisiana
Kristin M. Fitzgerald
Roy F. Weston, Inc.
Falls Church, Virginia
Bridget Flaherty
Pittsburgh Poison Center
Children's Hospital of Pittsburgh
Pittsburgh, Pennsylvania

Stephanie E. Foster
  College of Pharmacy
  Northeast Louisiana University
  Monroe, Louisiana
Arthur Furst
  Professor Emeritus, University of San Francisco
  San Francisco, California
Guilio Gabbiani
  Department of Pathology
  Centre Medical Universitaire
  Geneva, Switzerland
Shayne C. Gad
  Gad Consulting Services
  Raleigh, North Carolina
Donald E. Gardner
  Inhalation Toxicology Associates
  Raleigh, North Carolina
Gerald J. Gleich
  Department of Immunology
  Mayo Clinic
  Rochester, Minnesota
Carla M. Goetz
Terry Gordon
  Nelson Institute of Environmental Medicine
  New York University Medical Center
  Tuxedo Park, New York
Raymond A. Guilmette
  Lovelace Respiratory Research Institute
  Albuquerque, New Mexico
Ramesh C. Gupta
  Breathitt Veterinary Center
  Murray State University
  Hopkinsville, Kentucky
Robin Guy
  Monsanto Company
  Skokie, Illinois
William Halperin
  Deputy Director
  National Institute for Occupational Safety and
  Health
  Atlanta, Georgia
Jerry L. Hamelink
  Hameltronics
  Hudsonville, Michigan
Paul R. Harp
  College of Pharmacy
  Northeast Louisiana University
  Monroe, Louisiana

Linda Hart
  Pittsburgh Poison Center
  Children's Hospital of Pittsburgh
  Pittsburgh, Pennsylvania
Rolf Hartung
  School of Public Health (retired)
  University of Michigan
  Ann Arbor, Michigan
Donald Henderson
  Hearing Research Laboratory
  State University of New York at Buffalo
  Buffalo, New York
Rogene F. Henderson
  Lovelace Respiratory Research Institute
  Albuquerque, New Mexico
Cynthia Hess
  Albuquerque Public Schools
  Albuquerque, New Mexico
Michael T. Hodgman
  Toxicology Treatment Center
  University of Pittsburgh
  Pittsburgh, Pennsylvania
Michael Hodgson
  University of Connecticut Health Center
  Farmington, Connecticut
A. J. Hoffman-Kiefer
Michael P. Holsapple
  Health and Environmental Research Laboratory
  Dow Chemical Company
  Midland, Michigan
Mary Lee Hultin
  Michigan Department of Environmental Quality
  Lansing, Michigan
C. Lynn Humbertson
  Pittsburgh Poison Center
  Children's Hospital of Pittsburgh
  Pittsburgh, Pennsylvania
Nancy A. Jeter
  College of Pharmacy
  Northeast Louisiana University
  Monroe, Louisiana
Robert M. Joy[†]
  University of California, Davis
  Davis, California
Robert L. Judd
  College of Pharmacy
  Northeast Louisiana University
  Monroe, Louisiana

[†] Deceased.

Sam Kacew
  University of Ottawa
  Ottawa, Ontario, Canada
Norbert E. Kaminski
  Michigan State University
  East Lansing, Michigan
Michael A. Kamrin
  Institute of Environmental Toxicology
  Michigan State University
  East Lansing, Michigan
John Kao
  Division of Drug Safety and Metabolism
  Wyeth-Ayerst Research
  Kendall Park, New Jersey
Kathryn Kehoe
  Jacksonville University
  Jacksonville, Florida
David P. Kelly
  DuPont Haskell Laboratory
  Newark, Delaware
Edward Kerfoot
  BASF Corporation
  Wyandotte, Michigan
Janet E. Kester
  Dames & Moore Group
  Sedalia, Colorado
Wendy Khune
Daniel T. Kirkpatrick
  WIL Research Laboratories
  Ashland, Ohio
Paul Kleihues
  International Agency for Research on Cancer
  Lyon, France
Travis R. Kline
  Earth Tech, Inc.
  Grand Rapids, Michigan
David M. Krentz
  DuPont Haskell Laboratory
  Newark, Delaware
Edward P. Krenzelok
  Pittsburgh Poison Center
  Children's Hospital of Pittsburgh
  Pittsburgh, Pennsylvania
Gary R. Krieger
  School of Pharmacy
  University of Colorado
  Denver, Colorado
Swarupa G. Kulkarni
  Northeast Louisiana University
  Monroe, Louisiana

Denise L. Kurta
  Pittsburgh Poison Center
  Children's Hospital of Pittsburgh
  Pittsburgh, Pennsylvania
Denise A. Kuspis
  Pittsburgh Poison Center
  Children's Hospital of Pittsburgh
  Pittsburgh, Pennsylvania
Joseph R. Landolph
  Norris Comprehensive Cancer Center
  University of Southern California
  Los Angeles, California
Linda Larsen
  Michigan Department of Environmental Quality
  Lansing, Michigan
Bill L. Lasley
  Institute of Toxicology and Environmental Health
  University of California, Davis
  Davis, California
Virginia Lau
  Dames & Moore Group
  San Francisco, California
Gerald A. LeBlanc
  North Carolina State University
  Raleigh, North Carolina
Hon-Wing Leung
  Union Carbide Corporation
  Danbury, Connecticut
Barbara C. Levin
  Biotechnology Division
  National Institute of Standards and Technology
  Gaithersburg, Maryland
Jing Liu
  College of Pharmacy
  Northeast Louisiana University
  Monroe, Louisiana
Betty J. Locey
  Dames & Moore Group
  Farmington Hills, Michigan
Gaylord P. Lopez
  Georgia Poison Center
  Atlanta, Georgia
Frank C. Lu
  Consulting Toxicologist
  Miami, Florida
Harold MacFarland
  Consultant in Toxicology
  Victoria, British Columbia, Canada

Joseph A. Maga
   Colorado State University
   Fort Collins, Colorado
Linda Angevine Malley
   DuPont Haskell Laboratory
   Newark, Delaware
Raja S. Mangipudy
   University of Nebraska Medical Center
   Omaha, Nebraska
Anthony S. Manoguerra
   California Poison Control System
   San Diego, California
Robert R. Maronpot
   Laboratory of Experimental Pathology
   National Institute of Environmental Health Science
   Research Triangle Park, North Carolina
Arthur N. Mayeno
   Tsukuba Research Laboratories
   Upjohn Pharmaceuticals, Ltd.
   Ibaraki, Japan
Gordon P. McCallum
   University of Western Ontario
   London, Ontario, Canada
John A. McCants
   College of Pharmacy
   Northeast Louisiana University
   Monroe, Louisiana
Sandra L. McFadden
   State University of New York at Buffalo
   Buffalo, New York
Janice M. McKee
   Dames & Moore Group
   Farmington Hills, Michigan
Thomas E. McKone
   School of Public Health
   University of California, Berkeley
   Berkeley, California
Michele A. Medinsky
   Chemical Industry Institute of Toxicology
   Research Triangle Park, North Carolina
Harihara M. Mehendale
   College of Pharmacy
   Northeast Louisiana University
   Monroe, Louisiana
Peter G. Meier
   School of Public Health
   University of Michigan
   Ann Arbor, Michigan

Marion G. Miller
   Department of Environmental Toxicology
   University of California, Davis
   Davis, California
Ann D. Mitchell
   Genesys Research, Incorporated
   Durham, North Carolina
Rita Mrvos
   Pittsburgh Poison Center
   Children's Hospital of Pittsburgh
   Pittsburgh, Pennsylvania
Leyna Mulholland
   Transent Group, Inc.
   Indianapolis, Indiana
Lanita B. Myers
   Pittsburgh Poison Center
   Children's Hospital of Pittsburgh
   Pittsburgh, Pennsylvania
Lewis Nelson
   Emergency Medicine
   NYU Medical Center
   Bellevue Hospital Center
   New York, New York
Canice Nolan
   European Commission
   Science Research and Development
   Brussels, Belgium
Adrian J. Nordone
   Proctor & Gamble Company
   Strombeck-Bever, Belgium
Susan B. Norton
   National Center for Environmental Assessment
   U.S. Environmental Protection Agency
   Washington, DC
Frederick W. Oehme
   Comparative Toxicology Laboratories
   College of Veterinary Medicine
   Kansas State University
   Manhattan, Kansas
Juhani Paakkanen
   Food Quality and Standards Service
   Food and Agricultural Organization of the United
   Nations
   Rome, Italy
Stephanie Padilla
   Cellular and Molecular Toxicology Branch
   Neurotoxicology Division
   United States Environmental Protection Agency
   Research Triangle Park, North Carolina

438

Contributors

Richard A. Parent
Consultox Limited
Damariscotta, Maine
Colin Park
Dow Chemical Company
Midland, Michigan
Ralph Parod
Department of Toxicology, Ecology and Safety
BASF Corp.
Wyandotte, Michigan
Arthur Penn
Nelson Institute of Environmental Medicine
New York University Medical Center
Tuxedo, New York
Thuc Pham
College of Pharmacy
Northeast Louisiana University
Monroe, Louisiana
Gabriel L. Plaa
University of Montreal
Montreal, Quebec, Canada
Miriam C. Poirier
National Cancer Institute
National Institutes of Health
Bethesda, Maryland
Carey Pope
College of Pharmacy
Northeast Louisiana University
Monroe, Louisiana
Paolo Preziosi
Catholic University of the Sacred Heart
Rome, Italy
Shirley B. Radding
TETRAC, Inc.
Santa Clara, California
Shashi Kumar Ramaiah
College of Pharmacy
Northeast Louisiana University
Monroe, Louisiana
Priya Raman
University of Michigan
Ann Arbor, Michigan
Gary O. Rankin
Marshall University School of Medicine
Huntington, West Virginia
Vaman C. Rao
St. Xavier's College
Mumbai, India
Prathibha S. Rao
University of Nebraska Medical Center
Omaha, Nebraska

Sidhartha Ray
College of Pharmacy and Health Sciences
Long Island University
Brooklyn, New York
Janice Reeves
College of Pharmacy
Northeast Louisiana University
Monroe, Louisiana
Michele A. Reynolds
Placitas, New Mexico
Lorenz Rhomberg
Harvard Center for Risk Analysis
Harvard School of Public Health
Boston, Massachusetts
Rudy J. Richardson
University of Michigan
Ann Arbor, Michigan
Heriberto Robles
QST Environmental
Fountain Valley, California
Kathleen Rodgers
Livingston Research Institute
University of Southern California
Los Angeles, California
Regina M. Rogowski
Pittsburgh Poison Center
Children's Hospital of Pittsburgh
Pittsburgh, Pennsylvania
S. Rutherfoord Rose
Virginia Poison Center
Medical College of Virginia
Richmond, Virginia
Wilson K. Rumbeiha
Animal Health Diagnostic Laboratory
College of Veterinary Medicine
Michigan State University
East Lansing, Michigan
Barbara Salem
Institute for Environmental Technology
Michigan State University
East Lansing, Michigan
Harry Salem
Edgewood Research Development and Engineering
Center
Aberdeen Proving Ground, Maryland
Elizabeth J. Scharman
West Virginia Poison Center
Charleston, West Virginia

Lisa Scheuring-Mroz
Children's Hospital of Pittsburgh
Pittsburgh, Pennsylvania

Richard B. Schlesinger
New York University School of Medicine
New York, New York

Bobby R. Scott
Lovelace Respiratory Research Institute
Albuquerque, New Mexico

Michael Shannon
Harvard Medical School
Children's Hospital
Boston, Massachusetts

Dale E. Sharp
Consultox, Limited
Damariscotta, Maine

Frederick R. Sidell
Bel Air, Maryland

Christopher J. Sinal
University of Western Ontario
London, Ontario, Canada

Madhusudan G. Soni
College of Pharmacology
Northeast Louisiana University
Monroe, Louisiana

Henry A. Spiller
Kentucky Regional Poison Center
Louisville, Kentucky

Daniel Steinmetz
BASF Corp.
Wyandotte, Michigan

Bradford Strohm
Division of Chemical Risk Management
General Motors Corporation
Detroit, Michigan

Lester G. Sultatos
New Jersey Medical School
Newark, New Jersey

Brenda Swanson-Biearman
Pittsburgh Poison Center
Children's Hospital of Pittsburgh
Pittsburgh, Pennsylvania

Leonard I. Sweet
School of Public Health
University of Michigan
Ann Arbor, Michigan

Robert Tardif
Faculty of Medicine
University of Montreal
Montreal, Quebec, Canada

Kashyap N. Thakore
University of New Mexico Department of Internal
Medicine/Cardiology
Albuquerque, New Mexico

Melanie C. Thatcher
Nalco/Exxon Energy Chemicals Ltd
Southhampton, United Kingdom

Beatriz Tuchweber
Faculty of Medicine
Universite de Montreal
Montreal, Quebec, Canada

Rochelle W. Tyl
Center for Life Sciences and Toxicology
Research Triangle Institute
Research Triangle Park, North Carolina

William S. Utley
Nalco Chemical Company
Naperville, Illinois

Harri Vainio
Unit of Chemoprevention
International Agency for Research on Cancer
Lyon, France

William H. van der Schalie
National Center for Environmental Assessment
U.S. Environmental Protection Agency
Washington, DC

Robert Visser
Environmental Health and Safety Division
Organisation for Economic Cooperation and
Development (OECD)
Paris, France

William A. Watson
University of Missouri and Truman Medical
Center
Kansas City, Missouri

Elizabeth V. Wattenberg
Environmental and Occupational Health
University of Minnesota School of Public Health
Minneapolis, Minnesota

Gregory P. Wedin
Hennepin Regional Poison Center
Minneapolis, Minnesota

Bernard Weiss
University of Rochester Medical Center
Rochester, New York

Vittoria Werth
Pittsburgh Poison Control Center
Pittsburgh, Pennsylvania

Ainsley Weston
National Institute for Occupational Safety and
Health
Centers for Disease Control
Morgantown, West Virginia

Philip A. Wexler
Toxicology and Environmental Health Information
Program
National Library of Medicine
National Institutes of Health
Bethesda, Maryland

Carole Wezorek
Pittsburgh Poison Center
Children's Hospital of Pittsburgh
Pittsburgh, Pennsylvania

Gary Whitmyre
risksciences.com, L.L.C.
Arlington, Virginia

Regina Wiechelt
Pittsburgh Poison Center
Children's Hospital of Pittsburgh
Pittsburgh, Pennsylvania

C. F. Wilkinson
Jellinek, Schwartz & Connolly
Washington, DC

Calvin C. Willhite
Department of Toxic Substances Control
State of California
Berkeley, California

Barry W. Wilson
Department of Avian Sciences and Environmental
Toxicology
University of California, Davis
Davis, California

Hanspeter Witschi
Institute of Toxicology and Environmental
Health
University of California, Davis
Davis, California

Kathryn A. Wurzel
New Fields, Inc.
Atlanta, Georgia

Robert A. Young
Oak Ridge National Laboratory
Oak Ridge, Tennessee

Tim R. Zacharewski
Michigan State University
East Lansing, Michigan

Index

O

W

X

Y

Z

ISBN 0-12-227223-4

90038

9 780122 272233